Nonlinear Diffusion Equations

Nonlinear Diffusion Equations

Zhuoqun Wu
Jilin University, China

Junning Zhao
Xiamei University, China

Jingxue Yin, Huilai Li
Jilin University, China

World Scientific
New Jersey • London • Singapore • Hong Kong

Published by

World Scientific Publishing Co. Pte. Ltd.

P O Box 128, Farrer Road, Singapore 912805

USA office: Suite 1B, 1060 Main Street, River Edge, NJ 07661

UK office: 57 Shelton Street, Covent Garden, London WC2H 9HE

British Library Cataloguing-in-Publication Data
A catalogue record for this book is available from the British Library.

NONLINEAR DIFFUSION EQUATIONS

ISBN 981-02-4718-4

This book is printed on acid-free paper.

Printed in Singapore by World Scientific Printers (S) Pte Ltd

To Whom We Love

Preface

Nonlinear diffusion equations, as an important class of parabolic equations, come from a variety of diffusion phenomena appeared widely in nature. They are suggested as mathematical models of physical problems in many fields such as filtration, phase transition, biochemistry and dynamics of biological groups. In many cases the equations possess degeneracy or singularity. Comparing to linear equations and quasilinear equations without degeneracy and singularity, such equations, to a certain extent, reflect even more exactly the physical reality. For example, the solutions of such equations may possess the properties of finite speed of propagation of perturbations. On the other hand, the appearance of degeneracy or singularity makes the study more challenging and involved. Thus in the last three decades, especially in recent ten years or so, the study in this direction attracts a large number of mathematicians both in China and abroad. Remarkable progress has been achieved. Many new ideas and methods have been developed to overcome the special difficulties caused by the degeneracy and the singularity, which enrich enormously the theory of partial differential equations.

There have been a tremendous amount of papers on nonlinear diffusion equations with degeneracy or singularity. However, as we know, there are very little of monographs discussing such equations systematically and extensively, although some special topics have been explored in some books. For instance, part of the material is investigated in the books by A. Friedman [FR2] and by G. C. Dong [DO], and an exhaustive exploration on the evolutionary p-Laplacian equation is presented in a new book by E. DiBenedetto [D2]. The present book is devoted to a more comprehensive presentation of the basic problems, main results and typical methods for nonlinear diffusion equations with degeneracy. Some results for equations with singularity will be mentioned briefly in remarks or somewhere else. For convenience of the readers, a list of references is attached at the end of the book, which we hope to collect the related works as complete as possible.

The book contains four chapters. Chapter I and Chapter II deal with two kinds of equations respectively, which have clear physical significance and a rich mathematical theory, namely the Newtonian filtration equations with the typical example

$$\frac{\partial u}{\partial t} = \Delta u^m \tag{1}$$

and the non-Newtonian filtration equations with the typical example

$$\frac{\partial u}{\partial t} = \operatorname{div}(|\nabla u|^{p-2}\nabla u). \tag{2}$$

The equation (1) is degenerate if $m > 1$ and singular if $0 < m < 1$. The equation (2) is degenerate if $p > 2$ and singular if $1 < p < 2$. Only the cases $m > 1$ and $p > 2$ are considered with a few exceptions. In these two chapters it is not our intention to pursue the general form of the equations. In fact, we discuss mainly the equations (1) and (2) themselves, but a little part is devoted to the influence of strongly nonlinear sources to the properties of solutions. Because of the analogy of the theoretical frame and basic ideas in many aspects between the equation (1) and the equation (2) (in spite of the differences in crucial technique in the arguments and in the properties of the two different equations), to avoid the repetitions, sometimes we provide a detailed presentation in Chapter I whereas give a brief treatment in the corresponding part of Chapter II (for example, the discussion of free boundary) and sometimes we give a complete proof for the equation (2) whereas just mention the corresponding result for the equation (1) in Chapter I (for example, the existence of solutions under the optimal initial condition and the corresponding uniqueness). In Chapter I, we treat the one-dimensional case and multidimensional case separately and successively, so that the readers can be acquainted with the theory from easier situation to more difficult, compare the differences and check up what have been or what can be achieved in the study of these two cases.

The obvious character of the equations (1) and (2) is their simplicity in form. Another essential point is that either (1) or (2) has only one "point of degeneracy", namely either (1) with $m > 1$ or (2) with $p > 2$ degenerates only if $u = 0$ or $\nabla u = 0$ respectively. Chapter III of the book is devoted to general quasilinear degenerate parabolic equations of second order. Here by the word "general", we mean the two aspects in the following. The first is the generality in form of the equations, for example, instead of (1) we consider

$$\frac{\partial u}{\partial t} = \frac{\partial}{\partial x_i}\left(a^{ij}(x,t,u)\frac{\partial u}{\partial x_j}\right) + \frac{\partial}{\partial x_i} b^i(x,t,u) + c(x,t,u),$$

where $a^{ij} = a^{ji}$ and

$$a^{ij}(x,t,u) \geq 0, \qquad \forall \xi = (\xi_1, \cdots, \xi_N) \in \mathrm{R}^N.$$

Here and throughout the book we always adopt the summation convention on repeated indices. Also instead of (2), we consider

$$\frac{\partial u}{\partial t} = \operatorname{div} \vec{A}\,(\nabla B(u)),$$

where $\vec{A}\,(v) = (A^1(v), \cdots, A^N(v))$, and the more general equations, for example, in one dimensional case

$$\frac{\partial u}{\partial t} = \frac{\partial}{\partial x} F\left(\frac{\partial}{\partial x} A(u)\right) + \frac{\partial B(u)}{\partial x},$$

where $F'(s) > 0$, $A'(s) \geq 0$. The second and the more essential aspect is that the equations under consideration are allowed to have many "points of degeneracy" or even have arbitrary degeneracy. However, to keep the book in reasonable length we present our arguments in this chapter basically for equations of the form

$$\frac{\partial u}{\partial t} = \Delta A(u) + \operatorname{div} \vec{B}\,(u)$$

where $A'(s) \geq 0$. It should be noted that in equations with arbitrary degeneracy, the presence of a convection term leads to substantial difficulties, because in this case, equations turn out to be of hyperbolic–parabolic mixed type and their solutions might be discontinuous. As is well-known, a book by Oleinik and Radkevich [OR] deals with the theory of linear elliptic equations with arbitrary degeneracy (including degenerate parabolic equations), namely, equations with nonnegative characteristic form. The arguments in Chapter III might be regarded as a development of this theory to the quasilinear case. Some properties of BV functions used in the study of such equations are listed at the end of this chapter without proof.

Chapter IV entitled nonlinear diffusion equations of higher order is in fact devoted to the study of some special equations of fourth order with particular attention to the Cahn-Hilliard equation which arises from phase transition, dynamics of biological groups and the diffusion phenomena in fluids of high viscosity. The purpose of Chapter IV is to expose, to a certain degree, the common points and the differences between the degenerate quasilinear parabolic equations of higher order and those of second order through investigations of these typical equations.

It is impossible to include the study of systems of nonlinear diffusion equations in such a short volume. We will restrict ourselves to the discussion of equations.

The authors are very grateful to the Publishing House of Jilin University for its enthusiastic support for the publication of this book.

Wu Zhuoqun

Jilin University, P. R. China
Jan., 1996

Preface to the Second Edition

The first edition of this book published in 1996 was written in Chinese. The present edition is basically an English translation of the first edition. Corrections of the errors have been made throughout and part of the material has been entirely rewritten. A large number of new references are added in the bibliography. The authors wish to express their appreciation to World Scientific Publishing Company for its efficient handling of the publication of this book.

Wu Zhuoqun

Jilin University, P. R. China
May, 2001

Contents

Chapter 1

Newtonian Filtration Equations

1.1 Introduction

1.1.1 *Physical examples*

In this chapter we study a special class of quasilinear degenerate parabolic equations of second order, which come directly from a variety of diffusion phenomena, such as filtration.

Suppose that we are concerned with a homogeneous, isotropic and rigid porous medium filled with a fluid. Then the flow is governed by the continuity equation

$$\frac{\partial \theta}{\partial t} + \operatorname{div} \overrightarrow{V} = 0, \tag{1.1}$$

and Darcy's law

$$\overrightarrow{V} = -K(\theta)\nabla\varphi, \tag{1.2}$$

where $\overrightarrow{V}$ denotes the macroscopic velocity of the fluid, θ the volumetric moisture content, $K(\theta)$ the hydraulic conductivity and φ the total potential (cf. [BEA], [RHM], [SWA]).

If absorption and chemical, osmotic and thermal effects are ignored, then φ may be expressed as

$$\varphi = \Psi + z, \tag{1.3}$$

where Ψ is the hydrostatic potential due to capillary suction and z the gravitational potential. Here we choose the (x, y, z) coordinate system in such a way that the z-coordinate is vertical and pointing upwards.

Combining (1.1), (1.2) and (1.3) we obtain

$$\frac{\partial \theta}{\partial t} = \mathrm{div}(K(\theta)\nabla\Psi) + \frac{\partial K(\theta)}{\partial z}. \tag{1.4}$$

Between the variables θ and Ψ there exists an empirical relationship which might be quite complicated because of hysterical effects.

For many porous media, Ψ is a function of θ and $K(\theta)\frac{\mathrm{d}\Psi}{\mathrm{d}\theta}$ and $K(\theta)$ can be expressed reasonably as follows

$$K(\theta)\frac{\mathrm{d}\Psi}{\mathrm{d}\theta} = D_0\theta^{m-1}, \quad K(\theta) = K_0\theta^n,$$

where D_0, K_0, m, n are positive constants and $1 < m \leq n$. In this case, (1.4) becomes

$$\frac{\partial \theta}{\partial t} = \Delta\theta^m + \frac{\partial \theta^n}{\partial z}, \tag{1.5}$$

after necessary change of variables (cf.[BEA], [BRC], [IR], [SWA]). Thus we have, for horizontal flow,

$$\frac{\partial \theta}{\partial t} = \Delta\theta^m \tag{1.6}$$

and for vertical flow,

$$\frac{\partial \theta}{\partial t} = \frac{\partial^2 \theta^m}{\partial z^2} + \frac{\partial \theta^n}{\partial z}. \tag{1.7}$$

In another case, θ is a function of Ψ,

$$\theta = \theta(\Psi),$$

and $0 \leq \theta(\Psi) < \overline{\theta}$, with θ increasing strictly for $\Psi < 0$ and $\theta(\Psi) = \overline{\theta}$ for $\Psi > 0$. The equation (1.4) now becomes

$$\frac{\partial \theta(\Psi)}{\partial t} = \mathrm{div}(D(\Psi)\nabla\Psi) + \frac{\partial D(\Psi)}{\partial z}, \tag{1.8}$$

where $D(\Psi) = K(\theta(\Psi))$. In the saturated region, (1.4) reduces to

$$\Delta\Psi = 0.$$

Thus, the equation (1.8) is of elliptic type in the saturated region and of parabolic type in the unsaturated region.

If $K(\theta(\Psi)) \equiv 1$, then we obtain the following equation for one-dimensional flow

$$\frac{\partial \theta(\Psi)}{\partial t} = \frac{\partial^2 \Psi}{\partial x^2}. \tag{1.9}$$

There are a large number of other diffusion problems from which quasilinear degenerate parabolic equations are arisen. A typical example in dynamics of biological groups is the model for the space-diffusion of groups

$$\frac{\partial \rho}{\partial t} = \Delta A(\rho) + \sigma(\rho), \tag{1.10}$$

where ρ denotes the density of distribution of the species, $\sigma(\rho)$ the growth rate and $A'(\rho)$ the diffusion coefficient which is positive for $\rho > 0$ and vanishes for $\rho = 0$. Evolution of two classes of biological groups, for example, species with different levels of age, can be governed by the system

$$\frac{\partial u}{\partial t} = \operatorname{div}(u\nabla(u+v)),$$

$$\frac{\partial v}{\partial t} = k\operatorname{div}(v\nabla(u+v)),$$

where u and v denote the densities of the two different groups, $k > 0$ is a constant. In the extreme case, $k = 0$, v does not vary with time t and the equation for u turns out to be

$$\frac{\partial u}{\partial t} = \Delta\left(\frac{1}{2}u^2\right) + \operatorname{div}(u\nabla v). \tag{1.11}$$

Equation of the form (1.6) also appears in plasma physics, however, in which $0 < m < 1$, corresponding to the fast diffusion. Such kind of equation also arises in the study of phenomena occurring at the beginning of a nuclear explosion. At the very first stage, immediately following energy release, thermal waves are propagated in the as yet stationary gas. Heat conduction is then determined mainly by radiation, and the thermal conductivity is a function of temperature. Thus the equation to be considered is

$$\frac{\partial u}{\partial t} = \frac{\partial^2 A(u)}{\partial x^2}, \tag{1.12}$$

where u denotes the temperature and $A(u)$ satisfies the condition: $A'(u) > 0$ for $u > 0$, $A'(0) \geq 0$. The corresponding initial condition is

$$u(x,0) = E\delta(x),$$

where $E > 0$ is a constant and $\delta(x)$ the Dirac measure. The initial condition describes instant energy release at $t = 0$. Such kind of physical process motivates the interest in consideration of the problem with unbounded and measure initial data. (cf.[ZR])

Finally it should be pointed out that a lot of diffusion problems can be described by filtration equations with absorption term and convection term

$$\frac{\partial u}{\partial t} = \Delta A(u) + C \cdot \nabla B(u) + \sigma(u). \tag{1.13}$$

Equations of the form (1.5)-(1.13) are referred to as porous medium equations or Newtonian filtration equations. They are called briefly filtration equations very often.

1.1.2 *Definitions of generalized solutions*

Consider equations of the form

$$\frac{\partial u}{\partial t} = \Delta A(u), \tag{1.14}$$

where $\Delta = \sum_{i=1}^{N} \frac{\partial^2}{\partial x_i^2}$ and $A(u) \in C^1[0, \infty)$ satisfies

$$A'(0) = 0, A'(u) > 0, \quad \text{for } u > 0. \tag{1.15}$$

The equation (1.14) is of parabolic type when $u > 0$. However, it degenerates when $u = 0$. If we do not restrict ourselves to the study of nonnegative solutions, then it should be assumed that $A'(u) \in C^1(-\infty, \infty)$ and the condition (1.15) should be replaced by

$$A'(0) = 0, A'(u) > 0, \quad \text{for } u \neq 0. \tag{1.16}$$

We will pay more attention to the special case

$$\frac{\partial u}{\partial t} = \Delta u^m \tag{1.17}$$

with $m > 1$, corresponding to the slow diffusion. If we do not restrict ourselves to the study of nonnegative solutions, then (1.17) should be written as

$$\frac{\partial u}{\partial t} = \Delta(|u|^{m-1} u). \tag{1.18}$$

Basically no attention will be paid to the fast diffusion case, although it will be mentioned somewhere.

We will mainly consider the Cauchy problem with initial condition

$$u(x,0) = u_0(x), \qquad x \in \mathbb{R}^N, \tag{1.19}$$

where $u_0(x)$ is a nonnegative and locally integrable function.

Since quasilinear parabolic equations with degeneracy do not have classical solutions in general, it is necessary to generalize the notion of solutions.

Let $Q_T = \mathbb{R}^N \times (0,T)$ and G be a subdomain of Q_T.

Definition 1.1.1 A nonnegative function u is called a generalized solution of the equation (1.14) on G, if $u, A(u) \in L^1_{loc}(G)$ and u satisfies

$$\iint_G \left(u\frac{\partial \varphi}{\partial t} + A(u)\Delta\varphi \right) dxdt = 0 \tag{1.20}$$

for any $\varphi \in C_0^\infty(G)$.

Obviously, if u is a generalized solution of (1.14) on G, and the weak derivatives $\dfrac{\partial A(u)}{\partial x_i} \in L^1_{loc}(G)$ $(i = 1, 2, \cdots, N)$, then (1.20) can be transformed to the form

$$\iint_G \left(u\frac{\partial \varphi}{\partial t} - \nabla A(u) \cdot \nabla\varphi \right) dxdt = 0. \tag{1.21}$$

Definition 1.1.2 A nonnegative function u is called a generalized solution of the Cauchy problem (1.14), (1.19) on Q_T, if u is a generalized solution of the equation (1.14) on Q_T and u satisfies

$$\lim_{t\to 0^+} \int_{\mathbb{R}^N} u(x,t)h(x)dx = \int_{\mathbb{R}^N} u_0(x)h(x)dx \tag{1.22}$$

for any $h \in C_0^\infty(\mathbb{R}^N)$.

If $u, A(u) \in L^1_{loc}(\overline{Q}_T)$ (we mean $u, A(u) \in L^1(G)$ for any bounded subdomain G of Q_T.), then Definition 1.1.2 is equivalent to the following

Definition 1.1.3 A nonnegative function u is called a generalized solution of the Cauchy problem (1.14), (1.19) on Q_T, if $u, A(u) \in L^1_{loc}(\overline{Q}_T)$ and u satisfies

$$\iint_{Q_T} \left(u\frac{\partial \varphi}{\partial t} + A(u)\Delta\varphi \right) dxdt + \int_{\mathbb{R}^N} u_0(x)\varphi(x,0)dx = 0 \tag{1.23}$$

for any $\varphi \in C^\infty(\overline{Q}_T)$,which vanishes for large $|x|$ and $t = T$.

The function φ in the above definitions (and also in the definitions below)will be called a test function.

Proof of equivalence. Let u be a generalized solution of (1.14), (1.19) in Definition 1.1.2, and φ be a test function in Definition 1.1.3. Choose $\varphi\eta_\varepsilon$ as a test function in (1.20) where $\eta_\varepsilon \in C_0^\infty(0,T)$ such that $\eta_\varepsilon(t) = 1$ for $t \in (\varepsilon, T-\varepsilon)$, $|\eta'_\varepsilon(t)| \le \frac{C}{\varepsilon}$ for $t \in (0,T)$ with constants $\varepsilon \in (0,T)$ and C. Then we have

$$\iint_{Q_T}\left(u\frac{\partial\varphi}{\partial t} + A(u)\Delta\varphi\right)\eta_\varepsilon dxdt + \iint_{Q_T} u\varphi\eta'_\varepsilon dxdt = 0. \tag{1.24}$$

The first integral tends to the first term of (1.23) as $\varepsilon \to 0$. Moreover

$$\begin{aligned} &\iint_{Q_T} u\varphi\eta'_\varepsilon dxdt - \int_{\mathbb{R}^N} u_0(x)\varphi(x,0)dx \\ = &\int_{T-\varepsilon}^T\int_{\mathbb{R}^N} u\varphi\eta'_\varepsilon dxdt \\ &+\int_0^\varepsilon \eta'_\varepsilon(t)\left(\int_{\mathbb{R}^N} u(x,t)(\varphi(x,t)-\varphi(x,0))\right)dt \\ &+\int_0^\varepsilon \eta'_\varepsilon(t)\,(u(x,t)-u_0(x))\varphi(x,0)dx)\,dt \\ = &\; I_1 + I_2 + I_3. \end{aligned}$$

Since it is clear that $\lim\limits_{\varepsilon\to 0} I_2 = 0$, $\varphi(x,T) = 0$ implies $\lim\limits_{\varepsilon\to 0} I_1 = 0$ and (1.22) implies $lim_{\varepsilon\to 0} I_3 = 0$, we may assert that the second integral tends to the second term of (1.23) as $\varepsilon \to 0$. Thus u is a generalized solution of (1.14), (1.19) in Definition 1.1.3.

Conversely, let u be a generalized solution of (1.14), (1.19) in Definition 1.1.3,. Obviously, (1.23) implies that u is a generalized solution of (1.14) on Q_T. It remains to prove (1.22). For any $t_0 \in (0,T)$, choose $\varphi = h\eta_\varepsilon$ in (1.23) where $h \in C_0^\infty(\mathbb{R}^N)$, $\eta_\varepsilon(t) \in C^\infty[0,T]$ such that $\eta_\varepsilon(t) = 1$ for $t \in [0, t_0-\varepsilon]$, $\eta_\varepsilon(t) = 0$ for $t \in [t_0+\varepsilon, T]$, $|\eta'_\varepsilon(t)| \le \frac{C}{\varepsilon}$ for $t \in [0,T]$ and some constants C and $\varepsilon \in (0, t_0)$. Then we have

$$\int_{t_0-\varepsilon}^{t_0+\varepsilon}\int_{\mathbb{R}^N} uh\eta'_\varepsilon dxdt + \int_0^{t_0+\varepsilon}\int_{\mathbb{R}^N} A(u)\Delta h\eta_\varepsilon dxdt + \int_{\mathbb{R}^N} u_0(x)h(x)dx = 0,$$

namely

$$\int_{t_0-\varepsilon}^{t_0+\varepsilon} \eta_\varepsilon'(t)\left(\int_{\mathbb{R}^N} u(x,t)h(x)dx - \int_{\mathbb{R}^N} u(x,t_0)h(x)dx\right)dt$$

$$-\int_{\mathbb{R}^N} u(x,t_0)h(x)dx + \int_0^{t_0+\varepsilon}\int_{\mathbb{R}^N} A(u)\Delta h\eta_\varepsilon dxdt$$

$$+\int_{\mathbb{R}^N} u_0(x)h(x)dx = 0. \tag{1.25}$$

The absolute value of the first term of (1.25) is dominated by

$$\frac{C}{\varepsilon}\int_{t_0-\varepsilon}^{t_0+\varepsilon}\left|\int_{\mathbb{R}^N} u(x,t)h(x)dx - \int_{\mathbb{R}^N} u(x,t_0)h(x)dx\right|dt$$

which tends to zero as $\varepsilon \to 0$ for almost all t_0. In addition

$$\lim_{\varepsilon\to 0}\int_0^{t_0+\varepsilon}\int_{\mathbb{R}^N} A(u)\Delta h\eta_\varepsilon dxdt = \int_0^{t_0}\int_{\mathbb{R}^N} A(u)\Delta hdxdt.$$

Thus we may conclude from (1.25) that

$$\int_0^{t_0}\int_{\mathbb{R}^N} A(u)\Delta hdxdt - \int_{\mathbb{R}^N} u(x,t_0)h(x)dx + \int_{\mathbb{R}^N} u_0(x)h(x)dx = 0$$

for almost all $t_0 \in (0,T)$ and hence for all $t_0 \in (0,T)$, which implies (1.22). This means that u is a generalized solution in Definition 1.1.2.

If $\dfrac{\partial A(u)}{\partial x_i} \in L^1_{loc}(\overline{Q}_T)$ $(i = 1,\cdots,N)$, then (1.23) can be transformed to the form

$$\iint_{Q_T}\left(u\frac{\partial\varphi}{\partial t} - \nabla A(u)\cdot\nabla\varphi\right)dxdt + \int_{\mathbb{R}^N} u_0(x)\varphi(x,0)dx = 0. \tag{1.26}$$

□

Now we define generalized solutions of the boundary value problem.

Let $\Omega \subset \mathbb{R}^N$ be a domain with appropriately smooth boundary $\partial\Omega$. Denote $\Omega_T = \Omega\times(0,T)$, $\Gamma = \partial\Omega\times(0,T)$. Consider the first boundary value problem for the equation (1.14) on Ω_T, with boundary value condition

$$u(x,t)|_\Gamma = g(x,t) \quad \text{for } (x,t) \in \Gamma \tag{1.27}$$

and initial value condition

$$u(x,0) = u_0(x), \quad x \in \Omega. \tag{1.28}$$

Definition 1.1.4 A nonnegative function u is called a generalized solution of the boundary value problem (1.14), (1.27), (1.28) on Ω_T, if $u, A(u) \in L^1(\Omega_T)$ and u satisfies

$$\iint_{\Omega_T}\left(u\frac{\partial\varphi}{\partial t}+A(u)\Delta\varphi\right)dxdt+\int_\Omega u_0(x)\varphi(x,0)dx-\int_\Gamma A(g)\frac{\partial\varphi}{\partial n}ds=0 \tag{1.29}$$

for any $\varphi \in C^\infty(\overline{\Omega}_T)$ which vanishes for $(x,t) \in \Gamma$, $t = T$ and large $|x|$, where n denotes the outward normal to Γ.

If $\dfrac{\partial A(u)}{\partial x_i} \in L^1(\Omega_T)$ $i = 1, 2, \cdots, N$, then Definition 1.1.4 is equivalent to the following

Definition 1.1.5 A nonnegative function u is called a generalized solution of the boundary value problem (1.14), (1.27), (1.28) on Ω_T, if $u, A(u) \in L^1(\Omega_T)$ and u satisfies

$$\iint_{\Omega_T}\left(u\frac{\partial\varphi}{\partial t}+A(u)\Delta\varphi\right)dxdt+\int_\Omega u_0(x)\varphi(x,0)dx=0 \tag{1.30}$$

for any $\varphi \in C^\infty(\overline{\Omega}_T)$ which vanishes when $|x|$ is large enough, $t = T$ and in a neighborhood of Γ, and u has boundary value $u|_\Gamma$, namely, the limit $\lim\limits_{\substack{y\to x\\ y\in\Omega}} u(y,t)$ exists for almost all $(x,t) \in \Gamma$, and

$$\lim_{\substack{y\to x\\ y\in\Omega}} u(y,t) = g(x,t). \tag{1.31}$$

It is not difficult to check that if $\dfrac{\partial A(u)}{\partial x_i} \in L^1(\Omega_T)$ $(i = 1, 2, \cdots, N)$, then Definition 1.1.4 is equivalent to Definition 1.1.5.

To prove, first let u be a generalized solution of (1.14), (1.27), (1.28) in Definition 1.1.4. Then clearly (1.30) holds. Since $\dfrac{\partial A(u)}{\partial x_i} \in L^1(\Omega_T)$ $(i = 1, 2.\cdot, N)$ implies the existence of the boundary value of $A(u)$, and hence of u itself on Γ and $A(u) \in L^1(\Gamma)$. To prove (1.31), let φ be a test function in Definition 1.1.4 with $\varphi(x,0) = 0$. Set $\varphi = \varphi_1 + \varphi_2$, $\varphi_1 = \varphi\xi_\varepsilon$, $\varphi_2 = \varphi(1-\xi_\varepsilon)$, where $\xi_\varepsilon \in C^\infty(\Omega)$ satisfies $\text{supp}\xi_\varepsilon \subset \Omega$, $0 \le \xi_\varepsilon \le 1$, $\xi_\varepsilon(x) = 1$ for $x \in \Omega_\varepsilon = \{x \in \Omega; \text{dist}(x, \partial\Omega) \ge \varepsilon\}$ $(\varepsilon > 0)$, $|\nabla\xi_\varepsilon| \le \dfrac{C}{\varepsilon}$ for

$x \in \Omega$. Then from (1.29) we obtain

$$\iint_{\Omega_T} \left(u\frac{\partial \varphi_1}{\partial t} + A(u)\Delta\varphi_1 \right) dxdt + \iint_{\Omega_T} \left(u\frac{\partial \varphi_2}{\partial t} + A(u)\Delta\varphi_2 \right) dxdt + \int_\Gamma u_0(x)\frac{\partial \varphi}{\partial n} ds = 0. \quad (1.32)$$

Using (1.29) for $\varphi = \varphi_1$ shows that the first term of (1.32) equals zero. Moreover

$$\iint_{\Omega_T} u\frac{\partial \varphi_2}{\partial t} dxdt = \iint_{\Omega_T} u(1-\xi_\varepsilon)\frac{\partial \varphi}{\partial t} dxdt \longrightarrow 0 \qquad (\varepsilon \to 0),$$

$$\begin{aligned}
\iint_{\Omega_T} A(u)\Delta\varphi_2 dxdt &= \int_\Gamma A(u)\Big|_\Gamma \frac{\partial \varphi}{\partial n} ds - \iint_{\Omega_T} \nabla A(u) \cdot \nabla\varphi_\varepsilon dxdt \\
&= \int_\Gamma A(u)\Big|_\Gamma \frac{\partial \varphi}{\partial n} ds - \iint_{\Omega_T} \nabla A(u)(1-\xi_\varepsilon)\nabla\varphi dxdt \\
&\quad + \iint_{\Omega_T} \nabla A(u)\varphi \cdot \nabla\xi_\varepsilon dxdt \\
&\longrightarrow \int_\Gamma A(u)\Big|_\Gamma \frac{\partial \varphi}{\partial n} ds \qquad (\varepsilon \to 0).
\end{aligned}$$

Here we have used the properties of ξ_ε and the fact $\varphi|_\Gamma = 0$. Thus, from (1.32) we derive

$$\int_\Gamma (A(u)|_\Gamma - A(g)) \frac{\partial \varphi}{\partial n} ds = 0$$

which implies (1.31).

Conversely, let u be a generalized solution of (1.14), (1.27), (1.28) in Definition 1.1.5 and φ be a test function in Definition 1.1.4 Then

$$\iint_{\Omega_T} \left(u\xi_\varepsilon\frac{\partial \varphi}{\partial t} + A(u)\Delta(\varphi\xi_\varepsilon) \right) dxdt + \int_\Omega u_0(x)\varphi(x,0)\xi_\varepsilon(x)dx = 0. \quad (1.33)$$

Since $\varphi|_\Gamma = 0$, we have

$$\left| \iint_{\Omega_T} \varphi\nabla A(u) \cdot \nabla\xi_\varepsilon dxdt \right| \le \frac{C}{\varepsilon}\int_0^T \int_{\Omega\setminus\Omega_\varepsilon} |\varphi| \cdot |\nabla A(u)| dxdt \longrightarrow 0 \quad (\varepsilon \to 0).$$

Hence

$$\begin{aligned}
&\iint_{\Omega_T} \varphi \nabla A(u) \cdot \nabla(\varphi \xi_\varepsilon) dxdt \\
= & -\iint_{\Omega_T} \xi_\varepsilon \nabla A(u) \nabla \varphi dxdt - \iint_{\Omega_T} \varphi \nabla A(u) \cdot \nabla \xi_\varepsilon dxdt \\
& \longrightarrow -\iint_{\Omega_T} \nabla A(u) \cdot \nabla \varphi dxdt \qquad (\varepsilon \to 0)
\end{aligned}$$

and from (1.33),

$$\iint_{\Omega_T} \left(u \frac{\partial \varphi}{\partial t} - \nabla A(u) \nabla \varphi \right) dxdt + \int_\Omega u_0(x) \varphi(x, 0) = 0,$$

which implies (1.29) by integrating by parts.

The following assertion is valid obviously, which will be used in the sequel.

Proposition 1.1.1 *Let u be a generalized solution of the Cauchy problem (1.14), (1.19) and* $\dfrac{\partial A(u)}{\partial x_i} \in L^1_{loc}(Q_T)$, $(i = 1, 2, \cdots, N)$. *Then for any smooth domain* $\Omega \subset \mathbb{R}^N$, *u is a generalized solution of the boundary value problem for the equation (1.14) on* $\Omega_T = \Omega \times (0, T)$ *with boundary value*

$$g(x, t) = u(x, t)|_\Gamma$$

and initial value $u_0(x)$.

Remark 1.1.1 Sometimes we need to use the notion of generalized super (sub)-solutions. To define the generalized super-solutions(sub-solutions), it suffices to replace "=" in (1.20) by "$\le$" ("$\ge$") and require the test function φ to be nonnegative. To define the generalized super-solutions(sub-solutions) of the Cauchy problem (1.14), (1.19), it suffices to replace "=" in (1.23) by "$\le$"("$\ge$") and require φ to be nonnegative, or replace "=" in (1.20) and (1.22) by "$\le$"("$\ge$") and require $\varphi \ge 0, h \ge 0$. Similarly, to define the generalized super-solutions(sub-solutions) of the boundary value problem (1.14), (1.27), (1.28), it suffices to replace "=" in (1.29) by "$\le$"("$\ge$") and require $\varphi \ge 0$, or replace "=" in (1.30) and (1.31) by "$\le$"("$\ge$") and "$\ge$"("$\le$") respectively and require $\varphi \ge 0$.

Remark 1.1.2 If u is a generalized solution of the Cauchy problem

(1.14), (1.19) in Definition 1.1.3, then for any $\tau \in (0, T)$ there holds

$$\begin{aligned}\iint_{Q_\tau} \left(u\frac{\partial \varphi}{\partial t} + A(u)\Delta\varphi \right) dxdt - \int_{\mathbb{R}^N} u(x,\tau)\varphi(x,\tau)dx \\ + \int_{\mathbb{R}^N} u_0(x)\varphi(x,0)dx = 0\end{aligned} \tag{1.34}$$

for any $\varphi \in C^\infty(\overline{Q}_T)$ which vanishes when $|x|$ is large enough, where $Q_\tau = \mathbb{R}^N \times (0, \tau)$.

To prove, we choose $\varphi\eta_\varepsilon$ as a test function in (1.23), where $\eta_\varepsilon \in C^\infty[0, T]$ satisfies $\eta_\varepsilon(t) = 1$ for $t \in [0, \tau - \varepsilon]$, $\eta_\varepsilon(t) = 0$ for $t \in [\tau, T]$, $|\eta'_\varepsilon(t)| \leq \dfrac{C}{\varepsilon}$. Then we obtain

$$\begin{aligned}\iint_{Q_\tau} \eta_\varepsilon \left(u\frac{\partial \varphi}{\partial t} + A(u)\Delta\varphi \right) dxdt + \iint_{Q_\tau} u\varphi\eta'_\varepsilon dxdt \\ + \int_{\mathbb{R}^N} u_0(x)\varphi(x,0)dx = 0.\end{aligned} \tag{1.35}$$

Letting $\varepsilon \to 0$ and noticing that

$$\begin{aligned} & \left| \iint_{Q_\tau} u\varphi\eta'_\varepsilon dxdt - \int_{\mathbb{R}^N} u(x,\tau)\varphi(x.\tau)dx \right| \\ = & \left| \iint_{Q_\tau} \eta'_\varepsilon(t) \left(u(x,t)\varphi(x,t) - u(x,\tau)\varphi(x,\tau) \right) dxdt \right| \\ \leq & \frac{C}{\varepsilon} \int_{\tau-\varepsilon}^{\tau+\varepsilon} \left| \int_{\mathbb{R}^N} u(x,t)\varphi(x,t)dxdt - \int_{\mathbb{R}^N} u(x,\tau)\varphi(x,\tau)dx \right| dt \\ \longrightarrow & \ 0, \qquad (\varepsilon \to 0)\end{aligned}$$

from (1.35) we obtain (1.34).

It is clear that if for any $\tau \in (0, T)$ and any $\varphi \in C^\infty(\overline{Q}_T)$ vanishing when $|x|$ is large enough, (1.34) holds, then u is a generalized solution of the Cauchy problem (1.14), (1.19) in Definition 1.1.3. This means that we may use (1.34) to give another equivalent definition of generalized solutions of the Cauchy problem (1.14), (1.19).

Similarly it is easy to see that if u is a generalized solution of the boundary value problem (1.14), (1.27), (1.28) in Definition 1.1.4, then for any

$\tau \in (0,T)$, there holds

$$\begin{aligned}\iint_{\Omega_\tau}\left(u\frac{\partial\varphi}{\partial t}+A(u)\Delta\varphi\right)dxdt-\int_\Omega u(x,\tau)\varphi(x,\tau)dx\\+\int_\Omega u_0(x)\varphi(x,0)dx-\int_{\Gamma_\tau}A(g)\frac{\partial\varphi}{\partial n}dS=0\end{aligned} \tag{1.36}$$

for any $\varphi \in C^\infty(\overline{\Omega}_T)$ which vanishes when $|x|$ is large enough, $t = T$ and $(x,t) \in \Gamma$, where $\Omega_\tau = \Omega \times (0,\tau)$, $\Gamma_\tau = \partial\Omega \times (0,\tau)$.

1.1.3 *Special solutions*

We will give some special solutions of the equation (1.17) to conclude this section.

1. **Barenblatt solution**

$$B_m(x,t) = t^{-k}\left(\left(1-\frac{k(m-1)}{2mN}\frac{|x|^2}{t^{2k/N}}\right)_+\right)^{1/(m-1)}, \tag{1.37}$$

where $k = \left(m-1+\dfrac{2}{N}\right)^{-1}$, $u_+ = \max(u,0)$, N is the dimension of $\mathbb{R}^N$. Its initial value is the Dirac measure.

2. **Pressure solution of degree two**

$$u(x,t) = \left(\frac{T_0|x|^2}{T_0-t}\right)^{1/(m-1)}, \tag{1.38}$$

where $T_0 = \dfrac{m-1}{2m(2+N(m-1))}$. Its initial value is $|x|^{2/(m-1)}$.

3. **Pressure solution of degree one** $(N=1)$

$$u(x,t) = \left(\frac{m-1}{m}c\,(ct\pm x)_+\right)^{1/(m-1)}, \tag{1.39}$$

where $c > 0$ is an arbitrary constant. Its initial value is

$$\left(\frac{m-1}{m}c(\pm cx)_+\right)^{1/(m-1)}.$$

In fact, the function (1.38) is a classical solution of (1.17) on Q_{T_0}; this can be checked immediately.

For any fixed $0 < t < T$, the function (1.37) has a compact support

$$|x|^2 \leq \frac{2mNt^{2k/N}}{k(m-1)}.$$

To check that the function (1.37) is a generalized solution of the equation (1.17), it suffices to observe that $B_m(x,t)$ is a classical solution of (1.17) in the domain $\{(x,t); B_m(x,t) > 0\}$ and both $B_m(x,t)$ and $\frac{\partial}{\partial x_i} B_m^m(x,t)$ $(i = 1, 2, \cdots, N)$ vanish at the lateral boundary of the domain:

$$|x|^2 = \frac{2mNt^{2k/N}}{k(m-1)}.$$

Similarly, we can check that the function (1.39) is a generalized solution of the equation (1.17) in one dimension.

1.2 Existence and Uniqueness of Solutions: One Dimensional Case

In this section, we study the Cauchy problem for the filtration equation (1.14) in one dimension:

$$\frac{\partial u}{\partial t} = \frac{\partial^2 A(u)}{\partial x^2}, \tag{2.1}$$

$$u(x,0) = u_0(x), \tag{2.2}$$

where $u_0(x) \geq 0$ is a locally integrable function on $\mathbb{R}$ and $A(s) \in C^1[0, \infty)$ satisfies the conditions

$$A(s) > 0, A'(s) > 0, \text{for } s > 0, A(0) = A'(0) = 0. \tag{2.3}$$

Denote $Q_T = \mathrm{R} \times (0, T)$.

1.2.1 *Uniqueness of solutions*

We first discuss the uniqueness of generalized solutions of the Cauchy problem.

Theorem 1.2.1 *The Cauchy problem (2.1), (2.2) has at most one generalized solution u bounded together with the weak derivative* $\frac{\partial A(u)}{\partial x}$.

Proof. First notice that if both u and $\dfrac{\partial A(u)}{\partial x}$ are bounded, then by approximation, the integral identity in the definition of generalized solutions (Definition 1.1.3) holds for any $\varphi \in W^{1,\infty}(Q_T)$ vanishing when $|x|$ is large enough and $t = T$. Such functions will be also called test functions.

Now let u_1, u_2 be generalized solutions of (2.1), (2.2). Then u_1 and u_2 satisfy (1.26). Hence we have

$$\iint_{Q_T} \frac{\partial \varphi}{\partial t}(u_1 - u_2)dxdt = \iint_{Q_T} \frac{\partial \varphi}{\partial x}\left(\frac{\partial A(u_1)}{\partial x} - \frac{\partial A(u_2)}{\partial x}\right) dxdt \tag{2.4}$$

for any test function φ.

If

$$\varphi(x,t) = \int_T^t (A(u_1(x,\tau)) - A(u_2(x,\tau)))d\tau \tag{2.5}$$

could be chosen as a test function, then from (2.4) we would have

$$\begin{aligned}
&\iint_{Q_T} (A(u_1) - A(u_2))(u_1 - u_2)dxdt \\
= &\iint_{Q_T} \int_T^t \left(\frac{\partial A(u_1(x,\tau))}{\partial x} - \frac{\partial A(u_2(x,\tau))}{\partial x}\right) d\tau \\
&\cdot \left(\frac{\partial A(u_1(x,t))}{\partial x} - \frac{\partial A(u_2(x,t))}{\partial x}\right) dxdt \\
= &\frac{1}{2}\iint_{Q_T} \frac{\partial}{\partial t}\left(\int_T^t \left(\frac{\partial A(u_1)}{\partial x} - \frac{\partial A(u_2)}{\partial x}\right) d\tau\right)^2 dxdt \\
= &-\frac{1}{2}\int_{-\infty}^{\infty}\left(\int_T^0 \left(\left(\frac{\partial A(u_1)}{\partial x} - \frac{\partial A(u_2)}{\partial x}\right) d\tau\right)\right)^2 dx \le 0
\end{aligned}$$

and hence $u_1 = u_2$ a.e. on $\mathbb{R}$ due to the condition (2.3).

However the function φ defined by (2.5) can not play the role of a test function since in general it does not vanish for large $|x|$, although we have $\varphi \in W^{1,\infty}(Q_T)$ and $\varphi(x,T) = 0$.

A natural idea is to consider the following cut-off function instead of φ:

$$\varphi_n(x,t) = \alpha_n(x)\int_T^t (A(u_1) - A(u_2))d\tau$$

where $\alpha_n(x)$ is a smooth function such that

$$\alpha_n(x) = 1, \qquad \text{when } |x| \le n-1;$$

$$\alpha_n(x) = 0, \qquad \text{when } |x| \ge n;$$

$$0 \le \alpha_n(x) \le 1, \qquad \text{when } n-1 < |x| < n;$$

$$|\alpha_n'(x)| \text{ is bounded uniformly.}$$

Substituting $\varphi = \varphi_n$ into (2.4) yields

$$\begin{aligned}
&\iint_{Q_T} \alpha_n(x)(A(u_1) - A(u_2))(u_1 - u_2)dxdt \\
= \; & -\frac{1}{2}\int_{-\infty}^{\infty} \alpha_n(x)\left(\int_T^0 \left(\frac{\partial A(u_1)}{\partial x} - \frac{\partial A(u_2)}{\partial x}\right) d\tau\right)^2 dx \\
& + \iint_{Q_n} \alpha_n'(x)\left(\int_T^t (A(u_1) - A(u_2))\, d\tau\right) \\
& \cdot \left(\frac{\partial A(u_1)}{\partial x} - \frac{\partial A(u_2)}{\partial x}\right) dxdt \\
\le \; & \iint_{Q_n} \alpha_n'(x)\left(\int_T^t (A(u_1) - A(u_2))\, \mathrm{d}\tau\right) \\
& \cdot \left(\frac{\partial A(u_1)}{\partial x} - \frac{\partial A(u_2)}{\partial x}\right) dxdt,
\end{aligned} \tag{2.6}$$

where $Q_n = \{(x,t); n-1 < |x| < n, 0 < t < T\}$.

Since u_i and $\dfrac{\partial A(u_i)}{\partial x}$ $(i = 1,2)$ are bounded and $\alpha_n(x)$ is bounded uniformly in n, the right hand side of (2.6), denoted by I_{2n}, is bounded in n, so is the left hand side, denoted by I_{1n}. Since $\alpha_n(x)$ increases with n, so does I_{1n}. Thus $\lim\limits_{n\to\infty} I_{1n}$ exists and we have

$$\lim_{n\to\infty} I_{1n} = \iint_{Q_T} (A(u_1) - A(u_2))(u_1 - u_2)dxdt.$$

Furthermore we can prove

$$\lim_{n\to\infty} I_{2n} = 0. \tag{2.7}$$

In fact, from the boundedness of u_i and $\dfrac{\partial A(u_i)}{\partial x}$ and the uniform boundedness of $\alpha_n'(x)$, we have

$$\begin{aligned}|I_{2n}| &\le C\iint_{Q_n}|A(u_1)-A(u_2)|dxdt\\ &\le C\left(\iint_{Q_n}(A(u_1)-A(u_2))^2dxdt\right)^{1/2}\\ &\le C\left(\iint_{Q_n}(A(u_1)-A(u_2))(u_1-u_2)dxdt\right)^{1/2},\end{aligned}$$

where C is a constant independent of n. The finiteness of the integral $\iint_{Q_T}(A(u_1)-A(u_2))(u_1-u_2)dxdt$ implies that the right hand side of (2.7) tends to zero as $n\to\infty$.

Letting $n\to\infty$ in (2.6) yields

$$\iint_{Q_T}(A(u_1)-A(u_2))(u_1-u_2)dxdt=0,$$

and hence $u_1=u_2$ a.e. on Q_T. This completes the proof of our theorem.

1.2.2 *Existence of solutions*

Next we discuss the existence of generalized solutions of the Cauchy problem (2.1), (2.2). □

Theorem 1.2.2 *Assume that u_0 is a nonnegative, continuous and bounded function on $\mathbb{R}$, $A(u_0)$ satisfies the Lipschitz condition, $A(s)$ is appropriately smooth and $A(s)\to+\infty$ as $s\to+\infty$. Then for any $T>0$ the Cauchy problem (2.1), (2.2) admits a continuous and bounded generalized solution u on Q_T such that $\dfrac{\partial A(u)}{\partial x}$ is bounded. Moreover, the solution u is classical in the domain $\{(x,t)\in Q_T, u(x,t)>0\}$.*

Proof. Denote $v_0=A(u_0)$ and choose a sequence of smooth functions $\{v_0^n(x)\}$ such that $\{v_0^n(x)\}$ uniformly converges to $v_0(x)$ as $n\to\infty$ and

$$\left|\frac{d}{dx}v_0^n(x)\right|\le K_0,\quad 0<v_0^n(x)\le M\quad (n=1,2,\cdots)$$

with some constants K_0 and M. Construct a sequence of smooth functions $\{w_n(x)\}$ such that

$$\begin{cases} w_n(x) = v_0^n(x), & |x| \le n-2, \\ w_n(x) = M, & |x| \ge n-1, \\ 0 < w_n(x) \le M, & \\ \left|\dfrac{d}{dx} w_n(x)\right| \le N = \max\{K_0, M\}, & (n = 1, 2, \cdots). \end{cases} \tag{2.8}$$

Now denote $v = A(u), \Phi(v) = A^{-1}(v)$. Then (2.1) becomes

$$\Phi'(v)\frac{\partial v}{\partial t} = \frac{\partial^2 v}{\partial x^2}. \tag{2.9}$$

Consider the initial-boundary value problem for (2.9) with conditions

$$v(x,0) = w_n(x), \quad v(\pm n, t) = M, \tag{2.10}$$

which admits a classical solution $v_n(x,t)$ on $G_n = (-n,n) \times (0,T)$ by virtue of the standard theory for parabolic equations.

From the maximum principle for classical solutions, we have

$$0 < \inf w_n(x) \le v_n(x,t) \le M. \tag{2.11}$$

We will further prove that

$$\left|\frac{\partial v_n}{\partial x}\right| \le N \quad \text{on } G_n. \tag{2.12}$$

For this purpose, let $P_n = \dfrac{\partial v_n}{\partial x}$ which satisfies

$$\Phi'(v_n)\frac{\partial P_n}{\partial t} = \frac{\partial^2 P_n}{\partial x^2} - \frac{1}{\Phi'(v_n)}\frac{\partial}{\partial x}\Phi'(v_n)\frac{\partial P_n}{\partial x}.$$

The maximum principle shows that

$$\max_{G_n}\left|\frac{\partial v_n}{\partial x}\right| \le \max_{\Gamma_n}\left|\frac{\partial v_n}{\partial x}\right|,$$

where $\Gamma_n = \partial_p G_n$ denotes the parabolic boundary of G_n. Since $\left|\dfrac{\partial v_n}{\partial x}\right|_{t=0} = |w_n'(x)| \le N$, to prove (2.12) it suffices to prove $\left|\dfrac{\partial v_n}{\partial x}\right|_{x=\pm n} \le N$.

Notice that v_n achieves its maximum M on the lateral boundary $x = n$, $0 \le t \le T$. Hence

$$\left.\frac{\partial v_n}{\partial x}\right|_{x=n} \ge 0. \tag{2.13}$$

Consider the auxiliary function

$$z_n = v_n - M(x - n + 1),$$

which satisfies

$$\begin{aligned} &\Phi'(v_n)\frac{\partial z_n}{\partial t} = \frac{\partial^2 z_n}{\partial x^2}, \quad \text{on } D_n = \{n-1 < x < n, 0 < t < T\} \\ &z_n(x,0) = w_n(x) - M(x-n+1) = M(n-x) \ge 0 \quad \text{on } [n-1, n] \\ &z_n(n,t) = 0, \quad z_n(n-1,t) = v_n(n-1,t) > 0 \quad \text{on } (0,T]. \end{aligned}$$

z_n achieves its minimum $\min_{\overline{D}_n} z_n$ on $x = n$, $0 \le t \le T$. Hence

$$\left.\frac{\partial z_n}{\partial x}\right|_{x=n} = \left.\frac{\partial v_n}{\partial x}\right|_{x=n} - M \le 0,$$

which combining with (2.13) yields $0 \le \left.\frac{\partial v_n}{\partial x}\right|_{x=n} \le M \le N$.

Similarly we can prove $\left|\frac{\partial v_n}{\partial x}\right|_{x=-n} \le M$.

The estimate (2.12) implies the uniform Lipschitz continuity of v_n in x: for any $(x,t),(y,t) \in \overline{Q}_T$) and n large enough such that $(x,t),(y,t) \in \overline{G}_n$),

$$|v_n(x,t) - v_n(y,t)| \le N|x-y|. \tag{2.14}$$

Denote $u_n = A^{-1}(v_n)$. Note that the assumption $A(s) \to \infty$ as $s \to \infty$ and $A'(s) > 0$ imply the existence of the inverse function $A^{-1}(v_n)$. From (2.9) we have

$$\frac{\partial u_n}{\partial t} = \frac{\partial^2 v_n}{\partial x^2}. \tag{2.15}$$

For any $(x,t),(y,t) \in \overline{Q}_T)$, choose n large enough such that $(x,t),(y,t) \in \overline{G}_n)$, $x+|\Delta t|^{1/2} \in [-n,n]$ with $\Delta t = t-s$. Integrating (2.15) yields

$$\begin{aligned}
&\int_x^{x+|\Delta t|^{1/2}} (u_n(z,t)-u_n(z,s))dz \\
&= \int_s^t \int_x^{x+|\Delta t|^{1/2}} \frac{\partial u_n}{\partial t} dy d\tau \\
&= \int_s^t \int_x^{x+|\Delta t|^{1/2}} \frac{\partial^2 v_n}{\partial x^2} dy d\tau \\
&= \int_s^t \left(\frac{\partial v}{\partial x}(x+|\Delta t|^{1/2},\tau) - \frac{\partial v_n}{\partial x}(x,\tau) \right) d\tau.
\end{aligned}$$

Hence from (2.12) we obtain

$$\left| \int_x^{x+|\Delta t|^{1/2}} (u_n(z,t)-u_n(z,s))dz \right| \le 2N|\Delta t|.$$

Using the mean value theorem for integrals, we see that there exists $x^* \in [x, x+|\Delta t|^{1/2}]$ such that

$$\int_x^{x+|\Delta t|^{1/2}} (u_n(z,t)-u_n(z,s))dz = (u_n(x^*,t)-u_n(x^*,s))|\Delta t|^{1/2}.$$

Thus

$$|u_n(x^*,t)-u_n(x^*,s)| \le 2N|\Delta t|^{1/2}$$

and hence for some constant C,

$$\begin{aligned}
&|v_n(x^*,t)-v_n(x^*,s)| = |A(u_n(x^*,t)) - A(u_n(x^*,s))| \\
= \;& |A'(\xi_n)||u_n(x^*,t)-u_n(x^*,s)| \le C|\Delta t|^{1/2}.
\end{aligned} \tag{2.16}$$

Combining (2.14) with (2.16) deduces

$$\begin{aligned}
& |v_n(x,t)-v_n(y,s)| \\
\le\;& |v_n(x,t)-v_n(x^*,t)| + |v_n(x^*,t)-v_n(x^*,s)| \\
& + |v_n(x^*,s)-v_n(y,s)| \\
\le\;& N\left(|x-x^*| + |\Delta t|^{1/2} + |x^*-y|\right) \\
\le\;& \widetilde{C}(|x-y| + |\Delta t|^{1/2})
\end{aligned}$$

for some constant $\widetilde{C}$ and proves that $\{v_n\}$ is uniformly Hölder continuous on G_n with Hölder exponent $\{1, \frac{1}{2}\}$. This together with (2.11), (2.12) implies that there exists a subsequence of $\{v_n\}$, supposed to be $\{v_n\}$ itself, such that $\{v_n\}$ converges uniformly to a certain function v on any compact subset of $\overline{Q}_T$ and $\left\{\dfrac{\partial v_n}{\partial x}\right\}$ weak star converges to $\dfrac{\partial v}{\partial x}$ on any bounded domain of Q_T. Furthermore it is easy to see that $\{u_n\}$ $(u_n = A^{-1}(v_n))$ converges uniformly to $u = A^{-1}(v)$ on any compact subset of $\overline{Q}_T$.

Obviously u and $\dfrac{\partial A(u)}{\partial x}$ are bounded. Given any test function φ, i.e. $\varphi \in C^\infty(\overline{Q}_T)$ such that $\varphi = 0$ when $|x|$ is large enough and $t = T$. Let n be large enough such that $\operatorname{supp}\varphi \subset G_n$. Multiplying (2.15) by φ, integrating over Q_T and integrating by parts, we obtain

$$\iint_{Q_T} \left(u_n \frac{\partial \varphi}{\partial t} - \frac{\partial A(u_n)}{\partial x} \frac{\partial \varphi}{\partial x} \right) dxdt + \int_{-\infty}^{\infty} \varphi(x, 0) A^{-1}(w_n(x)) dx = 0$$

from which it follows by letting $n \to \infty$ and noticing that for large n, $w_n(x) = v_0^n(x)$ and $w_n(x)$ converges uniformly to $v_0 = A(u_0)$ on any finite interval, that u satisfies (1.24), i.e u is a generalized solution of (2.1), (2.2).

Finally, we prove that the solution u is classical in $\{(x,t) \in Q_T, u(x,t) > 0\}$. Let $(x_0, t_0) \in Q_T$, $u(x_0, t_0) > 0$. Then there is a neighborhood $U \subset Q_T$ of (x_0, t_0) and constant $\alpha_0 > 0$ such that

$$u(x,t) \geq \alpha_0 > 0, \quad \text{for } (x,t) \in U.$$

Hence

$$u_n(x,t) \geq \frac{\alpha_0}{2} > 0 \quad \text{for } (x,t) \in U \text{ and large } n.$$

This means that for large n the solution u_n satisfies

$$\frac{\partial u_n}{\partial t} = \frac{\partial}{\partial x} \left(a(x,t) \frac{\partial u_n}{\partial x} \right)$$

with $a(x,t) = A'(u_n)$ being uniformly parabolic on U. From the standard theory for parabolic equations, it follows that for large n, u_n is uniformly bounded and equi-continuous in $C^2(U)$. Thus $u \in C^2(U)$ and satisfies (2.1) in the classical sense. The proof of Theorem 1.2.2 is complete. □

1.2.3 *Comparison theorems*

First we have

Theorem 1.2.3 *Assume the conditions in Theorem 1.2.1 and Theorem 1.2.2. Let u_1 and u_2 be bounded generalized solutions of the equation (2.1) with bounded weak derivatives $\dfrac{\partial A(u_i)}{\partial x}$ and initial data $u_{0i}(x)$ $(i=1,2)$. If $u_{01}(x) \le u_{02}(x)$ a.e. on $\mathbb{R}$, then $u_1 \le u_2$ a.e. on Q_T.*

Proof. As in the proof Theorem 1.2.2, we construct a sequence of approximate smooth functions $\{u_{0i}^n(x)\}$ $(i=1,2)$ of $\{u_{0i}(x)\}$ $(i=1,2)$. However, an additional condition $u_{01}^n(x) \le u_{02}^n(x)$ is required. Then, by the comparison theorem for classical solutions, we get $u_1^n(x,t) \le u_2^n(x,t)$ on G_n. Letting $n \to \infty$ and using the uniqueness of generalized solutions yield $u_1 \le u_2$ a.e on Q_T and complete the proof of our theorem. □

Proving comparison theorem in this way seems to be too roundabout. Many approaches applied to prove uniqueness theorem are also adapted in establishing comparison theorem. The proof of Theorem 1.3.1 is an example in this respect. As we will point out in Remark 1.3.1, we can prove

Theorem 1.2.4 *Let $u_i \in L^1(Q_T) \cap L^\infty(Q_T)$ $(i=1,2)$ be generalized solutions of (2.1) with initial data u_{0i}, $(i=1,2)$. If $u_{01}(x) \le u_{02}(x)$ a.e on R, then $u_1 \le u_2$ a.e on Q_T.*

Sometimes we need to apply comparison theorem for generalized solutions of the boundary valuc problcm.

Theorem 1.2.5 *Let u_i $(i=1,2)$ be generalized solutions of the boundary value problem for (2.1) on $\Omega_T = (\alpha,\beta) \times (0,T)$ with initial data $u_{0i}(x)$ $(i=1,2)$ and boundary value conditions*

$$u_i(\alpha,t) = g_i(t), \quad u_i(\beta,t) = h_i(t), \quad (i=1,2).$$

Assume that $u_i \in L^\infty(\Omega_T)$ $(i=1,2)$ and $\dfrac{\partial A(u_i)}{\partial x} \in L^1(\Omega_T)$ $(i=1,2)$. If $u_{01}(x) \le u_{02}(x)$ a.e on (α,β), $g_1(t) \le g_2(t)$, $h_1(t) \le h_2(t)$ a.e. on $(0,T)$, then $u_1 \le u_2$ a.e on Ω_T.

The basic idea of the proof is just the same as in Theorem 1.3.1. We leave it as an exercise and suggest that the readers do it after reading the proof of Theorem 1.3.1.

Remark 1.2.1 Comparison theorem is still valid if (α, β) is replaced by (α, ∞) or $(-\infty, \beta)$. In this case we need to require the generalized solutions considered to be bounded and integrable on $(\alpha, \infty) \times (0, T)$ or $(-\infty, \beta) \times (0, T)$. The proof is similar to §1.1.3 Theorem 1.3.1, but appropriate adaptations should be made.

Checking the proof of Theorem 1.2.4 and Theorem 1.2.5 we see that in fact we have the following more general results.

Theorem 1.2.6 *Let $u_i \in L^\infty(Q_T) \cap L^1(Q_T)$ $(i = 1, 2)$ and u_1 (u_2) be a generalized subsolution (supersolution) of (2.1) with initial data $u_{01}(x)$ $(u_{02}(x))$ respectively. If $u_{01}(x) \leq u_{02}(x)$ a.e on$\mathbb{R}$, then $u_1 \leq u_2$ a.e on Q_T.*

Theorem 1.2.7 *Let u_1 and u_2 be a generalized subsolution and a generalized supersolution of (2.1) on $\Omega_T = (\alpha, \beta) \times (0, T)$ with initial data $u_{0i}(x)$ $(i = 1, 2)$ and boundary data*

$$u_i(\alpha, t) = g_i(t) \quad u_i(\beta, t) = h_i(t) \quad (i = 1, 2)$$

respectively. Assume that $u_i \in L^\infty(\Omega_T)$, $\dfrac{\partial A(u_i)}{\partial x} \in L^1(\Omega_T)$ $(i = 1, 2)$. If $u_{01}(x) \leq u_{02}(x)$ a.e on (α, β), $g_1(t) \leq g_2(t)$, $h_1(t) \leq h_2(t)$ a.e on $(0, T)$, then $u_1 \leq u_2$ a.e on Ω.

Remark 1.2.2 Since Proposition 1.1.1 implies that generalized solutions of the Cauchy problem on Q_T are generalized solutions of the boundary value problem on Ω_T, we can compare generalized solutions of the Cauchy problem on any domain of the form Ω_T. This fact will be used in the sequel.

1.2.4 *Some extensions*

The results presented in Theorem 1.2.1 and Theorem 1.2.2 were first obtained by Oleinik, Kalashnikov and Zhou. In their creative work [OKZ] they studied also the first boundary value problem. What they considered are equations of the form

$$\frac{\partial u}{\partial t} = \frac{\partial^2 A(x, t, u)}{\partial x^2} \tag{2.17}$$

where $A(x, t, u)$ satisfies

$$A(x, t, 0) = \frac{\partial A}{\partial u}(x, t, 0) = 0,$$

$$A(x,t,u) > 0, \qquad \frac{\partial A(x,t,u)}{\partial u} > 0, \quad \text{for } u > 0.$$

Following [OKZ], the existence and uniqueness theory for quasilinear degenerate parabolic equations in one dimension has been developed in several directions.

1. **Extension to unbounded initial value.**

Kalashnikov studied the Cauchy problem for (2.17) with unbounded initial value $u_0(x)$ in [KA3]. He discussed the existence and uniqueness of generalized solutions in a class of functions which are continuous and nonnegative but need not to be bounded. The uniqueness of generalized solutions of the Cauchy problem is established in a class of functions denoted by E, under the assumption that $\dfrac{\partial A(x,t,u)}{\partial u}$ increases with u and

$$A(x,t,u) \le M_1 \left(\frac{\partial A(x,t,u)}{\partial u} \right)^p$$

for some positive constants M_1 and p. By E he meant the class of all functions $u(x,t)$ satisfying the condition

$$\frac{\partial A(x,t,u(x,t))}{\partial u} \le M_2(1+x^2), \tag{2.18}$$

where $M_2 > 0$ is a constant.

The result in [BA1], [BA2] shows that for the uniqueness of generalized solutions to hold, the condition (2.18) defining the class E can not be replaced by

$$\frac{\partial A(x,t,u)}{\partial u} \le M_1(1+x^2)^{1+\varepsilon}, \quad 0 < \varepsilon < 1,$$

no matter how small ε is.

As for the existence, the following result is presented in [KA3]: the Cauchy problem for (2.17) admits a generalized solution in the class E_ϵ defined by

$$\frac{\partial A(x,t,u(x,t))}{\partial u} \le M_1(1+x^2)^{1-\varepsilon} \quad (\varepsilon > 0)$$

under the assumptions

$$A(x,t,u) = A(x,u),$$

$$m\left(\frac{\partial A(x,u)}{\partial u}\right)^p \le A(x,u) \le M\left(\frac{\partial A(x,u)}{\partial u}\right)^p,$$

$$\frac{\partial A(x,u_0(x))}{\partial u} \le M_2(1+x^2)^{1-\varepsilon}, \quad (\varepsilon > 0) \tag{2.19}$$

and other appropriate conditions, where m, M, p are positive constants.

It is indicated in [KA3] that if the condition (2.19) is replaced by

$$\frac{\partial A(x,u_0(x))}{\partial u} \le M_1(1+x^2), \tag{2.20}$$

then one can not obtain the global solution in general. The pressure solution of degree two given in §1.1.1 also shows this matter. In fact, in case $A(x,t,u) = u^m$ $(m > 1)$, the condition (2.20) becomes

$$u_0(x) \le \left(\frac{M_2}{m}\right)^{1/(m-1)} (1+x^2)^{1/(m-1)}.$$

The initial data of the solution (1.33) satisfy this condition, however (1.38) does exist locally.

For the further work in this direction, we mention [KAM] in which the author studied the Cauchy problem for (2.1) with measure initial value

$$u(x,0) = E\delta(x) \qquad (E > 0)$$

and proved the existence and uniqueness of the so-called source-type solutions under appropriate conditions on $A(u)$.

2.**Extension to equations of other form.**

Kalashnikov [KA2] studied equations with absorption term

$$\frac{\partial u}{\partial t} = \frac{\partial^2 A(u)}{\partial x^2} - \psi(u),$$

where $A(u)$ and $\psi(u)$ satisfy

$$A'(u) > 0, \psi'(u) > 0, \quad \text{for } u > 0,$$

$$A'(0) \ge 0, \psi(u) \ge 0, \quad \text{for } u \ge 0.$$

He proved the existence and uniqueness of generalized solutions of the Cauchy problem under some additional conditions on $A(u)$ and $\psi(u)$, the solution obtained is Hölder continuous.

Kershner [KE1] has studied this kind of equations too. His method can be applied to more general equations

$$\frac{\partial u}{\partial t} = \frac{\partial^2 A(u)}{\partial x^2} + \frac{\partial \psi(u)}{\partial x} + \gamma(u).$$

In another paper, published a little early [KE2], he studied the first boundary value problem for equations of the special form

$$\frac{\partial u}{\partial t} = \frac{\partial^2 u^m}{\partial x^2} - b\frac{\partial u^n}{\partial x} - cu^l$$

with some positive constants m, n, l.

Gilding and Peletier [GP2] considered the Cauchy problem for the equation

$$\frac{\partial u}{\partial t} = \frac{\partial^2 u^m}{\partial x^2} + \frac{\partial u^n}{\partial x} \quad (m > 1, n > 0)$$

and proved that it admits at most one generalized solution whenever

$$n \geq \frac{1}{2}(m+1)$$

and it admits a generalized solution if $u_0 \geq 0$ is bounded and continuous with ${u_0}^m$ Lipschitz continuous. Soon afterwards these results were extended by Gilding [GI] to more general equations

$$\frac{\partial u}{\partial t} = \frac{\partial}{\partial x}\left(a(u)\frac{\partial u}{\partial x}\right) + b(u)\frac{\partial u}{\partial x},$$

where $a(u)$, $b(u)$ are continuous and $a(u) > 0$ for $u > 0$, $a(0) = 0$. He proved the uniqueness for the Cauchy problem under the condition

$$b^2(u) = O(a(u)) \qquad (u \to 0_+) \tag{2.21}$$

and proved the existence under the conditions that $a'(u)$ and $b'(u)$ are locally integrable , $ua'(u), ub'(u) \in L^1(0,1)$ and

$$A(u_0(x)) = \int_0^{u_0(x)} a(\tau)d\tau$$

is Lipschitz continuous.

It should be pointed out that as a condition to ensure the uniqueness of generalized solutions, (2.21) is unnatural. Wu Dequan [WD] and Dong Guangchang, Ye Qixiao [DY] have ever made efforts to improve this condition. Chen Yazhe [CH1] removed any kind of conditions in which $b(u)$

is controlled by $a(u)$ and established the uniqueness by assuming $a(u)$ to satisfy the following condition: there exist constants δ, $m > 1$ such that

$$\frac{1}{m}\left(\frac{u_1}{u_2}\right)^{m-1} \leq \frac{a(u_1)}{a(u_2)} \leq m$$

for any $0 < u_1 < u_2 < \delta$. A substantial progress in the study of uniqueness was made by Zhao Junning [ZH1] who did not require $a(u)$ and $b(u)$ to have any relation and only assumed $a(u) \geq 0$ to satisfy the condition that the set $E = \{s, a(s) = 0\}$ does not include any interior point; uniqueness was proved in $L^\infty(Q_T)$. The uniqueness of solutions for equations without any additional assumption except $a(u) \geq 0$ is much more difficult to prove. Vol'pert and Hudjaev [VH1] tried to do that in the class of BV functions; however, as pointed out in [WZQ1], their proof is incorrect. Based on a deep investigation of functions in BV and in a more general class BV_x, Wu Zhuoqun and Yin Jingxue [WY1] completed the proof of uniqueness of solutions in BV. (For details see Chapter 3).

1.3 Existence and Uniqueness of Solutions: Higher Dimensional Case

We are ready to turn to the filtration equation in higher dimension. We will concentrate our attention on its typical case, i.e, equation of the form

$$\frac{\partial u}{\partial t} = \Delta u^m \tag{3.1}$$

with constant $m > 1$, where $\Delta = \sum_{i=1}^{N} \frac{\partial^2}{\partial x_i^2}$, and only discuss the Cauchy problem, whose initial value condition is

$$u(x, 0) = u_0(x). \tag{3.2}$$

1.3.1 *Comparison theorem and uniqueness of solutions*

Denote $Q_T = \mathbb{R}^N \times (0, T)$.

Theorem 1.3.1 *Let $u_i \in L^1(Q_T) \cap L^\infty(Q_T)$ $(i = 1, 2)$ be generalized solutions of the Cauchy problem for (3.1) with initial data $u_{0i}(x)$, $(i = 1, 2)$. If $0 \leq u_{01}(x) \leq u_{02}(x)$ a.e on $\mathbb{R}^N$, then $u_1(x,t) \leq u_2(x,t)$ a.e on Q_T.*

Proof. From Remark 1.1.2 we have

$$\int_{\mathbb{R}^N} u_i(x,\tau)\varphi(x,\tau)dx - \int_{\mathbb{R}^N} u_{i0}(x)\varphi(x,0)dx$$
$$= \iint_{Q_\tau} \left(u_i \frac{\partial \varphi}{\partial t} + u_i^m \Delta\varphi \right) dxdt \quad (i = 1, 2)$$

for any $\tau \in (0,T)$ and $\varphi \in C^\infty(\overline{Q}_T)$ which vanishes when $|x|$ is large enough.

Let $z = u_1 - u_2$, $z_0 = u_{01} - u_{02}$. Then

$$\begin{aligned} & \int_{\mathbb{R}^N} z(x,\tau)\varphi(x,\tau)dx - \int_{\mathbb{R}^N} z_0(x)\varphi(x,0)dx. \\ = & \iint_{Q_\tau} \left(z\frac{\partial \varphi}{\partial t} + (u_1^m - u_2^m)\Delta\varphi \right) dxdt. \\ = & \iint_{Q_\tau} z\left(\frac{\partial \varphi}{\partial t} + a\Delta\varphi \right) dxdt, \end{aligned} \tag{3.3}$$

where

$$a(x,t) = \begin{cases} \dfrac{u_1^m(x,t) - u_2^m(x,t)}{u_1(x,t) - u_2(x,t)}, & \text{if}\, u_1(x,t) \neq u_2(x,t), \\ m u_1^{m-1}(x,t), & \text{if}\, u_1(x,t) = u_2(x,t). \end{cases}$$

If for any nonnegative function $g \in C_0^\infty(\mathbb{R}^N)$, the problem

$$\begin{cases} \dfrac{\partial \varphi}{\partial t} + a\Delta\varphi = 0, \\ \varphi(x,\tau) = g(x), \end{cases} \tag{3.4}$$

had a solution $\varphi \in C^\infty(\overline{Q}_\tau)$ which vanishes when $|x|$ is large enough, then from (3.3) we would obtain

$$\int_{\mathbb{R}^N} z(x,t)g(x)dx \le 0. \tag{3.5}$$

Here we have used the assumption $z_0(x) \le 0$ and the fact $\varphi(x,t) \ge 0$ which follows from the maximum principle. Therefore by the arbitrariness of $g(x)$ we get $z(x,\tau) \le 0$ or $u_1(x,\tau) \le u_2(x,\tau)$ a.e for $x \in \mathbb{R}^N$ and this is what we want to prove.

However since the coefficient a in (3.4) is merely a nonnegative and locally integrable function, (3.4) does not admit any smooth solution in general and even if (3.4) does admit, the solution can not have compact support in x in general. In view of this point, we replace a by

$$a_n = \rho_n * a + \frac{1}{n},$$

where ρ_n is a mollifier on Q_T and consider the boundary value problem

$$\begin{cases} \dfrac{\partial \varphi}{\partial t} + a_n \Delta \varphi = 0 & \text{for } |x| < R, \ \ 0 < t < \tau, \\ \varphi = 0 & \text{for } |x| = R, \ \ 0 < t < \tau, \\ \varphi(x, \tau) = g(x) & \text{for } |x| \le R, \end{cases} \tag{3.6}$$

where $R > R_0 + 1$ such that $\operatorname{supp} g(x) \subset B_{R_0} = \{x \in \mathbb{R}^N; |x| < R_0\}$. We choose ρ_n such that

$$\int_0^\tau \int_{B_R} (a - \rho_n * a)^2 \, dxdt \le \frac{1}{n^2}. \tag{3.7}$$

Let φ_n be a solution of (3.6). Extend the definition of φ_n to the whole Q_τ by setting $\varphi_n = 0$ outside the domain $|x| \le R$, $0 \le t \le \tau$. Since the extended function φ_n may not necessarily be a sufficiently smooth function on Q_τ, we use a function $\xi_R \in C_0^\infty(\mathbb{R}^N)$ with the following properties to "cut-off" φ_n:

$$\begin{cases} 0 \le \xi_R(x) \le 1, \\ \xi_R(x) = 1, \quad \text{for } |x| < R - 1, \\ \xi_R(x) = 0, \quad \text{for } |x| > R - \dfrac{1}{2}, \\ |\nabla \xi_R(x)|, |\Delta \xi_R(x)| \le C. \end{cases} \tag{3.8}$$

Here and below, we always use C to denote a universal constant independent of R, n, which may take different values on different occasions. Choosing $\varphi = \xi_R \varphi_n$ in (3.3) yields

$$\begin{aligned} &\int_{\mathbb{R}^N} z(x,\tau) g(x) \xi_R(x) dx - \int_{\mathbb{R}^N} z_0(x) \xi_R(x) \varphi_n(x,0) dx \\ = &\iint_{Q_\tau} (u_1^m - u_2^m)(2 \nabla \xi_R \nabla \varphi_n + \varphi_n \Delta \xi_R) \, dxdt. \\ &+ \iint_{Q_\tau} z \xi_R (a - a_n) \Delta \varphi_n dxdt \equiv I_n + J_n. \end{aligned} \tag{3.9}$$

Now we are ready to estimate I_n and J_n. Multiplying the equation in (3.6) by $\Delta\varphi_n$, integrating over $B_R \times (t, \tau)$ and applying Green's formula, we obtain

$$\frac{1}{2}\int_{B_R} |\nabla\varphi_n(x,t)|^2 dx + \int_t^\tau \int_{B_R} a_n(\Delta\varphi_n)^2 dx ds = \frac{1}{2}\int_{B_R} |\nabla g|^2 dx,$$

from which it follows in particular

$$\int_t^\tau \int_{B_R} |\nabla\varphi_n|^2 dx ds \le C, \tag{3.10}$$

$$\int_t^\tau \int_{B_R} a_n(\Delta\varphi_n)^2 dx ds \le C. \tag{3.11}$$

Using (3.8), (3.10) and noting that $u_i \in L^\infty(Q_T)$ $(i = 1,2)$ and φ_n is uniformly bounded yield

$$\begin{aligned} |I_n| &\le C\int_0^\tau \int_{B_R\setminus B_{R-1}} (u_1^m + u_2^m)(|\nabla\varphi_n| + 1) dx dt \\ &\le C\int_0^\tau \int_{B_R\setminus B_{R-1}} (u_1 + u_2) dx dt. \end{aligned} \tag{3.12}$$

Using (3.11) and noting that $u_i \in L^\infty(Q_T)$ $(i = 1,2)$ yield

$$\begin{aligned} |J_n| &\le C\left(\int_0^\tau \int_{B_R} \frac{(a-a_n)^2}{a_n} dx dt\right)^{1/2} \left(\int_0^\tau \int_{B_R} a_n(\Delta\varphi_n)^2 dx dt\right)^{1/2} \\ &\le C\left(\int_0^\tau \int_{B_R} \frac{(a-a_n)^2}{a_n} dx dt\right)^{1/2}. \end{aligned}$$

By virtue of (3.7) we further obtain

$$|J_n| \le C\sqrt{n}\left(\int_0^\tau \int_{B_R} \left(a - \rho_n * a - \frac{1}{n}\right)^2 dx dt\right)^{1/2} \le \frac{C}{\sqrt{n}}. \tag{3.13}$$

Combining (3.12), (3.13) with (3.9) and noting that $z_0(x) \le 0$, $\varphi_n \ge 0$, $\xi_R \ge 0$, we finally arrive at

$$\begin{aligned} &\int_{\mathbb{R}^N} z(x,\tau)g(x)\xi_R(x) dx \\ &\le \int_{\mathbb{R}^N} z_0(x)\varphi_n(x,0)\xi_R(x) dx + |I_n| + |J_n| \end{aligned}$$

$$\leq |I_n| + |J_n| \leq C \int_0^T \int_{B_R \setminus B_{R-1}} (u_1 + u_2) dx dt + \frac{C}{\sqrt{n}}.$$

Letting $n \to \infty$ and then $R \to \infty$ and noting that $u_i \in L^1(Q_T)$ $(i = 1, 2)$, we see that the right hand side tends to zero and hence (3.5) holds. The proof is complete. □

Remark 1.3.1 Checking the proof of Theorem 1.3.1, we see that the method applied is adapted to more general equations of the form

$$\frac{\partial u}{\partial t} = \Delta A(u) \tag{3.14}$$

with $A(u) \in C^1[0, \infty)$ and

$$A(s) > 0, A'(s) > 0 \text{ for } s > 0, A(0) = A'(0) = 0.$$

. In other words, using the same method we can prove that for generalized solutions of (3.14) in $L^1(Q_T) \cap L^\infty(Q_T)$, the comparison theorem is valid.

Remark 1.3.2 It should be pointed out that requiring a generalized solution u to belong to $L^1(Q_T) \cap L^\infty(Q_T)$ is too restrictive, which means that u must be "small" at infinity and excludes even the nonzero constant solution. Fortunately those generalized solutions determined by the initial data with compact support satisfy such condition.

As an immediate corollary of Theorem 1.3.1, we have

Theorem 1.3.2 *Suppose $0 \leq u_0 \in L^1(\mathbb{R}^N) \cap L^\infty(\mathbb{R}^N)$. Then the Cauchy problem (3.1), (3.2) admits at most one generalized solution in $L^1(Q_T) \cap L^\infty(Q_T)$.*

Similar to the proof of Theorem 1.3.1, we may obtain the following comparison theorem for the boundary value problem.

Theorem 1.3.3 *Let $u_i \in L^1(\Omega_T) \cap L^\infty(\Omega_T)$ $(i = 1, 2)$ be generalized solutions of (3.1) satisfying the initial value condition and boundary value condition*

$$u_i(x, 0) = u_{i0}(x), \quad \textit{for } x \in \Omega \quad (i = 1, 2)$$

and

$$u_i(x, t)|_\Gamma = g_i(x, t) \quad \textit{for } (x, t) \in \Gamma \quad (i = 1, 2),$$

where $\Omega \subset \mathbb{R}^N$ *is a smooth domain,* $\Gamma = \partial\Omega \times (0,T)$, $\Omega_T = \Omega \times (0,T)$. *If* $u_1(x) \le u_2(x)$ *a.e on* Ω *and* $g_1(x,t) \le g_2(x,t)$ *a.e on* Γ, *then* $u_1(x,t) \le u_2(x,t)$ *a.e on* Ω_T.

Remark 1.3.3 Since the generalized solution u of the Cauchy problem for (3.1) on Q_T with locally integrable derivatives $\dfrac{\partial A(u)}{\partial x_i}$ $(i = 1, 2, \cdots, N)$ is also a generalized solution of the boundary value problem for (3.1) on any domain of the form $\Omega_T = \Omega \times (0,T)$, we can use the comparison theorem on any Ω_T for generalized solutions of the Cauchy problem.

1.3.2 *Existence of solutions*

Now we discuss the existence of generalized solutions of the Cauchy problem (3.1), (3.2). First we prove

Theorem 1.3.4 *Assume that* $u_0 \in L^1(\mathbb{R}^N) \cap L^\infty(\mathbb{R}^N)$ *and* $u_0(x) \ge 0$. *Then the Cauchy problem (3.1), (3.2) admits a generalized solution* $u \in L^1(Q_T) \cap L^\infty(Q_T)$.

Proof. The basic idea of the proof is to regularize the initial value and then to establish some estimates on the approximate solutions to obtain the desired compactness.

Choose a sequence of positive numbers R_n and η_n such that

$$\eta_n R_n^N \to 0 \text{ as } n \to \infty, \tag{3.15}$$

and then construct $u_{0n} \in C_0^\infty(B_{R_n})$ such that

$$\|u_{0n}\|_{L^\infty(\mathbb{R}^N)} \le \|u_0\|_{L^\infty(\mathbb{R}^N)}, \tag{3.16}$$

$$\|u_{0n} - u_0\|_{L^1(\mathbb{R}^N)} \to 0 \text{ as } n \to \infty. \tag{3.17}$$

To this purpose, it suffices to define $v_{0n} = u_0$ for $|x| \le B_{R_n - 1}$, $v_{0n} = 0$ elsewhere and then mollify it.

Now consider the boundary value problem

$$\begin{cases} \dfrac{\partial u_n}{\partial t} = \Delta u_n^m & \text{for } (x,t) \in B_{R_n} \times (0,T) \\ u = \eta_n & \text{for } (x,t) \in \partial B_{R_n} \times (0,T) \\ u = u_{0n} + \eta_n & \text{for } x \in B_{R_n}. \end{cases} \tag{3.18}$$

According to the classical theory, (3.18) admits a smooth solution u_n. By maximum principle and using (3.16), we obtain

$$\eta_n \le u_n \le \|u_0\|_{L^\infty(\mathbb{R}^N)} + \eta_n.$$

Multiplying the equation in (3.18) by pu_n^{p-1} $(1 \le p < \infty)$ we derive

$$\begin{aligned}\frac{\partial u_n^p}{\partial t} &= pu_n^{p-1}\Delta u_n^m = p\operatorname{div}(u_n^{p-1}\nabla u_n^m) - p\nabla u_n^{p-1}\cdot\nabla u_n^m \\ &= p\operatorname{div}(u_n^{p-1}\nabla u_n^m) - mp(p-1)u_n^{m+p-3}|\nabla u_n|^2.\end{aligned}$$

Integrating both sides of the above equality over $B_{r_n} \times (0,T)$ and noting that $\dfrac{\partial u_n}{\partial \nu} \le 0$ on $\partial B_{R_n} \times (0,T)$, where ν denotes the outward normal to $\partial B_{R_n} \times (0,T)$, we have

$$\begin{aligned}&\int_{B_{R_n}} u_n^p(x,t)dx - \int_{B_{R_n}} (u_{0n}(x)+\eta_n)^p dx \\ = \ & p\int_0^t\int_{\partial B_{R_n}} u_n^{p-1}\frac{\partial u_n^m}{\partial \nu}d\sigma d\tau - mp(p-1)\int_0^t\int_{B_{R_n}} u_n^{m+p-3}|\nabla u_n|^2 dxd\tau \\ \le \ & -mp(p-1)\int_0^t\int_{B_{R_n}} u_n^{m+p-3}|\nabla u_n|^2 dxd\tau\end{aligned}$$

or

$$\begin{aligned}&\int_{B_{R_n}} u_n^p(x,t)dx + mp(p-1)\int_0^t\int_{B_{R_n}} u_n^{m+p-3}|\nabla u_n|^2 dxd\tau \\ \le \ & \int_{B_{R_n}} (u_{0n}(x)+\eta_n)^p dx.\end{aligned}$$

Using (3.15) and (3.17) we see that the right hand side of the above inequality tends to $\int_{\mathbb{R}^N} u_0^p dx$ as $n \to \infty$, which implies that for any fixed $p \in [1,\infty)$ the left hand side is bounded. In particular we have

$$\int_{B_{R_n}} u_n(x,t)dx \le C, \tag{3.19}$$

and

$$\int_0^T\int_{B_{R_n}} |\nabla u_n^m| dxdt = m^2\int_0^T\int_{B_{R_n}} u_n^{2(m-1)}|\nabla u_n|^2 dxdt \le C. \tag{3.20}$$

Multiplying the equation in (3.18) by $mu_n^{m-1}\dfrac{\partial u_n}{\partial t}$ gives

$$\frac{4m}{(m+1)^2}\left(\frac{\partial u_n^{(m+1)/2}}{\partial t}\right)^2 = \operatorname{div}\left(\frac{\partial u_n^m}{\partial t}\nabla u_n^m\right) - \frac{1}{2}\frac{\partial}{\partial t}|\nabla u_n^m|^2.$$

Integrating over $B_{R_n}\times(0,T)$ and noting that $\dfrac{\partial u_n}{\partial t} = 0$ on $\partial B_{R_n}\times(0,T)$ we obtain

$$\begin{aligned}
&\frac{4m}{(m+1)^2}\int_t^T\int_{B_{R_n}}\left(\frac{\partial u_n^{(m+1)/2}}{\partial t}\right)^2 dxd\tau \\
=\ & -\frac{1}{2}\int_{B_{R_n}}|\nabla u_n^m(x,T)|^2dx + \frac{1}{2}\int_{B_{R_n}}|\nabla u_n^m(x,t)|^2dx \\
\le\ & \frac{1}{2}\int_0^T\int_{B_{R_n}}|\nabla u_n^m(x,t)|^2dx.
\end{aligned}$$

Integrating with respect to $t\in(0,T)$ we further obtain

$$\begin{aligned}
&\frac{4m}{(m+1)^2}\int_0^T\int_t^T\int_{B_{R_n}}\left(\frac{\partial u_n^{(m+1)/2}}{\partial t}\right)^2 dxd\tau dt \\
=\ & \frac{4m}{(m+1)^2}\int_0^T\int_{B_{R_n}} t\left(\frac{\partial u_n^{(m+1)/2}}{\partial t}\right)^2 dxdt \\
\le\ & \frac{1}{2}\int_0^T\int_{B_{R_n}}|\nabla u_n^m(x,t)|^2dxdt.
\end{aligned}$$

Using (3.20) and the uniform boundedness of u_n, we finally derive

$$\begin{aligned}
&\int_0^T\int_{B_{R_n}} t\left(\frac{\partial u_n^m}{\partial t}\right)^2 dxdt \\
=\ & m^2\int_0^T\int_{B_{R_n}} tu_n^{2(m-1)}\left(\frac{\partial u_n}{\partial t}\right)^2 dxdt \\
=\ & \frac{4m^2}{(m+1)^2}\int_0^T\int_{B_{R_n}} tu_n^{m-1}\left(\frac{\partial u_n^{(m+1)/2}}{\partial t}\right)^2 dxdt \le C.
\end{aligned}$$

The estimates (3.19), (3.20), and the Kolmogrov theorem imply that for any $\delta\in(0,T)$, $R>0$, $\{u_n^m\}$ is strongly compact in $L^2(B_R\times(\delta,T)$ and hence there exits a subsequence of $\{u_n^m\}$, supposed to be $\{u_n^m\}$ itself, which converges almost everywhere to a certain function v on Q_T:

$$u_n^m \to v \quad \text{as } n \to \infty \text{ a.e on } Q_T,$$

namely

$$u_n \to u = v^{1/m} \quad \text{as } n \to \infty \text{ a.e on } Q_T.$$

Obviously $u \in L^\infty(Q_T)$ which together with (3.19) implies $u \in L^1(Q_T)$.

Given any $\varphi \in C^\infty(\overline{Q}_T)$ which vanishes when $|x|$ is large enough and $t = T$. We have for large n

$$\iint_{Q_T} \left(u_n \frac{\partial \varphi}{\partial t} + u_n^m \Delta \varphi \right) dxdt + \int_{\mathbb{R}^N} (u_{0n}(x) + \eta_n)\varphi(x,0)dx = 0.$$

Here we regard u_n as zero outside $\{(x,t); |x| \leq R_n, t \in [0,T]\}$. Letting $n \to \infty$ we see that u is a generalized solution of (3.1), (3.2). The proof is complete. □

Remark 1.3.4 Different from one dimensional case, the above existence theorem does not provide a continuous solution. We will see in §1.5 that to prove the continuity of generalized solutions of the equation (3.1) in higher dimension is quite difficult.

1.3.3 *Some extensions*

We have discussed the existence and uniqueness of nonnegative generalized solutions of the Cauchy problem for the equation (3.1) under the assumption $u_0 \geq 0$ and $u_0 \in L^1(\mathbb{R}^N) \cap L^\infty(\mathbb{R}^N)$. The existence and uniqueness theory has been extended in several directions.

1. It is natural to consider the nonnegative solutions from the viewpoint of physics. However in mathematics we need not to restrict ourselves to this consideration. For signed solutions, the equation (3.1) should be changed into the form

$$\frac{\partial u}{\partial t} = \Delta(|u|^{m-1}u). \tag{3.21}$$

The argument stated in §1.3.1 and §1.3.2 is also available for (3.21) with signed initial data $u_0 \in L^1(\mathbb{R}^N) \cap L^\infty(\mathbb{R}^N)$.

2. All methods stated in §1.1.1 and §1.1.2 can be used to treat the first initial-boundary problem for (3.21) whose initial value condition and

boundary value condition are

$$u(x,0) = u_0(x), \qquad \text{for } x \in \Omega \tag{3.22}$$

and

$$u(x,t) = 0 \quad \text{for } (x,t) \in \partial\Omega \times (0,T), \tag{3.23}$$

where $\Omega \subset \mathbb{R}^N$ is a bounded domain with smooth boundary. The existence and uniqueness of generalized solutions of the problem (3.21)–(3.23) can be proved in a similar way [CE]. One can also treat the problem with nonzero boundary value [SA1].

3. Requiring u_0 to belong to both $L^\infty(\mathbb{R}^N)$ and $L^1(\mathbb{R}^N)$ is too harsh. Fortunately this condition can be relaxed. It is not very difficult to prove the following result.

Theorem 1.3.5 *Assume that $u_0 \in L^\infty(\mathbb{R}^N)$ and $u_0(x) \geq 0$. Then the Cauchy problem (3.1) and (3.2) admits a generalized solution $u \in L^\infty(Q_T)$.*

Proof. Choose $0 \leq u_{0n} \in C_0^\infty(\mathbb{R}^N)$ such that

$$\|u_{0n}\|_{L^\infty(\mathbb{R}^N)} \leq \|u_0\|_{L^\infty(\mathbb{R}^N)}$$

and for any fixed $R > 0$

$$\|u_{0n} - u_0\|\|_{L^1(B_R)} \to 0 \qquad (n \to \infty).$$

Letting $u_{n,R}$ be a classical solution of the problem

$$\begin{cases} \dfrac{\partial u}{\partial t} = \Delta u^m & \text{for } (x,t) \in B_R \times (0,T), \\ u = \dfrac{1}{n} & \text{for } (x,t) \in \partial B_R \times (0,T), \\ u = u_{0n} + \dfrac{1}{n} & \text{for } x \in B_R, \end{cases} \tag{3.24}$$

where $R > R_n$. The maximum principle implies

$$\frac{1}{n} \leq u_{n,R}(x,t) \leq \|u_0\|_{L^\infty(\mathbb{R}^N)} + 1,$$

$$u_{n,R}(x,t) \leq u_{n,\widetilde{R}}(x,t) \quad \text{for } \widetilde{R} > R, (x,t) \in B_R \times (0,T).$$

Thus the limit

$$u_n(x,t) = \lim_{R\to\infty} u_{n,R}(x,t)$$

exists and u_n is a classical solution of the problem

$$\begin{cases} \dfrac{\partial u}{\partial t} = \Delta u^m, \\ u = u_{0n} + \dfrac{1}{n} \quad \text{for } x \in \mathbb{R}^N. \end{cases} \tag{3.25}$$

Obviously we have

$$\frac{1}{n} \le u_n(x,t) \le \|u_0\|_{L^\infty(\mathbb{R}^N)} + 1.$$

Let

$$\omega(x) = \exp(-\sqrt{1+|x|^2}).$$

Clearly for some constant C,

$$|\nabla \omega(x)| \le C\omega(x).$$

Now we substitute $u = u_n$ into the equation in (3.25) and multiply both sides by $(m+1)u_n^m\omega$. Then we arrive at

$$\begin{aligned} \omega \frac{\partial u_n^{m+1}}{\partial t} = \; & (m+1)\text{div}(\omega u_n^m \nabla u_n^m) - (m+1)\omega|\nabla u_n^m|^2 \\ & -(m+1)u_n^m \nabla\omega \cdot \nabla u_n^m. \end{aligned}$$

Integrating over Q_T and using Young's inequality we obtain

$$\begin{aligned} & \int_{\mathbb{R}^N} \omega(x) u_n^{m+1}(x,T)dx + (m+1)\int_0^T \int_{\mathbb{R}^N} \omega|\nabla u_n^m|^2 dxdt \\ \le \; & \int_{\mathbb{R}^N} \omega(x)\left(u_{0n}^{m+1}(x) + \frac{1}{n}\right)dx + C\int_0^T\int_{\mathbb{R}^N} \omega|\nabla u_n^m|^2 dxdt \\ \le \; & C\int_{\mathbb{R}^N} \omega(x)dx + \varepsilon\int_0^T\int_{\mathbb{R}^N} \omega|\nabla u_n^m|^2 dxdt \end{aligned}$$

with small $\varepsilon > 0$. From this it follows that, in particular

$$\int_0^T \int_{\mathbb{R}^N} \omega|\nabla u_n^m|^2 dxdt \le C. \tag{3.26}$$

Multiplying both sides of the equation in (3.25) by $m\omega u_n^{m-1}\frac{\partial u_n}{\partial t}$, integrating over $\mathbb{R}^N \times (t,T)$ and using Young's inequality again yield

$$\begin{aligned}
& \frac{4m}{(m+1)^2}\int_t^T\int_{\mathbb{R}^N}\omega\left(\frac{\partial u_n^{(m+1)/2}}{\partial t}\right)^2 dxd\tau \\
= & -\int_t^T\int_{\mathbb{R}^N}\frac{\partial u_n^m}{\partial t}\nabla\omega\cdot\nabla u_n^m dxdt - \frac{1}{2}\int_{\mathbb{R}^N}\omega(x)|\nabla u_n^m(x,T)|^2dx \\
& +\frac{1}{2}\int_{\mathbb{R}^N}\omega(x)|\nabla u_n^m(x,t)|^2dx \\
\leq & C\int_t^T\int_{\mathbb{R}^N}\left|\frac{\partial u_n^m}{\partial t}\right|\omega|\nabla u_n^m|dxd\tau + \frac{1}{2}\int_{\mathbb{R}^N}\omega(x)|\nabla u_n^m(x,t)|^2dx \\
\leq & C\int_t^T\int_{\mathbb{R}^N}\left|\frac{\partial u_n^{(m+1)/2}}{\partial t}\right|\omega|\nabla u_n^m|dxd\tau \\
& +\frac{1}{2}\int_{\mathbb{R}^N}\omega(x)|\nabla u_n^m(x,t)|^2dx \\
\leq & \varepsilon\int_t^T\int_{\mathbb{R}^N}\omega\left(\frac{\partial u_n^{(m+1)/2}}{\partial t}\right)^2 dxd\tau + C(\varepsilon)\int_t^T\int_{\mathbb{R}^N}\omega|\nabla u_n^m|^2dxd\tau \\
& +\frac{1}{2}\int_{\mathbb{R}^N}\omega(x)|\nabla u_n^m(x,t)|^2dx
\end{aligned}$$

with small $\varepsilon > 0$. Hence

$$\begin{aligned}
& \int_t^T\int_{\mathbb{R}^N}\omega\left(\frac{\partial u_n^{(m+1)/2}}{\partial t}\right)^2 dxd\tau \\
\leq & C\int_t^T\int_{\mathbb{R}^N}\omega|\nabla u_n^m|^2dxd\tau + \frac{1}{2}\int_{\mathbb{R}^N}\omega(x)|\nabla u_n^m(x,t)|^2dx.
\end{aligned}$$

Integrating over $(0,T)$ we further obtain

$$\begin{aligned}
& \int_0^T\int_{\mathbb{R}^N}t\omega\left(\frac{\partial u_n^{(m+1)/2}}{\partial t}\right)^2 dxdt \\
\leq & C\int_0^T\int_{\mathbb{R}^N}t\omega|\nabla u_n^m|^2dxdt + \frac{1}{2}\int_0^T\int_{\mathbb{R}^N}\omega|\nabla u_n^m(x,t)|^2dxdt \\
\leq & C\int_0^T\int_{\mathbb{R}^N}\omega|\nabla u_n^m(x,t)|^2dxdt.
\end{aligned}$$

This together with (3.26) yields

$$\int_0^T \int_{\mathbb{R}^N} t\omega \left(\frac{\partial u_n^m}{\partial t} \right)^2 dxdt \le C. \tag{3.27}$$

The estimates (3.26), (3.27) and the uniform boundedness of $\{u_n\}$ imply the compactness of $\{u_n^m\}$ in $L^2_{loc}(Q_T)$. From this it follows that there exists a subsequence of $\{u_n\}$, which converges to a certain bounded function almost everywhere on Q_T and the limit function is a generalized solution of (3.1), (3.2). The proof of Theorem 1.3.5 is thus completed. □

In [BEN], [BFU], the existence of a generalized solution u of the Cauchy problem (3.21), (3.2) is proved for $u_0 \in L^1(\mathbb{R}^N)$ by using the semigroup method. u satisfies the following estimate: for $t \in (0, T)$

$$\|u(\cdot, t)\|_{L^\infty(\mathbb{R}^N)} \le \frac{C}{t^{N/(2+(m-1)N)}} \|u_0\|_{L^1(\mathbb{R}^N)}^{2/(2+(m-1)N)} \tag{3.28}$$

with a constant C independent of u_0. This estimate can also be obtained by using the regularization method [FR1]. A similar estimate holds for the boundary value problem [EV]. The estimate like (3.28) expresses a kind of regularization effect.

As pointed out in §1.2.4, in some cases, the initial data might be a measure. [PIE] has proved the existence of generalized solutions for the case that the initial value is a nonnegative finite Radon measure. In [BCP] it is noticed that the finiteness of the measure can be relaxed; the authors have proved in [BCP] that there exists a nonnegative continuous generalized solutions of (3.1) if and only if the nonnegative Radon measure initial data μ satisfies the condition

$$\sup_{R>\tau} \frac{1}{R^{N+2/(m-1)}} \int_{|x|<R} d\mu < \infty, \quad \text{for any } R > 0.$$

(see §1.8.3).

4. It should also be pointed out that the uniqueness of generalized solutions of the Cauchy problem(3.1), (3.2) is valid not only in $L^1(Q_T) \cap L^\infty(Q_T)$ but also in a more general class of functions, especially in the class of all bounded and measurable functions. (see §1.8.3).

5. All what we have discussed above is the case $m > 1$ which corresponds to the slow diffusion. In this case, the equation (3.1) or (3.21) degenerates at $u = 0$. Many authors have studied another important case $0 < m < 1$ which corresponds to the fast diffusion. In this case, another kind of singularity

occurs in (3.1) or (3.21) at $u = 0$ [VA3], [DD1], [DD2]. Some authors have studied (3.1) or (3.21) in the case $m < 0$ [ERV], [DD3], [DD4] and the equation

$$\frac{\partial u}{\partial t} = \Delta \ln u,$$

[HU1], [HU2], [DD5] which correspond to the super-fast diffusion. Both the slow diffusion equations and the fast diffusion equations are examples of equations of the form

$$\frac{\partial u}{\partial t} = \Delta A(u),$$

where $A(u)$ is increasing and continuous. Many people have studied equations under such general conditions. For details, see §3.5.

6. Many authors have been devoted to degenerate equations of the form

$$\frac{\partial u}{\partial t} = \Delta A(u) + \Psi(u)$$

and

$$\frac{\partial u}{\partial t} = \Delta A(u) + \operatorname{div} \overrightarrow{B}(u)$$

which involve some lower order terms. In Chapter 3 we will discuss more general quasilinear degenerate parabolic equations of second order, which may have many points of degeneracy, even have a set of points of degeneracy including interior points.

1.4 Regularity of Solutions: One Dimensional Case

1.4.1 *Lemma*

Generalized solutions of the filtration equation in one dimension obtained in §1.1.2, are all continuous, which satisfy the equation in classical sense in a neighborhood of any "point of non-degeneracy".

Early in [KA1], Kalashnikov noticed that solutions of the equation

$$\frac{\partial u}{\partial t} = \frac{\partial^2 u^m}{\partial x^2} \tag{4.1}$$

with $m > 1$, may have discontinuous derivative $\frac{\partial u}{\partial x}$, even if its initial data

$$u(x,0) = u_0(x) \qquad (x \in \mathbb{R}) \tag{4.2}$$

are sufficiently smooth.

In this section a thorough investigation on the regularity of generalized solutions of the equation (4.1) will be made.

To this purpose we first prove

Lemma 1.4.1 *Let $m > 1$ and u be a positive and smooth solution of (4.1) on $G = (a,b) \times (0,T]$. Here by "smooth" we mean $u \in C^2(G) \cap C(\overline{G})$, such that $\frac{\partial^3 u}{\partial x^3} \in C(G)$. Then for any $\tau \in (0,T)$, a_1, $b_1 \in (a,b)$ $(a_1 < b_1)$, we have on $G_1 = (a_1, b_1) \times [\tau, T]$,*

$$\left| \frac{\partial u^{m-1}(x,t)}{\partial x} \right| \le C, \tag{4.3}$$

where $C = C(\tau, m, M, a_1 - a, b_1 - b)$, $M = \sup_G u$. If

$$M_1 = \sup_{(a,b)} \left| \frac{du_0^{m-1}(x)}{dx} \right| < \infty,$$

then (4.3) holds on $(a_1, b_1) \times [0,T]$ where C depends on M_1 instead of τ.

Proof. Let $v = u^{m-1}$. Then v satisfies

$$\frac{\partial v}{\partial t} = mv\frac{\partial^2 v}{\partial x^2} + \frac{m}{m-1}\left(\frac{\partial v}{\partial x}\right)^2.$$

Here we have used the assumption $u > 0$, otherwise $\frac{\partial v}{\partial x}$ and $\frac{\partial^2 v}{\partial x^2}$ may not exist everywhere.

Since our goal is to estimate $\frac{\partial v}{\partial x}$, a natural approach is to derive and use the equation for $\frac{\partial v}{\partial x}$. However it seems difficult to achieve our goal in this way. So we consider a transformation of v: $v = \varphi(w)$ with φ to be determined later. We first require that

$$\varphi : [0,1] \to [0, M^*], \quad \varphi'(r) > 0, \tag{4.4}$$

where $M^* = M^{m-1}$. Then w satisfies

$$\frac{\partial w}{\partial t} = m\varphi \frac{\partial^2 w}{\partial x^2} + m\varphi \frac{\varphi''}{\varphi'}\left(\frac{\partial w}{\partial x}\right)^2 + \frac{m}{m-1}\varphi'\left(\frac{\partial w}{\partial x}\right)^2.$$

Let $p = \dfrac{\partial w}{\partial x}$. Then p satisfies

$$\begin{aligned}\frac{\partial p}{\partial t} = \quad & m\varphi \frac{\partial^2 p}{\partial x^2} + \left(\frac{m^2}{m-1}\varphi'' + m\varphi\left(\frac{\varphi''}{\varphi'}\right)'\right)p^3 \\ & + \left(\frac{m(m+1)}{m-1}\varphi' + wm\varphi\frac{\varphi''}{\varphi'}\right)p\frac{\partial p}{\partial x}\end{aligned}$$

or after multiplying both sides by p,

$$\begin{aligned}\frac{1}{2}\frac{\partial p^2}{\partial t} = \quad & m\varphi p \frac{\partial^2 p}{\partial x} + \left(\frac{m^2}{m-1}\varphi'' + m\varphi\left(\frac{\varphi''}{\varphi'}\right)'\right)p^4 \\ & + \left(\frac{m(m+1)}{m-1}\varphi' + 2m\varphi\frac{\varphi''}{\varphi'}\right)p^2\frac{\partial p}{\partial x}.\end{aligned} \tag{4.5}$$

Now we choose $\zeta = \zeta(x,t) \subset C^2(\overline{G})$ satisfying

$$\begin{aligned}&\zeta = 1 \quad \text{for } (x,t) \in \overline{G}_1, \\ &\zeta = 0 \quad \text{for } x = a, b \text{ and } t = 0, \\ &0 \le \zeta \le 1 \quad \text{for } (x,t) \in \overline{G}, \\ &\left|\frac{\partial \zeta}{\partial t}\right| \le \frac{C}{\tau}, \left|\frac{\partial \zeta}{\partial x}\right| \le C\max\left\{\frac{1}{a_1 - a}, \frac{1}{b_1 - b}\right\}, \\ &\left|\frac{\partial^2 \zeta}{\partial x^2}\right| \le C\max\left\{\frac{1}{(a_1-a)^2}, \frac{1}{(b_1-b)^2}\right\}\end{aligned}$$

and use $\zeta(x,t)$ to cut off the function p^2, namely to consider the function

$z = \zeta^2 p^2$ instead of p^2. Multiplying (4.5) by ζ^2, we can check that z satisfies

$$\begin{aligned}
&\frac{1}{2}\left(\frac{\partial z}{\partial t} - m\varphi\frac{\partial^2 z}{\partial x^2}\right) \\
= \; &\frac{p}{2}\left(\frac{m(m+1)}{m-1}\varphi' + 2m\varphi\frac{\varphi''}{\varphi'}\right)\frac{\partial z}{\partial x} \\
&+\left(\zeta\frac{\partial \zeta}{\partial t} - m\varphi\left(\frac{\partial \zeta}{\partial x}\right)^2 - m\varphi\zeta\frac{\partial^2 \zeta}{\partial x^2}\right)p^2 \\
&-\left(\frac{m(m+1)}{m-1}\varphi' + 2m\varphi\frac{\varphi''}{\varphi'}\right)\zeta\frac{\partial \zeta}{\partial x}p^3 \\
&+\left(\frac{m^2}{m-1}\varphi'' + m\varphi\left(\frac{\varphi''}{\varphi'}\right)'\right)\zeta^2 p^4 \\
&-m\varphi\zeta^2\left(\frac{\partial p}{\partial x}\right)^2 - 4m\varphi\zeta\frac{\partial \zeta}{\partial x}p\frac{\partial p}{\partial x}.
\end{aligned} \tag{4.6}$$

Let $(x_0, t_0) \in G$ be a point where z achieves its maximum. Then at this point

$$\frac{\partial z}{\partial t} \geq 0, \frac{\partial z}{\partial x} = 0, \frac{\partial^2 z}{\partial x^2} \leq 0. \tag{4.7}$$

Since

$$\frac{\partial z}{\partial x} = 2\zeta^2 p\frac{\partial p}{\partial x} + 2\zeta\frac{\partial \zeta}{\partial x}p^2,$$

at (x_0, t_0) we have

$$\zeta\frac{\partial p}{\partial x} = -p\frac{\partial \zeta}{\partial x}. \tag{4.8}$$

Using (4.7) and (4.8), from (4.6) we obtain

$$\begin{aligned}
&\left(\zeta\frac{\partial \zeta}{\partial t} - m\varphi\left(\frac{\partial \zeta}{\partial x}\right)^2 - m\varphi\zeta\frac{\partial^2 \zeta}{\partial x^2}\right)p^2 + 3m\varphi\left(\frac{\partial \zeta}{\partial x}\right)^2 p^2 \\
&\quad+\left(\frac{m^2}{m-1}\varphi'' + m\varphi\left(\frac{\varphi''}{\varphi'}\right)'\right)\zeta^2 p^4 \\
&\quad-\left(\frac{m(m+1)}{m-1}\varphi' + 2m\varphi\frac{\varphi''}{\varphi'}\right)\zeta\frac{\partial \zeta}{\partial x}p^3 \geq 0,
\end{aligned}$$

namely

$$
\begin{aligned}
&-\left(\frac{m^2}{m-1}\varphi'' + m\varphi\left(\frac{\varphi''}{\varphi'}\right)'\right)\zeta^2 p^4 \\
\leq\ &\left(\zeta\frac{\partial\zeta}{\partial t} + 2m\varphi\left(\frac{\partial\zeta}{\partial x}\right)^2 - m\varphi\zeta\frac{\partial^2\zeta}{\partial x^2}\right)p^2 \qquad (4.9)\\
&-\left(\frac{m(m+1)}{m-1}\varphi' + 2m\varphi\frac{\varphi''}{\varphi'}\right)\zeta\frac{\partial\zeta}{\partial x}p^3.
\end{aligned}
$$

We hope that the coefficient of the left side is positive. It suffices to require

$$
\begin{aligned}
&\varphi'' < 0, \left(\frac{\varphi''}{\varphi'}\right)' = \frac{\varphi'''\varphi' - (\varphi'')^2}{(\varphi')^2} < 0, \qquad (4.10)\\
&\text{or } \varphi'''\varphi' - (\varphi'')^2 < 0.
\end{aligned}
$$

Clearly $\varphi(r) = \alpha r^2 + (M^* - \alpha)r$ with $-M^* < \alpha < 0$ satisfies both (4.4) and (4.10). In particular, we choose

$$\varphi(r) = \frac{M^* r}{3}(4 - r).$$

Substituting this function into (4.9) and using Young's inequality, we obtain

$$\zeta^2 p^2 \leq C_1 + C_2\zeta|p| \leq C_1 + \varepsilon\zeta^2 p^2 + \frac{c_2^2}{4\varepsilon} \qquad (\varepsilon > 0)$$

or

$$(1-\varepsilon)\zeta^2 p^2 \leq C_1 + \frac{c_2^2}{4\varepsilon}$$

with $C_i = C_i(\tau, m, M, a_1 - a, b_1 - b)$ $(i = 1, 2)$. Taking $\varepsilon = \frac{1}{2}$ yields

$$\zeta^2 p^2 \leq C.$$

Hence

$$\left|\frac{\partial v}{\partial x}\right| = \left|\varphi'(w)\frac{\partial w}{\partial x}\right| = \varphi'(w)|p| \leq C \quad \text{on } \overline{G}_1.$$

This complete the proof of the first conclusion. □

Similarly we can prove the second conclusion. To do this we choose $\zeta(x) \in C_0^2(a, b)$ such that $0 \leq \zeta(x) \leq 1$, $\zeta = 1$ for $x \in (a_1, b_1)$.

1.4.2 *Regularity of solutions*

Theorem 1.4.1 *Suppose that $u_0(x)$ is nonnegative and bounded, $u_0^m(x)$ is Lipschitz continuous and $u(x,t)$ is a generalized solution of (4.1), (4.2). Then*

1. for any $\tau \in (0,T)$, $u \in C^{\alpha,\alpha/2}(\mathbb{R} \times [\tau, T])$ with $\alpha = \min\left\{1, \dfrac{1}{m-1}\right\}$ and $u \in C^{\alpha,\alpha/2}(\mathbb{R} \times [0,T])$ provided $u_0^{m-1}(x)$ is Lipschitz continuous;

2. $\dfrac{\partial u^m}{\partial x}$ exists and is continuous in x, in particular, $\dfrac{\partial u^m}{\partial x} = 0$ whenever $u(x,t) = 0$;

3. $\dfrac{\partial u}{\partial x}$ exists and is continuous in x if $1 < m < 2$, in particular, $\dfrac{\partial u}{\partial x} = 0$ whenever $u(x,t) = 0$.

Remark 1.4.1 The Barenblatt solution (1.37) shows that $u \in C^{\alpha,\alpha/2}(\mathbb{R}\times [\tau, T])$ is the best possible global result and the Hölder exponent $\alpha = \min\left\{1, \dfrac{1}{m-1}\right\}$ cannot be increased.

Proof of Theorem 1.4.1. The case $u_0 \equiv 0$ is trivial. We suppose that $\sup u_0 > 0$.

1. From the proof of Theorem 1.2.2 and also using the uniqueness theorem we assert that $u(x,t)$ can be obtained as the limit of the classical solution $u_n(x,t)$ of the problem

$$\begin{cases} \dfrac{\partial u}{\partial t} = \dfrac{\partial^2 u^m}{\partial x^2} & |x| < n, t > 0, \\ u(x,0) = u_{0n}(x) & |x| \le n, \\ u(\pm n, t) = M = \sup u_0 & 0 < t < T, \end{cases}$$

and $u_n(x,t)$ uniformly converges to $u(x,t)$ on any compact subset of Q_T. Here $u_{0n}(x) > 0$ and $\{u_{0n}(x)\}$ uniformly converges to $u_0(x)$ on any finite interval.

Given any $\tau \in (0,T)$ and $a < b$. By Lemma 1.4.1, for any large n and $(x,t) \in G^\tau = [a,b] \times [\tau, T]$, we have

$$\left|\frac{\partial u_n^{m-1}}{\partial x}\right| \le C \tag{4.11}$$

with some constant $C = C(\tau, m, M)$ independent of a, b.

It follows by virtue of the uniform boundedness of u_n

$$\left|\frac{\partial u_n^m}{\partial x}\right| \le C \quad \text{on } G^\tau. \tag{4.12}$$

Checking the proof of Theorem 1.2.2 we see that in fact this estimate has been established there.

Let $(x,t),(y,s) \in G^\tau$. Based on (4.12) we see from the proof of Theorem 1.2.2 that for some $x^* \in [x, x+|t-s|^{1/2}]$,

$$|u_n(x^*,t) - u_n(x^*,s)| \le C|t-s|^{1/2}. \tag{4.13}$$

Therefore if $m > 2$, then

$$|u_n^{m-1}(x^*,t) - u_n^{m-1}(x^*,s)| \le C|t-s|^{1/2}.$$

This and (4.11) imply

$$\begin{aligned}
& |u_n^{m-1}(x,t) - u_n^{m-1}(y,s)| \\
\le\ & |u_n^{m-1}(x,t) - u_n^{m-1}(x^*,t)| + |u_n^{m-1}(x^*,t) - u_n^{m-1}(x^*,s)| \\
& + |u_n^{m-1}(x^*,s) - u_n^{m-1}(y,s)| \\
\le\ & C(|x-x^*| + |t-s|^{1/2} + |x^*-y|) \le C(|x-y| + |t-s|^{1/2}).
\end{aligned}$$

Using the inequality $|a-b|^\alpha \le |a^\alpha - b^\alpha|$ $(a>0, b>0, \alpha \ge 1)$ we further obtain

$$|u_n(x,t)-u_n(y,s)|^{m-1} \le |u^{m-1}(x,t)-u^{m-1}(y,s)| \le C(|x-y|+|t-s|^{1/2})$$

or

$$|u_n(x,t) - u_n(y,s)| \le C(|x-y|^\alpha + |t-s|^{\alpha/2})$$

with $\alpha = \dfrac{1}{m-1}$.

If $1 < m < 2$, then using (4.11) again yields

$$\left|\frac{\partial u_n}{\partial x}\right| \le \left|\frac{1}{m-1} u_n^{2-m} \frac{\partial u_n^{m-1}}{\partial x}\right| \le \frac{1}{m-1} u_n^{2-m} \left|\frac{\partial u_n^{m-1}}{\partial x}\right| \le C$$

and hence

$$|u_n(x,t) - u_n(y,t)| \le C|x-y|.$$

This and (4.13) give

$$\begin{aligned}
&|u_n(x,t)-u_n(y,s)| \\
\le\ & |u_n(x,t)-u_n(x^*,t)|+|u_n(x^*,t)-u_n(x^*,s)| \\
&\quad +|u_n(x^*,s)-u_n(x,s)| \\
\le\ & C(|x-x^*|+|t-s|^{1/2}+|x^*-y|)\le C(|x-y|+|t-s|^{1/2}).
\end{aligned}$$

Summing up, in either case we have

$$|u_n(x,t)-u_n(y,s)|\le C(|x-y|^{\alpha}+|t-s|^{\alpha/2})$$

with $\alpha=\min\left\{1,\dfrac{1}{m-1}\right\}$, which implies that $u\in C^{\alpha,\alpha/2}(\mathbb{R}\times[\tau,T])$ by letting $n\to\infty$.

If u_0^{m-1} is Lipschitz continuous, then by Lemma 1.4.1, (4.11) holds for $(x,t)\in[a,b]\times[0,T]$. Based on this fact we can prove $u\in C^{\alpha,\alpha/2}(\mathbb{R}\times[0,T])$.

2. If $u(x_0,t_0)>0$, then from Theorem 1.2.2, we see that u is a classical solution in some neighborhood of (x_0,t_0), in particular, $\dfrac{\partial u^m}{\partial x}$ exists and is continuous near (x_0,t_0).

If $u(x_0,t_0)=0$ $(t_0>0)$, then for any given $\delta>0$, from the first conclusion of our theorem

$$0\le u(x,t_0)=u(x,t_0)-u(x_0,t_0)\le C\delta^{\alpha}$$

for $x\in I_\delta=[x_0-\delta,x_0+\delta]$. Hence for large n, we have

$$u_n(x,t_0)\le C_1\delta^{\alpha}\quad \text{for } x\in I_\delta,$$

and

$$\begin{aligned}
|u_n^m(x,t_0)-u_n^m(y,t_0)| &= \left|\int_x^y \frac{\partial u_n^m(\xi,t_0)}{x}d\xi\right| \\
&= \frac{m}{m-1}\left|\int_x^y u_n(\xi,t_0)\frac{\partial u_n^{m-1}}{\partial x}(\xi,t_0)d\xi\right| \\
&\le \frac{m}{m-1}C_1\delta^{\alpha}|x-y|
\end{aligned}$$

for $x,y\in I_\delta$. Here we have used (4.11) again. Letting $n\to\infty$ gives

$$|u^m(x,t_0)-u^m(y,t_0)|\le\frac{mC_1}{m-1}\delta^{\alpha}|x-y|\quad\text{for } x,y\in I_\delta$$

with some constant C_1, which implies that $\dfrac{\partial u^m(x,t_0)}{\partial x}$ is continuous at $x = x_0$.

3. Suppose $1 < m < 2$. Since u is a classical solution of (4.1) whenever $u(x,t) > 0$, we need to treat only the case that $u(x_0,t_0) = 0$ for some (x_0,t_0). We have

$$\left|\frac{\partial u_n^{m-1}(x,t_0)}{\partial x}\right| \le C_1 u_n^{2-m}(x,t_0) \le C_1 \delta^{\alpha(2-m)} \quad \text{for } x \in I_\delta.$$

From this it is easily seen that $\dfrac{\partial u(x,t_0)}{\partial x}$ is continuous at $x = x_0$. □

1.4.3 *Some extensions*

Kalashnikov [KA3] extended the above result of Aronson in two directions. On the one hand, he proved that if u is a generalized solution of the problem

$$\frac{\partial u}{\partial t} = \frac{\partial^2 u^m}{\partial x^2} - cu^n \quad (c > 0, m > 1, n > 0)$$

with initial data u_0 which are nonnegative, continuously differentiable and bounded together with $\dfrac{\mathrm{d}}{\mathrm{d}x} u_0^m(x)$, then for any $\tau \in (0,T]$, the derivative $\dfrac{\partial u^\sigma}{\partial x}$ is bounded on $\mathbb{R} \times [\tau, T]$, where $\sigma = \max\left\{m-1, \dfrac{m-n}{2}\right\}$. Examples show that $\dfrac{\partial u^{\sigma-\varepsilon}}{\partial x}$ is unbounded no matter how small $\varepsilon > 0$ is. (cf. [MP3]).

On the other hand, he proved that if u is a generalized solution of the equation

$$\frac{\partial u}{\partial t} = \frac{\partial^2 A(u)}{\partial x^2} \tag{4.14}$$

with initial data u_0, which are nonnegative, continuously differentiable and bounded together with $\dfrac{d}{dx} A(u_0)(x)$, where $A(u)$ is appropriately smooth, $A'(0) = A(0) = 0$ and

$$\Psi(u) = \int_0^u \frac{A'(\sigma)}{\sigma} \mathrm{d}\sigma < \infty,$$

then for any $\tau \in (0,T)$, $\dfrac{\partial \Psi(u)}{\partial x}$ is bounded on $\mathbb{R} \times [\tau,])$ provided $\dfrac{d\Psi(u_0)}{dx}$ is bounded. As a corollary, one has

$$|u(x,t) - u(y,t)| \leq \Psi^{-1}(K|x - x'|)$$

provided $\Psi''(u) > 0$ for $u > 0$, where Ψ^{-1} is the inverse function of Ψ and K is a constant. Examples show that this result is sharp.

Gilding and Peletier [GP2] studied the Cauchy problem for the equation

$$\frac{\partial u}{\partial t} = \frac{\partial^2 u^m}{\partial x^2} + \frac{\partial u^n}{\partial x} \quad (m > 1, n > 1)$$

and proved that the problem admits a generalized solution $u \in C^{\alpha,\alpha/2}(\mathbb{R} \times [\tau, T]) \cap C^{1/m,1/2m}(\mathbb{R} \times [0,T])$ for any $\tau \in (0,T)$ with $\alpha = \left\{1, \dfrac{1}{m-1}\right\}$, provided that the initial data u_0 are nonnegative, continuous and bounded and u_0^m is Lipschitz continuous. The exponent α is the best possibility. In addition, he obtained a result similar to Theorem 1.4.1.

1.5 Regularity of Solutions: Higher Dimensional Case

In this section we discuss the regularity of generalized solutions of the Cauchy problem

$$\frac{\partial u}{\partial t} = \Delta u^m \quad (m > 1), \tag{5.1}$$

$$u(x,0) = u_0(x). \tag{5.2}$$

Compared with one dimensional case, the investigation of regularity of generalized solutions in higher dimension is a very difficult problem. Even the proof of continuity of generalized solutions of (5.1), (5.2) had been remained as an open problem in a long period. In 1979 Caffarelli and Friedmann [CF1]proved the interior Hölder continuity and global continuity; their argument is based on the comparison principle. As shown in [LSU], it is effective to apply the properties of the so-called class B_2 of functions to investigate the regularity of solutions of uniformly parabolic equations. Chen Yazhe [CH2] introduced a class of functions, called generalized class B_2; using the properties of such class, he extended the result of [CF1] to more general quasilinear degenerate parabolic equations and proved the Hölder continuity of generalized solutions.

In this section we will use Chen's method to prove the Hölder continuity of generalized solutions. Assume that u is a generalized solution of the Cauchy problem (5.1), (5.2), which is bounded on $Q_T = \mathbb{R}^N \times (0,T)$. As we have shown in Theorem 1.3.5, we can obtain a bounded generalized solution as the limit of a sequence of positive and smooth solutions with positive initial data. We have also indicated in §1.3.3 that the uniqueness is valid in a general class of functions,especially in the class of bounded and measurable functions. By virtue of uniqueness, we may regard the given generalized solution u as the one obtained by regularization in Theorem 1.3.5. To prove the Hölder continuity of u, it suffices to establish the uniformly Hölder estimate for its approximate positive smooth solution. We simply assume that u is just the later one.

1.5.1 *Generalized class B_2*

Let $\Omega \subset \mathbb{R}^N$ be a bounded domain, $w \in W^{1,p}(\Omega)$ $(p \geq 1)$. For any real number k, denote

$$A_k = \{x \in \Omega; w(x) > k\}.$$

Then the set A_k is measurable. The function defined by

$$w^{(k)}(x) = (w(x) - k)^+$$

is called a cut-off function of $w(x)$. It is easily seen that $w^{(k)}(x) \in W^{1,p}(\Omega)$ and

$$\frac{\partial w^{(k)}(x)}{\partial x} = \frac{\partial w(x)}{\partial x} \qquad \text{for } x \in A_k,$$
$$\frac{\partial w^{(k)}(x)}{\partial x} = 0 \qquad \text{elsewhere.}$$

Let $w = u^m$, where u is the given positive smooth solution of (5.1) on Q_T. Then w satisfies

$$\frac{\partial \Phi(w)}{\partial t} = \Delta w \tag{5.3}$$

where

$$\Phi(s) = s^{1/m}.$$

Let $(x_0, t_0) \in Q_T$, $B_\rho = B_\rho(x_0) = \{x \in \mathbb{R}^N; |x - x_0| < \rho\}$ and $\zeta(x)$ be a function on $\overline{B}_\rho$ which is continuous and piecewise smooth such that $0 \le \zeta(x) \le 1$ and $\zeta(x) = 0$ for $x \in \partial B_\rho$ (also called a cut-off function). Denote

$$A_{k,\rho}(t) = \{x \in B_\rho; w(x,t) > k\}, B_{k,\rho} = \{x \in B_\rho; w(x,t) < k\}.$$

Multiplying (5.3) by $\zeta^2(w-k)^+$ and integrating with respect to x over B_ρ yield

$$\int_{B_\rho} \zeta^2 (w-k)^+ \frac{\partial \Phi(w)}{\partial t} dx = \int_{B_\rho} \zeta^2 (w-k)^+ \Delta w dx.$$

It is clear that

$$\begin{aligned}
&\int_{B_\rho} \zeta^2 (w-k)^+ \frac{\partial \Phi(w)}{\partial t} dx \\
= \quad & \int_{B_\rho} \zeta^2 (w-k)^+ \Phi'(w) \frac{\partial (w-k)^+}{\partial t} dx \\
= \quad & \int_{B_\rho} \zeta^2 \frac{\partial}{\partial t} \int_0^{(w-k)^+} \Phi'(k+\tau)\tau d\tau dx \\
= \quad & \frac{\partial}{\partial t} \int_{B_\rho} \zeta^2 \int_0^{(w-k)^+} \Phi'(k+\tau)\tau d\tau dx \\
= \quad & \frac{\partial}{\partial t} \int_{A_{k,\rho}(t)} \zeta^2 \chi_k(w-k) dx,
\end{aligned}$$

where

$$\chi_k(s) = \int_0^s \Phi'(k+\tau)\tau d\tau. \tag{5.4}$$

Integrating by parts gives

$$\begin{aligned}
&\int_{B_\rho} \zeta^2 (w-k)^+ \Delta w dx \\
= \quad & -\int_{B_\rho} \zeta^2 \nabla (w-k)^+ \nabla w dx - 2\int_{B_\rho} \zeta (w-k)^+ \nabla \zeta \cdot \nabla w dx \\
= \quad & -\int_{A_{k,\rho}(t)} \zeta^2 |\nabla w|^2 - 2\int_{A_{k,\rho}(t)} \zeta (w-k) \nabla \zeta \cdot \nabla w dx.
\end{aligned}$$

Therefore we have

$$\begin{aligned}
&\frac{\partial}{\partial t}\int_{A_{k,\rho}(t)}\zeta^2\chi_k(w-k)dx+\int_{A_{k,\rho}(t)}\zeta^2|\nabla w|^2dx\\
=\;&-2\int_{A_{k,\rho}(t)}\zeta(w-k)\nabla\zeta\cdot\nabla w dx\\
\le\;&\frac{1}{2}\int_{A_{k,\rho}(t)}\zeta^2|\nabla w|^2dx+2\int_{A_{k,\rho}(t)}|\nabla\zeta|^2(w-k)^2dx.
\end{aligned}$$

Hence

$$\begin{aligned}
&\frac{\partial}{\partial t}\int_{A_{k,\rho}(t)}\zeta^2\chi_k(w-k)dx+\frac{1}{2}\int_{A_{k,\rho}(t)}\zeta^2|\nabla w|^2dx\\
\le\;&2\int_{A_{k,\rho}(t)}|\nabla\zeta|^2(w-k)^2dx.
\end{aligned}\tag{5.5}$$

If we multiply (5.3) by $\zeta^2(w-k)^-$ instead of $\zeta^2(w-k)^+$, then, similarly we can obtain

$$\begin{aligned}
&\frac{\partial}{\partial t}\int_{B_{k,\rho}(t)}\zeta^2\widetilde{\chi}_k(w-k)dx+\frac{1}{2}\int_{B_{k,\rho}(t)}\zeta^2|\nabla w|^2dx\\
\le 2\;&\int_{B_{k,\rho}(t)}|\nabla\zeta|^2(w-k)^2dx,
\end{aligned}\tag{5.6}$$

where

$$\widetilde{\chi}_k(s)=\int_0^s\Phi'(k-\tau)\tau d\tau.\tag{5.7}$$

Definition 1.5.1 $w\in B_2(Q_T,M,m,\gamma)$ if $0<w(x,t)\le M$, $|\nabla w|\in L^2_{loc}(Q_T)$ and w satisfies

$$\frac{\partial}{\partial t}\int_{A_{k,\rho}(t)}\zeta^2\chi_k(w-k)dx+\frac{1}{2}\int_{A_{k,\rho}(t)}\zeta^2|\nabla w|^2dx\tag{5.8}$$

$$\frac{\partial}{\partial t}\int_{B_{k,\rho}(t)}\zeta^2\widetilde{\chi}_k(w-k)dx+\frac{1}{2}\int_{B_{k,\rho}(t)}\zeta^2|\nabla w|^2dx\tag{5.9}$$

for any cut-off function $\zeta(x)$ where $\chi_k(s)$ and $\widetilde{\chi}_k(s)$ are defined by (5.4) and (5.7).

The above derivation shows the following proposition.

Proposition 1.5.1 *Assume that u is a positive, smooth solution of (5.1) on Q_T, which is bounded by a constant M. Then $w=u^m\in B_2(Q_T,M,m,2)$.*

1.5.2 *Some lemmas*

We need the following lemmas.

Lemma 1.5.1 *[LU1] Let $\Omega \subset \mathbb{R}^N$ be a bounded domain, $w \in H_0^1(\Omega)$. Then*

$$\int_S |w|^2 dx \le C(mes\, S)^{2/N} \int_S |\nabla w|^2 dx,$$

where $S = \{x \in \Omega; w(x) > 0\}$ and C is a constant depending only on N.

Proof. By Hölder's inequality we have

$$\int_S |w|^2 dx \le (\text{mes}\, S)^{1-(N-2)/N} \left(\int_S |w|^{2N/(N-2)} dx \right)^{(N-2)/2N}.$$

Using the embedding theorem yields

$$\begin{aligned} & \left(\int_S |\nabla w|^{2N/(N-2)} dx \right)^{(N-2)/2N} \\ = & \left(\int_\Omega |w^+|^{2N/(N-2)} dx \right)^{(N-2)/2N} \\ \le & \ C \left(\int_\Omega |\nabla w^+|^2 \right)^{1/2} \le C \left(\int_S |\nabla w|^2 \right)^{1/2} \end{aligned}$$

which combining with the above inequality gives the desired conclusion. □

Lemma 1.5.2 *[LU2] For any $w \in W^{1,p}(B_\rho)$ $(p \ge 1)$, there holds*

$$(\lambda - k)(mes\, A_{\lambda,\rho})^{1-1/N} \le \frac{\beta \rho^N}{mes\,(B_\rho \backslash A_{k,\rho})} \int_{A_{k,\rho} \backslash A_{\lambda,\rho}} |\nabla w| dx,$$

where $\lambda > k$, $A_{k,\rho} = \{x \in B_\rho; w(x) > k\}$ and β is a constant depending only on N.

Proof. Consider the following nonnegative function

$$\begin{aligned} \overline{w}(x) &= 0 && \text{for } x \in B_\rho \backslash A_{k,\rho}, \\ \overline{w}(x) &= w - k && \text{for } x \in A_{k,\rho} \backslash A_{\lambda,\rho}, \\ \overline{w}(x) &= \lambda - k && \text{for } x \in A_{\lambda,\rho}. \end{aligned}$$

For almost all $s, y \in B_\rho$, we have

$$\overline{w}(y) - \overline{w}(x) = \int_0^{|y-x|} \frac{\partial \overline{w}}{\partial r}(x + r\omega)dr,$$

where (r, ω) is the spherical coordinates centered at x. Integrating this formula with respect to $y \in B_\rho \backslash A_{k,\rho}$ and noting that $\overline{w}(y) = 0$, we obtain

$$\text{mes}\,(B_\rho \backslash A_{k,\rho})\overline{w}(x) = -\int_{B_\rho \backslash A_{k,\rho}} \int_0^{|y-x|} \frac{\partial \overline{w}}{\partial r}(x + r\omega)drdy.$$

Now we estimate the integral on the right hand side:

$$\begin{aligned}
&\left| \int_{B_\rho \backslash A_{k,\rho}} dy \int_0^{|y-x|} \frac{\partial \overline{w}}{\partial r}(x + r\omega)dr \right| \\
\le\ & \int_{B_\rho} dy \int_0^{|y-x|} \left| \frac{\partial \overline{w}}{\partial r}(x + r\omega) \right| dr \\
\le\ & \int_{y \in B_\rho} |x - y|^{N-1} d_y|x - y| d\omega \int_0^{|y-x|} \left| \frac{\partial \overline{w}}{\partial r}(x + r\omega) \right| dr \\
\le\ & \int_0^{2\rho} |x - y|^{N-1} d_y|x - y| \int_{B_\rho} \frac{|\nabla \overline{w}(\xi)|}{|x - \xi|^{N-1}} d\xi \\
=\ & \frac{(2\rho)^N}{N} \int_{B_\rho} \frac{|\nabla \overline{w}(\xi)|}{|x - \xi|^{N-1}} d\xi.
\end{aligned}$$

Therefore

$$\text{mes}\,(B_\rho \backslash A_{k,\rho})\overline{w}(x) \le \frac{(2\rho)^N}{N} \int_{B_\rho} \frac{|\nabla \overline{w}(\xi)|}{|x - \xi|^{N-1}} d\xi.$$

Integrating both sides over $A_{\lambda,\rho}$ and estimating the right hand side yield

$$\begin{aligned}
&\text{mes}\,(B_\rho \backslash A_{k,\rho})(\lambda - k)\text{mes}\,A_{\lambda,\rho} \\
\le\ & \frac{(2\rho)^N}{N} \int_{B_\rho} |\nabla \overline{w}(\xi)| d\xi \int_{A_{\lambda,\rho}} \frac{dx}{|x - \xi|^{N-1}} \\
\le\ & C(N)\rho^N (\text{mes}\,A_{\lambda,\rho})^{1/N} \int_{B_\rho} |\nabla \overline{w}(\xi)| d\xi.
\end{aligned}$$

Here we have used the estimate

$$\int_{A_{\lambda,\rho}} \frac{dx}{|x - \xi|^{N-1}} \le (\omega_N + 1)(\text{mes}\,A_{\lambda,\rho})^{1/N}$$

which can be proved by noting

$$\int_{A_{\lambda,\rho}} \frac{dx}{|x-\xi|^{N-1}} \le \int_{|x-\xi|\le\delta} \frac{dx}{|x-\xi|^{N-1}} + \int_{\substack{x\in A_{\lambda,\rho}\\ |x-\xi|>\delta}} \frac{dx}{|x-\xi|^{N-1}},$$

$$\int_{|x-\xi|\le\delta} \frac{dx}{|x-\xi|^{N-1}} = \omega_N\delta,$$

$$\int_{\substack{x\in A_{\lambda,\rho}\\ |x-\xi|>\delta}} \frac{dx}{|x-\xi|^{N-1}} \le \delta^{-N+1}\mathrm{mes}\, A_{\lambda,\rho}$$

and choosing $\delta = (\mathrm{mes}\, A_{\lambda,\rho})^{1/N}$. □

Lemma 1.5.3 *$\chi_k(s)$ and $\widetilde{\chi}_k(s)$ possess the following properties:*

(i) if $\mu > k > \frac{\mu}{2} > 0$, $0 < h \le \mu - k$, $0 < \beta < 1$, then

$$\frac{h^2}{2} \le \frac{\chi_k(h)}{\Phi'(\mu)} \le h^2, \quad \frac{\chi_k(h)}{\chi_k(\beta h)} \le \frac{1}{\beta^2};$$

(ii) if $k > h > 0$, $0 < \beta < 1$, then

$$\frac{h^2}{2} \le \frac{\widetilde{\chi}_k(h)}{\Phi'(k)} \le mh^2,$$

$$\frac{\widetilde{\chi}_k(h)}{\widetilde{\chi}_k(\beta h)} \le 1 + \max\left\{\frac{4m(1-\beta)^{1/m}}{\beta^2}, \frac{2(1-\beta^2)}{\beta^2}\right\}.$$

The proof is easy. We leave it to the readers.

We need also the following iteration lemma which can be proved by induction.

Lemma 1.5.4 *Let $\{y_n\}$ $(n = 0, 1, 2, \cdots)$ be a sequence of positive numbers satisfying*

$$y_{n+1} \le cb^n y_n^{1+\alpha},$$

where $c > 0$, $b > 1$, $\alpha > 0$ are constants. If

$$y_0 \le c^{-1/\alpha} b^{-1/\alpha^2},$$

then $\lim_{n\to\infty} y_n = 0$.

1.5.3 *Properties of functions in the generalized class* B_2

First let us introduce some notations. For any fixed $\rho \in (0,1]$, $\varepsilon \in (0,1]$, denote

$$Q_\rho = \left\{(x,t); |x-x_0| < \rho, t_0 - aB\rho^2 < t < t_0\right\},$$

where $B = \Phi'(\rho^\varepsilon)$, a is a constant to be determined later. Let

$$\kappa_N = \operatorname{mes} B_1,$$

$$\mu = \sup_{Q_\rho} w(x,t),\ \widetilde{\mu} = \inf_{Q_\rho} w(x,t),\ \omega = \mu - \widetilde{\mu},$$

$$\sigma = 2^{-1/N},\ \rho_1 = \frac{2+\sigma}{3}\rho,\ \rho_2 = \frac{1+2\sigma}{3}\rho,\ \rho_3 = \sigma\rho,$$

$$A = \Phi'(\mu), \quad B_k = \Phi'(k).$$

The fact $\Phi''(s) < 0$ implies that

$$A \le B_k \le B, \quad \text{if } \rho^\varepsilon \le k \le \mu.$$

In this section, we will always assume that $w \in B_2(Q_T, M, m, 2)$ and

$$\mu \ge \rho^\varepsilon. \tag{5.10}$$

In addition, we require that $t_0 - aB\rho^2 > 0$. The assumption (5.10) implies that $t_0 - aB\rho^2 \le t_0 - aA\rho^2$.

Lemma 1.5.5 *There exist constants β, $a = a(\gamma, N) \in (0,1)$, $b = b(\gamma, N) \in (0,1)$ such that*

(i) if $\mu > k > \mu/2$, $h = \mu - k > 0$ and

$$mes\, A_{k,\rho_1}(t_0 - aA\rho^2) \le \frac{1}{2}\kappa_N \rho_1^N, \tag{5.11}$$

then for any $t \in [t_0 - aA\rho^2, t_0]$

$$mes\ (B_{\rho_1} \setminus A_{k+\beta h,\rho_1}(t)) \ge b\kappa_N \rho_1^N;$$

(ii) if $h = k - \widetilde{\mu} \ge \rho^\varepsilon$ and

$$mes\, B_{k,\rho_1}(t_0 - aB_k\rho^2) \le \frac{1}{2}\kappa_N \rho_1^N, \tag{5.12}$$

then for any $t \in [t_0 - aB_k\rho^2, t_0]$

$$mes\ (B_{\rho_1} \setminus B_{k-\beta h,\rho_1}(t)) \ge b\kappa_N \rho_1^N.$$

Proof. For $\tau_1 > \tau_2 > 0$, let

$$\zeta(x;\tau_1,\tau_2) = \begin{cases} 1, & \text{if } |x-x_0| < \tau_2, \\ \dfrac{\tau_1 - |x-x_0|}{\tau_1-\tau_2}, & \text{if } \tau_2 \le |x-x_0| < \tau_1, \\ 0, & \text{if } |x-x_0| \ge \tau_1. \end{cases}$$

Choose $\zeta(x) = \zeta(x;\rho_1,\rho_1-\sigma_1\rho_1)$ in (5.8) with $\sigma_1 \in (0,1)$ to be determined, integrate over $(t_0 - aA\rho^2, t)$ and use the condition (5.11). Then we obtain

$$\begin{aligned} &\int_{A_{k,\rho_1}(t)} \zeta^2\chi_k(w-k)dx \\ \le\ & \int_{A_{k,\rho_1}(t_0-aA\rho^2)} \zeta^2\chi_k(w-k)dx + \frac{\gamma aA\rho^2h^2}{(\sigma_1\rho_1)^2}\kappa_N\rho_1^N \\ \le\ & \frac{1}{2}\kappa_N\rho_1^N\chi_k(h) + \frac{2\gamma aAh^2}{\sigma_1^2}\kappa_N\rho_1^N. \end{aligned}$$

On the other hand,

$$\begin{aligned} \int_{A_{k,\rho_1}(t)} \zeta^2\chi_k(w-k)dx \ge\ & \int_{A_{k+\beta h,\rho_1-\sigma_1\rho_1}(t)} \zeta^2\chi_k(w-k)dx \\ \ge\ & \chi_k(\beta h)\text{mes}\, A_{k+\beta h,\rho_1-\sigma_1\rho_1}(t). \end{aligned}$$

Therefore for any $t \in [t_0 - aA\rho^2, t_0]$

$$\text{mes}\, A_{k+\beta h,\rho_1-\sigma_1\rho_1}(t) \le \frac{\chi_k(h)}{\chi_k(\beta h)}\cdot\frac{1}{2}\kappa_N\rho_1^N + \frac{2a\gamma Ah^2}{\sigma_1^2\chi_k(\beta h)}\kappa_N\rho_1^N.$$

Using Lemma 1.5.4 we obtain

$$\text{mes}\, A_{k+\beta h,\rho_1-\sigma_1\rho_1}(t) \le \left(\frac{1}{2\beta^2} + \frac{4a\gamma}{\beta^2\sigma_1^2}\right)\kappa_N\rho_1^N.$$

Choose $\beta = \sqrt{2/3}$ and small b_1, σ_1 and a such that

$$\frac{1}{2\beta^2} + \frac{4a\gamma}{\beta^2\sigma_1^2} \le (1-b_1)(1-\sigma_1)^N.$$

Then for $t \in [t_0 - aA\rho^2, t_0]$ we further have

$$\text{mes}\, A_{k+\beta h,\rho_1-\sigma_1\rho_1}(t) \le (1-b_1)(1-\sigma_1)^N\kappa_N\rho_1^N.$$

Hence

$$
\begin{aligned}
& \text{mes } (B_{\rho_1} \setminus A_{k+\beta h,\rho_1}(t)) \\
\geq\ & \text{mes } (B_{\rho_1-\sigma_1\rho_1} \setminus A_{k+\beta h,\rho_1-\sigma_1\rho_1}(t)) \\
\geq\ & (1-\sigma_1)^N \kappa_N \rho_1^N - (1-b_1)(1-\sigma_1)^N \kappa_N \rho_1^N \\
=\ & b_1(1-\sigma_1)^N \kappa_N \rho_1^N,
\end{aligned}
$$

and (i) is proved by choosing $b = b_1(1-\sigma_1)^N$.

(ii) can be proved similarly. □

Lemma 1.5.6 *For any $\theta_1 > 0$, there exists $s(\theta_1) > 0$ such that*

(i) if $\mu > k > \mu/2$, $h = \mu - k$ and

$$mes\, A_{k,\rho_1}(t_0 - aA\rho^2) \leq \frac{1}{2}\kappa_N \rho_1^N,$$

then

$$\int_{t_0-aA\rho^2}^{t_0} mes\, A_{\mu-h/2^{s+1},\rho_1}(t)dt \leq \theta_1 A \rho_1^{N+2}; \tag{5.13}$$

(ii) if

$$\max_{t\in[t_0-aB\rho^2,t_0-aA\rho^2]} mes, B_{\widetilde{\mu}+\omega/2,\rho_1}(t) \leq \frac{1}{2}\kappa_N \rho_1^N, \tag{5.14}$$

then

$$\omega \leq 2^{s+2}\rho^\varepsilon$$

or

$$\int_{t_0-aB_{k_{s+2}}\rho^2}^{t_0} mes\, B_{k_{s+1},\rho_1} \leq \theta_1 B_{k_{s+2}} \rho_1^{N+2}, \tag{5.15}$$

where $k_s = \widetilde{\mu} + \omega/2^s$, $B_{k_s} = \Phi'(k_s)/m$.

Proof. First we prove (i). From Lemma 1.5.5 (i), for any $t \in [t_0 - aA\rho^2, t_0]$ we have

$$\text{mes } (B_{\rho_1} \setminus A_{k+\beta h,\rho_1}(t)) \geq b\kappa_N \rho_1^N. \tag{5.16}$$

Choose r_0 such that

$$1 - \beta \geq \frac{1}{2^{r_0}},$$

where β is the constant in Lemma 1.5.5. Let $k_l = \mu - \frac{h}{2^l}$. Noting that

$$k + \beta h \leq k + \left(1 - \frac{1}{2^{r_0}}\right) h = \mu - \frac{h}{2^{r_0}} \leq \mu - \frac{h}{2^l} = k_l, \quad \text{for } l \geq r_0,$$

from (5.16) we have, for $t \in [t_0 - aA\rho^2, t_0]$,

$$\begin{aligned} \text{mes } (B_{\rho_1} \setminus A_{k_l,\rho_1}(t)) &\geq \text{mes } (B_{\rho_1} \setminus A_{k+\beta h,\rho_1}(t)) \\ &\geq b\kappa_N \rho_1^N, \quad \text{for } l \geq r_0. \end{aligned} \tag{5.17}$$

Using Lemma 1.5.2 we obtain

$$(k_{l+1} - k_l)(\text{mes } A_{k_{l+1},\rho_1}(t))^{1-1/N} \leq \frac{\widetilde{\beta}\rho_1^N}{\text{mes } (B_{\rho_1} \setminus A_{k_l,\rho_1}(t))} \int_{D_l(t)} |\nabla w| dx,$$

where $\widetilde{\beta} > 0$ is a constant and

$$D_l(t) = A_{k_l,\rho_1}(t) \setminus A_{k_{l+1},\rho_1}(t).$$

Thus from (5.17) we see that for $t \in [t_0 - aA\rho^2, t_0]$,

$$\frac{h}{2^{l+1}} \text{mes } A_{k_{l+1},\rho_1}(t) \leq \frac{\widetilde{\beta}}{b\kappa_N} \int_{D_l(t)} |\nabla w| dx.$$

Noting that $(\text{mes } A_{k_{l+1},\rho_1}(t))^{1/N} \leq \kappa_N^{1/N} \rho_1$ and applying the Schwartz inequality to the right hand side of the above inequality, we further see that for $t \in [t_0 - aA\rho^2, t_0]$,

$$\frac{h}{2^{l+1}} \text{mes } A_{k_{l+1},\rho_1}(t) \leq \frac{\widetilde{\beta}\rho_1}{b\kappa_N^{1-1/N}} \left(\int_{D_l(t)} |\nabla w|^2 dx \right)^{1/2} (\text{mes } D_l(t))^{1/2}.$$

Integrating the square of both sides of this inequality over $[t_0 - aA\rho^2, t_0]$ and using the Schwartz inequality once again we arrive at

$$\begin{aligned} &\frac{h^2}{2^{2(l+1)}} \left(\int_{t_0 - aA\rho^2}^{t_0} \text{mes } A_{k_{l+1},\rho_1}(t) dt \right)^2 \\ \leq &\left(\frac{\widetilde{\beta}\rho_1}{b\kappa_N^{1-1/N}} \right)^2 \int_{t_0 - aA\rho^2}^{t_0} \int_{A_{k_l},\rho_1(t)} |\nabla w|^2 dx dt \int_{t_0 - aA\rho^2}^{t_0} \text{mes } D_l(t) dt. \end{aligned} \tag{5.18}$$

Now we use the properties of functions in the generalized class B_2 to estimate the integral of the right hand side of (5.18). Take $k = k_l$, $\zeta(x) = \zeta(x; \rho, \rho_1)$ in (5.8). Then we have

$$\begin{aligned} &\frac{1}{2}\int_{t_0-aA\rho^2}^{t_0}\int_{A_{k_l,\rho_1(t)}} |\nabla w|^2 dxdt \\ \leq\ & \int_{A_{k_l,\rho_1(t_0-aA\rho^2)}} \zeta^2 \chi_{k_l}(w-k_l)dx + \frac{9a\gamma A}{(1-\sigma)^2}\frac{h^2}{2^{2l}}\kappa_N\rho_1^N \\ \leq\ & \left(1+\frac{9a\gamma}{(1-\sigma)^2}\right)\frac{Ah^2}{2^{2l}}\kappa_N\rho_1^N, \end{aligned}$$

where Lemma 1.5.3 has been used at the last step of the above derivation. Substituting this into (5.18) yields

$$\left(\int_{t_0-aA\rho^2}^{t_0} \operatorname{mes} A_{k_{l+1},\rho_1}(t)dt\right)^2 \leq C_1 A\rho_1^{N+2}\int_{t_0-aA\rho_1^2}^{t_0} \operatorname{mes} D_l(t)dt$$

with constant C_1 depending only on N, γ. Summing up the above inequalities for l from r_0 to s, gives

$$(s-r_0+1)\left(\int_{t_0-aA\rho^2}^{t_0} \operatorname{mes} A_{k_{s+1},\rho_1}(t)dt\right)^2 \leq C_1\kappa_N aA^2\rho_1^{2N+4}.$$

Choosing s such that

$$\sqrt{\frac{C_1\kappa_N a}{s-r_0+1}} \leq \theta_1\sigma^{N+2},$$

we finally obtain (5.13) and complete the proof of (i).

Next we prove (ii). Suppose $\omega \geq 2^{s+2}\rho^\varepsilon$. We are ready to prove that (5.15) must hold. To this purpose, we take $k = \tilde{\mu} + \dfrac{\omega}{2}$, $hh = k - \tilde{\mu} = \dfrac{\omega}{2}$. It is evidently that (5.14) implies

$$\operatorname{mes} B_{\tilde{\mu}+\omega/2,\rho_1}(t_0 - aB_k\rho^2) \leq \frac{1}{2}\kappa_N\rho_1^N$$

and the assumption $\omega \geq 2^{s+2}\rho^\varepsilon$ implies $h = \dfrac{\omega}{2} ge\rho^\varepsilon$. Hence from (ii) in Lemma 1.5.5, we have, for $t \in [t_0 - aB\rho^2, t_0]$,

$$\operatorname{mes}\left(B_{\rho_1} \setminus B_{\tilde{\mu}+(\omega(1-\beta))/2,\rho_1}(t)\right) \geq b\kappa_N\rho_1^N. \tag{5.19}$$

Now we choose r_0 such that for $t \in [t_0 - aB\rho^2, t_0]$,

$$\text{mes}\left(B_{\rho_1} \setminus B_{\widetilde{k}_l,\rho_1}(t)\right) \geq b\kappa_N \rho_1^N, \quad \text{if } l \geq r_0,$$

which is possible from (5.19), where $\widetilde{k}_l = \widetilde{\mu} + \dfrac{\omega}{2^l}$. Then similar to the derivation from (5.16) to (5.18), we can verify that for $s \geq l \geq r_0$,

$$\begin{aligned}
&\frac{\omega^2}{2^{2l+2}}\left(\int_{t_0-aB_{\widetilde{k}_{s+2}}\rho^2}^{t_0} \text{mes}\, B_{\widetilde{k}_l,\rho_1}(t)dt\right)^2 \\
\leq &\left(\frac{\widetilde{\beta}\rho_1}{b\kappa_N^{1-1/N}}\right)^2 \int_{t_0-aB_{\widetilde{k}_{s+2}}\rho^2}^{t_0}\int_{B_{\widetilde{k}_l},\rho_1(t)} |\nabla w|^2 dxdt \int_{t_0-aB_{\widetilde{k}_{s+1}}\rho^2}^{t_0} \text{mes}\, \widetilde{D}_l(t)dt,
\end{aligned} \tag{5.20}$$

where

$$\widetilde{D}_l(t) = B_{\widetilde{k}_l,\rho_1}(t) \setminus B_{\widetilde{k}_{l+1},\rho_1}(t).$$

Based on (5.20) we can obtain (5.15) and complete the proof of (ii). □

Lemma 1.5.7 *For any $\theta_2 > 0$, there exists $\theta_1 = \theta_1(\theta_2) > 0$ such that*
(i) if $\mu > k > \mu/2$, $h = \mu - k > 0$ and

$$\int_{t_0-aA\rho^2}^{t_0} mes\, A_{k,\rho_1}(t)dt \leq \theta_1 A\rho_1^{N+2}, \tag{5.21}$$

then for any $t \in [t_0 - aA\rho^2/4, t_0]$,

$$mes\, A_{k+h/2,\rho_2}(t) \leq \theta_2\rho_2^N; \tag{5.22}$$

in addition, if $mes\, A_{k,\rho_1}(t_0 - aA\rho^2) = 0$, then for any $t \in [t_0 - aA\rho^2, t_0]$, (5.22) holds;
(ii) if $h = k - \widetilde{\mu} \geq \rho^\varepsilon$,

$$\int_{t_0-a\rho^2 B_{k-h/2}}^{t_0} mes\, B_{k,\rho_1}(t)dt \leq \theta_1 B_{k-h/2}\rho_1^{N+2}, \tag{5.23}$$

then for any $t \in [t_0 - aB_{k-h/2}\rho^2/4, t_0]$,

$$mes\, B_{k-h/2,\rho_2}(t) \leq \theta_2\rho_2^N; \tag{5.24}$$

in addition, if $mes\, B_{k,\rho_1}(t_0 - aB_k\rho^2) = 0$, then for $t \in [t_0 - aB_{k-h/2}\rho^2, t_0]$, (5.24) holds.

Proof. We only prove (ii); the proof of (i) is similar. To this purpose, we use the property (5.9) of functions in the generalized class B_2. Take $\zeta(x) = \zeta(x; \rho_1, \rho_2)$, assume that $t_0 \geq t > \tau \geq t_0 - a\rho^2 B_{k-h/2}\rho^2$ and then integrate (5.9) over $[\tau, t]$. We easily see that

$$\begin{aligned} &\widetilde{\chi}_k(\frac{h}{2})\text{mes}\, B_{k-h/2,\rho_2}(t) - \widetilde{\chi}_k(h)\text{mes}\, B_{k,\rho_1}(\tau) \\ \leq\ &\gamma\left(\frac{h^2}{(\rho_1-\rho_2)^2}+1\right)\int_\tau^t \text{mes}\, B_{k,\rho_1}(t)dt. \end{aligned} \tag{5.25}$$

From (5.23), we have, in particular,

$$\int_{t_0 - a\rho^2 B_{k-h/2}/4}^{t_0 - a\rho^2 B_{k-h/4}} \text{mes}\, B_{k,\rho_1}(t)dt \leq \theta_1 B_{k-h/2}\rho_1^{N+2},$$

which implies that there exists $\tau \in [t_0 - a\rho^2 B_{k-h/2}, t_0 - a\rho^2 B_{k-h/2}/4]$ such that

$$\text{mes}\, B_{k,\rho_1}(\tau) \leq \frac{4}{3a}\theta_1\rho_1^N.$$

Substituting this into (5.25) and using the condition (5.23), we have, for $t \in [t_0 - a\rho^2 B_{k-h/2}/4, t_0]$,

$$\text{mes}\, B_{k-h/2,\rho_2}(t) \leq \left\{\frac{\widetilde{\chi}_k(h)}{\widetilde{\chi}_k(h/2)}\frac{4}{3a} + \frac{\gamma B_{k-h/2}}{\widetilde{\chi}_k(h/2)}\left(\frac{9h^2}{(1-\sigma)^2}+\rho^2\right)\right\}\theta_1\rho_1^N.$$

Since Lemma 1.5.3 implies

$$\frac{\widetilde{\chi}_k(h)}{\widetilde{\chi}_k(h/2)} \leq 1 + 16m, \quad \frac{\widetilde{\chi}_k(h/2)}{B_{k-h/2}} \geq \frac{h^2}{8},$$

we can conclude the existence of θ_1 satisfying (5.24).

If $\text{mes}\, B_{k,\rho_1}(t_0 - a\rho^2 B_{k-h/2}) = 0$, then we take $\tau = t_0 - a\rho^2 B_{k-h/2}$ in (5.25) and assert that (5.24) holds for all $t \in [t_0 - a\rho^2 B_{k-h/2}, t_0]$. □

Lemma 1.5.8 *There exists $\theta_2 > 0$ such that*

(i) if $\mu > k \geq \mu/2$, $h = \mu - k > 0$ and

$$\max_{t\in[t_0 - aA\rho^2, t_0]} \textit{mes}\, A_{k,\rho_2}(t) \leq \theta_2\rho_2^N, \tag{5.26}$$

then for any $t \in [t_0 - aA\rho^2/4, t_0]$,

$$\textit{mes}\, A_{k+h/2,\rho_2}(t) = 0; \tag{5.27}$$

in addition, if $mes\, A_{k,\rho_2}(t_0 - aA\rho^2) = 0$, *then for any* $t \in [t_0 - aA\rho^2, t_0]$, *(5.27) holds;*

(ii) if $h = k - \widetilde{\mu} \geq \rho^{\varepsilon}$,

$$\max_{t\in[t_0 - aB_k\rho^2, t_0]} mes\, B_{k,\rho_2}(t) \leq \theta_2 \rho_2^N, \tag{5.28}$$

then for any $t \in [t_0 - aB_k\rho^2/4, t_0]$,

$$mes\, B_{k-h/2,\rho_2}(t) = 0; \tag{5.29}$$

in addition, if $mes\, B_{k,\rho_2}(t_0 - aB_k\rho^2) = 0$, *then for any* $t \in [t_0 - aB_k\rho^2, t_0]$, *(5.29) holds.*

Proof. We only prove (ii); the proof of (i) is similar. Let

$$k_j = k - \frac{h}{2} + \frac{h}{2^{j+1}}, \quad t_j = t_0 - \frac{1}{4}aB_k\rho^2 - \frac{3}{2^{j+2}}aB_k\rho^2,$$

$$\rho_j = \rho_3 + \frac{\rho_2 - \rho_3}{2^j}, \quad \mu_j = \max_{t\in[t_j,t_0]} \text{mes}\, B_{k_j,\rho_j}(t),$$

$$\zeta_j(x) = \zeta(x; \rho_j, \rho_{j+1}), \quad I_j(t) = \int_{B_{k_j,\rho_j}(t)} \widetilde{\chi}_{k_j}(k_j - w)\zeta_j^2 dx.$$

Since $k_j \geq k/2$, by Lemma 1.5.3, we have

$$\widetilde{\chi}_{k_j}(k_j - w) \leq m\Phi'(k_j)(k_j - w)^2 \leq 2mB_k(k_j - w)^2.$$

Hence

$$I_j(t) \leq 2mB_k \int_{B_{k_j,\rho_j}(t)} (k_j - w)^2 \zeta_j^2 dx. \tag{5.30}$$

Using Lemma 1.5.1, we further derive

$$I_j(t) \leq CB_k\mu_j^{2/N} \int_{B_{k_j,\rho_j}(t)} |\nabla w|^2 \zeta_j^2 dx + CB_k\mu_j^{2/N} \cdot \frac{h^2}{(\rho_j - \rho_{j+1})^2}\mu_j. \tag{5.31}$$

On the other hand, (5.4)implies

$$I_j'(t) + \frac{1}{2}\int_{B_{k_j,\rho_j}(t)} |\nabla w|^2 \zeta_j^2 dx \leq \gamma \left(\frac{h^2}{(\rho_j - \rho_{j+1})^2} + 1 \right) \mu_j. \tag{5.32}$$

Given $t \in [t_{j+1}, t_0]$.

(a) If $I_j'(t) \geq 0$, then from (5.32), we have

$$\int_{B_{k_j,\rho_j}(t)} |\nabla w|^2 \zeta_j^2 dx \leq 2\gamma \left(\frac{h^2}{(\rho_j - \rho_{j+1})^2} + 1 \right) \mu_j.$$

Substituting it into (5.31) yields

$$I_j(t) \leq CB_k \mu_j^{2/N+1} \left((2\gamma+1) \frac{h^2}{(\rho_j - \rho_{j+1})^2} + 2\gamma \right). \tag{5.33}$$

(b) If $I_j'(t) < 0$ and for some $\tau \in [t_j, t]$, $I_j'(\tau) = 0$, then we can take τ such that $I_j'(s) < 0$ for $s \in (\tau, t]$ and $I_j(t) \leq I_j(\tau)$. Since (5.31), (5.32) imply (5.33) for $t = \tau$, (5.33) holds for t too.

(c) If for any $\tau \in [t_j, t]$, $I_j'(\tau) < 0$, then from (5.32) we have

$$\frac{1}{2} \int_{t_j}^t \int_{B_{k_j,\rho_j}(t)} |\nabla w|^2 \zeta_j^2 dx dt \leq I_j(t_j) + \gamma(t - t_j) \left(\frac{h^2}{(\rho_j - \rho_{j+1})^2} + 1 \right) \mu_j.$$

Integrating (5.31) over $[t_j, t]$ and using the above inequality yield

$$\int_{t_j}^t I_j(\tau) \mathrm{d}\tau \leq CB_k \mu_j^{2/N} \left[2I_j(t_j) + (2\gamma+1) \frac{h^2 (t - t_j) \mu_h}{(\rho_j - \rho_{j+1})^2} + 2\gamma(t - t_j) \mu_j \right].$$

Substituting (5.30) into this inequality and noting that $I_j(\tau)$ is increasing on $[t_j, t]$, we derive that for $t \in [t_{j+1}, t_0]$,

$$I_j(t) \leq CB_k \mu_j^{2/N+1} \left(\frac{4mh^2 B_k}{t_{j+1} - t_j} + \frac{(2\gamma+1)h^2}{(\rho_j - \rho_{j+1})^2} + 2\gamma \right). \tag{5.34}$$

Thus in all cases (5.34) holds for $t \in [t_{j+1}, t_0]$.

On the other hand, using (5.3) we have

$$\begin{aligned} I_j(t) \geq \; & \widetilde{\chi}_{k_j}(k_j - k_{j+1}) \operatorname{mes} B_{k_{j+1},\rho_{j+1}}(t) \\ \geq \; & \frac{1}{2} B_k (k_j - k_{j+1})^2 \operatorname{mes} B_{k_{j+1},\rho_{j+1}}(t), \end{aligned}$$

which combining with (5.34) yields

$$\mu_{j+1} \leq 2C \mu_j^{2/N+1} 2^{2(j+1)} \left(\frac{4mB_k}{t_{j+1} - t_j} + \frac{2\gamma+1}{(\rho_j - \rho_{j+1})^2} + \frac{2\gamma}{h^2} \right).$$

Hence from the definition of t_j and ρ_j, we see that the above inequality can be simplified as

$$\mu_{j+1} \leq \frac{C_1 2^{4j} \mu_j^{2/N+1}}{\rho^2},$$

where C_1 is a constant depending only on N, m, γ, T. If we set $y_j = \mu_j/\rho^N$, then the above inequality can be written as

$$y_{j+1} \leq C_1 2^{4j} y_j^{2/N+1}.$$

The assumption (5.28) implies that

$$y_0 \leq \theta_2.$$

Therefore we can use Lemma 1.5.4 to conclude that $\lim_{j\to\infty} y_j = 0$ provided θ_2 is chosen small enough and complete the proof of (5.29).

If $\operatorname{mes} B_{k,\rho_2}(t_0 - aB_k\rho^2) = 0$, then we take $t_j = t_0 - aB_k\rho^2$ instead of $t_j = t_0 - \frac{1}{4}aB_k\rho^2 - \frac{3}{2^{j+2}}ab_k\rho^2$. The desired conclusion can be easily obtained. □

Lemma 1.5.9 *There exists $s > 0$ such that*

(i) if $\mu > k \geq \mu/2$, $h = \mu - k > 0$ and

$$mes\, A_{k,\rho_1}(t_0 - aA\rho^2) \leq \frac{1}{2}\kappa_N \rho_1^N,$$

then for any $t \in [t_0 - aA\rho^2/16, t_0]$,

$$mes\, A_{\mu-h/2^{s+3},\rho_3}(t) = 0; \tag{5.35}$$

(ii) if

$$\max_{t\in[t_0-aB\rho^2, t_0-aA\rho^2]} mes\, B_{\widetilde{\mu}+\omega/2,\rho_1}(t) \leq \frac{1}{2}\kappa_N \rho_1^N, \tag{5.36}$$

then

$$\omega \leq 2^{s+2}\rho^{\varepsilon} \tag{5.37}$$

or

$$osc\,\{w, \widetilde{Q}_{\rho/4}\} \leq \left(1 - \frac{1}{2^{s+3}}\right) osc\,\{w, \widetilde{Q}_\rho\}, \tag{5.38}$$

where

$$\widetilde{Q}_{\rho/4} = \left\{(x,t); x \in B_{\rho/4}, t_0 - aA(\rho/4)^2 < t < t_0\right\}.$$

Proof. Let θ_2, θ_1 and s be constants determined by Lemma 1.5.8, Lemma 1.5.7 and Lemma 5.6 successively. Then (5.35) holds. Similarly from the assumption (5.36) we can derive (5.37) or

$$\text{mes}\, B_{\widetilde{\mu}+\omega/2^{s+3},\rho_3}(t) = 0$$

for any $t \in [t_0 - aB_{k_{s+2}}(\rho/4)^2, t_0]$, where $B_{k_s} = \Phi'(k_s)$, $k_s = \widetilde{\mu} + \omega/2^s$. In the later case, obviously we have

$$\text{osc}\,\{w, \widetilde{Q}_{\rho/4}\} \leq \mu - \left(\widetilde{\mu} + \frac{\omega}{2^{s+3}}\right) \leq \left(1 - \frac{1}{2^{s+3}}\right)\omega,$$

which is just what we want to prove. □

1.5.4 *Hölder continuity of solutions*

Proposition 1.5.2 *Let $\rho_0 \in (0,1]$, $\varepsilon \in (0,1]$, $Q_{\rho_0} \subset Q_T$ and $u(x,t)$ be a classical solution of (5.1) on Q_{ρ_0} such that $0 < u \leq M$. Denote*

$$\mu_0 = \sup_{Q_{\rho_0}} w(x,t), \quad \widetilde{\mu}_0 = \inf_{Q_{\rho_0}} w(x,t), \quad \omega_0 = \mu_0 - \widetilde{\mu}_0,$$

where $w = u^m$. If for some constant $C_0 \geq 1$,

$$\omega_0 \leq C_0\rho_0^{\varepsilon}, \quad \mu_0 \geq 2C_0\rho_0^{\varepsilon}, \tag{5.39}$$

then there exist positive constants α, C depending only on N, m, T, M, such that for $(x,t) \in Q_{\rho_0}$,

$$|w(x,t) - w(x_0,t_0)| \leq C(|x-x_0|^{\alpha} + |t-t_0|^{\alpha/2}).$$

Proof. Let

$$x' = \frac{x - x_0}{\rho_0}, \quad t' = \frac{t - t_0}{a\rho_0^2\Phi'(\mu_0)}.$$

Then Q_{ρ_0} is transformed into

$$Q_1' = \left\{(x',t'); |x'| < 1, -\frac{\Phi'(\rho_0^{\varepsilon})}{\Phi'(\mu_0)} < t' < 0\right\}$$

and $v(x',t')=u(x,t)$ satisfies

$$\frac{\partial v}{\partial t'}=\mathrm{div}_{x'}(A(x',t')\nabla_{x'}v),$$

where

$$A(x',t')=am\Phi'(\mu_0)u^{m-1}(x,t).$$

From (5.39), $\widetilde{\mu}_0\geq\frac{\mu_0}{2}$ and hence

$$a\left(\frac{1}{2}\right)^{(m-1)/m}\leq A(x',t')\equiv a\left(\frac{w}{\mu_0}\right)^{(m-1)/m}\leq a.$$

Thus using the estimate which holds for parabolic equations without degeneracy, we have

$$|v(x',t')-v(0,0)|\leq C_1(|x'|^{\alpha_1}+|t'|^{\alpha_1/2})$$

with constants $C_1>0$ and $\alpha_1\in(0,1)$ depending only on N, m, M, T, which holds for any $(x',t')\in Q'_{1/2}=\{(x',t');|x'|<1/2,-1<t'<0\}$.

Returning to the original variables (x,t), we get

$$|u(x,t)-u(x_0,t_0)|\leq C_1\left(\frac{|x-x_0|^{\alpha_1}}{\rho_0^{\alpha_1}}+\frac{|t-t_0|^{\alpha_1/2}}{(a\rho_0^2\Phi'(\mu_0))^{\alpha_1/2}}\right) \tag{5.40}$$

for any $(x,t)\in\widetilde{Q}_{\rho_0/2}$, where

$$\widetilde{Q}_{\rho_0/2}=\left\{(x,t);|x-x_0|<\frac{\rho_0}{2},t_0-\frac{a}{2}\Phi'(\mu_0)\rho_0^2<t<t_0\right\}.$$

Let

$$\Sigma=\left\{(x,t);|x-x_0|\leq\frac{\rho_0^2}{4},t_0-\left(\frac{a}{2}\Phi'(\mu_0)\rho_0^2\right)^2<t\leq t_0\right\}.$$

Then from (5.40), we see that, for $(x,t)\in\Sigma$,

$$|u(x,t)-u(x_0,t_0)|\leq C_1\left(|x-x_0|^{\alpha_1/2}+|t-t_0|^{\alpha_1/4}\right). \tag{5.41}$$

If $(x,t) \in Q_{\rho_0} \setminus \Sigma$, then using the assumption (5.39) again, we have

$$\begin{aligned}
&\left(|x-x_0|^2+|t-t_0|\right)^{1/4} \\
\geq\ &\min\{\rho_0/2, [a\Phi'(\mu_0)\rho_0^2/2]^{1/2}\} \\
\geq\ &\min\left\{\frac{1}{2}, \left(\frac{a\Phi'(M^m)}{2}\right)^{1/2}\right\}\rho_0 \\
\geq\ &\min\left\{\frac{1}{2}, \left(\frac{a\Phi'(M^m)}{2}\right)^{1/2}\right\}(C_0^{-1}\omega_0)^{1/2}.
\end{aligned}$$

Hence

$$\omega_0 \leq C_0\left(\min\left\{\frac{1}{2}, \left(\frac{a\Phi'(M^m)}{2}\right)^{1/2}\right\}\right)^{-\varepsilon}(|x-x_0|^2+|t-t_0|)^{\varepsilon/4},$$

which shows that (5.41) holds on $Q_{\rho_0} \setminus \Sigma$ for some other $\alpha_1 \in (0,1)$. The proof is complete. □

Theorem 1.5.1 *Let u be a generalized solution of (5.1) on Q_T such that $0 \leq u(x,t) \leq M$. Then there exist constants $C>0$, $\alpha \in (0,1)$ depending only on N, m, M, T such that*

$$|w(x,t)-w(x_0,t_0)| \leq C\rho_0^{-1}\left(|x-x_0|^\alpha+|t-t_0|^{\alpha/2}\right), \quad \forall (x_0,t_0) \in Q_T,\ (x,t) \in Q^*_{\rho_0},$$

provided $t_0 - a\rho_0^2/m^2 > 0$, where $w = u^m$,

$$\rho_0 = \min\left\{1, \left(\frac{t_0}{a}\right)^{m/(m+1)}\right\},$$

$$Q^*_{\rho_0} = \left\{(x,t); |x-x_0| < \rho_0, t_0 - \frac{a}{m^2}\rho_0^2 < t < t_0\right\},$$

and $a = a(2,N)$ is the constant determined in Lemma 1.5.5.

Obviously, Theorem 1.5.1 implies that for any $\eta \in (0,T)$, the generalized solution u is Hölder continuous in $\mathbb{R}^N \times (\eta, T)$.

Proof. As we have indicated at the beginning of this section we may simply assume that$u(x,t)$ is a positive classical solution of (5.1).

Given $(x_0, t_0) \in Q_T$. Denote

$$\hat{C} = \max\left\{(2^{s+8}m^3)^m, \frac{4M^m}{\rho_0}\right\},$$
$$\eta = 64m^3\hat{C}^{1-1/m},$$

where s is the constant determined in Lemma 1.5.9. Choose $\varepsilon \in (0,1)$ so small that

$$\left(\frac{1}{\eta}\right)^{\varepsilon} \geq 1 - 2^{-1-(s+3)N_c}, \quad N_c = 4m^3\hat{C}^{1-1/m},$$

and let

$$\rho_l = \frac{\rho_0}{\eta^l},$$
$$Q_l = Q_{\rho_l} = \left\{(x,t); |x - x_0| < \rho_l, t_0 - a\Phi'(\rho_l^\varepsilon)\rho_l^2 < t < t_0\right\},$$
$$\mu_l = \sup_{Q_l} w(x,t), \quad \widetilde{\mu}_l = \inf_{Q_l} w(x,t) \quad (l = 0, 1, 2, \cdots),$$
$$\omega_l = \mu_l - \widetilde{\mu}_l, \quad l^* = \inf\{l; \mu_l \geq 2\hat{C}\rho_l^\varepsilon\}.$$

From the definition of $\hat{C}$, it is clear that

$$\omega_0 \leq \hat{C}\rho_0^\varepsilon.$$

If $l^* = 0$, then the conclusion for our theorem follows from Proposition 1.5.2. Now we suppose $l^* > 0$ and prove

$$\omega_l \leq \hat{C}\rho_l^\varepsilon \quad \text{when } l \leq l^* \tag{5.42}$$

by induction.

Suppose (5.42) holds for $l < l^*$. We want to prove that (5.42) holds for $l+1$, provided $l + 1 \leq l^*$.

If $\mu_l \leq 2\rho_l^\varepsilon$, then it is easy to see that

$$\omega_{l+1} \leq \mu_l \leq \hat{C}\rho_{l+1}^\varepsilon$$

which means that (5.42) holds for $l+1$. Now we suppose $\mu_l \geq 2\rho_l^\varepsilon$.

If

$$\max_{t\in[t_0 - aB\rho_l^2, t_0 - aA\rho_l^2]} \operatorname{mes} B_{\widetilde{\mu}_l + \omega_l/2, \rho_{1l}}(t) \leq \frac{1}{2}\kappa_N \rho_{1l}^N, \tag{5.43}$$

where

$$B = \Phi'(\rho_l^\varepsilon), \quad A = \Phi'(\mu_l), \quad \rho_{1l} = \frac{2+\sigma_0}{3}\rho_l, \quad \sigma_0 = 2^{-1/N},$$

then from the definition of $\hat{C}$ and Lemma 1.5.9, we see that

$$\begin{aligned}\omega_l &\le 2^{s+2}\rho_l^\varepsilon \le 2^{s+2}\eta^\varepsilon\rho_{l+1}^\varepsilon \\ &\le \frac{2^{-6}}{m^3}\hat{C}^{1/m}(64m^3\hat{C}^{1-1/m})^\varepsilon\rho_{l+1}^\varepsilon \le \hat{C}\rho_{l+1}^\varepsilon\end{aligned}$$

or

$$\operatorname{osc}\{w;\widetilde{Q}_{\rho_l/4}\} \le \left(1-\frac{1}{2^{s+3}}\right)\operatorname{osc}\{w;Q_{\rho_l}\} \le \left(\frac{1}{\eta}\right)^\varepsilon \hat{C}\rho_l^\varepsilon \le \hat{C}\rho_{l+1}^\varepsilon.$$

From the definition of η and noting that $\mu_l < 2\hat{C}\rho_l^\varepsilon$, we can verify that $Q_{l+1} \subset \widetilde{Q}_{\rho_l/4}$. In fact, $\rho_{l+1} \le \dfrac{\rho_l}{4}$ and

$$\begin{aligned}\frac{aA\,(\rho_l/4)^2}{a\Phi'(\rho_{l+1}^\varepsilon)\rho_{l+1}^2} &= \frac{\Phi'(\mu_l)\rho_l^2}{16\Phi'(\rho_{l+1}^\varepsilon)\rho_{l+1}^2} \\ &\ge \frac{\eta^2}{16m^3}\left(\frac{1}{2\hat{C}\eta^\varepsilon}\right)^{1-1/m} \ge \frac{\eta^{1+1/m}}{32m^3\hat{C}^{1-1/m}} \ge 1.\end{aligned}$$

Thus $Q_{l+1} \subset \widetilde{Q}_{\rho_l/4}$ and hence $\omega_{l+1} \le \hat{C}\rho_{l+1}^\varepsilon$.

If (5.43) does not hold, then there exists $\tau \in [t_0 - aB\rho_l^2, t_0 - aA\rho_l^2]$, such that $\operatorname{mes} B_{\widetilde{\mu}_l+\omega_l/2,\rho_{1l}}(\tau) > \dfrac{1}{2}\kappa_N\rho_{1l}^N$ and hence

$$\operatorname{mes} A_{\widetilde{\mu}+\omega_l/2,\rho_{1l}} \le \frac{1}{2}\kappa_N\rho_{1l}^N$$

or

$$\operatorname{mes} A_{\mu_l-\omega_l/2,\rho_{1l}}(\tau) \le \frac{1}{2}\kappa_N\rho_{1l}^N. \tag{5.44}$$

Now we divide the interval $[\tau, t_0]$ into K equal parts, such that

$$\frac{1}{2}aA\rho_l^2 \le \Delta t = \frac{t_0-\tau}{K} \le aA\rho_l^2.$$

Clearly

$$K \le \frac{2a\Phi'(\rho_l^\varepsilon)\rho_l^2}{aA\rho_l^2} \le 2\left(\frac{2\hat{C}\rho_l^\varepsilon}{\rho_l^\varepsilon}\right)^{1-1/m} \le 4\hat{C}^{1-1/m} = K_0.$$

Let $t_p = \tau + (p-1)\Delta t$ $(p = 1, 2, \cdots, K)$, $t_{K+1} = t_0$. Using Lemma 1.5.9 on $[t_1, t_2]$, from (5.44) we obtain

$$\text{mes}\, A_{\mu_l - \omega_l/2^{s+4}, \rho_{3l}}(t_2) = 0,$$

where $\rho_{3l} = \sigma_0 \rho_l$, $\sigma_0 = 2^{-1/N}$. Therefore

$$\begin{aligned} & \text{mes}\, A_{\mu_l - \omega_l/2^{s+4}, \rho_{3l}}(t_2) \\ \leq\ & \text{mes}\, A_{\mu_l - \omega_l/2^{s+4}, \rho_{3l}}(t_2) + \kappa_N(\rho_{1l}^N - \rho_{3l}^N) \\ \leq\ & \frac{1}{2}\kappa_N \rho_{1l}^N. \end{aligned}$$

By induction, we finally arrive at

$$\text{mes}\, A_{\mu_l - \omega_l/2^{(s+3)(K-1)+1}, \rho_{1l}}(t_K) \leq \frac{1}{2}\kappa_N \rho_{1l}^N$$

and for $t \in (t_0 - \frac{a}{32} A\rho_l^2, t_0)$,

$$\text{mes}\, A_{\mu_l - \omega_l/2^{(s+3)K+1}, \rho_{3l}}(t) = 0.$$

Therefore

$$\begin{aligned} \omega_{l+1} = \text{osc}\,\{w; Q_{l+1}\} \leq\ & \left(1 - \frac{1}{2^{(s+3)K+1}}\right) \text{osc}\,\{w; Q_l\} \\ \leq\ & \eta^{-\varepsilon} \hat{C}\rho_l^\varepsilon \leq \hat{C}\rho_{l+1}^\varepsilon. \end{aligned}$$

Summing up, we have proved (5.42) for all $l \leq l^*$. In particular, we have $\omega_{l^*} \leq \hat{C}\rho_{l^*}^\varepsilon$. On the other hand, by the definition of l^*, $\mu_{l^*} \geq 2\hat{C}\rho_{l^*}^\varepsilon$. Thus, by Proposition 1.5.2, there exist positive constants α, C depending only on N, m, T, M, such that

$$|w(x,t) - w(x_0, t_0)| \leq C\left(|x - x_0|^\alpha + |t - t_0|^{\alpha/2}\right) \tag{5.45}$$

holds on $Q_{\rho_{l^*}}$. For any $\rho \in [\rho_{l^*}, \rho_0)$, there must be an $l < l^*$ such that $\rho_{l+1} \leq \rho \leq \rho_l$. Thus by the definition of l^*, we have

$$\begin{aligned} & \text{osc}\{w; Q_\rho\} = \sup_{Q_\rho} w(x,t) - \inf_{Q_\rho} w(x,t) \leq \omega_l \\ \leq\ & \hat{C}\rho_l^\varepsilon = \hat{C}(\eta\rho_{l+1})^\varepsilon \leq \hat{C}(\eta\rho)^\varepsilon = \hat{C}\eta^\varepsilon\rho^\varepsilon, \end{aligned}$$

from which it is easy to see that (5.45) holds on Q_{ρ_0} for some positive constants α, C depending only on N, m, T, M. The proof is complete. □

Remark 1.5.1 If the initial data $u_0 \geq 0$ are nonnegative, bounded and Hölder continuous, then by a similar argument we can prove that the corresponding generalized solution is Hölder continuous down to $t = 0$.

1.6 Properties of the Free Boundary: One Dimensional Case

1.6.1 *Finite propagation of disturbances*

Consider the equation

$$\frac{\partial u}{\partial t} = \frac{\partial^2 A(u)}{\partial x^2} \tag{6.1}$$

with initial data

$$u(x,0) = u_0(x) \quad x \in \mathbb{R}. \tag{6.2}$$

We always assume that $A(u) \in C^1[0,\infty) \cap C^2(0,\infty)$,

$$A(u) > 0, A'(u) > 0, A''(u) > 0 \quad \text{for } u > 0,$$

$$A(0) = A'(0) = 0,$$

and u_0 is nonnegative, bounded and continuous on $\mathbb{R}$ with $A(u_0)$ satisfying the Lipschitz condition.

By virtue of Theorem 1.2.1 and Theorem 1.2.2, the Cauchy problem (6.1), (6.2) admits exactly one nonnegative, bounded and continuous generalized solution u on $Q = \mathbb{R} \times (0,\infty)$ with bounded weak derivative $\dfrac{\partial A(u)}{\partial x}$. In §1.4, we have discussed the regularity of generalized solutions for the equation

$$\frac{\partial u}{\partial t} = \frac{\partial^2 u^m}{\partial x^2} \tag{6.3}$$

with $m > 1$ which corresponds to the slow diffusion.

In physics, slow diffusion should imply that the speed of propagation of disturbances is finite. The mathematical description of this fact is that if $\operatorname{supp} u_0$ is bounded, then for any $t > 0$, $\operatorname{supp} u(x,t)$ is also bounded.

We have the following general result.

Theorem 1.6.1 *Assume that for any* $u > 0$,

$$\Psi(u) = \int_0^u \frac{A'(s)}{s} ds < +\infty. \tag{6.4}$$

Let u *be a generalized solution of the Cauchy problem (6.1), (6.2) on* Q. *If* $\operatorname{supp} u_0(x)$ *is bounded, then for any* $t > 0$, $\operatorname{supp}\ u(\cdot, t)$ *is also bounded.*

Proof. Consider the function of the form

$$\overline{u}(x,t) = \Psi^{-1}(c(ct + t_1) - x)_+).$$

It is easy to verify that for any $t_1 > 0$, $c > 0$, $\overline{u}(x,t)$ is a generalized solution of (6.1) with initial data

$$\overline{u}(x,0) = \Psi^{-1}(c(ct_1 - x)_+).$$

Let $x_0 = \sup\{\operatorname{supp} u_0(x)\}$. Then

$$u(x,0) = u_0(x) = 0 \le \overline{u}(x,0) \quad \text{for } x \ge x_0.$$

Besides, since Ψ and Ψ^{-1} are increasing, we have

$$\begin{aligned} \overline{u}(x,t) &= \Psi^{-1}(c(c(t+t_1) - x_0)_+) \\ &\ge \Psi^{-1}(c(ct_1 - x_0)_+) \ge M = \sup_Q u, \end{aligned}$$

provided $ct_1 - x_0 > 0$ and $c(ct_1 - x_0) \ge \Psi(M)$. Hence

$$u(x_0,t) \le M \le \overline{u}(x_0,t) \quad \text{for } t \ge 0.$$

Now we apply the comparison theorem on $G_T = (x_0, \infty) \times (0,T)$ (see Theorem 1.2.4 and Remark 1.2.2) and then obtain

$$u(x,t) \le \overline{u}(x,t), \quad \text{on } G_T,$$

from which it follows that $u(x,t) = 0$ when $x \ge X_t = c(t + t_1)$, since $\overline{u}(x,t) = 0$ for $x \ge X_t = c(t+t_1)$.

Notice that in applying the comparison theorem on G_T, we need to check $u \in L^1(G_T)$ (see Remark 1.2.2). From Definition 1.1.3 and Remark 1.1.2, for any $\tau \in (0,T)$ and $\varphi \in C^\infty(\overline{Q}_T)$ which vanishes when $|x|$ is large enough, we have

$$\int_{\mathbb{R}} u(x,\tau)\varphi(x,\tau)dx - \int_{\mathbb{R}} u_0(x)\varphi(x,0)dx$$
$$= \iint_{Q_\tau} \left(u\frac{\partial \varphi}{\partial t} + A(u)\frac{\partial^2 \varphi}{\partial x^2} \right) dxdt.$$

In particular, we take $\varphi = \varphi_X \in C^\infty(\mathbb{R})$ such that

$$0 \le \varphi_X(x) \le 1,\ \varphi_X(x) = 1, \text{ for } |x| \le X,$$

$$\varphi_X(x) = 0, \text{ for } |x| \ge X+1,\ |\varphi''_X(x)| \le C,$$

where the constant C is independent of X. Then we obtain

$$\int_{|x|\le X} u(x,\tau)dx + \iint_{\substack{X\le|x|\le X+1 \\ 0\le t\le T}} A(u)\varphi''_X(x)dxdt \le \int_{\mathbb{R}} u_0(x)dx.$$

Since the right hand side is bounded uniformly in X, letting $X \to \infty$, we see that $\int_{\mathbb{R}} u(x,\tau)dx$ is bounded and hence $u \in L^1(Q_T)$.

Similarly we can prove that there exists X'_t such that $u(x,t) = 0$ when $x \le X'_t$. □

Remark 1.6.1 Theorem 1.6.1 shows that (6.4) is a sufficient condition for (6.1) to possess the property of finite speed of disturbances. One can prove that (6.4) is also necessary for (6.1) to possess such property. Oleinik guessed the necessity of this condition at a symposium held in Moscow university. Since then people verified this supposition for some special cases or under some additional condition. Finally Peletier [PE3] gave a satisfactory result.

When $A(u) = u^m$, the condition (6.4) is equivalent to $m > 1$, the slow diffusion case.

Assume that $\operatorname{supp} u_0 = [x_1, x_2]$ $(-\infty < x_1 < x_2 < +\infty)$ and u is a generalized solution of (6.3), (6.2) on $Q = \mathbb{R} \times (0,\infty)$. Denote

$$\Omega = \{(x,t); u(x,t) > 0, t > 0\}$$

$$\Omega(t) = \{x; u(x,t) > 0\}$$

$$\zeta_1(t) = \inf \Omega(t), \quad \zeta_2(t) = \sup \Omega(t).$$

By the continuity of u, Ω is an open set. Theorem 1.6.1 implies that for any $t > 0$, $\zeta_i(t)$ $(i = 1, 2)$ is finite. We call $x = \zeta_i(t)$ $(i = 1, 2)$ the **free boundary** or **interface** of u.

Theorem 1.6.2 *Assume that $m > 1$. Then $(-1)^i\zeta_i(t)$ $(i = 1, 2)$ is increasing and*

$$\lim_{t\to\infty}(-1)^i\zeta_i(t) = +\infty, \quad (i = 1, 2). \tag{6.5}$$

Theorem 1.6.2 means that in case of slow diffusion, disturbances will be propagated to infinite scope, although the speed of propagation is finite.

Before proving Theorem 1.6.2, we first prove the following proposition which is also very useful in the sequel.

Proposition 1.6.1 *Assume that $m > 1$ and u is the generalized solution of the Cauchy problem (6.3), (6.2)on $Q = \mathbb{R} \times (0, \infty)$. Then*

$$\frac{\partial u}{\partial t} \geq -\frac{ku}{t}, \tag{6.6}$$

$$\frac{\partial v}{\partial t} \geq -\frac{(m-1)kv}{t}, \tag{6.7}$$

$$\frac{\partial^2 v}{\partial x^2} \geq -\frac{k}{t} \tag{6.8}$$

in the sense of distributions.

Remark 1.6.2 As will be seen from the proof stated below, this proposition is valid for any initial data $u_0 \in L^\infty(\mathbb{R}^N)$.

Proof of Proposition 1.6.1. From the proof of Theorem 1.3.5 and point 4 in §1.3.3, the given generalized solution u can be obtained as the limit of a sequence of classical solutions which are positive, and uniformly bounded on Q and whose derivatives up to second order are bounded. To prove the proposition, we may simply suppose that u is just its approximate smooth solution.

We first verify (6.8). Notice that v satisfies

$$\frac{\partial v}{\partial t} = (m-1)v\frac{\partial^2 v}{\partial x^2} + \left(\frac{\partial v}{\partial x}\right)^2 \tag{6.9}$$

and $w = \dfrac{\partial^2 v}{\partial x^2}$ satisfies

$$Lw = \frac{\partial w}{\partial t} - (m-1)v\frac{\partial^2 w}{\partial x^2} - 2m\frac{\partial v}{\partial x}\cdot\frac{\partial w}{\partial x} - (m+1)w^2 = 0. \qquad (6.10)$$

Clearly the function $\widetilde{w} = -\dfrac{k}{t}$ also satisfies the equation (6.10) on Q. Since $w = \dfrac{\partial^2 v}{\partial x^2}$ is bounded, we have

$$w(x,\varepsilon) > -\frac{k}{\varepsilon}$$

provided $\varepsilon > 0$ is small enough. Thus we may use the comparison theorem on $\mathrm{R}\times(\varepsilon,\infty)$ to assert

$$w(x,t) \geq \widetilde{w}(x,t) = -\frac{k}{t} \quad \text{for } (x,t) \in \mathbb{R}\times(\varepsilon,\infty),$$

from which (6.8) follows by virtue of the arbitrariness of ε.

(6.7) follows from (6.8) and (6.9).

To verify (6.6), it suffices to note that

$$\begin{aligned}\frac{\partial u}{\partial t} &= \left(\frac{m-1}{m}\right)^{1/(m-1)}\left(\frac{\partial^2 v}{\partial x^2} + \frac{1}{m-1}v^{(2-m)/(m-1)}\left(\frac{\partial v}{\partial x}\right)^2\right)\\ &\geq u\frac{\partial^2 v}{\partial x^2} \geq -\frac{ku}{t}.\end{aligned}$$

□

Proof of Theorem 1.6.2. From (6.6) we have

$$\frac{\partial}{\partial t}(t^k u) \geq 0.$$

From this and the continuity of u, it is easy to see that for any $x \in \mathbb{R}$, $t^k u$ is increasing in t. Therefore

$$\Omega(t_1) \subset \Omega(t_2), \quad \text{for } t_1 < t_2,$$

in other words, $(-1)^i\zeta_i(t)$ $(i = 1, 2)$ is increasing.

Next we prove (6.5). For simplicity we suppose that $x_1 < 0 < x_2$, $u_0(0) > 0$. Then there exist $\delta > 0$, $\varepsilon_0 > 0$ such that

$$u_0(x) \geq \varepsilon_0 \quad \text{for } |x| \leq \delta.$$

Consider the function

$$B_{\tau,L}(x,t) = L^{1/m-1} B_m(x, L(t+\tau)) \qquad (L, \tau > 0)$$

where $B_m(x,t)$ is the Barenblatt solution of (6.3) (see (1.37)). It is easy to check by direct calculation that for any $L, \tau > 0$, $B_{\tau,L}(x,t)$ is a generalized solution of (6.3) with initial data

$$B_{\tau,L}(x,0) = L^{1/(m-1)} B(x, L\tau)$$

$$= \frac{L^{1/(m-1)}}{(L\tau)^{1/(m+1)}} \left(\left(1 - \frac{m-1}{2m(m+1)} \frac{x^2}{(L\tau)^{2/(m+1)}} \right)_+ \right)^{1/(m-1)}.$$

Obviously

$$0 \le B_{\tau,L}(x,0) \le \frac{L^{2/(m^2-1)}}{\tau^{1/(m+1)}},$$

$$\operatorname{supp} B_{\tau,L}(x,0) = \left\{ x : |x|^2 \le \frac{2m(m+1)}{m-1} (L\tau)^{2/(m+1)} \right\}.$$

Choose $L, \tau > 0$ such that

$$L^{2/(m^2-1)} = \varepsilon_0 \tau^{1/(m+1)}, \qquad \frac{2m(m+1)}{m-1} (L\tau)^{2/(m+1)} = \delta^2.$$

It is easy to see that this is possible. For such L, $\tau > 0$, we have

$$B_{\tau,L}(x,0) \le u(x,0),$$

and thus by the comparison theorem (Theorem 1.2.3)

$$B_{\tau,L}(x,t) \le u(x,t).$$

In particular

$$\Omega(t) \supset \left\{ x : |x|^2 \le \frac{2m(m+1)}{m-1} (L(t+\tau))^{2/(m+1)} \right\}.$$

This completes the proof of (6.5). □

1.6.2 *Localization and extinction of disturbances*

We will indicate in the sequel that for filtration equations with appropriate absorption term, the support of generalized solutions might be included in a bounded domain forever. In this case we say that the disturbances possess the property of localization.

Consider equations of the form

$$\frac{\partial u}{\partial t} = \frac{\partial^2 A(u)}{\partial x^2} - c(u) \quad (x,t) \in Q, \tag{6.11}$$

where $A(u)$, $c(u)$ are appropriately smooth. In addition to the assumptions on $A(u)$ and u_0 given at the beginning of this section, we assume that $c'(u) > 0$ for $u > 0$ and $c(0) = 0$.

Similar to the argument stated in §1.1.2, we can prove the existence and uniqueness of generalized solutions (which can be defined in an obvious way) of the Cauchy problem for (6.11). Also the comparison theorem is valid (cf. [KA2]). In the sequel, we will use these results without proof.

Theorem 1.6.3 *Let u be a generalized solution of the Cauchy problem (6.11), (6.2) on Q. If $u_0(x) = 0$ for $|x| \geq X > 0$ and*

$$\int_0^1 \left(\int_0^v \Psi(\xi) d\xi \right)^{-1/2} dv < +\infty, \tag{6.12}$$

where $\Psi(v) = c(\Phi(v))$, $\Phi(v) = A^{-1}(v)$, then there exists $X_1 > X$ such that $u(x,t) = 0$ for $|x| \geq X$, $t \geq 0$.

Proof. To prove that for some $X_1 \geq X$, $u(x,t) = 0$ for $x \geq X$, $t \geq 0$, it suffices to construct a generalized solution $w(x,t)$ on the domain $G_X = \{(x,t); x > X, t > 0\}$ such that

$$u(x,t) \leq w(x,t) \quad \text{for } (x,t) \in G_X \tag{6.13}$$

and $w(x,t) = 0$ for $x \geq X_1$, $t \geq 0$.

We try to seek such generalized solution $w(x,t)$ among functions which depend only on x. First we require $w(x)$ to satisfy (6.11) on $G_X \backslash \overline{G}_{X_1}$, namely, for $X < x < X_1$,

$$\frac{d^2}{dx^2} A(w(x)) = c(w(x))$$

or

$$\frac{d^2 j(x)}{dx^2} = c(\Phi(j(x))) = \Psi(j(x)), \tag{6.14}$$

where $j(x) = A(w(x))$ and $X_1 > X$ is a constant to be determined later. Denote the inverse function of $j(s)$ by $J(v)$. Then

$$\frac{dj(x)}{dx} = \frac{1}{J'(v)},$$

$$\frac{d^2j(x)}{dx^2} = -\frac{J''(v)}{(J'(v))^2}\cdot\frac{dj(x)}{dx} = -\frac{J''(v)}{(J'(v))^3} = \left(\frac{1}{2(J'(v))^2}\right)',$$

and (6.14) turns out to be

$$\left(\frac{1}{2(J'(v))^2}\right)' = \Psi(v).$$

Integrating this equality yields

$$J(v) = \int_0^v \left(2\int_0^\eta \Psi(\xi)d\xi\right)^{-1/2} d\eta. \tag{6.15}$$

From the above analysis, it is natural to define $J(v)$ by (6.15) and then to consider its inverse function $j(x)$. The condition (6.12) ensures the definition of $J(v)$ for all $v \geq 0$. Since $J(v)$ is increasing and $J(+\infty) = +\infty$, $j(x)$ is well-defined for all $x \geq 0$. $j(x)$ is a solution of (6.14) for $x > 0$; so is $j(X_1 - x)$ for any X_1 and $x < X_1$. Hence for any $X_1 \geq X$, $\Phi(j(X_1 - x))$ is a (classical) solution for $x < X_1$. We choose X_1 such that

$$u(x,t) \leq w(X). \tag{6.16}$$

This is possible; for example, we may take $X_1 = X + J(A(M))$, where $M = \sup u$.

Now we define

$$w(x,t) = w(x) = \begin{cases} \Phi(j(X_1 - x)) & \text{for } X \leq x \leq X_1, \\ 0 & \text{for } x \geq X_1. \end{cases}$$

Since w and $\dfrac{dA(w)}{dx}$ equal zero at $x = X_1$, it is easy to check that w is a generalized solution of (6.11) on G_X. Using the comparison theorem for u and w on G_X and noticing (6.16) and that $u_0(x) = 0$ for $x \geq X$, we arrive at (6.15) and that $u(x,t) = 0$ for $x \geq X$, $t > 0$.

Similarly we can prove that $u(x,t) = 0$ for $x \leq -X$, $t > 0$. □

Remark 1.6.3 For equations of the form

$$\frac{\partial u}{\partial t} = \frac{\partial^2 u^m}{\partial x^2} - cu^n \quad (m \geq 1, n > 0, c > 0),$$

the condition (6.12) becomes

$$\left(\frac{m+n}{mc}\right)^{1/2} \int_0^1 v^{-(m+n)/(2m)} dv < +\infty$$

which is equivalent to $n < m$.

Theorem 1.6.4 *Let $u(x,t)$ be a generalized solution of the Cauchy problem (6.11), (6.2) on $Q = \mathbb{R} \times (0, \infty)$. If*

$$\int_0^1 \frac{dy}{c(y)} < \infty, \tag{6.17}$$

then there exists $T \in (0, \infty)$ such that $u(x,t) \equiv 0$, for $x \in \mathbb{R}$, $t \geq T$.

In this case we will say that **extinction** *occurs for the solution $u(x,t)$ at the time $t = T$.*

Proof. To prove our theorem, we first choose a function $w(t)$ such that

$$\frac{dw(t)}{dt} = -c(w(t)) \quad \text{for } 0 < t < T, \tag{6.18}$$

$$w(T) = 0 \tag{6.19}$$

with $T > 0$ to be determined later. Integrating (6.18) and using (6.19) we obtain

$$\int_0^{w(t)} \frac{dy}{c(y)} = T - t, \quad \text{for } 0 < t < T. \tag{6.20}$$

Here we notice that the condition (6.17) ensures that the integral $\displaystyle\int_0^v \frac{dy}{c(y)}$ is defined for any $v \geq 0$ and is increasing in v.

Extend the function $w(t)$ defined by (6.20) to $(0, \infty)$ with $w(t) = 0$ for $t \in [T, \infty)$. Then it is easy to check that $w(t)$ is a generalized solution of (6.11) on $(0, \infty)$.

If we have

$$u(x, 0) = u_0(x) \leq w(0), \tag{6.21}$$

then the comparison theorem gives

$$u(x,t) \le w(t) \quad \text{for } (x,t) \in Q$$

and hence $u(x,t) = 0$ for $x \in \mathbb{R}$, $t \ge T$. Since

$$\int_0^{w(0)} \frac{dy}{c(y)} = T,$$

for (6.21) to be held, it suffices to take $T = \int_0^M \frac{dy}{c(y)}$ where $M \ge \sup u_0(x)$. The proof is thus completed. □

Remark 1.6.4 Kalashnikov proved in [KA1] that for the generalized solution of (6.11)to have the property of extinction, the condition (6.17) is also necessary.

1.6.3 *Differential equation on the free boundary*

In what follows we will further investigate the properties of free boundaries $x = \zeta_i(t)$ $(i = 1, 2)$ of the generalized solution u of the equation (6.3).

Naturally we expect the free boundaries to move with the local velocity. Set

$$v = \frac{m}{m-1} u^{m-1}.$$

Then, in view of the equation of state, which we have used to derive the equation (6.3), v is essentially the pressure, and, by Darcy's law, we expect

$$\zeta_i'(t) = - \lim_{\substack{(x,t)\in\Omega \\ x \to \zeta_i(t)}} \frac{\partial v(x,t)}{\partial x},$$

where $\Omega = \{(x,t) \in Q; u(x,t) > 0\}$. The next result shows that this is almost true.

Theorem 1.6.5 *The limits*

$$\begin{aligned} v_x(\zeta_i(t), t) &\equiv \lim_{\substack{(x,t)\in\Omega \\ x \to \zeta_i(t)}} \frac{\partial v(x,t)}{\partial x} \quad (i = 1, 2), \\ \zeta_i'(t+0) &\equiv \lim_{\Delta t \to 0_+} \frac{\zeta_i(t+\Delta t) - \zeta_i(t)}{\Delta t} \quad (i = 1, 2) \end{aligned}$$

exist for all $t > 0$ *and*

$$\zeta_i'(t+0) = -v_x(\zeta_i(t), t) \quad (i = 1, 2).$$

Proof. From Theorem 1.2.2, $u \in C^\infty(\Omega)$. By Proposition 1.6.1, for any $\tau > 0$, there exists a constant β depending only on τ such that

$$\frac{\partial^2 v}{\partial x^2} \geq -\beta \quad \text{for } (x,t) \in \Omega,\ t \geq \tau$$

in the sense of distributions. Hence the function $f(x,t) = v(x,t) + \beta x^2$ satisfies $\dfrac{\partial^2 f}{\partial x^2} \geq 0$. This means that $\dfrac{\partial f}{\partial x}$ is an increasing function of x for each fixed $t \geq \tau$. Since for any $t \geq \tau$, u has compact support, it follows from Lemma 1.4.1 that for any finite $T > \tau$, $\dfrac{\partial v}{\partial x}$ and $\dfrac{\partial f}{\partial x}$ are bounded on $\Omega \cap \{(x,t); t \in [\tau, T]\}$. Thus the limits $\lim\limits_{\substack{x \to \zeta_i(t) \\ (x,t) \in \Omega}} \dfrac{\partial f(x,t)}{\partial x}$ $(i = 1, 2)$ and $\lim\limits_{\substack{x \to \zeta_i(t) \\ (x,t) \in \Omega}} \dfrac{\partial v(x,t)}{\partial x}$ $(i = 1, 2)$ exist for any $t \in [\tau, T]$. Since T, τ are arbitrary, these limits exist for any $t > 0$.

We are ready to prove the rest part of the theorem for any t. For simplicity we take $t = 0$ and treat $\zeta_2(t)$ only. Set $a = \zeta_2(0), v_x(a, 0) = \alpha$. Then either $\alpha = 0$ or $\alpha < 0$.

Case 1. $\alpha < 0$.

We will show that for any sufficiently small $\varepsilon > 0$, there exists a $\delta > 0$ depending only on ε such that

$$\left| \frac{\zeta_2(\Delta t) - a}{\Delta t} + \alpha \right| < \varepsilon \tag{6.22}$$

whenever $0 < \Delta t < \delta$. Since $\lim\limits_{x \to a^-} \dfrac{\partial v(x,0)}{\partial x} = \alpha$, it follows that for any $0 < \varepsilon < -\alpha$, there exists a $\delta_0 > 0$ such that

$$\alpha - \varepsilon < \frac{\partial v(x,0)}{\partial x} < \alpha + \varepsilon, \quad (a - \delta_0 < x < a).$$

Using the mean value theorem we get

$$(\alpha + \varepsilon)(x - a) < v(x, 0) < (\alpha - \varepsilon)(x - a)$$

and hence

$$\begin{cases} u(x,0) > \left(\dfrac{m-1}{m}(\alpha+\varepsilon)(x-a)\right)^{1/(m-1)}, \\ u(x,0) < \left(\dfrac{m-1}{m}(\alpha-\varepsilon)(x-a)\right)^{1/(m-1)} \end{cases} \tag{6.23}$$

wherever $a - \delta_0 < x < a$.

We will use the following pressure solutions of (6.3)

$$u_1(x,t) = \left(\frac{m-1}{m}(\alpha+\varepsilon)((\alpha+\varepsilon)t + (x-a))_-\right)^{1/(m-1)},$$

$$u_2(x,t) = \left(\frac{m-1}{m}(\alpha-\varepsilon)((\alpha-\varepsilon)t + (x-a))_-\right)^{1/(m-1)},$$

which are some modification of (4.37). From (6.23) it follows that

$$u_1(x,0) < u(x,0) < u_2(x,0), \quad \text{for } x \ge a - \delta_0.$$

By continuity there exists $\tau > 0$ such that,

$$u_1(a-\delta_0,t) < u(a-\delta_0,t) < u_2(a-\delta_0,t), \quad \text{for } 0 < t < \tau.$$

Therefore the comparison theorem gives

$$u_1(x,t) \le u(x,t) \le u_2(x,t), \quad \text{for } x \ge a-\delta_0, 0 < t < \tau$$

and from this it is easy to see that

$$a - (\alpha+\varepsilon)t \le \zeta_2(t) \le a - (a-\varepsilon)t, \quad \text{for } 0 < t < \tau$$

or

$$\left|\frac{\zeta_2(t) - a}{t} + \alpha\right| \le \varepsilon, \quad \text{for } 0 < t < \tau,$$

which completes the proof in Case 1.

2^0. $\alpha = 0$.

From the proof in Case 1, we see that we still have

$$u_1(x,t) \le u(x,t), \quad \text{for } x \ge a-\delta_0, 0 < t < \tau$$

and hence

$$\frac{\zeta_2(t) - a}{t} \le \varepsilon, \quad \text{for } 0 < t < \tau.$$

On the other hand, by the monotonicity of $\zeta_2(t)$,

$$\frac{\zeta_2(t) - a}{t} \geq 0 \quad \text{for } t > 0.$$

Combining these two inequalities completes the proof in Case 2. □

1.6.4 *Continuously differentiability of the free boundary*

Proposition 1.6.2 *For any $0 < \delta < 1$, $\zeta_i(t)$ $(i = 1, 2)$ is Lipschitz continuous on $\left[\delta, \frac{1}{\delta}\right]$.*

Proof. Let $t_1 \in \left[\delta, \frac{1}{\delta}\right]$. Denote $\alpha_0 = \zeta_2(t_1)$ and $C = \max\limits_{\substack{\delta \leq t \leq 1/\delta \\ x \in \mathbb{R}}} \left| \frac{\partial v(x,t)}{\partial x} \right|$ which is finite by Lemma 1.4.1.

Consider the pressure solution of (6.3),

$$U(x,t) = \left(\frac{m-1}{m} c(c(t - t_1) + \alpha_0 - x)_+ \right)^{1/(m-1)}.$$

Since $v(x,t) \leq C(\alpha_0 - x)$ for $x \leq \alpha_0$, we have

$$u(x,t_1) \leq U(x,t_1) \quad \text{for } x \in \mathbb{R}.$$

Therefore by the comparison theorem,

$$u(x,t) \leq U(x,t) \quad \text{for } x \in \mathbb{R}, t_1 \leq t \leq \frac{1}{\delta}.$$

From this it follows that

$$\zeta_2(t) \leq \alpha_0 + c(t - t_1) \quad \text{for } t_1 \leq t \leq \frac{1}{\delta},$$

and hence, since $\zeta_2(t)$ is increasing, we obtain

$$0 \leq \zeta_2(t) - \zeta_2(t_1) \leq C(t - t_1), \qquad \text{for } t_1 \leq t \leq \frac{1}{\delta}.$$

The Lipschitz continuity of ζ_1 can be proved similarly.

In the sequel we will use the Barenblatt solutions (see (1.37))

$$w(x,t) = B_m(x,t) = \frac{1}{\lambda(t)} \left(\left(1 - \frac{x^2}{\lambda^2(t)} \right)_+ \right)^{1/(m-1)},$$

where

$$\lambda(t) = C_m(t+1)^{1/(m+1)}, \quad C_m = \left(\frac{2m(m+1)}{m-1}\right)^{1/(m+1)}.$$

An immediate calculation shows that for any $c, L > 0$,

$$w_{c,L}(x,t) = cw(Lx, c^{m-1}L^2t)$$

is a generalized solution of (6.3). Denote

$$v_{c,L}(x,t) = \frac{m}{m-1} w_{c,L}^{m-1}(x,t),$$

or

$$v_{c,L}(x,t) = \frac{m}{m-1} \frac{c^{m-1}}{\lambda_{c,L}^{m-1}(t)} \left(1 - \frac{L^2x^2}{\lambda_{c,L}^2(t)}\right)_+, \tag{6.24}$$

where

$$\lambda_{c,L}(t) = C_m \left(c^{m-1}L^2t + 1\right)^{1/(m+1)}.$$

Let $x = \zeta_{c,L}^{(i)}(t)$ $(i = 1, 2)$ be the free boundaries of $v_{c,L}$. Then

$$\begin{aligned} \zeta_{c,L}^{(1)}(t) &= \frac{C_m}{L} \left(c^{m-1}L^2t + 1\right)^{1/(m+1)}, \\ \zeta_{c,L}^{(2)}(t) &= -\frac{C_m}{L} \left(c^{m-1}L^2t + 1\right)^{1/(m+1)}. \end{aligned}$$

□

Proposition 1.6.3 *For any $\delta > 0$ there exists a convex function $\xi_i(t)$ $(t > 0)$ and a $C^{1,1}$ function $\eta_i(t)$ such that*

$$(-1)^i \zeta_i(t) = \xi_i(t) + \eta_i(t), \quad (t > \delta, i = 1, 2). \tag{6.25}$$

Here $C^{1,1}$ denotes the class of all functions whose derivatives are Lipschitz continuous.

Proof. We only prove (6.25) for $i = 2$. We set $\zeta(t) = \zeta_2(t)$ and want to study the free boundary $x = \zeta(t)$ near a point (x_0, t_0) with $t_0 > \delta$. For the sake of simplicity, we perform a translation of x, t variables so that, in new x, t variables,

$$t_0 = 0, \quad x_0 = \zeta(t_0) = \frac{c_m}{L}.$$

Then $(x_0, t_0) = \left(\frac{c_m}{L}, 0\right)$ lies on the free boundary of both $v(x,t)$ and $v_{c,L}(x,t)$ defined by (6.24). We wish to choose c, L so that

$$\zeta'(0+0) = \widetilde{\zeta}'(0), \tag{6.26}$$

$$v_{c,L}(x,0) \le v(x,0), \tag{6.27}$$

where $\widetilde{\zeta}(t) = \zeta^{(2)}_{c,L}(t)$. In view of Theorem 1.6.5, (6.26) is equivalent to

$$\frac{\partial}{\partial x} v\left(\frac{c_m}{L}, 0\right) = \frac{\partial}{\partial x} v_{c,L}\left(\frac{c_m}{L}, 0\right), \tag{6.28}$$

$$\frac{\partial^2 v(x,t)}{\partial x^2} \ge -2P, \quad \text{for } x \in \mathbb{R}, t \ge \delta. \tag{6.29}$$

If c, L are chosen so that

$$\frac{\partial^2 v_{c,L}(x,0)}{\partial x^2} = -2P, \text{ whenever } v_{c,L}(x,0) > 0, \tag{6.30}$$

then since also

$$v_{c,L}\left(\frac{C_m}{L}, 0\right) = 0 = v\left(\frac{C_m}{L}, 0\right),$$

we then deduce the inequality (6.27).

By an immediate calculation, (6.26), (6.30) turn out to be

$$\zeta'(0+0) = \frac{1}{m+1} C_m c^{m-1} L, \tag{6.31}$$

$$\frac{m}{m-1} \frac{2L^2 c^{m-1}}{C_m^{m+1}} = 2p, \tag{6.32}$$

which clearly have a unique positive solution (c, L). For such (c, L), (6.27) holds and hence by the comparison theorem,

$$v_{c,L}(x,t) \le v(x,t),$$

from which it follows that

$$\widetilde{\zeta}(t) \le \zeta(t).$$

Using this inequality and recalling (6.26) we obtain

$$\zeta(h) - \zeta(0) - h\zeta'(0+0) \ge \widetilde{\zeta}(h) - \widetilde{\zeta}(0) - h\widetilde{\zeta}'(0). \tag{6.33}$$

By a direct calculation, using (6.26), (6.32) yields

$$\widetilde{\zeta}''(0) = -C_m \frac{m}{(m+1)^2} c^{2(m-1)} L^3$$
$$= -\frac{(m-1)C_m^{m+2}}{(m+1)^2} P c^{m-1} L = -\gamma P \zeta'(0+0),$$

where $\gamma = \dfrac{(m-1)C_m^{m+1}}{m+1}$. It is clear that

$$\widetilde{\zeta}''(t) = \widetilde{\zeta}''(0) + O(t) \qquad (t \to 0_+).$$

Therefore the right hand side of (6.33) is equal to

$$\frac{1}{2} h^2 \widetilde{\zeta}''(0) + O(h^3) = -\frac{1}{2}\gamma h^2 P \zeta'(0+0) + O(h^3),$$

and (6.33) gives

$$\zeta(h) - \zeta(0) - h\zeta'(0+0) \geq -\frac{1}{2} h^2 \gamma \zeta'(0+0) + O(h^3).$$

Going back to the original (x,t) coordinates, we obtain

$$\Phi_h(t) \geq -\gamma P \zeta'(t+0) + O(h) \tag{6.34}$$

at any point $t = t_0 > 0$, where

$$\Phi_h(t) = \frac{\zeta(t+h) - \zeta(t) - h\zeta'(t+0)}{h^2/2} \qquad (h > 0).$$

Now, in any interval $\left[\delta, \frac{1}{\delta}\right]$ and for any $h \in \left(0, \frac{\delta}{2}\right)$, $\Phi_h(t) \in L^\infty\left[\delta, \frac{1}{\delta}\right]$, since by Proposition 1.6.2, $\zeta(t)$ is Lipschitz continuous on $\left[\delta, \frac{1}{\delta}\right]$. Also, for any $0 < s < T < \infty$,

$$\int_s^T \Phi_h(t)dt = \frac{2}{h^2}\left(\int_T^{T+h} \zeta(t)dt - \int_s^{s+h} \zeta(t)dt - h(\zeta(T) - \zeta(s))\right)$$
$$= \frac{2}{h^2}\int_T^{T+h} (\zeta(t) - \zeta(T))dt - \frac{2}{h^2}\int_s^{s+h} (\zeta(t) - \zeta(s))dt \leq C.$$

Here and below $C > 0$ is a universal constant independent of h. Since, by (6.34), $\zeta(t)$ is Lipschitz continuous,

$$\Phi_h(t) \geq -C, \quad \text{for } t \in (s,T),$$

we conclude that

$$\int_s^T |\Phi_h(t)|dt \le C.$$

We can therefore choose a sequence $h = h_n \to 0$ such that Φ_{h_n} weakly converges to a measure μ_0. Using (6.34) we then have

$$\Phi_{h_n}(t) + \gamma P\zeta'(t+0) \to \mu, \quad \text{weakly as} n \to \infty,$$

where μ is a nonnegative measure. Since the weak convergence implies the convergence in the sense of distributions, we conclude from the definition of Φ_h, that

$$\zeta'' + \gamma P\zeta' = \mu \tag{6.35}$$

in the sense of distributions. Set

$$\xi(t) = \int_\delta^t \int_\delta^\tau d\mu(s)d\tau,$$

$$\eta(t) = \zeta(\delta) + \zeta'(\delta+0)(t-\delta) - \gamma P \int_\delta^t (\zeta(s) - \zeta(\delta))ds.$$

Then $\xi(t)$ is convex, $\eta \in C^{1,1}$, and $\hat{\zeta}(t) = \xi(t) + \eta(t)$ is a solution of (6.35) with $\hat{\zeta}(\delta) = \xi(\delta)$, $\hat{\zeta}'(\delta) = \zeta(\delta+0)$. By uniqueness we must have $\hat{\zeta} \equiv \zeta$. The proof is complete. □

Corollary 1.6.1 *For any $t > 0$, $\zeta_i'(t \pm 0)$ exists and*

$$(-1)^i \zeta_i'(t-0) \le (-1)^i \zeta_i'(t+0). \tag{6.36}$$

Furthermore, for any $\delta > 0$, there exists $P > 0$ such that for $t_2 > t_1 > \delta$,

$$(-1)^i \zeta_i'(t_2 - 0)e^{\gamma P t_2} \ge (-1)^i \zeta_i'(t_1 + 0)e^{\gamma P t_1}, \tag{6.37}$$

which implies that there exists a constant $t_i^* \ge 0$ such that $\zeta'(t)$ is strictly increasing for $t > t^*$ and $\zeta_i(t) \equiv x_i$ for $0 \le t \le t^*$.

If $t^* > 0$, then the free boundary $x = \zeta(t)$ remains stationary (vertical) for t^* units of time after which it begins to move without further stops. We will call t_i^* the **waiting time***.*

Proof. (6.36) follows from the convexity of $\xi(t)$ and $\eta \in C^{1,1}$. The existence of $\zeta_i(t-0)$ can be proved similarly.

From (6.35), we get

$$(\zeta' e^{\gamma P t})' = e^{\gamma P t}\mu \geq 0,$$

which implies (6.37).

If $\zeta(t) \equiv$const. on the interval $0 < s_1 < t < s_2$, then $\zeta'(t) \equiv 0$ on this interval. Using(6.37) we deduce $\zeta'(t) \equiv 0$ on $0 < t < s_1$. This completes the proof.

Now we are ready to discuss the continuous differentiability of $\zeta_i(t)$ for $t > t_i^*$. We will treat $\zeta(t) = \zeta_2(t)$ only.

Let $(x_0, t_0) = (\zeta(t_0), t_0)$ with $t_0 > t_2^*$ and N_δ be the intersection of a δ-neighborhood of (x_0, t_0) and $\Omega = \{(x,t) \in Q; u(x,t) > 0\}$. Take $\delta > 0$ so small that $\overline{N}_{2\delta}$ stays away from $t = t_2^*$ and from the free boundary $x = \zeta_1(t)$. Denote by $d(\overline{x}, \overline{t})$ the distance from $(\overline{x}, \overline{t})$ to the free boundary. □

Lemma 1.6.1 *There exist positive constants C_1, C_2 depending only on m, δ such that*

$$C_1 \leq \frac{v(x,t)}{d(x,t)} \leq C_2 \quad \textit{in } N_\delta.$$

Proof. By Corollary 1.6.1 and our assumption on N_δ, there exist constants $c > 0$,$C > 0$, depending only on δ, such that for all t with $(\zeta(t), t) \in \partial N_\delta$,

$$c \leq \zeta'(t \pm 0) \leq C. \tag{6.38}$$

Hence there exist constants $\overline{C}_1 > 0$, $\overline{C}_2 > 0$ depending only on δ , such that

$$\overline{C}_1 \leq \frac{d(x,t)}{|x - \zeta(t)|} \leq \overline{C}_2, \quad \text{for } (x,t) \in N_\delta.$$

Since $v(x,t)$ is Lipschitz continuous,

$$v(x,t) = v(x,t) - v(\zeta(t),t) \leq C'|x - \zeta(t)| \leq C_2 d(x,t).$$

Next, since $\dfrac{\partial^2 v}{\partial x^2} \geq -2P$ (see (6.29)), using the Taylor formula, Theorem

1.6.5 and (6.38), we see that for $\delta > 0$ small enough,

$$\begin{aligned} v(x,t) &\geq |v_x(\zeta(t),t)||x-\zeta(t)| - P|x-\zeta(t)|^2 \\ &= |\zeta'(t+0)||x-\zeta(t)| - P|x-\zeta(t)|^2 \geq c'|x-\zeta(t)| \\ &\geq C_1 d(x,t) \end{aligned}$$

with a constant $C_1 > 0$ depending only on δ . The rest part of the lemma can be proved similarly. □

Lemma 1.6.2 *There exists a constant C depending only on δ, such that*

$$\left|\frac{\partial v}{\partial t}\right| \leq C, \qquad \left|\frac{\partial^2 v}{\partial t^2}\right| \leq \frac{C}{d(x,t)}, \quad \text{for } (x,t) \in N_\delta.$$

Proof. Let $(\overline{x},\overline{t}) \in N_\delta$ and S be a square with center $(\overline{x},\overline{t})$ and side $\gamma = \frac{1}{2}d(\overline{x},\overline{t})$. For simplicity, we suppose $\overline{x} = 0, \overline{t} = 0$. Consider the function

$$w(x,t) = \gamma^{-1} v(\gamma x, \gamma t),$$

where (x,t) varies in a unit square S_0 (so that $(\gamma x, \gamma t) \in S$). Clearly, in S_0

$$\frac{\partial w}{\partial t} = (m-1)w\frac{\partial^2 w}{\partial x^2} + \left(\frac{\partial w}{\partial x}\right)^2, \tag{6.39}$$

and in view of Lemma 1.6.1, for some constants $C_1 > 0$, $C_2 > 0$,

$$C_1 \leq (m-1)w \leq C_2.$$

Thus (6.39) is uniformly parabolic. Noting that $\left(\frac{\partial v}{\partial x}\right)^2$ is a bounded function, we can apply the Nash-Moser estimate [LSU] and conclude that for some $\alpha \in (0,1)$,

$$|w|_{C^\alpha(S')} \leq C,$$

where S' is a square concentric with S_0 and with side $\frac{3}{4}$, and $C > 0$ is a constant depending only on $C_1, C_2, \operatorname{dist}(S', \partial S_0)$. We can now construct a fundamental solution for the equation

$$\frac{\partial W}{\partial t} = (m-1)w\frac{\partial^2 W}{\partial x^2}$$

and use it to express w (see [FR1]), from which we see that $\frac{\partial w}{\partial x}$ is Hölder continuous. Applying Schauder's estimates [FR1] we conclude that $\frac{\partial w}{\partial t}$, $\frac{\partial^2 w}{\partial t^2}$ (in fact, any derivative of w) are bounded by a constant C^*, depending only on δ, in a square S'' concentric with S_0 and with side $\frac{1}{2}$. Going back to the function v we deduce that $\frac{\partial v(\overline{x},\overline{t})}{\partial t}$, $\frac{\partial^2 v(\overline{x},\overline{t})}{\partial t^2}$ are bounded by C^*; this completes the proof. □

Lemma 1.6.3 *Let* $x_0 = \zeta(t_0), t_0 > 0$, $\frac{\partial v(x_0,t_0)}{\partial x} = -\zeta'(t_0+0) = -b < 0$. *Then there exists a neighborhood* N_δ *of* (x_0, t_0)*, such that for any constant* a*,*

$$v(x,t) = L_b(x - x_0, t - t_0) + o(|x - x_0| + |t - t_0|)$$

for $(x,t) \in N_\delta$ *with* $\frac{x - x_0}{t - t_0} = a$*, where* $L_b(x,t) = b(bt - x)_+$.

Proof. For $\eta > 0$, define

$$v_\delta(x,t) = \eta^{-1} v(\eta x - x_0, \eta t - t_0).$$

Fix $\alpha \in (0, t_0)$. Then v is defined in the $\frac{\alpha}{\eta}$-neighborhood of the origin and v is also a generalized solution of the equation (6.39). (This means that $\left(\frac{m-1}{m} v_0\right)^{1/(m-1)}$ is a generalized solution of (6.3)). By Proposition 1.6.1, for $t \geq -\frac{\alpha}{\eta}$,

$$\frac{\partial^2 v_\eta}{\partial x^2} = \eta \frac{\partial^2 v_\eta}{\partial x^2}(\eta x + x_0, \eta t + t_0) \geq -\frac{\eta k}{\eta t + t_0} \geq -\frac{\eta k}{t_0 - \eta}. \tag{6.40}$$

From the proof of Theorem 1.4.1, we see that $v_\eta \in C^{1,1/2}$ with the Hölder coefficient independent of η. Therefore there exists a subsequence $\{v_{\eta_n}\}$ with $\eta_n \to 0$ such that $v_{\eta_n} \to \overline{v}$ in $\mathbb{R}^2$ uniformly on compact sets and $\overline{v}$ is a generalized solution of (6.39). We may simply suppose that $v_\eta \to \overline{v}$ in R^2 as $\eta \to 0$ uniformly on compact sets. It is easily seen that all of the derivatives of v_η converge to the corresponding derivatives of $\overline{v}$.

In view of our hypothesis, for each fixed $x < 0$,

$$\frac{\partial \overline{v}}{\partial x}(x,0) = \lim_{\eta\to 0} \frac{\partial v_\eta}{\partial x}(x,0) = \frac{\partial v}{\partial x}(x_0,t_0) = -b;$$

and for each fixed $x \geq 0$,

$$\overline{v}(x,0) = \lim_{\eta\to 0} v_\eta(x_0,0) = \lim_{\eta\to 0} \frac{1}{\eta} v(\eta x + x_0, t_0) = 0.$$

Thus

$$\overline{v}(x,0) = \begin{cases} -bx & \text{for } x < 0 \\ 0 & \text{for } x \geq 0. \end{cases}$$

Moreover,

$$\begin{aligned} v_\eta(x,t) &\leq \frac{C}{\eta}|\eta x + x_0 - \zeta(\eta t + t_0)| \\ &\leq \frac{C}{\eta}(\eta|x| + |x_0 - \zeta(\eta t + t_0)|) \leq C(|x| + 2bt) \end{aligned}$$

provided $x \in \mathbb{R}$ and $0 < t < \varepsilon$ with $\varepsilon > 0$ small enough. Therefore we can use a uniqueness theorem in [KA3] mentioned in §1.2.4 to conclude that for $x \in \mathbb{R}$, $0 < t < \varepsilon$,

$$\overline{v}(x,t) = L_b(x,t). \tag{6.41}$$

Repeating the argument we can further prove that (6.41) holds on the whole Q.

By Lemma 1.6.2, it is easy to see that there exists a constant C depending only on δ, such that

$$\left|\frac{\partial^2 v_\eta(x,t)}{\partial t^2}\right| = \eta \left|\frac{\partial^2 v}{\partial t^2}(\eta x + x_0, \eta t + t_0)\right| \leq \frac{C_1}{-x}$$

for $x < 2bt < 0$ with $t^2 + x^2 < \dfrac{\delta^2}{\eta^2}$. Therefore

$$\left|\frac{\partial^2 \overline{v}_\eta(x,t)}{\partial t^2}\right| \leq \frac{C_1}{-x} \qquad \text{for } x < 2bt < 0. \tag{6.42}$$

For $x < 2bt < 0$,

$$\begin{aligned}\overline{v}(x,t) &= \overline{v}(x,0) + t\frac{\partial \overline{v}(x,0)}{\partial t} + \frac{t^2}{2}\frac{\partial^2 \overline{v}(x,\widetilde{t})}{\partial t^2} \\ &= L_b(x,0) + t\frac{\partial L_b(x,0)}{\partial t} + \frac{t^2}{2}\frac{\partial^2 \overline{v}(x,\widetilde{t})}{\partial t^2}\end{aligned}$$

where $t < \widetilde{t} < 0$. Thus

$$\overline{v}(x,t) - L_b(x,t) = \frac{t^2}{2}\frac{\partial^2 \overline{v}(x,\widetilde{t})}{\partial t}$$

and it follows by using (6.42) that

$$\lim_{x\to-\infty} (\overline{v}(x,t) - L_b(x,t)) = 0 \quad \text{for } t < 0. \tag{6.43}$$

Next we show that

$$\overline{v}(x,t) \geq L_b(x,t) \quad \text{for } t < 0. \tag{6.44}$$

Since $L_b(x,t) = 0$ for $x \geq bt$, it suffices to prove (6.44) for $x < bt$, $t < 0$. In view of (6.40), we have $\dfrac{\partial^2 \overline{v}}{\partial x^2} \geq 0$. Thus for $x < bt$, $t < 0$,

$$\frac{\partial \overline{v}(x,t)}{\partial x} \leq \frac{\partial \overline{v}(bt,t)}{\partial x} = -b,$$

$$\begin{aligned}\overline{v}(x,t) &= \overline{v}(bt,t) + \int_{bt}^{x} \frac{\partial \overline{v}}{\partial x}(y,t)dy \\ &\geq \overline{v}(x,t) - b(x - bt) = \overline{v}(bt,t) + L_b(x,t)\end{aligned}$$

from which (6.44) follows.

Now we prove that (6.41) holds for $t < 0$. Consider a fixed point $(\overline{x},\overline{t})$ with $\overline{x} < b\overline{t}$. Let S be a rectangle with center $(\overline{x},\overline{t})$ and boundaries parallel to the coordinate axis, such that $(x,t) \in S$ implies that $x < bt$ and S contains points with $t > 0$. In S, $\overline{v} - L_b \geq 0$ and achieves its minimum value, 0, on $S \cap \{t > 0\}$. By the strong minimum principle, we must have $\overline{v} \equiv L_b$ in S. Therefore (6.41) holds for $x < bt$, $t < 0$.

Suppose that $\overline{v}(\overline{x},\overline{t}) > 0$ for some $(\overline{x},\overline{t})$ with $\overline{x} > b\overline{t}$, $\overline{t} < 0$. Then from Theorem 1.6.2, we have $\overline{v}(\overline{x},\overline{t}) > 0$ for all $t > \overline{t}$. However this contradicts the fact that the line $x = \overline{x}$ must intersect the line $x = bt$ for some $t > \overline{t}$ and $\overline{v} = L_b = 0$ at that intersection. Therefore (6.41) also holds for $x > bt$, $t < 0$.

Given $a > 0$. Let $\eta = t - t_0 \neq 0$. Since $v_{|\eta|} \to \overline{v} \equiv L_b$ as $|\eta| \to 0$, given $\varepsilon > 0$ there exists a $\eta_0 = \eta_0(\varepsilon) > 0$ such that

$$\left| \frac{1}{|\eta|} v(\eta a + x_0, \eta + t_0) - L_b(a \operatorname{sgn} \eta, \operatorname{sgn} \eta) \right| < \varepsilon$$

provided $|\eta| = |t - t_0| \in (0, \eta_0)$, $\dfrac{x - x_0}{t - t_0} = a$. Namely

$$|v(x,t) - L_b(x - x_0, t - t_0)| < \varepsilon |t - t_0|.$$

The proof of Lemma 1.6.3 is complete. □

Theorem 1.6.6 $\zeta_i(t)$ $(i = 1, 2)$ *is continuously differentiable at any* $t \neq t_i^*$ $(i = 1, 2)$, *where* t_i^* $(i = 1, 2)$ *is defined in Corollary 1.6.1.*

Proof. We will prove the assertion only for $\zeta(t) = \zeta_2(t)$. Since a convex function which is differentiable everywhere on an interval is necessarily continuously differentiable there, by Proposition 1.6.3, it suffices to prove that $\zeta(t)$ is differentiable everywhere at any $t = t_0 > t_2^*$, i.e

$$\zeta'(t_0 - 0) = \zeta'(t_0 + 0) = b. \tag{6.45}$$

If $b = 0$, then by Corollary 1.6.1, (6.45) is trivial. We suppose $b > 0$. Denote $x_0 = \zeta(t_0)$. From the definition of t_2^*, there must be a $\tau \in (t_2^*, t_0)$ such that $\zeta'(\tau + 0) > 0$.

First we prove that for any $a > b$, there exists $\varepsilon > 0$ such that

$$\zeta(t) > x_0 + a(t - t_0), \quad t_0 - \varepsilon < t < t_0. \tag{6.46}$$

If (6.46) is false, then there exists a sequence ε_n with $\varepsilon_n \downarrow 0$ as $n \to \infty$, such that

$$\zeta(t_0 - \varepsilon_n) \leq x_0 - \varepsilon_n a$$

and hence, by the definition of $\zeta(t)$

$$v(x_0 - \varepsilon_n a, t_0 - \varepsilon_n) = 0.$$

However, from Lemma 1.6.3 we have

$$v(x_0 - \varepsilon_n a, t_0 - \varepsilon_n) = L_b(-\varepsilon_n a, -\varepsilon_n) + o(\varepsilon_n)$$

$$= \quad -b^2 \varepsilon_n + ba\varepsilon_n + o(\varepsilon_n).$$

Hence

$$-b^2 + ba + o(1) = -b(b-a) + o(1) = 0 \quad (n \to \infty)$$

which contradicts the fact that $a > b > 0$. Therefore (6.46) holds and we conclude that

$$\lim_{t \to t_0 - 0} \frac{x_0 - \zeta(t)}{t_0 - t} \le a.$$

Since $a > b$ is arbitrary, it follows that

$$\lim_{t \to t_0 - 0} \frac{x_0 - \zeta(t)}{t_0 - t} \le b. \tag{6.47}$$

Next we prove that for any $c < b$, there exists $\eta > 0$ such that

$$\zeta(t) < x_0 + c(t - t_0), \quad \text{for } t_0 - \eta < t < t_0. \tag{6.48}$$

If (6.48) is false, then there exists a sequence $\{\eta_n\}$ with $\eta_n \downarrow 0$ as $n \to \infty$, such that

$$\zeta(t_0 - \eta_n) \ge x_0 - c\eta_n > x_0 - b\eta_n.$$

By the mean value theorem there exists $\widetilde{x} \in (x_0 - b\eta_n, x_0 - c\eta_n)$, such that

$$\frac{\partial v}{\partial x}(\widetilde{x}, t_0 - \eta_n) = \frac{v(x_0 - c\eta_n, t_0 - \eta_n) - v(x_0 - b\eta_n, t_0 - \eta_n)}{(b-c)\eta_n}.$$

Since $c < b$, we have, by Lemma 1.6.3,

$$\frac{\partial v}{\partial x}(\widetilde{x}, t_0 - \eta_n) = \frac{o(\eta_n)}{(b-c)\eta_n} = o(1) \quad (n \to \infty).$$

According to Proposition 1.6.1,

$$v(x,t) + \frac{k}{2t}(\zeta(t) - x)^2$$

is a convex function of x. Thus

$$\begin{aligned} &\frac{\partial v}{\partial x}(\zeta(t_0 - \eta_n), t_0 - \eta_n) \\ \ge \; &\frac{\partial v}{\partial x}(\widetilde{x}, t_0 - \eta_n) + \frac{k(\zeta(t_0 - \eta_n) - \widetilde{x})}{t_0 - \eta_n} = o(1) \quad (n \to \infty), \end{aligned}$$

so that

$$\limsup_{t\to t_0-0} \frac{\partial v}{\partial x}(\zeta(t),t) \geq 0.$$

On the other hand, by Corollary 1.6.1,

$$\begin{aligned}\frac{\partial v}{\partial x}(\zeta(t),t) &= -\zeta'(t_0+0)\\ &\leq -\zeta'(\tau+0)e^{\gamma P(\tau-t)}\\ &\leq -\zeta'(\tau+0)e^{\gamma P(\tau-t_0)} < 0 \quad \text{for } t\in(\tau,t_0).\end{aligned}$$

This contradiction shows that (6.48) holds. Since $c < b$ is arbitrary, we conclude that

$$\liminf_{t\to t_0-0} \frac{x_0-\zeta(t)}{t_0-t} \geq b.$$

This, combining with (6.47) gives (6.45). □

1.6.5 *Some further results*

1. The property about the finite speed of propagation of disturbances is valid for more general equations. Gilding [GI] studied equations with convection term

$$\frac{\partial u}{\partial t} = \frac{\partial}{\partial x}\left(a(u)\frac{\partial u}{\partial x}\right) + b(u)\frac{\partial u}{\partial x}$$

and proved that if

$$b^2(u) = O(a(u)) \quad (u \to 0_+), \quad ua'(u), \;\; ub'(u) \in L^1(0,1),$$

then the disturbances possess the property of finite speed of propagation if and only if

$$\int_0^1 \frac{a(u)}{u}du < \infty.$$

2. The property about the localization of disturbances was first discovered by Martinson and Pavlov [MP1], [MP2], [MP3], who studied equations of the form

$$\frac{\partial u}{\partial t} = \frac{\partial^2 u^m}{\partial x^2} - cu(m>1, c>0),$$

$$\frac{\partial u}{\partial t} = \frac{\partial^2 u}{\partial x^2} - cu^n \quad (0 < n < 1, c > 0).$$

Theorem 1.6.2 and Theorem 1.6.3 were obtained by Kalashnikov [KA1]. Afterwards, Kershner [KE2] studied the property of localization and also the property of extinction for the equation

$$\frac{\partial u}{\partial t} = \frac{\partial^2 u^m}{\partial x^2} - b\frac{\partial u^n}{\partial x} - cu^l$$

with $m > 1$, $n \geq 1$, $l > 0$.

3. The result given in Theorem 1.6.6 is the best possible. Aronson, Caffarelli and Vazquez [ACV] pointed out that, for some initial value,

$$\zeta_i'(t_i^* + 0) \neq \zeta_i'(t_i^* - 0).$$

This shows that in general, $\zeta_i(t)$ is impossible to be continuously differentiable on the whole $(0, \infty)$.

Knerr [KN] discussed the problem that when t^* will be zero or not.

Quantitative investigation on the waiting time for some equations was done by Aronson, Caffarelli and Kamin [ACK], Lacey, Ockendon and Tayler [LOT] and others. They gave some sufficient conditions for the waiting time to be positive although the free boundary is continuously differentiable on $(0, \infty)$.

4. Using the iterative weighted norm, Höllig and Pilant [HPI], Höllig and Kreiss [HKR] proved the C^∞ smoothness of free boundaries after the waiting time. Such result was also obtained by Aronson and Vazquez [AVA]. Semi-group method was applied by Angenent [ANG] to prove the analyticity of free boundaries.

For recent works concerning propagation of disturbances and free boundaries, see for example, [AB1], [DS], [GSV1], [GSV2], [GSV3], [GV], [KAM3].

1.7 Properties of the Free Boundary: Higher Dimensional Case

1.7.1 *Monotonicity and Hölder continuity of the free boundary*

Consider the Cauchy problem

$$\frac{\partial u}{\partial t} = \Delta u^m, \tag{7.1}$$

$$u(x,0) = u_0(x) \quad x \in \mathbb{R}^N, \tag{7.2}$$

where $m > 1$ and $u_0 \geq 0$ is Hölder continuous on $\mathbb{R}^N$ with compact support.

From the discussion in §1.3 and §1.5, the Cauchy problem (7.1), (7.2) admits a unique generalized solution u on $Q = \mathbb{R}^N \times (0, \infty)$ which is Hölder continuous down to $t = 0$.

Denote

$$\Omega = \{(x,t); u(x,t) > 0,\ t > 0\},$$

$$\Omega(t) = \{x; u(x,t) > 0\},$$

$$\Gamma = \partial\Omega \cap \{t > 0\},$$

$$\Gamma(t) = \partial\Omega(t).$$

By continuity, Ω and $\Omega(t)$ are open sets of Q and $\mathbb{R}^N$ respectively. In §1.1.6 the condition for disturbances to possess the property of finite speed of disturbances is presented for one dimensional case (Theorem 1.6.1). It is not difficult to prove that a similar result is valid for higher dimensional case, from which it follows that for any $t > 0$, $\Omega(t)$ is a bounded domain and $\Gamma(t)$ is a bounded set. We will call Γ the **free boundary** or **interface** of the generalized solution u.

This section is devoted to a study of properties of the free boundary. To this purpose we first point out that we can extend Proposition 1.6.1 to higher dimensional case.

Proposition 1.7.1 *Assume that $m > 1$ and u is the generalized solution of the Cauchy problem (7.1), (7.2) on $Q = \mathbb{R}^N \times (0, \infty)$. Then*

$$\frac{\partial u}{\partial t} \geq -\frac{ku}{t},$$

$$\frac{\partial v}{\partial t} \geq -\frac{(m-1)kv}{t},$$

$$\Delta v \geq -\frac{k}{t}$$

in the sense of distributions, where $v = \dfrac{m}{m-1}u^{m-1}$, $k = (m-1+2/N)^{-1}$.

Below we will always assume that $m > 1$ and u is the generalized solution of the Cauchy problem (7.1), (7.2) on $Q = \mathbb{R}^N \times (0, \infty)$.

Theorem 1.7.1 *For any* $t_2 > t_1 > 0$,

$$\Omega(t_1) \subset \Omega(t_2),$$

and for any $t > 1$,

$$\inf\{|x|; x \in \Gamma(t)\} \geq Ct^{k/N},$$

where $C > 0$ *is a constant and* $k = (m - 1 + 2/N)^{-1}$.

We will omit the proof, since the first conclusion follows from Proposition 1.7.1 immediately and the proof of the second conclusion is similar to that of Theorem 1.6.2.

Theorem 1.7.2 *Assume that* $\partial\Omega \in C^1$. *If there exist positive constants* C_0, δ, $\gamma \in (0, 1]$ *such that for* $x \in D$ *with* $dist(x, \partial D) < \delta$,

$$u_0(x) \geq C_0(dist(x, \partial D))^{\gamma/(m-1)}, \tag{7.3}$$

then for any $t > 0$,

$$\overline{\Omega(0)} = \overline{D} \subset \Omega(t).$$

In other words, the support of u *expands from the very beginning, or* $x \in \overline{D} = \overline{\Omega(0)}$ *implies that for any* $t > 0$, $u(x, t) > 0$.

Proof. Since from Theorem 1.7.1, $\Omega(0) \subset \Omega(t)$ $(t > 0)$, it remains to prove that $\partial D \subset \Omega(t)$ $(t > 0)$.

Let $x_0 \in \partial D$. Since $\partial D \in C^1$, D satisfies the inner ball condition, there exist $\rho > 0$ and $\overline{x}_0 \in D$ such that $B_\rho(\overline{x}_0) \subset D$, $\overline{B_\rho(\overline{x}_0)} \cap \partial D = \{x_0\}$. We wish to prove that $x_0 \in \Omega(t)$ or $u(x_0, t) > 0$ $(t > 0)$. To this purpose, we use the Barenblatt solution of (7.1):

$$\begin{aligned}
& w_{c,\tau}(x, t) \\
= \quad & cB_m(x - \overline{x}_0, c^{m-1}(t + \tau)) \\
= \quad & c(c^{m-1}(t+\tau))^{-k}\left(1 - \frac{k(m-1)}{2mN} \cdot \frac{|x - \overline{x}_0|^2}{(c^{m-1}(t+\tau))^{2k/N}}\right)_+^{1/(m-1)} \\
= \quad & \left(\frac{k(m-1)}{2mN}\right)^{1/(m-1)} (t+\tau)^{-1/(m-1)} \\
& \cdot \left(\frac{2mN}{k(m-1)} c^{2k(m-1)/N} (t+\tau)^{2k/N} - |x - \overline{x}_0|^2\right)_+^{1/(m-1)},
\end{aligned}$$

where $k = (m-1+2/N)^{-1}$.

Observe that

$$w_{c,\tau}(x,0) = \left(\frac{k(m-1)}{2mN}\right)^{1/(m-1)} \tau^{-1/(m-1)}.$$
$$\left(\frac{2mN}{k(m-1)} c^{2k(m-1)/N} \tau^{2k/N} - |x-x_0|^2\right)_+^{1/(m-1)}.$$

Choose $\tau > 0$ large enough such that

$$\left(\frac{k(m-1)}{2mN}\right)^{1/(m-1)} \tau^{-1/(m-1)} \leq C_0,$$

and then choose $c > 0$ such that

$$\frac{2mN}{k(m-1)} c^{2k(m-1)/N} \tau^{2k/N} = \rho^2.$$

Then

$$w_{c,\tau}(x,0) \leq C_0(\rho^2 - |x-\overline{x}_0|^2)_+^{1/(m-1)}.$$

On the other hand, from (7.3),

$$\begin{aligned} u_0(x) &\geq C_0(\rho^2 - |x-\overline{x}_0|^2)_+^{\gamma/(m-1)} \\ &\geq C_0(\rho^2 - |x-\overline{x}_0|^2)^{1/(m-1)} \end{aligned}$$

provided that ρ is small enough. Therefore

$$w_{c,\tau}(x,0) \leq u_0(x) \quad \text{for } x \in \mathbb{R}^N.$$

By the comparison theorem,

$$w_{c,\tau}(x,t) \leq u(x,t) \quad \text{for } (x,t) \in Q.$$

In particular

$$w_{c,\tau}(x_0,t) \leq u(x_0,t) \quad \text{for } t > 0.$$

Since

$$|x_0 - \overline{x}_0|^2 = \rho^2 < \frac{2mN}{k(m-1)} c^{2k(m-1)/N}(t+\tau)^{2k/N},$$

from the definition of $w_{c,\tau}$, we have $w_{c,\tau}(x_0,t) > 0$ and hence $u(x_0,t) > 0$ for $t > 0$. □

Now we are ready to discuss the Hölder continuity of the free boundary. To this purpose we need some auxiliary propositions.

Proposition 1.7.2 *Given $\eta_0 > 0$. Assume that the constants η, C satisfy*

$$36mN exp\left(\frac{mk\eta}{\eta_0}\right)\left(2^N C + \frac{mk\eta}{2N\eta_0}\right) < 1, \tag{7.4}$$

and $x_0 \in \mathbb{R}^N$, $t_0 \geq \eta_0$, $R > 0$, $0 < \sigma \leq \eta$. If

$$v(x, t_0) = 0, \quad \text{for } x \in B_R(x_0) \tag{7.5}$$

and

$$⨍_{B_R(x_0)} v(x, t_0 + \sigma)dx \leq \frac{CR^2}{\sigma}, \tag{7.6}$$

then

$$v(x, t_0 + \sigma) = 0, \quad \text{for } x \in B_{R/6}(x_0). \tag{7.7}$$

Here

$$v = \frac{m}{m-1}u^{m-1}, \quad ⨍_\Omega f(x)dx = \frac{1}{|\Omega|}\int_\Omega f(x)dx.$$

The physical interpretation of the proposition is that if the gas reached the ball $B_{R/6}(x_0)$ at time $t_0 + \sigma$ and there was no gas in $B_R(x_0)$ at time t_0, then a considerable amount of gas has entered the ball $B_R(x_0)$ at time $t_0 + \sigma$.

Proof. 1) Let

$$x' = \frac{1}{R}(x - x_0), \quad t' = \frac{1}{\sigma}(t - t_0),$$
$$\overline{v}(x', t') = \frac{\sigma}{R^2}v(x_0 + Rx', t_0 + \sigma t').$$

It is easy to check that

$$\overline{u} = \left(\frac{m-1}{m}\overline{v}(x', t')\right)^{1/(m-1)} = \left(\frac{\sigma}{R^2}\right)^{1/(m-1)} u(x_0 + Rx', t_0 + \sigma t')$$

is a generalized solution of (7.1) on Q.

Dropping the mark "′", we obtain

$$\overline{v}(x, t) = \frac{\sigma}{R^2}v(x_0 + Rx, t_0 + \sigma t_0),$$

$$\overline{u}(x,t) = \left(\frac{m-1}{m}\overline{v}(x,t)\right)^{1/(m-1)} = \left(\frac{\sigma}{R^2}\right)^{1/(m-1)} u(x_0 + Rx, t_0 + \sigma t).$$

Thus after changing variables, (7.5), (7.6) turn out to be

$$\overline{v}(x,0) = 0 \quad \text{for } x \in B_1 = B_1(0), \tag{7.8}$$

$$\int_{B_1} \overline{v}(x,1)dx \leq C \tag{7.9}$$

and (7.7) becomes

$$\overline{v}(x,1) = 0 \quad \text{for } x \in B_{1/6}(0). \tag{7.10}$$

2) Set

$$U = \left(\frac{m-1}{m}V\right)^{1/(m-1)}, \quad V = V(r,t) = \lambda\left(\frac{t}{6} + r - \frac{1}{3}\right)^+,$$

where $r = |x|$, $\lambda = 1/(6mN)$. It is easy to verify that U is a generalized super-solution of (7.1), i.e U satisfies

$$\frac{\partial U}{\partial t} \geq \Delta U^m \quad \text{on } \mathbb{R}^N \times (0,1)$$

in the sense of distributions.

3) The basic idea of the following argument is to compare u with U on $B_{1/2} \times (0,1)$. If we can assert that

$$\overline{u}(x,t) \leq U(r,t) \quad \text{for } (x,t) \in B_{1/2} \times (0,1),$$

i.e

$$\overline{v}(x,t) \leq V(r,t) \quad \text{for } (x,t) \in B_{1/2} \times (0,1), \tag{7.11}$$

then, since clearly we have

$$V(r,t) = 0 \quad \text{for } (x,t) \in B_{1/6} \times (0,1),$$

(7.10) follows immediately.

To prove (7.11), by the comparison theorem which is valid also for generalized super-solutions (sub-solutions), it suffices to verify

$$\overline{v}(x,0) = 0 < V(r,0) \quad \text{for } x \in B_{1/2}, \tag{7.12}$$

$$\overline{v}(x,t) < V(r,t) \quad \text{for } (x,t) \in \partial B_{1/2} \times (0,1) \tag{7.13}$$

4) (7.12) follows from the assumption (7.8). However the proof of (7.13) is a little tortured. By Proposition 1.7.1 and an immediate calculation, we can check

$$\frac{\partial \overline{v}}{\partial t} \geq -\varepsilon \overline{v}, \tag{7.14}$$

$$\Delta \overline{v} \geq -\varepsilon, \tag{7.15}$$

where $\varepsilon = \dfrac{mk\eta}{\eta_0}$. (7.15) implies that

$$\Delta \left(\overline{v} + \frac{\varepsilon}{2N}|x|^2 \right) \geq 0,$$

that is, $\overline{v} + \dfrac{\varepsilon}{2N}|x|^2$ is subharmonic in x. Hence by the mean value theorem for subharmonic functions, for any $x \in B_{1/2}$,

$$\begin{aligned} \overline{v}(x,1) + \frac{\varepsilon}{2N}|x|^2 &\leq \fint_{B_{1/2}(x)} \left(\overline{v}(\xi,1) + \frac{\varepsilon}{2N}|\xi|^2 \right) d\xi \\ &\leq 2^N \fint_{B_1} \overline{v}(\xi,1) d\xi + \frac{\varepsilon}{2N}. \end{aligned}$$

Using (7.9), we conclude that

$$\overline{v}(x,1) \leq 2^N C + \frac{\varepsilon}{2N}, \quad \text{for } x \in B_{1/2}. \tag{7.16}$$

From (7.14) it follows that

$$\overline{v}(x,t) \leq e^{\varepsilon(1-t)} \overline{v}(x,1) \quad \text{for } t \in (0,1).$$

Combining this with (7.16) and using the assumption (7.4) we obtain

$$\overline{v}(x,t) \leq e^{\varepsilon} \left(2^N C + \frac{\varepsilon}{2N} \right) \leq \frac{1}{36mN} \quad \text{for } x \in B_{1/2}, t \in (0,1).$$

Therefore (7.13) holds. The proof of our proposition is completed. □

Corollary 1.7.1 *Under the assumptions of Proposition 1.7.2, if*

$$v(x,t_0) = 0 \quad \textit{for } x \in B_R(x_0),$$

and $(x_0, t_0 + \sigma) \in \Gamma$, *then*

$$\fint_{B_R(x_0)} v(x, t_0 + \sigma) dx \geq \frac{CR^2}{\sigma}.$$

Proposition 1.7.3 *Let $x_0 \in \mathbb{R}^N$, $t_0 \geq \eta_0$, $R > 0$, $\nu > 0$, $0 < \sigma \leq \eta$. If*

$$\fint_{B_R(x_0)} u^m(x, t_0)dx \geq \nu \left(\frac{R^2}{\sigma}\right)^{m/(m-1)}, \tag{7.17}$$

then there exists $\lambda_0 = \lambda_0(m, N, \nu) > 0$ such that

$$u^m(x_0, t_0 + \lambda\sigma) \geq c \left(\frac{R^2}{\sigma}\right)^{m/(m-1)}, \tag{7.18}$$

provided that $\lambda > \lambda_0$, $\eta < \dfrac{\eta_0}{mk\lambda}$, where $c = \dfrac{\nu^{1/m}}{\lambda}$.

The proposition expresses the physical fact that if a large mass of gas was in $B_R(x_0)$ at time t_0, then the gas covers a neighborhood of x_0 at time $t_0 + \lambda\sigma$.

Proof.

1) For simplicity we replace $x - x_0$ by x and $t - t_0$ by t. Let

$$\overline{u}(x, t) = \left(\frac{\sigma}{R^2}\right)^{1/(m-1)} u(Rx, \sigma t).$$

Then (7.17), (7.18) turn out to be

$$\fint_{B_1} \overline{u}^m(x, 0)dx \geq \nu, \tag{7.19}$$

$$\overline{u}^m(0, \lambda) \geq c. \tag{7.20}$$

2) Let

$$\Phi(t) = \int_{B_1} \overline{u}^m(x, t)dx.$$

We wish to prove

$$\int_0^t \Phi(s)ds \leq C_1 \int_0^t \overline{u}^m(0, s)ds + C_2\varepsilon^\delta + C_3(\Phi(t))^{1/m} \tag{7.21}$$

with positive constants C_1, C_2, C_3, δ depending only on m, M and $\varepsilon = \dfrac{mk\eta}{\eta_0}$.

To prove (7.21), we introduce Green's function

$$u^m(0, \lambda) \geq c.$$

$$G_\rho(r) = \begin{cases} r^{2-N} - \rho^{2-N} - \dfrac{N-2}{2\rho^N}(\rho^2 - r^2), & \text{for } N \neq 2, 0 < r \leq \rho, \\ \log \dfrac{\rho}{r} - \dfrac{1}{2\rho^2}(\rho^2 - r^2), & \text{for } N = 2, 0 < r \leq \rho, \end{cases}$$

$$G_\rho(r) = 0 \qquad \text{for } r > \rho.$$

Obviously

$$\begin{aligned} &G_\rho(\rho) = G'_\rho(\rho) = 0, \quad G_\rho(r) > 0, \\ &\Delta G_\rho = \alpha_N \chi_{B_\rho} - \gamma_N \delta(x), \end{aligned} \qquad \text{for } 0 < r < \rho \tag{7.22}$$

where $\alpha_N = N(N-2)$, $r_N > 0$ is a constant depending only on N, χ_A denotes the characteristic function of the set A, and $\delta(x)$ denotes the Dirac measure. Note that (7.22) is valid in the sense of distributions.

If $0 < t < 1$ and $\overline{u}$ is smooth, then

$$\begin{aligned} \int_{B_1} G_1(r)\overline{u}(x,t)dx &\geq \int_0^t \int_{B_1} G_1(r) \frac{\partial \overline{u}(x,s)}{\partial s} dxds \\ &= \int_0^t \int_{B_1} G_1(r) \Delta \overline{u}^m(x,s) dxds \\ &= -\gamma_N \int_0^t \overline{u}^m(0,s)ds + \alpha_N \int_0^t \int_{B_1} \overline{u}^m(x,s)dxds. \end{aligned}$$

Hence for $t \in (0,1)$,

$$\int_s^t \Phi(s)ds \leq \frac{\gamma_N}{\alpha_N} \int_0^T \overline{u}^m(0,s)ds + \frac{1}{\alpha_N} \int_{B_1} G_1(r)\overline{u}(x,t)dx. \tag{7.23}$$

In general case, we substitute the approximate smooth solution of $\overline{u}$ into (7.23) and then pass to the limit to conclude (7.24) for $\overline{u}$.

We will consider the following three cases separately.

Case 1

$$N \leq 2 \quad \text{or} \quad \frac{N}{N-2} > \frac{m}{m-1}. \tag{7.24}$$

We have, for $q = \dfrac{m}{m-1}$,

$$\begin{aligned} &\int_{B_1} G_1(r)\overline{u}(x,t)dx \\ \leq\ &\left(\int_{B_1} G_1^q(r)dx \right)^{1/q} \left(\int_{B_1} \overline{u}^m(x,t)dx \right)^{1/m}. \end{aligned} \tag{7.25}$$

Using (7.24) it is easy to verify

$$\int_{B_1} G_1^q(r)dx < \infty.$$

Combining this with (7.23), (7.25) yields (7.21) with $C_2 = 0$.

Case 2 (7.24) is not satisfied, but

$$N \le 4 \quad \text{or} \quad \frac{N}{N-4} > \frac{m}{m-1}. \tag{7.26}$$

Without loss of generality we suppose that $\overline{u}$ is smooth as above. For $x \in B_{1/2}$, using (7.22) we can express $\overline{u}^m$ by Green's function $G_{1/4}$:

$$\begin{aligned}\gamma_N \overline{u}^m(x,t) = \; & \alpha_N 4^N \int_{B_{1/4}(x)} \overline{u}^m(y,t)dy \\ & - \int_{B_{1/4}(x)} G_{1/4}(|y-x|)\Delta \overline{u}^m(y,t)dy.\end{aligned}$$

By Proposition 1.7.1

$$\Delta \overline{u}^m = \frac{\partial \overline{u}}{\partial t} \ge -\varepsilon \overline{u}.$$

Substituting this into the previous formula and using Young's inequality yield

$$\begin{aligned} & \overline{u}(x,t) = (\overline{u}^m(x,t))^{1/m} \\ \le \; & 2\beta_N^{1/m}(\Phi(t))^{1/m} + 2\beta_N^{1/m}\varepsilon^{1/m}\left(\int_{B_{1/4}(x)} G_{1/4}(|y-x|)\overline{u}(y,t)dy\right)^{1/m} \\ \le \; & 2\beta_N^{1/m}(\Phi(t))^{1/m} + C_4\varepsilon^{q/m} + \int_{B_{1/4}(x)} G_{1/4}(|y-x|)\overline{u}(y,t)dy, \end{aligned} \tag{7.27}$$

where $\beta_N = \max\left\{\frac{4^N\alpha_N}{\gamma_N}, \frac{1}{\gamma_N}\right\}$, $q = \frac{m}{m-1}$.

Now we calculate the second integral on the right hand side of (7.23). Using (7.27) we have

$$\begin{aligned} & \int_{B_{1/2}} G_1(|x|)\overline{u}(x,t)dx \\ \le & C_5(\Phi(t))^{1/m} + C_6\varepsilon^{q/m} \end{aligned}$$

$$
\begin{aligned}
&+\int_{B_{1/2}}\int_{B_{1/4}(|x|)} G_{1/4}(|y-x|)G_1(|x|)\overline{u}(y,t)dydx \\
\leq & C_5(\Phi(t))^{1/m} + C_6\varepsilon^{q/m} \\
&+\int_{B_{1/2}}\int_{B_{3/4}(|x|)} G_{1/4}(|y-x|)G_1(|x|)\overline{u}(y,t)dydx \\
= & C_5(\Phi(t))^{1/m} + C_6\varepsilon^{q/m} + \int_{B_{3/4}} \overline{G}(y)\overline{u}(y,t)dy, \qquad (7.28)
\end{aligned}
$$

where

$$\overline{G}(y) = \int_{B_{1/2}} G_{1/4}(|y-x|)G_1(|x|)dx.$$

It is easy to see that $\overline{G}(y)$ is continuous on $B_{3/4}$ when $N < 4$, $|\overline{G}(y)| \leq \alpha \log|y| + \beta$ with some constants α, β, when $N = 4$ and $|\overline{G}(y)| \leq \dfrac{\alpha}{|y|^{N-4}}$ with some constant α when $N > 4$. Thus under the condition (7.26), we have

$$\int_{B_{3/4}} |\overline{G}(y)|^q dy < \infty.$$

Similar to the proof of (7.25) we can use Hölder's inequality to estimate $\int_{B_{3/4}} \overline{G}(y)\overline{u}(y,t)dy$ and then substitute into (7.28). Therefore we obtain

$$\int_{B_{1/2}} G(|x|)\overline{u}(x,t)dx \leq C_7(\Phi(t))^{1/m} + C_6\varepsilon^{q/m}. \qquad (7.29)$$

Since $G_1(r)$ is bounded away from $r = 0$,

$$\int_{B_1\setminus B_{1/2}} G_1(|x|)\overline{u}(x,t)dx \leq C_8 \int_{B_1\setminus B_{1/2}} \overline{u}(x,t)dx \leq C_9(\Phi(t))^{1/m},$$

which combining with (7.29) yields

$$\int_{B_1} G_1(|x|)\overline{u}(x,t)dx \leq C_{10}(\Phi(t))^{1/m} + C_6\varepsilon^{q/m}.$$

Substituting this into (7.23) we obtain (7.21).

Case 3 (7.26) is not satisfied, but

$$N \leq 6 \quad \text{or} \quad \frac{N}{N-7} > \frac{m}{m-1}. \qquad (7.30)$$

In this case, we can proceed as above, but in evaluating the integral $\int_{b_{3/4}} \overline{G}(y)\overline{u}(y,t)dy$, we first express $\overline{u}(y,t)$ in terms of Green's function $G_{1/8}$. Finally we also obtain (7.21).

If (7.30) is not satisfied, then we can repeat the above procedure step by step until (7.21) is proved for given positive integer N.

3) Now we come back to the proof of (7.20). By the assumption (7.19),

$$\Phi(0) \geq \nu_0 = \nu|B_1|.$$

Since from Proposition 1.7.1,

$$\frac{\partial \overline{u}^m}{\partial t} \geq -\varepsilon \overline{u}^m, \tag{7.31}$$

we have $\Phi'(t) \geq -\varepsilon\Phi(t)$. Hence

$$\Phi(t) \geq \nu_0 e^{-\varepsilon\lambda} \geq e^{-1}\nu_0 \quad \text{for } t \in (0,\lambda). \tag{7.32}$$

Suppose that (7.20) is not satisfied, that is

$$\overline{u}^m(0,\lambda) \leq c.$$

Then using (7.31) gives

$$\overline{u}^m(0,t) \leq \overline{u}^m(0,\lambda)e^{\varepsilon(\lambda-t)} \leq ce \quad \text{for } t \in (0,\lambda), \varepsilon\lambda \leq 1.$$

Combining this with (7.21), (7.32) we obtain

$$\begin{aligned}
&\int_0^t \Phi(s)ds \\
&\leq C_1ect + C_2\varepsilon^\delta + C_3(\Phi(t))^{1/m} \\
&\leq C_1ec\lambda + C_2\varepsilon^\delta + C_3(\Phi(t))^{1/m} \\
&= C_1e\nu^{1/m} + C_2\varepsilon^\delta + C_3(\Phi(t))^{1/m} \\
&= (C_1e\nu^{1/m} + C_2\varepsilon^\delta)\cdot\frac{e^{1/m}}{\nu_0^{1/m}}\cdot(e^{-1}\nu_0)^{1/m} + C_3(\Phi(t))^{1/m} \\
&\leq \overline{C}(\Phi(t))^{1/m} + C_3(\varphi(t))^{1/m} = \frac{1}{B}(\Phi(t))^{1/m},
\end{aligned} \tag{7.33}$$

where

$$\overline{C} = (C_1e\nu^{1/m} + C_2\varepsilon^\delta)\cdot\frac{e^{1/m}}{\nu_0^{1/m}}, \quad \frac{1}{B} = \overline{C} + C_3. \tag{7.34}$$

Set

$$\psi(t) = \int_0^t \Phi(s)ds.$$

Then (7.33) can be written as

$$\psi'(t) \geq B^m(\Phi(t))^m. \tag{7.35}$$

Notice that from (7.32),

$$\Psi(t) \geq At, \quad A = e^{-1}\nu_0. \tag{7.36}$$

We will compare $\Psi(t)$ with the solution of the equation

$$\chi'(t) = B^m \chi(t)^m, \tag{7.37}$$

satisfying

$$\chi(\frac{\lambda}{2}) = A \cdot \frac{\lambda}{2}. \tag{7.38}$$

From (7.35), (7.36) we can conclude that

$$\psi(t) \geq \chi(t), \quad \text{for } t \in \left[\frac{\lambda}{2}, \lambda\right]. \tag{7.39}$$

On the other hand, the solution $\chi(t)$ satisfying the condition (7.38) is determined by

$$(m-1)\chi^{m-1}(t) = \frac{1}{C - B^m t},$$

with C satisfying

$$\frac{1}{C - B^m\lambda/2} = (m-1)A^{m-1}(\lambda/2)^{m-1}. \tag{7.40}$$

Since $\chi(t) \to \infty$ as $t \to \dfrac{C}{B^m}$, we also have, by (7.39),

$$\psi(t) \to \infty, \quad \text{as } t \to \frac{C}{B^m},$$

provided $\lambda > \dfrac{C}{B^m}$. However this is impossible. The contradiction shows that if $\lambda \leq \dfrac{C}{B^m}$, then (7.20) must be satisfied.

From (7.40), noting $A = e^{-1}\nu_0$ and (7.34) it is easy to check

$$\frac{C}{B^m} = \frac{\lambda}{2} + \frac{(2e)^{m-1}\left[(C_1\nu^{1/m}e + C_2\varepsilon^{\delta})\left(\frac{e}{\nu_0}\right)^{1/m} + C_5\right]^m}{(m-1)\nu_0^{m-1}\lambda^{m-1}}.$$

Thus we can find a $\lambda_0 = \lambda_0(m, N, \nu) > 0$, such that (7.2) and hence (7.18) hold provided $\lambda \geq \lambda_0$, $\varepsilon = \frac{\eta m k_0}{\eta_0} \leq \frac{1}{\lambda}$. The proof is complete. □

Proposition 1.7.4 *Under the assumptions of Proposition 1.7.3, if*

$$\fint_{B_R(x_0)} v(x, t_0)dx \geq \nu_0 \cdot \frac{R^2}{\sigma}$$

with $\nu_0 = \nu^{(m-1)/m}$, *then*

$$v(x_0, t_0 + \lambda\sigma) \geq c_0 \cdot \frac{R^2}{\sigma}$$

with $c_0 = c^{(m-1)/m}$.

Proof. Since

$$\begin{aligned}\fint_{B_R(x_0)} v(x, t_0)dx = & \frac{m}{m-1}\fint_{B_R(x_0)} u^{m-1}(x, t_0)dx \\ \leq & C^*\left(\fint_{B_R(x_0)} u^m(x, t_0)dx\right)^{(m-1)/m},\end{aligned}$$

where C^* is a constant depending only on m, N, Proposition 1.7.3 can immediately be applied to deduce our conclusion. □

Denote

$$\begin{aligned}\sigma(x, t) = & \{(x, \tau); 0 < \tau < t\}, \\ \Gamma_m(x, t) = & \{(x, t) \in \Gamma; \sigma(x, t) \cap \Gamma = \emptyset\}, \\ \Gamma_s(x, t) = & \{(x, t) \in \Gamma; \sigma(x, t) \subset \Gamma\}.\end{aligned}$$

Clearly $\Gamma_m \cap \Gamma_s = \emptyset$.

Theorem 1.7.3 $\Gamma = \Gamma_m \cup \Gamma_s$, *that is, for any* $(x_0, t_0) \in \Gamma$, *either (i)* $\sigma(x_0, t_0) \subset \Gamma$ *or (ii)* $\sigma(x_0, t_0) \cap \Gamma = \emptyset$.

Thus if the free boundary contains a vertical segment, then the extension of this segment down to $t = 0$ *also belongs to the free boundary.*

Combining Theorem 1.7.3 with Theorem 1.7.2 shows that under the condition (7.3), either $\Gamma = \Gamma_m$ or $\Gamma_s = \emptyset$.

Proof. If the assertion is not true, then there exist t_1, t_2 with $0 < t_1 < t_2 < t_0$ such that

$$(x_0, t_2) \in \Gamma, (x_0, t_1) \notin \Gamma.$$

Without loss of generality we may assume that $\sigma = t_2 - t_1$ is sufficiently small and $\lambda = \dfrac{t_0 - t_2}{t_2 - t_1}$ is sufficiently large. (otherwise we may replace (x_0, t_2) by another point at Γ). Theorem 1.7.1 implies that $x_0 \notin \Omega(t_1)$. However $(x_0, t_1) \notin \Gamma$. Thus $u(x_0, t_1) = 0$ and there exists $R > 0$ such that

$$u(x, t_1) = 0, \quad \text{for } x \in B_R(x_0).$$

Applying Corollary 1.7.1 with $\sigma = t_2 - t_1$ we obtain

$$\fint_{B_R(x_0)} v(x, t_2)dx \geq \frac{cR^2}{t_2 - t_1}.$$

Next, we apply Proposition 1.7.4 with $\nu_0 = c$, $\sigma = t_2 - t_1$ and $\sigma = t_0 - t_2$, and then obtain

$$v(x_0, t_0) = v(x_0, t_2 + \lambda\sigma) > 0,$$

which contradicts the fact $(x_0, t_0) \in \Gamma$. □

Theorem 1.7.4 *Given any $\eta_0 > 0$, there exist positive constants C, γ, h, such that for any $(x_0, t_0) \in \Gamma_m$, $t_0 \geq \eta_0$,*

$$u(x,t) = 0, \qquad \begin{array}{l} if |x - x_0| \leq C(t_0 - t)^\gamma \\ t_0 - h < t < t_0, \end{array} \tag{7.41}$$

$$u(x,t) > 0, \qquad \begin{array}{l} if |x - x_0| \leq C(t_0 - t)^\gamma \\ t_0 < t < t_0 + h, \end{array} \tag{7.42}$$

Proof. First we prove (7.41). Fix $\tau \in [0, t_0)$. Let $h = t_0 - \tau$. Since $(x_0, t_0) \in \Gamma_m$, there must be an $R > 0$ such that $u(x, \tau) = 0$ for $x \in B_R(x_0)$. Let

$$t_1 = \tau + (1 - \lambda)h = t_0 - \lambda h,$$

where $\lambda \in (0,1)$ is to be determined later. Suppose

$$\operatorname{dist}(x_0, \Gamma(t_1)) < \alpha R, \tag{7.43}$$

where $\alpha \in (0,1)$ is also to be determined later. Applying Corollary 1.7.1, we deduce

$$\fint_{B_{(1-\alpha)R}(x_1)} v(x,t_1)dx \geq \frac{c(1-\alpha)^2 R^2}{(1-\lambda)h},$$

where $c > 0$, $x_1 \in \Gamma(t_1)$ with $|x_1 - x_0| \leq \alpha R$. Hence

$$\fint_{B_R(x_0)} v(x,t_1)dx \geq \frac{c(1-\alpha)^2 R^2}{(1-\lambda)h}(1-\alpha)^N.$$

Given any large $C > 0$ we can choose α, $\lambda > 0$ so that

$$\frac{c(1-\alpha)^2 R^2}{(1-\lambda)h}(1-\alpha)^N \geq C\frac{R^2}{\lambda h}. \tag{7.44}$$

But if C is sufficiently large (depending on m, N, η_0), then Proposition 1.7.4 implies $u(x_0, t_0) > 0$, which is of course impossible. Thus, if (7.44) holds, then (7.43) cannot be valid, that is,

$$\operatorname{dist}(x_0, \Gamma(t_1)) \geq \alpha R. \tag{7.45}$$

Notice that (7.44) is implied by

$$(1-\alpha)^{N+2} \geq C(1-\lambda)$$

with another large constant C, provided $\lambda > \frac{1}{2}$. Taking $\alpha = \lambda^\gamma$, we reduce

$$(1-\lambda^\gamma)^{N+2} \geq C(1-\lambda).$$

This inequality is valid for some λ near 1 and γ sufficiently large (for instance, $\lambda = 1 - \frac{1}{k}$, $\gamma = k$, k sufficiently large).

From (7.45) it follows that

$$u(x,t_1) = 0, \quad \text{for } x \in B_{\alpha R}(x_0),\ \alpha = \lambda^\gamma.$$

We can now repeat the previous argument with R replaced by αR, τ replaced by t_1 and h replaced by λh. Thus we deduce that

$$\operatorname{dist}(x_0, \Gamma(t_2)) \geq \alpha^2 R, \quad t_2 = t_0 - \lambda^2 h.$$

Proceeding step by step, we may obtain

$$\operatorname{dist}(x_0,\Gamma(t)) \ge \lambda^{\gamma n} R, \quad t = t_0 - \lambda^n h.$$

Now we take $\overline{h} = \dfrac{\lambda\eta_0}{2}$ and let h vary on $\left[\overline{h}, \dfrac{\overline{h}}{\lambda}\right]$. Then $t_0 - h$ varies on $\left[t_0 - \dfrac{\eta_0}{2}, t_0 - \dfrac{\lambda\eta_0}{2}\right]$. It is easily seen that the corresponding values of R are bounded from below by a positive number $R_0 > 0$. Thus for any $h \in \left[\overline{h}, \dfrac{\overline{h}}{\lambda}\right]$ and $n = 1, 2, \cdots$, we have

$$\operatorname{dist}(x_0,\Gamma(t)) \ge \lambda^{\gamma n} R_0, t = t_0 - \lambda^n h. \tag{7.46}$$

Taking C such that

$$C\left(\frac{\overline{h}}{\lambda}\right)^{\gamma} \le R_0, \tag{7.47}$$

we must have

$$u(x,t) = 0, \quad \text{if } |x - x_0| \le C(t_0 - t)^{\gamma}, t_0 - \overline{h} < t < t_0.$$

In fact, for any $t \in (t_0 - \overline{h}, \overline{t}_0)$, we may choose a positive integer $n(t)$ such that

$$t_0 - t \in \left(\lambda^{n(t)}\overline{h}, \lambda^{n(t)} \cdot \frac{\overline{h}}{\lambda}\right).$$

Set $h(t) = \lambda^{-n(t)}(t_0 - t)$, that is,

$$t_0 - t = \lambda^{n(t)} h(t).$$

Then $h(t) \in \left(\overline{h}, \dfrac{\overline{h}}{\lambda}\right)$. Therefore from (7.46), (7.47) we conclude that if $|x - x_0| \le C(t_0 - t)^{\gamma}$, then

$$\begin{aligned} |x - x_0| &\le C(\lambda^{n(t)} h(t))^{\gamma} \\ &\le C\left(\frac{\overline{h}}{\lambda}\right)^{\gamma} \lambda^{\gamma n(t)} \\ &\le R_0 \lambda^{\gamma n(t)} \\ &\le \operatorname{dist}(x_0, \Gamma(t)), \end{aligned}$$

which implies (7.41).

Next we prove (7.42). Consider the set

$$E = \{(x, t); |x - x_0| < C(t - t_0)^\gamma, t_0 < t < t_0 + h\}.$$

For any $(x, t) \in E$, if $(x, t) \in \Gamma$, then the segment $\sigma(x, t)$ must contain interior points of the set defined by

$$|x - x_0| \le C(t - t_0)^\gamma, t_0 - h < t < t_0,$$

that is, $\sigma(x, t)$ contains points not belonging Γ. Thus, by Theorem 1.7.3, $\sigma(x, t) \cap \Gamma = \emptyset$. Similar to the proof of (7.41), with (x_0, t_0) replaced by (x, t), we can derive that $u(x_0, t_0) > 0$, which contradicts $(x_0, t_0) \in \Gamma$. Therefore $(x, t) \notin \Gamma$,that is, E does not contain any point of Γ. This shows that on E either $u \equiv 0$ or $u > 0$. Since $(x_0, t_0) \in \Gamma$, we must have, for any $(x, t) \in E$, $u > 0$. Thus (7.42) is proved. □

As a corollary of Theorem 1.7.3, we have

Theorem 1.7.5 *For any $(x_0, t_0) \in \Gamma_m$, there exists a neighborhood in which Γ can be expressed by a Hölder continuous function*

$$t = S(x) \quad \textit{for } x \in U_x,$$

where U_x is the projection on $t = 0$ of Γ in the neighborhood.

Proof. Applying Theorem 1.7.3, it is easy to see that there exists a neighborhood, such that any point of Γ in this neighborhood must be a point of Γ_m. Hence the portion $U(x_0, t_0)$ of Γ in this neighborhood can be expressed by a function $t = S(x)$.

For any $x \in U_x$, since $(x, S(x)) \in \Gamma$, we see from Theorem 1.7.4 that $(x, S(x))$ must satisfy

$$|x - x_0| \ge C|S(x) - S(x_0)|^\gamma$$

or

$$|S(x) - S(x_0)| \le \left(\frac{1}{C}\right)^{1/\gamma} |x - x_0|^{1/\gamma}.$$

□

Theorem 1.7.6 *Under the assumptions of Theorem 1.7.2, Γ can be expressed by a function*

$$t = S(x) \quad \textit{for } x \in \mathbb{R}^N \backslash D.$$

Moreover, $S(x)$ is uniformly continuous on any compact subset K which does not intersect with $\overline{D}$.

Proof. By Theorem 1.7.2 and Theorem 1.7.3, Γ does not contain any vertical segment. Hence the whole Γ can be expressed by a function $t = S(x)$, which is defined on $\mathbb{R}^N \setminus D$ in virtue of Theorem 1.7.1.

From the continuity of u it is easy to check that $S(x)$ is uniformly continuous on K, that is, for any $h > 0$, there exists $\eta(h) > 0$ such that

$$|S(x) - S(x_0)| < h, \quad \text{for } x, x_0 \in K, |x - x_0| < \eta(h). \tag{7.48}$$

Since $K \cap \overline{D} = \emptyset$, there exists $\eta_0 > 0$ such that $S(x) > \eta_0$ for $x \in K$. From the proof of Theorem 1.7.4 we see that C, h, γ appeared there can be chosen to be independent of $x_0 \in K$, $t_0 = S(x_0)$ $(\geq \eta_0)$. Applying Theorem 1.7.4, we conclude that for $x, x_0 \in K$, $|x - x_0| < \eta(h)$,

$$|x - x_0| \geq C(t - t_0)^\gamma,$$

or

$$|S(x) - S(x_0)| \leq \left(\frac{1}{C}\right)^{1/\gamma} |x - x_0|^{1/\gamma}.$$

□

1.7.2 *Lipschitz continuity of the free boundary*

Without any additional condition besides those stated at the beginning of this section, one can prove the Lipschitz continuity of the free boundary for large time. Precisely, one can prove that the free boundary Γ is Lipschitz continuous for $t \geq T_0$ where $T_0 > 0$ is such a time that for $t \geq T_0$, $B_{R_0}(0) \subset \Omega(t)$ where $B_{R_0}(0)$ is the smallest ball containing $D = \Omega(0)$. For the proof of this result, see [CVW], in which the authors also proved that under certain conditions on the initial value $u_0(x)$, the free boundary Γ is globally Lipschitz continuous.

Theorem 1.7.7 *Assume that $u_0 \in \mathcal{C}$. Then the free boundary Γ can be expressed by a function*

$$t = S(x) \quad \textit{for } x \in \mathbb{R}^N \setminus D$$

and $S(x)$ is locally Lipschitz continuous on $\mathbb{R}^N \setminus D$.

We say that $u_0 \in \mathcal{C}$, if $D = \{x; u_0(x) > 0\}$ is a bounded C^1 domain and

(i) $v_0 = \dfrac{m}{m-1} u_0^{m-1} \in C^1(\overline{D})$;
(ii) $v_0 + |\nabla v_0| \geq K_1 > 0$, *for* $x \in D$;
(iii) in the sense of distributions,

$$\Delta v_0 \geq -K_2,$$

where K_1, K_2 *are positive constants.*

It is easy to check that the condition (i), (ii) imply that v_0 satisfies the Lipschitz condition near ∂D and

$$v_0(x) \geq C^* \text{dist}\,(x, \partial D) \quad \text{for } x \in D$$

with a constant $C^* > 0$. This shows that functions in $\mathcal{C}$ satisfy the condition (7.3).

To prove Theorem 1.7.7, we first prove some auxiliary propositions.

Proposition 1.7.5 *Assume that* $u_0 \in \mathcal{C}$. *Then there exist constants* A, B *depending only on* K_1, K_2, R, *such that*

$$(A-2)v(x,t) + x \cdot \nabla v(x,t) + (At+B)\frac{\partial v(x,t)}{\partial t} \geq 0 \tag{7.49}$$

in the sense of distributions. Here $R > 0$ *is a constant such that* $D \subset B_R(0)$ *and* $v = \dfrac{m}{m-1} u^{m-1}$.

Proof. Let us first sketch the basic idea of the proof. Suppose that u is smooth and $u > 0$. Then

$$\frac{\partial v}{\partial t} = (m-1)v\Delta v + |\nabla v|^2. \tag{7.50}$$

For $\varepsilon > 0$, (7.50) admits a family of the following solutions

$$v_\varepsilon(x,t) = \frac{1+A\varepsilon}{(1+\varepsilon)^2} v((1+\varepsilon)x, (1+A\varepsilon)t + B\varepsilon),$$

where $A > 0$, $B > 0$ are constants. In fact one can easily check that for any constants $a \neq 0$, b, α, β, $\dfrac{\alpha}{a^2} v(ax+b, \alpha t + \beta)$ is a solution of (7.50).

If we can prove that $v_\varepsilon(x,t) \geq v(x,t)$, then

$$\left.\frac{d}{d\varepsilon} v_\varepsilon(x,t)\right|_{\varepsilon=0} = \lim_{\varepsilon \to 0} \frac{v_\varepsilon(x,t) - v(x,t)}{\varepsilon} \geq 0.$$

However

$$\frac{d}{d\varepsilon}v_\varepsilon(x,t)\bigg|_{\varepsilon=0} = (A-2)v(x,t) + x_0\nabla v(x,t) + (At+B)\frac{\partial v(x,t)}{\partial t}.$$

Therefore (7.49) holds (in classical sense).

Unfortunately, u is not smooth and positive everywhere in general. It is natural to consider its approximate smooth and positive solution and then pass to the limit.

For any $\delta > 0$, let

$$v_{0\delta}(x) = (v_0 * \rho_\delta)(x) + \delta^\alpha,$$

where $\alpha \in (0,1)$, ρ_δ is the kernel of a mollifier with compact support contained in $B_\delta(0)$. Since $v_{0\delta} \in C^\infty(\mathbb{R}^N)$ and $v_{0\delta}(x) \geq \delta^\alpha$, there must be a smooth solution $v_\delta(x,t)$ with initial data $v_{0\delta}(x)$, which satisfies $v_\delta \geq \delta^\alpha$, and $v_\delta \to v$ as $\delta \to 0$.

To prove that v_δ satisfies (7.49), as stated above, it suffices to prove $v_{\varepsilon\delta}(x,t) \geq v_\delta(x,t)$, where

$$v_{\varepsilon\delta}(x,t) = \frac{1+A\varepsilon}{(1+\varepsilon)^2}v_\delta((1+\varepsilon)x, (1+A\varepsilon)t + B\varepsilon).$$

By the comparison theorem, to prove $v_{\varepsilon\delta} \geq v_\delta(x,t)$, it suffices to check $v_{\varepsilon\delta}(x,0) \geq v_\delta(x,0)$. To do this, we need a careful discussion.

For simplicity below we will drop the mark "δ". The readers need to keep in mind that v, v_0 and v_ε are just v_δ, $v_{0\delta}$ and $v_{\varepsilon\delta}$ respectively.

Denote

$$I_\varepsilon = \frac{1}{\varepsilon}(v_\varepsilon(x,0) - v(x,0)).$$

Then we have

$$\begin{aligned} I_\varepsilon &= \frac{1}{\varepsilon}\left(\frac{1+A\varepsilon}{(1+\varepsilon)^2}v((1+\varepsilon)x, B\varepsilon) - v(x,0)\right) \\ &\geq \frac{1}{\varepsilon}\left(\left(1+\frac{1}{2}(A-2)\varepsilon\right)v((1+\varepsilon)x, B\varepsilon) - v(x,0)\right) \\ &= \frac{1}{2}(A-2)v((1+\varepsilon)x, B\varepsilon) \\ &\quad + B\frac{\partial}{\partial t}v((1+\varepsilon)x, \theta\varepsilon) + x\cdot\nabla v(\xi, 0) \end{aligned}$$

with some $\theta \in (0, B)$ and some ξ on the segment joining the points x and $(1+\varepsilon)x$, provided that $A > 2$ and $\varepsilon > 0$ is small enough. Since ε is sufficiently small, there exists a constant C depending only on m, N and $v_{0\delta}$ such that

$$\begin{aligned} I_\varepsilon \ &\geq \frac{1}{2}(A-2)v(x,0) + B\frac{\partial}{\partial t}v(x,0) \\ &+x \cdot \nabla v(x,0) - C\varepsilon. \end{aligned}$$

Notice that here we have used the fact that for given $\delta > 0$, the derivatives of v up to second order are bounded. Since v satisfies (7.50), we have

$$\begin{aligned} I_\varepsilon \ &\geq \left(\frac{1}{2}(A-2) + B(m-1)\Delta v(x,0)\right) v(x,0) \\ &+B|\nabla v(x,0)|^2 + x \cdot \nabla v(x,0) - C\varepsilon, \end{aligned} \tag{7.51}$$

provided that $A > 2$ and $\varepsilon > 0$ is small enough. If $|x| \leq R + \delta$, then

$$\begin{aligned} I_\varepsilon \geq \ &\left(\frac{1}{2}(A-2) - B(m-1)K_0\right) v_0(x) \\ &+|\nabla v_0(x)|(B|\nabla v_0(x)| - R - \delta) - C\varepsilon, \end{aligned} \tag{7.52}$$

where we have used the fact $\Delta v_0 \geq -K_2$, which follows from the condition (iii). Denote

$$S = \{x \in D; \mathrm{dist}\,(x, \partial D) < \delta\}.$$

The condition (ii) implies that for small δ, we have

$$|\nabla v_0| \geq \frac{K_1}{4}$$

on S. In fact, for small δ, this inequality holds also on

$$U_{c,\delta} = \{x \in \mathbb{R}^N; \mathrm{dist}\,(x, \partial D) < c\delta\},$$

where $c \in \left(0, \frac{1}{2}\right)$ is independent of δ.

Now we are ready to prove $I_\varepsilon \geq 0$ on $\mathbb{R}^N$. We will divide $\mathbb{R}^N$ into several parts and prove $I_\varepsilon \geq 0$ on each part of $\mathbb{R}^N$.

Denote

$$D_1 = \{x \in \mathbb{R}^N; \mathrm{dist}\,(x, D) \geq \delta\}, \quad D_2 = \left\{s \in \mathbb{R}^N; |\nabla v_0| \geq \frac{K_1}{4}\right\},$$

$$D_3 = D \backslash S, \qquad D_4 = \{x \in \mathbb{R}^N; c\delta \leq \mathrm{dist}\,(x, D) < \delta\}.$$

Since $S \cup U_{c,\delta} \subset D_2$ (but D_2 may contain points outside $S \cup U_{c,\delta}$), we have $\mathbb{R}^N = D_1 \cup D_2 \cup D_3 \cup D_4$.

Case 1 On $D_1 = \{x \in \mathbb{R}^N; \text{dist}\,(x, D) \geq \delta\}$.

Since $v(x,t) \geq \delta^\alpha$, we have

$$v_\varepsilon(x,t) \geq \frac{1 + A\varepsilon}{(1+\varepsilon)^2}\delta^\alpha > \delta^\alpha.$$

Then noting that on D_1, $v(x,0) = \delta^\alpha$, we have $v_\varepsilon(x,0) \geq v(x,0)$ or $I_\varepsilon \geq 0$ on D_1.

Case 2 On $D_2 = \left\{x \in \mathbb{R}^N; |\nabla v_0| \geq \dfrac{K_1}{4}\right\}$.

On D_2 we must have $|x| \leq R + \delta$. Hence, since on D_2, $|\nabla v_0(x)| \leq \dfrac{K_1}{4}$, from (7.52),

$$\begin{aligned} I_\varepsilon &\geq |\nabla v_0(x)|(B|\nabla v_0(x)| - R - \delta) - C\varepsilon \\ &\geq \frac{K_1}{4}\left(\frac{BK_1}{4} - R - \delta\right) - C\varepsilon > 0, \end{aligned}$$

provided that $B > \dfrac{4R}{K}$, $A > 2 + 2B(m-1)K_2$.

Case 3 On $D_3 = D \backslash S$.

It suffices to consider those points in D_3 such that $|\nabla v_0(x)| \leq \dfrac{K_1}{4}$. Since $v_0(x)$ is bounded below by a constant $a > 0$, from (7.52) we have

$$\begin{aligned} I_\varepsilon &\geq \left(\frac{1}{2}(A-2) - B(m-1)K_2\right) a \\ &\quad - \frac{K_1}{4}(R + \delta) - C\varepsilon > 0, \end{aligned}$$

provided that $A > 2 + 2B(m-1)K_2 + \dfrac{2RK_1}{a}$ and ε, δ are small enough.

Case 4 On $D_4 = \{x \in \mathbb{R}^N; c\delta \leq \text{dist}\,(x, D) \leq \delta\}$.

This case is a little difficult to treat. In order to prove $I_\varepsilon \geq 0$ on D_4, we need to choose the kernel ρ_δ of the mollifier such that

$$0 \leq \rho_\delta(x) \leq \rho_\delta(0), \rho_\delta(x) = \rho_\delta(0), \quad \text{for } |x| \leq \delta - \delta^{\gamma+1},$$

$$\rho_\delta(x) = 0, \quad \text{for } |x| \geq \delta,$$

where $0 < \gamma < 1$ is to be determined. Of course, we also assume that $\int_{\mathbb{R}^N} \rho_\delta(x)dx = 1$.

We will further divide D_4 into two parts:

$$D_4^{(1)} = \{x \in \mathbb{R}^N; \delta - \delta^{\gamma+1} < \text{dist}\,(x, D) < \delta\},$$

$$D_4^{(2)} = \{x \in \mathbb{R}^N; c\delta \le \text{dist}\,(x, D) \le \delta - \delta^{\gamma+1}\}.$$

Since $\partial D \in C^1$, from the condition (i) we can verify (but not clearly) that for δ small enough,

$$|\nabla v_{0\delta}(x)| \le CK_3\delta^{\gamma(1+(N-1)/2)},$$

where K_3 is the upper bound of $|\nabla v_0|$ near ∂D and $C > 0$ is a constant depending only on N. Using this we obtain from (7.52) that

$$\begin{aligned} I_\varepsilon \ge & \left(\frac{1}{2}(A-2) - B(m-1)K_2\right)\delta^\alpha \\ & -(R+\delta)CK_3\delta^{\gamma(1+(N-1)/2} - C\varepsilon. \end{aligned}$$

Hence, if we choose $\dfrac{\alpha}{1+(N-1)/2} < \gamma < 1$, $A > 2 + 2B(m-1)K_2$ and δ small enough, then for small ε we have $I_\varepsilon \ge 0$ on $D_4^{(1)}$.

Finally, we prove that $I_\varepsilon \ge 0$ on $D_4^{(2)}$. Since we have proved the conclusion on D_2, it remains to consider those points in $D_4^{(2)}$ such that $|\nabla v_0(x)| \le \dfrac{K_1}{4}$. For such points, from (7.51) we have

$$\begin{aligned} I_\varepsilon \ge & \; B(m-1)v(x,0)\Delta v(x,0) \\ & -(R+\delta)|\nabla v(x,0)| - C\varepsilon \end{aligned} \tag{7.53}$$

provided $A > 2$.

It can be proved (but not easily) that for $0 < \gamma < \dfrac{2}{N-2}$ and small δ,

$$\Delta v(x,0) \ge CK_1\delta^{-(1-(N-1)\gamma/2)},$$

where $C > 0$ is a certain constant. Using this in (7.53) and noting that $v(x,0) \ge \delta^\alpha$, we derive

$$I_\varepsilon \ge \; B(m-1)CK_1\delta^{\alpha-(1-(N-1)\gamma/2)} - (R+\delta)\frac{K_1}{4} - C\varepsilon,$$

provided $\gamma < \dfrac{2}{N-2}$, $0 < \alpha < 1 - \dfrac{(N-1)\gamma}{2}$ and we first choose δ and then choose ε small enough, then $I_\varepsilon > 0$ follows.

The proof of our proposition is thus completed. $\square$

Proposition 1.7.6 *Under the assumptions of Proposition 1.7.5, if we denote*

$$x(s) = x_0 e^s, \quad t(s) = \frac{1}{A}\left((At_0 + B)s^{As} - B\right), \quad s \geq 0$$

then the function $v(x(s), t(s))e^{(A-2)s}$ *is decreasing.*

Proof. Since $x'(s) = x(s)$, $t'(s) = At(s) + B$, from Proposition 1.7.5 we have

$$\begin{aligned} & \frac{d}{ds}\left(v(x(s), t(s))e^{(A-2)s}\right) \\ = \; & e^{(A-2)s}((A-2)v + x \cdot \nabla v + (At + B)v_t) \\ \geq \; & 0 \end{aligned}$$

in the sense of distributions. $\square$

Theorem 1.7.8 *Assume that* $v_0 \in \mathcal{C}$. *Then for any* $(\overline{x}, \overline{t}) \in Q$, *there exist positive constants* A, B, C *depending only on* v_0, $\overline{t}$, $R_1 = \sup\{dist(\overline{x}, y);\, y \in D\}$ *such that*

$$v(x,t) \geq v(\overline{x}, \overline{t})e^{-C(t-\overline{t})}, \tag{7.54}$$

where $(x,t) \in Q$, $\overline{t} < t < \overline{t} + \varepsilon$ *with* ε *depending only on* A, B *and*

$$\frac{|x - \overline{x}|}{t - \overline{t}} \leq \frac{R_1}{A\overline{t} + B}. \tag{7.55}$$

Proof. Let A, B be the constants in Proposition 1.7.5 corresponding to $R = 3R_1$ and $C = A - 2$. To verify (7.54), it suffices to prove that any point (ξ, τ) in the cone (7.55) can be joined with $(\overline{x}, \overline{t})$ by a curve in Proposition 1.7.6. Since Proposition 1.7.6 is based on Proposition 1.7.5, we have to check $D \subset B_R(0)$.

Suppose (ξ, τ) is a point in the cone (7.55), $\overline{t} < \tau < \overline{t} + \varepsilon$. Choose x_0 such that

$$\xi - x_0 = (\overline{x} - x_0)\left(\frac{A\tau + B}{A\overline{t} + B}\right)^{1/A}.$$

For simplicity, we change coordinates so that x_0 becomes the origin but the notations $(\overline{x}, \overline{t})$ and (ξ, τ) do not change. Then the above formula turns out to be

$$\xi = \overline{x}\left(\frac{A\tau + B}{A\overline{t} + B}\right)^{1/A}.$$

Consider the curve

$$\xi = \overline{x}\left(\frac{At + B}{A\overline{t} + B}\right)^{1/A},$$

which can be expressed in the form of parameter

$$x(s) = \overline{x}e^s, \quad t(s) = \frac{1}{A}((A\overline{t} + B)e^{As} - B) \quad (s \geq 0).$$

This curve joins $(\overline{x}, \overline{t})$ and (ξ, τ). By Proposition 1.7.6, if we can verify that $D \subset B_R(0)$, then we obtain the conclusion of our theorem.

To prove $D \subset B_R(0)$, notice that

$$\begin{aligned} & |\overline{x}|\left(\frac{A\tau + B}{A\overline{t} + B}\right)^{1/A} \\ = \quad & |\xi| \leq |\xi - \overline{x}| + |\overline{x}| \\ \leq \quad & R_1 \frac{\tau - \overline{t}}{A\overline{t} + B} + |\overline{x}|, \end{aligned}$$

that is,

$$|\overline{x}|\left(\left(\frac{A\tau + B}{A\overline{t} + B}\right)^{1/A} - 1\right) \leq R_1 \frac{\tau - \overline{t}}{A\overline{t} + B}.$$

Clearly, there exists $\varepsilon = \varepsilon(A, B) > 0$, such that for $\overline{t} < \tau < \overline{t} + \varepsilon$,

$$\left(\frac{A\tau + B}{A\overline{t} + B}\right)^{1/A} - 1 \geq \frac{1}{2}\frac{\tau - \overline{t}}{A\overline{t} + B}.$$

Substituting this into the previous formula yields

$$\frac{1}{2}|\overline{x}|\frac{\tau - \overline{t}}{A\overline{t} + B} \leq R_1 \frac{\tau - \overline{t}}{A\overline{t} + B},$$

that is, $|\overline{x}| \leq 2R_1$. Thus for $y \in D$, we have

$$|y| \leq |y - \overline{x}| + |\overline{x}| \leq 3R_1 = R.$$

Hence $D \subset B_R(0)$. The proof is complete. □

Proof of Theorem 1.7.7. By Theorem 1.7.8, if $(\overline{x}, \overline{t}) \in \Gamma$, then

$$u(x,t) = 0, \qquad \text{when } |x - \overline{x}| \le \frac{R_1}{A\overline{t} + B}(\overline{t} - t), \overline{t} - \varepsilon < t < \overline{t},$$

$$u(x,t) > 0, \qquad \text{when } |x - \overline{x}| \le \frac{R_1}{A\overline{t} + B}(t - \overline{t}), \overline{t} < t < \overline{t} + \varepsilon,$$

provided that $\varepsilon = \varepsilon(A, B) > 0$ is small enough. From this, it follows that Γ can be expressed by a function $t = S(x)$. Theorem 1.7.1 implies that $S(x)$ is defined on $\mathbb{R}^N \backslash D$. Furthermore, since for $(x,t), (\overline{x}, \overline{t}) \in \Gamma$ with $\overline{t} - \varepsilon < t < \overline{t} + \varepsilon$,

$$|x - \overline{x}| \ge \frac{R_1}{A\overline{t} + B}|t - \overline{t}|,$$

we have

$$|S(x) - S(\overline{x})| \le C|x - \overline{x}|$$

for some constant $C > 0$. Thus we obtain the conclusion of Theorem 1.7.7. □

Remark 1.7.1 Under certain conditions, Caffarelli and Wolanski [CW] further proved that the free boundary is a $C^{1,\alpha}$ surface. C^∞-regularity of the free boundary was proved for two dimensional case by P. Daskalopoulos and R. Hamilton [DH1]. See also [DH2].

1.7.3 *Differential equation on the free boundary*

We have proved that on the free boundary $x = \zeta_i(t)$ there holds

$$\zeta_i'(t+0) = -\frac{\partial v(\zeta_i(t), t)}{\partial x}, \tag{7.56}$$

where $v = \dfrac{m}{m-1}u^{m-1}$ and u is a generalized solution of the equation (6.3) with compact support for any $t > 0$. This result can be extended to the higher dimensional case.

Let $\tau > 0$, $y \notin \overline{\Omega(\tau)}$ be fixed. Denote

$$h(t) = \max\{R; v(x,t) = 0, x \in B_R(y)\}.$$

From the monotonicity and continuity of the free boundary, it follows that $h(t)$ is decreasing and there exists $\delta_0 > 0$ such that

$$h(t) > 0, \quad \text{for } 0 < t \leq \tau + \delta_0.$$

The following theorem is an extension of the formula (7.56) to the higher dimensional case.

Theorem 1.7.9 *For $0 < t < \tau + \delta_0$, limits*

$$\gamma_t(0+0) = \lim_{\varepsilon \to 0+} \gamma_t(\varepsilon) = \lim_{\varepsilon \to 0+} \max_{x \in B_{h(t)+\varepsilon}} \frac{v(x,t)}{\varepsilon}$$

and

$$h'(t+0) = \lim_{\Delta t \to 0+} \frac{h(t+\Delta t) - h(t)}{\Delta t}$$

exist and

$$h'(t+0) = -\gamma_t(0+0). \tag{7.57}$$

For the proof, see [FR2].

If $N = 1$, then $h(t) = y - \zeta(t)$, where $x = \zeta(t)$ is the free boundary, for example, the right free boundary, then

$$\gamma_t(0+0) = -\frac{\partial v(\zeta(t),t)}{\partial t},$$
$$h'(t+0) = \zeta'(0+0)$$

and (7.57) is just (7.56).

Remark 1.7.2 Topics related to propagation of disturbances and the free boundary for (7.1) and equations with convection term and absorption term are studied by many authors. For example, the readers may refer to [AB2], [AB3] for the property of instantaneous shrinking and [AA1], [AA2], [BBA] for the focusing problem.

1.8 Initial Trace of Solutions

The Widder-type theorem shows that , if $u(x,t)$ is a nonnegative solution of the heat equation

$$\frac{\partial u}{\partial t} = \Delta u \tag{8.1}$$

on $Q_T = \mathbb{R}^N \times (0,T]$ $(T>0)$, then there exists a unique nonnegative Radon measure μ, such that

$$\lim_{t\to 0+}\int_{\mathbb{R}^N} u(x,t)\varphi(x)dx = \int_{\mathbb{R}^N}\varphi(x)d\mu \tag{8.2}$$

for any $\varphi \in C_0^\infty(\mathbb{R}^N)$. In other words, any nonnegative solution u of (8.1) has a unique Radon measure μ as its initial trace in the sense of (8.2); μ satisfies the growth condition

$$\int_{\mathbb{R}^N} e^{-C|x|}d\mu < +\infty \tag{8.3}$$

for some constant $C > 0$.

In this section, we extend this result to the porous medium equation

$$\frac{\partial u}{\partial t} = \Delta u^m \tag{8.4}$$

where $m > 1$ (cf. [AC]).

1.8.1 *Harnack inequality*

Proposition 1.8.1 *Assume that u is a continuous generalized solution of (8.4) on Q_T. For any $\zeta \in C_0^\infty(\mathbb{R}^N)$, $0 \le \zeta \le 1$ and $\tau \in (0,T)$, let w be the generalized solution of (8.4) on $\mathbb{R}^N \times [\tau,T]$ with initial data $w(x,\tau) = \zeta u(x,\tau)$. Then*

$$u(x,t) \ge w(x,t) \quad \text{for } (x,t) \in \mathbb{R}^N \times (\tau,T].$$

Proof. Choose $R > 0$ large enough such that $\operatorname{supp}\zeta \subset B_R(0)$. Let v_R be a generalized solution of the problem

$$\begin{aligned} &\frac{\partial v_R}{\partial t} = \Delta v_R^m && \text{for } (x,t) \in B_R(0)\times(\tau,T),\\ &v_R(x,\tau) = \zeta(x)u(x,\tau) && \text{for } t\in(\tau,T],\\ &v_R(x,t) = 0 && \text{for } (x,t)\in \partial B_R(0)\times[\tau,T]. \end{aligned}$$

Using the comparison theorem to u and v_R on $B_R(0)\times(\tau,T)$ (Theorem 1.3.3) gives

$$u \ge v_R \quad \text{for } (x,t) \in B_R(0)\times(\tau,T).$$

Since w has compact support, for sufficiently large R, w is also a generalized solution of the above problem, we have $v_R \equiv w$ on $B_R(0) \times (\tau, T)$ by uniqueness. □

Lemma 1.8.1 *Let u be a generalized solution of (8.4) on Q_T, which is continuous on $\overline{Q}_T$ and has compact support for any $t \in [0,T]$. If $K = \operatorname{supp} u(x,0) \subset \{x \in \mathbb{R}^N; x_N > 0\}$, then*

$$u(y, x_N, t) \geq u(y, -x_N, t) \quad \textit{for } y \in R^{N-1}, x_N > 0, t \in [0,T].$$

Proof. Let $v(x,t) = u(y, -x_N, t)$, $x = (y, x_N)$. Then v is a generalized solution of (8.4) and

$$v(y,0,t) = u(y,0,t) \quad \text{for } y \in R^{N-1}, t \in [o,T],$$

$$v(y,x_N,0) = u(y,-x_N,t) = 0$$

$$\leq u(y,x_N,0) \quad \text{for } y \in R^{N-1}, x_N > 0.$$

Using the comparison theorem to v and u on the domain $\{(x,t); x_N > 0, t \in (0,T)\}$ yields the conclusion of our lemma. □

Proposition 1.8.2 *Let u be a generalized solution of (8.4) on Q_T, which is continuous on $\overline{Q}_T$. If $\operatorname{supp} u(x,0) \subset B_1(0)$, then*

$$\inf_{x \in B_r(0)} u(x,t) \geq \max_{x \in \partial B_{r+2}(0)} u(x,t) \quad \textit{for } r > 0, t \in [0,T].$$

Proof. For fixed $x_0 \in B_r(0)$ and $x_1 \in \partial B_{r+2}(0)$, let Π denote the hy perplane consisting of points in $\mathbb{R}^N$ which are equidistant from x_0 and x_1. Thus

$$\Pi = \{x \in \mathbb{R}^N; < x, x - x_0 > = < \frac{1}{2}(x_0 + x_1), x_0 - x_1 >\},$$

where $< \cdot, \cdot >$ denotes the usual product in $\mathbb{R}^N$. It is easy to verify that

$$\operatorname{dist}(\Pi, \{0\}) = \frac{1}{2} \frac{|x_1|^2 + |x_0|^2}{|x_1 - x_0|} > \frac{(r+2)^2}{4(r+1)} \geq 1.$$

Therefore x_0 and $\operatorname{supp} u(x,0)$ are in the same half-space with respect to Π. Moreover, x_1 is the reflection of x_0 in Π. Thus, by Lemma 1.8.1, $u(x_0,t) \geq u(x_1,t)$ and the assertion follows since $x_0 \in B_r(0)$ and $x_1 \in \partial B_{r+2}(0)$ are arbitrary. □

Theorem 1.8.1 *Let u be a generalized solution of (8.4), which is continuous on $\overline{Q}_1$. Then there exist positive constants $M_0 = M_0(m, N)$, $E = E(m, N)$ such that*

$$u(0,1) \geq EM^{2/k}, \tag{8.5}$$

provided that $M > M_0$, where $k = N(m-1)+2$ and

$$M = \int_{B_1(0)} u(x,0)dx.$$

Proof. In the proof of this proposition we make extensive use of the Barenblatt solution of (8.4),

$$\begin{aligned} B(x,t) &= B_m(x,t) = t^{-N/k}\left(1 - \frac{(m-1)|x|^2}{2mkt^{2/k}}\right)_+^{1/(m-1)} \\ &= t^{-N/k}\left(1 - \frac{F|x|^2}{t^{2/k}}\right)_+^{1/(m-1)}, \quad F = \frac{m-1}{2mk} \end{aligned}$$

(see (1.37)). Denote

$$G = \int_{\mathbb{R}^N} B(x,t)dx.$$

G is a constant depending only on m, N. For any $\mu > 0$, define

$$u_\mu(x,t) = \mu^{-2/(m-1)} B(\mu x, t).$$

It is easy to verify that u_μ is a generalized solution of (8.4). Moreover

$$\int_{\mathbb{R}^N} u_\mu(x,t)dx = \mu^{-k/(m-1)}G.$$

Choose μ, t_0 such that

$$\int_{\mathbb{R}^N} u_\mu(x,t)dx = \frac{M}{2}, \tag{8.6}$$

$$\operatorname{supp} u_\mu(x,t_0) = \overline{B_4(0)}. \tag{8.7}$$

It suffices to take

$$\mu = \left(\frac{2G}{M}\right)^{(m-1)/k}, \quad t_0 = (4\sqrt{F}\mu)^k.$$

Note that with these choices of μ and t_0, $\mu^{-2/(m-1)}t_0^{-N/k} = \dfrac{(4\sqrt{F})^{-N}M}{2G}$. Moreover, M large implies that both μ and t_0 are small.

Assume temporarily that u has compact support with

$$\operatorname{supp} u(x,0) \subset B_1(0). \tag{8.8}$$

Given $\delta > 0$, we choose $r(t)$ such that

$$u_\mu(x, t+t_0) = \delta u_\mu(0, t+t_0) \quad \text{for } x \in \partial B_{r(t)}(0).$$

This is possible, in fact

$$r(t) = \frac{(1-\delta^{m-1})^{1/2}(t+t_0)^{1/k}}{\mu F^{1/2}}.$$

In view of our choice of μ and t_0, we have

$$r(0) = 4(1-\delta^{m-1})^{1/2}.$$

Thus we can fix δ so that $3 < r(0) < 4$. Since $\operatorname{supp} u(x,0) \subset B_1(0) \subset B_3(0)$, it follows from (8.7) that for small $t > 0$,

$$u(x,t) < u_\mu(x, t+t_0) \quad \text{for } x \in \partial B_{r(t)}(0).$$

We now distinguish two cases.

Case 1 There exist $t_1 \in (0,1)$, $x_1 \in \partial B_{r(t)}$ such that

$$u(x_1, t_1) = u_\mu(x_1, t_1 + t_0) = \delta u_\mu(0, t_1 + t_0).$$

Let $\rho(t) = r(t) - 2$. By Proposition 1.8.2, for $t \in (0, T)$,

$$\inf_{B_{\rho(t_1)}} u(x,t) \geq \max_{\partial B_{r(t_1)}} u(x,t).$$

In particular,

$$\inf_{B_{\rho(t_1)}} u(x,t) \geq u(x_1,t_1) = \delta u_\mu(0, t_1 + t_0).$$

Choose $L = L(m, N) > 0$ so large that

$$\operatorname{supp} u_{\mu L}(x, t_1 + t_0) = L^{-1} \operatorname{supp}_\mu(x, t_1 + t_0) \subset B_{\rho(t_1)}$$

and $L^{(m-1)/2} \geq \delta^{-1}$. Then

$$\begin{aligned}\max_{B_{\rho(t_1)}} u_{\mu L}(x, t_1 + t_0) &= L^{-(m-1)/2} u_\mu(0, t_1 + t_0) \\ &\leq \delta u_\mu(0, t_1 + t_0) \leq \inf_{B_{\rho(t_1)}} u(x, t_1), \\ u_{\mu L}(x, t_1 + t_0) &= 0 \leq u(x, t_1), \quad \text{for } x \in \mathbb{R}^N \backslash B_{\rho(t_1)}.\end{aligned}$$

Using the comparison theorem to $u_{\mu L}(x, t + t_0)$ and $u(x, t)$ on $\mathbb{R}^N \times (t_1, 1)$ gives

$$u_{\mu L}(x, t + t_0) \leq u(x, t), \quad \text{for}(x, t) \in \mathbb{R}^N \times (t_1, 1).$$

In particular,

$$\begin{aligned}u(0, 1) &\geq u_{\mu L}(0, 1 + t_0) = (L\mu)^{-2/(m-1)}(1 + t_0)^{N/k} \\ &\geq 2^{-(N-2)/k} G^{-2/k} L^{-2/(m-1)} M^{2/k}.\end{aligned}$$

Case 2 For all $t \in [0, 1]$, on $\partial B_{r(t)}$

$$u(x, t) < u_\mu(x, t + t_0).$$

In this case, since

$$u_\mu(x, t_0) \geq 0 = u(x, 0), \quad \text{for } |x| \geq r(0) > 3,$$

we can use the comparison theorem (Theorem 1.3.3) to $u_\mu(x, t + t_0)$ and $U(x, t)$ on $\mathbb{R}^N \backslash B_{r(t)}$ to obtain

$$u_\mu(x, t + t_0) \geq u(x, t), \quad \text{for } |x| \geq r(t), t \in [0, 1]. \tag{8.9}$$

Hence, in view of (8.6), we have

$$\begin{aligned}M &\leq \int_{\mathbb{R}^N} u(x, 0) dx = \int_{\mathbb{R}^N} u\left(x, \frac{1}{2}\right) dx \\ &= \int_{B_{r(1/2)}} u\left(x, \frac{1}{2}\right) dx + \int_{\mathbb{R}^N \backslash B_{r(1/2)}} u\left(x, \frac{1}{2}\right) dx \\ &\leq \int_{B_{r(1/2)}} u\left(x, \frac{1}{2}\right) dx + \frac{M}{2}.\end{aligned}$$

Notice that here we have also used the fact

$$\int_{\mathbb{R}^N} u(x, t) dx = \int_{\mathbb{R}^N} u(x, 0) dx \quad \text{for } t \in (0, 1),$$

which can be easily verified from the definition of generalized solutions, since u has compact support. Therefore we obtain

$$\int_{B_{r(1/2)}} u\left(x, \frac{1}{2}\right) dx \geq \frac{M}{2}. \tag{8.10}$$

We now distinguish two subcases.

Case 2a There exists $x \in \mathbb{R}^N \backslash B_3$ such that $u\left(x, \frac{1}{2}\right) \geq \gamma M^{2/k}$ with $\gamma > 0$ to be determined below.

By Proposition 1.8.2,

$$u\left(0, \frac{1}{2}\right) \geq \gamma M^{2/k}. \tag{8.11}$$

It is shown in Proposition 1.7.1 that $\dfrac{\partial u}{\partial t} \geq -\dfrac{ku}{t}$, that is, $\dfrac{\partial}{\partial t}(ut^k) \geq 0$ with $k = (m - 1 + 2/N)^{-1}$. Consequently, (8.11) implies

$$u(0,1) \geq u\left(0, \frac{1}{2}\right) e^{-k} \geq \gamma e^{-k} M^{2/k}.$$

Case 2b For all $x \in \mathbb{R}^N \backslash B_3$, $u\left(x, \frac{1}{2}\right) < \gamma M^{2/k}$.

In this case, in view of the choice of μ, t_0 and $r(t)$, there exists a constant $J = J(m, N)$ such that

$$\int_{B_{r(1/2)} \backslash B_3} u\left(x, \frac{1}{2}\right) dx < \gamma M^{2/k} \omega_N r^N \left(\frac{1}{2}\right) = \gamma JM. \tag{8.12}$$

In particular, we take $\gamma = \dfrac{1}{4J}$. Then since from (8.9), $u(x,t) \leq u_\mu(x, t+t_0)$ on $\mathbb{R}^N \backslash B_{r(1/2)}$, it follows from (8.6) and (8.12) that

$$\begin{aligned}
M &\leq \int_{\mathbb{R}^N} u(x,0)dx = \int_{\mathbb{R}^N} u\left(x, \frac{1}{2}\right) dx \\
&= \left(\int_{B_3} + \int_{B_{r(1/2)} \backslash B_3} + \int_{\mathbb{R}^N \backslash B_{r(1/2)}}\right) u\left(x, \frac{1}{2}\right) dx \\
&\leq \int_{B_3} u\left(x, \frac{1}{2}\right) dx + \frac{M}{4} + \frac{M}{2} \\
&= \int_{B_3} u\left(x, \frac{1}{2}\right) dx + \frac{3M}{4}.
\end{aligned}$$

Hence

$$\int_{B_3} u\left(x, \frac{1}{2}\right) dx \geq \frac{M}{4}.$$

This and Hölder's inequality yield

$$\int_{B_3} u^m\left(x, \frac{1}{2}\right) dx \geq KM^m$$

for some constant $K = K(m, N) > 0$, or

$$\fint_{B_3} u^m\left(x, \frac{1}{2}\right) dx \geq \nu\left(\frac{3^2}{\sigma}\right)^{m/(m-1)}$$

$$\nu = K\pi^{-1}3^{-N-2m/(m-1)}, \quad \sigma = \frac{1}{M^{m-1}}.$$

Thus, by Proposition 1.7.3, for some constant $\lambda > 0$, we have

$$u^m\left(0, \frac{1}{2} + \sigma\lambda\right) \geq c\left(\frac{3^2}{\sigma}\right)^{m/(m-1)} = c3^{2m/(m-1)}M^m,$$

$$c = \frac{\nu^{1/m}}{\lambda}.$$

If $M^{m-1} \geq 2\lambda$, then $\sigma\lambda \leq \frac{1}{2}$. Therefore, using the first inequality in Proposition 1.7.1 which implies that $t^{mk}u^m(x,t)$ is increasing, we obtain

$$u^m(0,1) \geq CM^m$$

provided $M^{m-1} \geq 2\lambda$. This will be the case if $M > M_0$ and $M_0 = (2\lambda)^{1/(m-1)}$.

To complete the proof of Theorem 1.8.1, we must remove the assumption (8.8).

For $l \in (0,1)$, let $\psi_l(r)$ be a monotone $C^\infty(R^+)$ function such that

$$\psi_l(r) = \begin{cases} 1 & \text{for } 0 \leq r \leq l, \\ 0 & \text{for } r \geq \dfrac{l+1}{2}. \end{cases}$$

Denote by w_l the generalized solution of (8.4) with $w_l(x,0) = \psi_l(|x|)u(x,0)$. Since w_l has compact support and satisfies (8.8), as we have proved above, there holds

$$w_l(0,1) \geq EM_l^{2/k} \tag{8.13}$$

provided that

$$M_l = \int_{B_1(0)} w_l(x,0)dx = \int_{B_1(0)} \psi_l(|x|)u(x,0)dx > M_0.$$

Since

$$\int_{B_1(0)} \psi(|x|)u(x,0)dx \to \int_{B_1(0)} u(x,0)dx = M \quad (l \to 1-),$$

it follows that for $l < 1$ sufficiently close to 1, $M > M_0$ implies that $M_l > M_0$ and hence (8.13) holds.

By Proposition 1.8.1, we have $u(x,t) \geq w_l(x,t)$; in particular, $u(0,1) \geq w_l(0,1)$, which together with (8.13)completes the proof of our theorem. □

Theorem 1.8.2 *Let u be a generalized solution of (8.4), which is continuous on $\overline{Q}_T$. Then there exists a constant C depending only on m, N, such that for any $\xi \in \mathbb{R}^N$,*

$$\int_{B_R(\xi)} u(x,0)dx \leq C(R^{k/(m-1)}T^{-1/(m-1)} + T^{N/2}u^{k/2}(\xi,T)), \tag{8.14}$$

where $k = N(m-1)+2$.

Proof. Fix $R > 0$. Define

$$u^*(x,t) = R^{-2/(m-1)}T^{1/(m-1)}u((x+\xi)R, tT).$$

It is easy to check that $u^*(x,t)$ is a continuous generalized solution of (8.4).

Without loss of generality, we suppose that ξ is the origin, Then

$$u^*(x,t) = R^{-2/(m-1)}T^{1/(m-1)}u(xR, tT).$$

It is easy to see that when $\xi = 0$, (8.14) turns out to be

$$\int_{B_1(0)} u^*(x,0)dx \equiv C(1 + u^*(0,1)^{k/2}).$$

Denote

$$M^* = \int_{B_1(0)} u^*(x,0)dx.$$

If $M^* > M_0$, then by Theorem 1.8.1, there exists $E = E(m,N) > 0$ such that

$$u^*(0,1) \geq E(M^*)^{2/k},$$

or

$$\int_{B_1(0)} u^*(x,0)dx \le E^{-k/2} \left(u^*(0,1)\right)^{k/2},$$

which means that (8.14) holds in this case. If

$$M^* = \int_{B_1(0)} u^*(x,0)dx \le M_0,$$

then we see that (8.14) holds for any $C \ge M$. Therefore if we take $C = \max\{M_0, E^{-k/2}\}$, then (8.14) holds in all cases. □

1.8.2 *Main result*

As an immediate corollary of Theorem 1.8.2, we can prove the existence of initial trace of generalized solutions.

Theorem 1.8.3 *Let u be a continuous solution of (8.4) on $Q_T = \mathbb{R}^N \times (0,T]$. Then there exists a unique nonnegative Radon measure μ on $\mathbb{R}^N$, such that for any $\varphi \in C_0(\mathbb{R}^N)$,*

$$\lim_{t\to 0^+} \int_{\mathbb{R}^N} \varphi(x)u(x,t)dx = \int_{\mathbb{R}^N} \varphi(x)d\mu.$$

Moreover, there exists a constant $C > 0$ depending only on m, N, such that for any $R > 0$

$$\int_{B_R(0)} d\mu \le C(R^{k/(m-1)}T^{-1/(m-1)} + T^{N/2}u^{k/2}(0,T)), \tag{8.15}$$

where $k = N(m-1)+2$.

Proof. The uniqueness is clearly valid. To prove the existence, we use Theorem 1.8.2. Then for any $\varepsilon \in (0,T)$,

$$\begin{aligned} &\int_{B_R} u(x,\varepsilon)dx \\ \le\ & C\left(R^{k/(m-1)}(T-\varepsilon)^{1/(m-1)} + T^{N/2}u^{k/2}(0,T)\right), \end{aligned} \tag{8.16}$$

where the constant C is independent of ε. From this it follows that there exists $\varepsilon_n \downarrow 0$, such that $u(x,\varepsilon_n)$ converges weakly to a measure μ, that is

$$\lim_{n\to\infty} \int_{B_R} \varphi(x)u(x,\varepsilon_n)dx = \int_{\mathbb{R}^N} \varphi(x)d\mu,$$

for all $\varphi \in C_0(\mathbb{R}^N)$. The fact that μ satisfies (8.15) is an easy consequence of(8.16). □

The measure μ whose existence has just been proved, may, of course depend on the sequence $\{\varepsilon_n\}$. The remainder of our argument is to prove that this is not the case. We will first prove the following elementary lemma.

Lemma 1.8.2 *Let u be a generalized solution of (8.4), which is continuous on $\overline{Q}_T$. If*

$$supp\, u(x,t) \subset B_1(0), \qquad \int_{\mathbb{R}^N} u(x,0)dx \le \nu,$$

then there exists a constant $P > 0$, depending only on N, such that for any $\eta > 0$, $u(x,t) \le \dfrac{\nu}{P\eta^{N-1/2}}$ in $\{\mathbb{R}^N \backslash B_{1+\eta}(0)\} \times [0,T]$.

Proof. For any two points x_0 and x_1 in $\mathbb{R}^N$, let $\Pi(x_0, x_1)$ be the hyperplane consisting of points which are equi-distant from x_0, and x_1. According to Lemma 1.8.1, we have

$$u(x_1,t) \ge u(x_0,t)$$

provided that x_0 and $B_1(0)$ lie in the same half-space, with respect to $\Pi(x_0, x_1)$.

Fix $x_0 \in \mathbb{R}^N \backslash B_{1+\eta}(0)$ and define

$$S(x_0) = \left\{ x; |x - x_0| < \frac{\eta}{2}, \cos\theta \ge \frac{1+\eta/4}{1+\eta} \right\},$$

where θ denotes the angle between $x_0 - x$ and x_0. For $x \in S(x_0)$,

$$\begin{aligned} \text{dist}\,(\Pi(x_0,x),\{0\}) &= |x_0|\cos\theta - \frac{1}{2}|x - x_0| \\ &\ge (1+\eta)\frac{1+\eta/4}{1+\eta} - \frac{\eta}{4} = 1. \end{aligned}$$

Thus, $x \in S(x_0)$ implies that x and $B_1(0)$ lie in the same half-plane with respect to $\Pi(x_0, x_1)$. We therefore have the estimate

$$\begin{aligned} \nu &\ge \int_{\mathbb{R}^N} u(x,0)dx = \int_{\mathbb{R}^N} u(x,t)dx \\ &\ge \int_{S(x_0)} u(x,t)dx \ge \text{mes}\, S(x_0) u(x_0,t) \\ &\ge P\eta^{3N/2-1} u(x_0,t) \end{aligned}$$

and the assertion follows from the fact that $|S(x_0)| \geq P(N)\eta^{N-1/2}$ with some constant $P(N) > 0$, which can be checked by calculation. The lemma is thus proved. □

Lemma 1.8.3 *Let u satisfy the hypothesis of Lemma 1.8.2. Then for any $\varepsilon > 0$, there exists a $\tau = \tau(m, N, \varepsilon, \nu, T) > 0$ such that $\operatorname{supp} u(x,t) \subset B_{1+\varepsilon}(0)$ for all $t \in [0, \tau]$.*

Proof. Let $u_\mu(x, t + t_0)$ be the Barenblatt solution of (8.4) used in the proof of Theorem 1.8.1. Given $\varepsilon > 0$, one can verify from the formula for u_μ that there exist $t_0 = t_0(m, N, \varepsilon, \nu, T) \in (0, T])$ and $\mu = \mu(m, N, \varepsilon, \nu, T) > 0$ such that

$$\operatorname{supp} u_\mu(x, t_0) = \overline{B_{1+\varepsilon/2}(0)},$$

and

$$\min_{t\in[0,T]} u_\mu(x, t + t_0)\Big|_{|x|=1+\varepsilon/4} \geq \frac{4^{2N-1}\nu}{P\varepsilon^{N-1/2}},$$

where P is the constant defined in Lemma 1.8.2. By hypothesis, $u(x, 0) = 0$ in $\mathbb{R}^N \backslash B_{1+\varepsilon/4}(0)$. Therefore

$$u(x, 0) = 0 \leq u_\mu(x, t_0), \quad \text{in } \mathbb{R}^N \backslash B_{1+\varepsilon/4}(0).$$

Moreover, by Lemma 1.8.2 and the construction of u_μ,

$$u(x, t) \leq \frac{4^{N-1/2}}{P\varepsilon^{N-1/2}} \leq u_\mu(x, t + t_0) \quad \text{on } \partial B_{1+\varepsilon/4}(0) \times [0, T].$$

It follows by the comparison theorem that

$$u(x, t) \leq u_\mu(x, t + t_0), \quad \text{in } \{\mathbb{R}^N \backslash B_{1+\varepsilon/4}(0)\} \times [0, T].$$

Thus the assertion holds for $\tau = \tau(m, N, \varepsilon, \nu, T)$ defined by

$$\operatorname{supp} u_\mu(\cdot, \tau + t_0) = \overline{B_{1+\varepsilon}(0)}.$$

□

Proposition 1.8.3 *Let u be a generalized solution of (8.4), which is continuous on $\overline{Q}_T$. If*

$$0 < \lambda < \int_{B_1(0)} u(x, 0)dx \leq \nu,$$

then for any $\varepsilon > 0$, there exists a $\tau = \tau(m, N, \varepsilon, \nu, T)$ such that

$$\int_{B_{1+\varepsilon}} u(x,t)dx \geq \lambda \quad \text{for } t \in [0, \tau].$$

Proof. Choose δ_1, δ_2, $0 < \delta_1 < \delta_2 < 1$, such that

$$\int_{B_{\delta_1}(0)} u(x,0)dx = \lambda.$$

Let w be the generalized solution of (8.4) with initial data

$$w(x,0) = \zeta(|x|)u(x,0),$$

where $\zeta \in C_0^\infty(R^+)$ satisfies $0 \leq \zeta(r) \leq 1$ and

$$\zeta(r) = \begin{cases} 1 & \text{for } 0 \leq r \leq \delta_1 \\ 0 & \text{for } r \geq \delta_2. \end{cases}$$

By Proposition 1.8.1, we have

$$w(x,t) \leq u(x,t) \quad \text{in } \overline{Q}_T.$$

Since $\operatorname{supp} w(x,0) \subset B_1(0)$ and

$$\int_{\mathbb{R}^N} w(x,0)dx = \int_{\mathbb{R}^N} \zeta(|x|)u(x,0)dx \leq \int_{B_1(0)} u(x,0) \leq \nu,$$

by Lemma 1.8.3, there exists a $\tau = \tau(m, N, \varepsilon, \nu, T) \in (0, T)$ such that

$$\operatorname{supp} w(x,t) \subset B_{1+\varepsilon}(0) \quad \text{for } t \in [0, \tau].$$

Therefore

$$\begin{aligned} \int_{B_{1+\varepsilon}(0)} u(x,t)dx &\geq \int_{B_{1+\varepsilon}(0)} w(x,t)dx = \int_{\mathbb{R}^N} w(x,t)dx \\ &= \int_{\mathbb{R}^N} w(x,0)dx \geq \int_{B_{\delta_1}(0)} w(x,0)dx \\ &= \int_{B_{\delta_1}(0)} u(x,0)dx = \lambda. \end{aligned}$$

The proof of our proposition is complete. □

Now we come back to proving that the measure μ does not depend on the sequence $\{\varepsilon_n\}$.

Suppose that u converges weakly to a measure μ along the sequence $\{\varepsilon_n\}$ and to a measure μ' along a sequence $\{\varepsilon'_n\}$, (8.15) must hold for both of these measures so that, in particular, they are regular.

If for any $r > 0$, there exists $t_0(r) \in (0, T)$ such that

$$u(x,t) = 0 \quad \text{for } (x,t) \in B_r(0) \times (0, t_0),$$

that is,

$$\int_{B_r(0)} u(x,t)dx = 0 \quad \text{for } t \in (0, t_0),$$

then clearly we have $\mu = \mu' = 0$. We need to treat only the case that for some $r_0 > 0$, there exists another sequence $\{t_n\}$ with $t_n \downarrow 0$ such that

$$u(x,t_n) \not\equiv 0 \quad \text{for } x \in B_{r_0}(0). \tag{8.17}$$

It is easy to prove that for any $t \in (0, T]$ there must hold

$$u(x,t) \not\equiv 0 \quad \text{for } x \in B_{r_0}(0). \tag{8.18}$$

To this purpose, we choose $\zeta(x) \in C_0^\infty(R^+)$ such that $0 < \zeta(r) \leq 1$ for $r \in [0, r_0)$ and $\zeta(r) = 0$ for $r \geq r_0$, Let w be the generalized solution with initial data $w(x,t_n) = \zeta(|x|)u(x,t_n)$. Applying Proposition 1.7.1 to w gives

$$w(x,t)t^k \geq w(x,t_n)t_n^k \quad \text{for } t \in (t_n, T).$$

This combining with (8.17) deduces that for any $t \in (t_n, T)$,

$$w(x,t) \not\equiv 0 \quad \text{for } x \in B_{r_0}(0).$$

Since by Proposition 1.8.1,

$$w(x,t) \leq u(x,t) \quad \text{for } (x,t) \in \mathbb{R}^N \times (t_n, T],$$

noting that $t_n \downarrow 0$, we conclude that (8.18) holds for $t \in (0, T]$. Thus

$$\int_{B_r} u(x,t)dx \geq \int_{B_{r_0}} u(x,t)dx > 0 \quad \text{for } t \in (0, T]$$

provided that $r \geq r_0$.

Now we prove that, for any $r \geq r_0$,

$$\mu(B_r(0)) = \mu'(B_r(0)).$$

First consider the case $r = 1$. Our task is to prove

$$\mu(B_r(0)) = \mu'(B_r(0)) \tag{8.19}$$

under the condition

$$\int_{B_1(0)} u(x,t)dx \neq 0 \quad \text{for } t \in (0,T].$$

Fix $s \in \left(0, \dfrac{T}{2}\right)$ and choose $\delta > 0$ sufficiently small so that

$$\lambda = \int_{B_1(0)} u(x,s)dx - \delta > 0.$$

For any $\varepsilon > 0$, by Proposition 1.8.3, there exists a $\tau = \tau(m, N, \varepsilon, \nu, T) > 0$ with $\nu = \displaystyle\int_{B_1(0)} u(x,s)dx$, such that

$$\int_{B_{1+\varepsilon}(0)} u(x,t+s)dx \geq \lambda = \int_{B_1(0)} u(x,s)dx - \delta \quad \text{for } t \in \left(0, \min\left\{\frac{T}{2}, \tau\right\}\right).$$

Taking $s = s_n$ and letting $n \to \infty$, we obtain

$$\int_{B_{1+\varepsilon}(0)} u(x,t)dx \geq \mu(B_1(0)) - \delta.$$

Now set $t = \varepsilon_n'$ and pass to the limit to get

$$\mu'(B_{1+\varepsilon}(0)) \geq \mu(B_1(0)) - \delta.$$

However, since both ε and δ are arbitrary we conclude that

$$\mu'(B_1(0)) \geq \mu(B_1(0)).$$

On the other hand, if we repeat the argument first taking the limit with $s = \varepsilon_n'$ followed by the limit with $t = \varepsilon_n$, we obtain

$$\mu(B_1(0)) \geq \mu'(B_1(0)).$$

Therefore (8,19) is valid. By a simple scaling argument, μ and μ' must agree on every ball in $\mathbb{R}^N$ so that $\mu = \mu'$.

1.8.3 *Extension of existence and uniqueness theorem*

Theorem 1.8.3 implies that if for some $T > 0$, the equation (8.4) admits a continuous generalized solution u on Q_T, satisfying the initial value condition

$$u(x,0) = u_0(x) \quad \text{for } x \in \mathbb{R}^N,$$

then $u_0(x)$ must satisfy the condition that for some $r_0 > 0$

$$\sup_{r \geq r_0} r^{-(N+2/(m-1))} \int_{B_r(0)} u_0(x)dx < \infty. \tag{8.20}$$

If the initial value is a measure μ, then Theorem 1.8.3 gives a necessary condition for (8.4) to have a continuous generalized solution on Q_T for some constant $T > 0$, which is similar to (8.20), that is, for some $r_0 > 0$

$$\sup_{r \geq r_0} r^{-(N+2/(m-1))} \int_{B_r(0)} d\mu < \infty. \tag{8.21}$$

It is natural to ask: is the condition (8.21) also sufficient for (8.4) to have a generalized solution with initial value μ? Bénilan, Crandall and Pierre [BCP] gave a positive answer to this problem; they proved that if the measure μ satisfies the condition (8.21), then for some $T > 0$, there exists a generalized solution of (8.4) on Q_T with μ as its initial value. The solution obtained is unbounded on Q_T in general, but is locally bounded. Dahlberg and Kenig [DK] further proved that the generalized solution obtained in [BCP] is the unique one of the Cauchy problem considered.

Remark 1.8.1 Some people have studied the initial trace of solutions for porous medium equations with convection term and absorption term, see for example, [DA], [HU3], [LHF2].

1.9 Other Problems

We will introduce briefly some other problems and related results to close the discussions of this chapter.

1.9.1 *Equations with strongly nonlinear sources*

In a variety of diffusion phenomena, people are required to study equations of the form

$$\frac{\partial u}{\partial t} = \Delta u^m + \lambda u^p, \tag{9.1}$$

where $m > 0$, $p > 0$, λ are constants. The nonlinear term λu^p in the equation describes the nonlinear source in the diffusion process; it is called "heat source" whenever $\lambda > 0$ and "cold source" or absorption term whenever $\lambda < 0$. Just as shown in the diffusion process, the appearance of nonlinear sources will exert a great influence to the properties of solutions. For instance, when the "heat source" occurs, the solutions of the equation (9.1) might be blowing-up, namely, the solutions might be unbounded at finite time. In this case, in order that the equation (9.1) has a generalized solution (its definition is similar to the case $\lambda = 0$), the condition on the growth of initial data

$$u(x, 0) = u_0(x) \geq 0 \quad \text{for } x \in \mathbb{R}^N \tag{9.2}$$

should be more restrictive, compared with the case $\lambda = 0$. Contrary, if the "cold source" occurs, then in order that the Cauchy problem (9.1), (9.2) has a generalized solution, the condition on the growth of $u_0(x)$ might be less restrictive, compared with the case $\lambda = 0$; in some circumstances the solutions even exist for any $u_0 \in L^1_{loc}(\mathbb{R}^N)$.

Let us first consider the case $\lambda > 0$.

For $f \in L^h_{loc}(\mathbb{R}^N)$ $(h \geq 1)$, denote

$$[f]_h = \sup_{x \in \mathbb{R}^N} \left(\int_{B_1(x)} |f(y)|^h dy \right)^{1/h},$$

where $B_1(x) = \{y \in \mathbb{R}^N; |x - y| < 1\}$. We have the following result on the local existence of solutions: if $m > 1$, $u_0 \in L^h_{loc}(\mathbb{R}^N)$ with $[u_0]_h < \infty$, where $h = 1$ whenever $1 < p < m + \frac{2}{N}$, $h > \frac{N}{2}(p - m)$ whenever $p \geq m + \frac{2}{N}$, then there exists a $T_0 = T_0(N, m, p, h)$, such that the Cauchy problem (9.1), (9.2) admits a generalized solution u on Q_{T_0} possessing the following properties:

$$[u(\cdot, t)]_h \leq \gamma [u_0]_h,$$

$$u(x, t) \leq \gamma t^{-N/\kappa_h} [u_0]_h^{2h/\kappa_h},$$

$$\int_0^t \int_{B_\rho(x_0)} |\nabla u^m|^\sigma dx d\tau$$
$$\leq \gamma t^{(1+(Nm+1)(1-\sigma))/\kappa} \left(\sup_{0<\tau<t} \int_{B_\rho(x_0)} u(x,\tau) dx \right)^{1+(\sigma(m+1)-2)/\kappa},$$
$$\text{for } 1 \leq \sigma \leq 1 + \frac{1}{Nm+1}$$

provided that T_0 satisfies

$$T_0 [u_0]_h^{m-1} + T_0^{1-N(p-m)/2h} [u_0]_h^{p-1} = \gamma_0^{-1}.$$

Here $\kappa_h = N(m+1) + 2h$, $\kappa = \kappa_1$, $\gamma = \gamma(N, m, p, h)$. The condition on the initial data u_0 depends on p; for the Cauchy problem.(9.1), (9.2) to have a local solution, the larger p is, the more restrictive the condition on u_0 should be. The basic idea of the proof is to consider the approximate problem

$$\frac{\partial u}{\partial t} = \Delta u^m + \min\{n, u^p\},$$
$$u(x,0) = u_{0n},$$

where $u_{0n} \in C_0^\infty(\mathbb{R}^N)$ with $[u_{0n}]_h \leq \gamma [u_0]_h$ and

$$\lim_{n\to\infty} \int_{B_\rho} |u_{0n} - u_0|^h dx = 0, \quad \text{for } \rho > 0,$$

and establish the locally uniform estimate on the bound of the approximate solution u_n. Such estimate can be obtained only for small t, because of the occurrence of the "heat source". From the proof it can be seen that the above result is also valid for the case that the initial value is a nonnegative Radon measure with

$$\sup_{x\in\mathbb{R}^N} \int_{B_1(x)} d\mu < \infty,$$

provided $1 < p < m + \dfrac{2}{N}$, $m > 1$.

The same method can be used to prove that if $p > m + \dfrac{2}{N}$, then for "small" initial value, the Cauchy problem (9.1), (9.2) admits a global solution. Precisely, we have the following result: if $p > m + \dfrac{2}{N}$, $h > \dfrac{N}{2}(p-m)$, then there exists a $\gamma_0 = \gamma_0(N, m, p, h)$, such that the Cauchy problem

(9.1), (9.2) admits a generalized solution on $Q = \mathbb{R}^N \times (0, \infty)$ provided that $u_0 \in L^1(\rho^N) \cap L^h(\mathbb{R}^N)$ and

$$\|u_0\|_{L^1(\mathbb{R}^N)} + \|u_0\|_{L^h(\mathbb{R}^N)} \le \gamma_0.$$

Based on the a priori estimate on the solutions of (9.1), it can be proved that if $1 < p < m + \dfrac{2}{N}$, then (9.1) does not admit any nontrivial global solution.

The uniqueness of solutions of the Cauchy problem (9.1), (9.2) is established in the class S of functions. Here we say that $u \in S$, if for some constants C, $\delta < (\max\{p, m\} - 1)^{-1}$,

$$[u(\cdot, t)]_1 \le C \quad \text{for } t \in (0, T),$$

$$\sup_{x \in \mathbb{R}^N} u(x, t) \le C t^{-\delta} \quad \text{for } t \in (0, T).$$

For the proofs of the above results, we refer to [AD].

Some results have been obtained in [ZL] for the case $\lambda < 0$. First, the existence of generalized solutions of the Cauchy problem (9.1), (9.2) on $Q = \mathbb{R}^N \times (0, \infty)$ is established in the following cases:

(i) $1 < m < p < m + \dfrac{2}{N}$ and the initial value is a Radon measure on $\mathbb{R}^N$;

(ii) $p > m > 1$ and $u_0 \in L^1_{loc}(\mathbb{R}^N)$.

It can be proved that if $p \ge m + \dfrac{2}{N}$, $u_0 = \delta(x)$, then the Cauchy problem (9.1), (9.2) does not admit any generalized solution. This means that if the initial value is a measure, then for the Cauchy problem (9.1), (9.2) to have a solution, the condition $p < m + \dfrac{2}{N}$ is not only sufficient, but also necessary. It is worth pointing out that if $p > m > 1$, then the equation (9.1) admits a global solution for any $u_0 \in L^1_{loc}(\mathbb{R}^N)$. This situation is completely different to the case $\lambda = 0$.

If $p \le m$, then for the Cauchy problem (9.1), (9.2) to have a solution, some restriction on the growth of the initial value u_0 should be made. Moreover the smaller p is, the more restrictive the condition on u_0 should be, which is

$$\int_{\mathbb{R}^N} \exp\{-\sqrt{1 + |x|^2}\} u_0(x) dx < \infty,$$

whenever $p = m$ and

$$u_0 \in L^\infty_{loc}(\mathbb{R}^N),\ u_0(x) \le C_1(C_2 + |x|^2)^{1/(m-p)} \quad \text{a.e on } \mathbb{R}^N \tag{9.3}$$

whenever $1 < p < m$, where C_1, C_2 are positive constants with

$$C_1 \le \left(\frac{(m-p)^2}{2Nm(m-p) + 4mp}\right)^{1/(m-p)}.$$

Under such conditions, it is proved in [ZL] that the Cauchy problem (9.1), (9.2) admits a generalized solution on $Q = \mathbb{R}^N \times (0, \infty)$.

The condition (9.3) requires the growth of u_0 not exceeding $|x|^{2/(m-p)}$, which is less restrictive compared with the case $\lambda = 0$. In later case, the condition on u_0 is that for some $r > 0$,

$$\sup_{\rho \ge r} \rho^{-N-2/(m-1)} \int_{B_\rho} u_0(x)dx < \infty,$$

roughly speaking, this means that the growth of u_0 is required to not exceed $|x|^{2/(m-1)}$.

It is also pointed out in [ZL] that for the Cauchy problem (9.1), (9.2)to have a solution, the condition (9.3) is almost the best possible. In fact, if $u_0 \in L^\infty_{loc}(\mathbb{R}^N)$ and for some constant $\alpha > \dfrac{2}{m-p}$,

$$\liminf_{|x| \to \infty} \frac{u_0(x)}{|x|^\alpha} > 0,$$

then the Cauchy problem (9.1), (9.2) does not admit any generalized solution.

In the case $\lambda = 0$, according to a general theorem on the existence and uniqueness (see §1.1.3, §1.1.8, which is indeed valid for $m > \left(1 - \dfrac{2}{N}\right)_+$), the equation (9.1) admits a unique generalized solution u satisfying the initial condition

$$u(x, 0) = E\delta(x)$$

with some constant $E > 0$. Such kind of solutions are called **source-type solutions**, which possess singularity near $(0,0)$. This result can be extended to the case $\lambda < 0$, $0 < p < m + \dfrac{2}{N}$, namely, in this case, the equation (9.1) also admits a source-type solution.

In the case $\lambda < 0$, the equation (9.1) not only admits source-type solutions, but also admits very singular solutions if $m > \left(1 - \frac{2}{N}\right)_+$. By a very singular solution U, we mean a solution U possessing the following properties:

$$U \in C(\overline{Q}_T \backslash \{0,\}), \quad U(x,0) = 0 \quad \text{for } x \in \mathbb{R}^N \backslash \{0\},$$

$$\lim_{t \to 0} \int_{B_R} U(x,t)dx = +\infty \quad \text{for } R > 0.$$

This result shows that in the present case, the conclusion on the initial trace of solutions in §1.1.8 is no longer valid. In other words, to ensure the existence of generalized solutions of (9.1), it is possible to relax the condition on u_0.

To prove the existence of very singular solutions, one considers the generalized solution u_k of the equation (9.1) with initial data

$$u(x,0) = k^{N+1}h(kx)$$

where $h(x) \geq 0$ such that $\int_{\mathbb{R}^N} h(x)dx = 1$. The proof is based on the uniform estimate on the bound of u_k on any compact subset of $\overline{Q} \backslash \{0,0\}$. In the proof, the absorption term λu^p and the condition $m < p < m + \frac{2}{N}$ plays key roles. For details, see [KP1], [PT], [PZ1]. A similar problem for more general equation has been discussed in [Z] by means of different method. For other related works, see [LS1], [LS2], [LS3], [SBC] in one dimensional case and [CC], [KM1], [LE], [ZH12] in multi-dimensional case.

1.9.2 *Asymptotic properties of solutions*

In the study of asymptotic properties of solutions, one of the basic problems is to analyse the relationship between the behavior of solutions for large time and the asymptotic behavior of the initial value u_0 as $|x| \to \infty$.

Assume that

$$\lim_{|x| \to \infty} |x|^\alpha u_0(x) = A,$$

where $\alpha > 0$ and $A > 0$ are constants. Denote

$$\gamma = \alpha(m-1) + 2, \quad \mu = m - 1 + \frac{2}{N}, \quad \beta = \frac{2(p-1)}{p-m}.$$

Again assume that $m > \left(1 - \frac{2}{N}\right)_+$.

In the case $\lambda = 0$, we have the following results (see [FK]):

(1) If $0 \le \alpha < N$, then the solution u of (9.1), (9.2) satisfies

$$t^{\alpha/\gamma}|u(x,t) - w_A(x,t)| \to 0 \quad \text{as } t \to \infty$$

uniformly on $\{x \in \mathbb{R}^N; |x| \le bt^{1/\gamma}\}$ with some constant $b > 0$, where $w_A(x,t)$ is the solution of (9.1) with initial data

$$u(x,0) = A|x|^{-\alpha}. \tag{9.4}$$

By the uniqueness of solutions of (9.1),(9.4),

$$W_A = t^{-\alpha/\gamma} f(\eta, A),$$

where $\eta = |x|t^{-1/\gamma}$ and f is the solution of the problem

$$\begin{cases} (f^m)'' + \dfrac{N-1}{\eta}(f^m)' + \dfrac{1}{\eta} f' + \dfrac{\alpha}{\gamma} f = 0, \ \eta > 0, \\ f'(0) = 0, \ f \ge 0, \ \lim\limits_{\eta \to \infty} \eta^\alpha f(\eta) = A. \end{cases}$$

(2) If $\alpha > N$, then the solution of (9.1),(9.2) satisfies

$$t^{1/\mu}|u(x,t - E_c(x,t))| \to 0, \quad \text{as } t \to \infty$$

uniformly on $\{x \in \mathbb{R}^N; |x| \le bt^{1/N\mu}\}$ with some constant $b > 0$, where $c = \|u_0\|_{L^1(\mathbb{R}^N)}$, E_c is the Barenblatt solution of (9.1),

$$E_c = t^{-1/\mu}\left(c_0 - \frac{(m-1)|x|^2}{2mN\mu t^{2/\mu N}}\right)_+^{1/(m-1)} \tag{9.5}$$

with a constant $c_0 > 0$ such that $\int_{\mathbb{R}^N} E_c(x,t)dx = c$.

In the case $\lambda < 0$, in view of the occurrence of the absorption, the behavior of the solution of (9.1), (9.2) as $t \to \infty$ depends not only on the asymptotic behavior of u_0 as $|x| \to \infty$, but also on the "competition" of diffusion and absorption. We have the following results (see [KP2], [PZ2]):

(1)If $p > \max\{m, 1\}, 0 < \alpha < \frac{2}{p-m}$, then the solution u of (9.1), (9.2) satisfies

$$t^{1/(p-1)}u(x,t) \to \left(\frac{1}{p-1}\right)^{1/(p-1)} \quad \text{as } t \to \infty$$

uniformly on $\{x \in \mathbb{R}^N; |x| < bt^{1/\beta}\}$ with some constant $b > 0$.

(2) If $p > m + \frac{2}{N}$, $\frac{2}{(p-m)} < \alpha < N$, then the solution u of (9.1),(9.2) satisfies

$$t^{\alpha/\gamma}|u(x,t) - w_A(x,t)| \to 0 \quad \text{as } t \to \infty$$

uniformly on $\{x \in \mathbb{R}^N; |x| \le bt^{1/\gamma}\}$ with some constant $b > 0$, where w_A is the solution of (9.1), (9.4) with $\lambda = 0$.

(3) If $p > m + \frac{2}{N}$, $\alpha > N$, then the solution u of (9.1), (9.2) satisfies

$$t^{1/\nu}|u(x,t) - E_c(x,t)| \to 0 \quad \text{as } t \to \infty$$

uniformly on $\{x \in \mathbb{R}^N; |x| \le bt^{1/N\nu}\}$ with some constant $b > 0$, where $c = \int_{\mathbb{R}^N} u_0(x,t)dx + \lambda \int_{\mathbb{R}^N} u^p(x,t)dxdt$, E_c is given by (9.5) with c_0 such that $c_0 \int_{\mathbb{R}^N} E_c(x,t)dx = C$.

(4) If $\max\{m, 1\} < p < m + \frac{2}{N}$, $\alpha > \frac{2}{p-m}$, then the solution u of (9.1), (9.2) satisfies

$$t^{1/(p-m)}|u(x,t) - U(x,t)| \to 0 \quad \text{as } t \to \infty$$

uniformly on $\{x \in \mathbb{R}^N; |x| \le bt^{1/\beta}\}$ with some constant $b > 0$, where U is a very singular solution of (9.1).

Another basic problem in the study of asymptotic properties is to discuss whether the solutions of the evolution equation considered "tend" and in what sense "tend" to the solutions of the corresponding stationary equation.

As an example, we introduce a result in [ACP]. Consider the boundary value problem

$$\begin{cases} \dfrac{\partial u}{\partial t} = \dfrac{\partial^2 u^m}{\partial x^2} + \psi(u), \quad (m > 1) \\ u|_{x=0,1} = 0, \\ u|_{t=0} = u_0(x,t), \end{cases} \tag{9.6}$$

where $0 \le u_0 \le 1$. Let $u = u(t; u_0)$ be its solution. Define the distance

$$d(u,v) = \|u - v\|_{L^1(0,1)} + \left\| \frac{\partial u^m}{\partial x} - \frac{\partial v^m}{\partial x} \right\|_{L^2(0,1)}$$

in the space

$$X = \{u \in L^\infty(0,1); 0 \le u \le 1, \frac{\partial u^m}{\partial x} \in L^2(0,1)\}.$$

Let $\omega(u_0)$ be the ω-limit set of $u(t,u_0)$:

$$\omega(u_0) = \{\omega \in X; \text{there exists } t_n \to \infty \text{ such that } u(t_n,u_0) \to \omega \text{ in } X\}.$$

Denote

$$\gamma_\tau(u_0) = \{u(t,u_0); t \ge \tau\} \quad (\tau \ge 0).$$

It is proved in [ACP] that for each $\tau > 0$, $\gamma_\tau(u_0)$ is precompact in X and $\omega(u_0) \subset E$, where E is the set of solutions of the stationary problem

$$\begin{cases} \dfrac{d^2u^m}{dx^2} + \psi(u) = 0, \\ u|_{x=0,1} = 0. \end{cases}$$

The proof is based on a usage of the Liapunov functional

$$V(w) = \frac{1}{2}\int_0^1 (w^m)'^2 dx - \int_0^1 F^*(w)dx,$$

where $F^*(w) = \dfrac{1}{m}\displaystyle\int_0^w \rho^{m-1}\psi(\rho)d\rho$. Under some conditions on u_0, the ω-limit set of $u(t,u_0)$ contains only one point.

Remark 1.9.1 In addition to those discussed above, there are many interesting problems which have been studied by many authors, among them are various kinds of singular limits, see for example, [AB], [DI], [GP], [HU4]–[HU9].

Chapter 2

Non-Newtonian Filtration Equations

2.1 Introduction Preliminary Knowledge

2.1.1 *Introduction Physical example*

This chapter is devoted to a study of the non-Newtonian filtration equation

$$\frac{\partial u}{\partial t} = \operatorname{div}\left(|\nabla u|^{p-2}\nabla u\right), \tag{1.1}$$

which is also called evolution p-Laplacian equation. Sometimes we also talk about the so-called polypropic filtration equation

$$\frac{\partial u}{\partial l} = \operatorname{div}\left(|\nabla u^m|^{p-2}\nabla u^m\right). \tag{1.2}$$

Here $m > 0$, $p > 1$, $\nabla u = \left(\frac{\partial u}{\partial x_1}, \frac{\partial u}{\partial x_2}, \cdots, \frac{\partial u}{\partial x_N}\right)$. If we do not restrict ourselves to the nonnegative solutions, then (1.2) should be written as

$$\frac{\partial u}{\partial t} = \operatorname{div}\left(|\nabla(|u|^{m-1}u)|^{p-2}\nabla(|u|^{m-1}u)\right). \tag{1.3}$$

Just as the Newtonian filtration equation, equations (1.1), (1.2), (1.3) have been the subject of intensive study in the last three decades.

We will first illustrate the physical background of these equations with an example.

Suppose a compressible fluid flows in a homogeneous isotropic rigid porous medium. Then the volumetric moisture content θ, the seepage velocity $\overrightarrow{V}$ and the density of the fluid are governed by the continuity equation

$$\theta\frac{\partial\rho}{\partial t} + \mathrm{div}(\rho\overrightarrow{V}) = 0. \tag{1.4}$$

For non-Newtonian fluid, the linear Darcy's law is no longer valid, because the influence of many factors such as the molecular and ion effects needs to be concerned. Instead, one has the following nonlinear relation

$$\rho\overrightarrow{V} = -\lambda|\nabla P|^{\alpha-1}\nabla P, \tag{1.5}$$

where $\rho\overrightarrow{V}$ and P denote the momentum velocity and pressure respectively, $\lambda > 0$ and $\alpha > 0$ are some physical constants.

If the fluid considered is the polytropic gas, then the pressure and density satisfy the following equation of state

$$P = c\rho^\gamma,$$

where $c > 0$, $\gamma > 0$ are some constants. Thus from (1.4) and (1.5)we obtain

$$\theta\frac{\partial\rho}{\partial t} = c^\alpha\lambda\mathrm{div}\left(|\nabla\rho^\gamma|^{\alpha-1}\nabla\rho^\gamma\right),$$

which is just (1.2) after changing variables and notations.

The equation (1.2) can be written as

$$\frac{\partial u}{\partial t} = a^{ij}\frac{\partial^2 u}{\partial x_i\partial x_j} + m^{p-1}(m-1)(p-1)u^{mp-p-m}|\nabla u|^p,$$

where

$$a^{ij} = m^{p-1}u^{(m-1)(p-1)}|\nabla u|^{p-2}\left(\delta_{ij} + (p-2)|\nabla u|^{-2}\frac{\partial u}{\partial x_i}\frac{\partial u}{\partial x_j}\right).$$

Here and in the sequel, double indices imply a summation from 1 to N. Clearly, for $\xi \in \mathbb{R}^N$,

$$\min\{1, p-1\}a_0(u, \nabla u)|\xi|^2 \le a^{ij}\xi_i\xi_j \le \max\{1, p-1\}a_0(u, \nabla u)|\xi|^2,$$

where

$$a_0(u, \nabla u) = m^{p-1}u^{(m-1)(p-1)}|\nabla u|^{p-2}.$$

If $a_0(u,\nabla u)\neq 0$, then

$$\lim_{\kappa\to 0} a_0(\kappa u,\nabla(\kappa u))=\begin{cases} 0, & \text{for } p>1+1/m,\\ \infty, & \text{for } p<1+1/m.\end{cases}$$

The case $p>1+\frac{1}{m}$ will be called the slow diffusion and the case $p<1+\frac{1}{m}$, the fast diffusion. Since the equations (1.1) and (1.2) possess degeneracy in the slow diffusion case and possess singularity in the fast diffusion case, they do not admit classical solutions in general.

The study of the p-Laplacian equation (1.1) started twenty years ago (see [L], [MP1],[MP2]). In recent years, rapid development has been achieved for the study of this equation along with the deep going investigation of the Newtonian equation. Theory on the uniqueness and existence of solutions, regularity of solutions and their interfaces, the initial trace problem and so on is being perfected. (see [AE], [CH3], [CH4], [CD1], [CD2], [DF], [DH1], [DH2], [ZH6], [ZH7], [ZY3]). Progress has been made also for the polytropic filtration equation (1.2) ([YU1], [ZY1], [ZX]). In this chapter we will introduce part of the basic content of the theory for the equations (1.1) and (1.2). In fact, more attention is paid only to the equation (1.1). As for the equation (1.2), we merely mention something briefly about its study. To save space, we almost restrict ourselves to the Cauchy problem, however, the initial data

$$u(x,0)=u_0(x),\quad \text{for } x\in\mathbb{R}^N, \tag{1.6}$$

might be even a Radon measure, usually assumed to be nonnegative.

2.1.2 *Basic spaces and some lemmas*

Let $\Omega\subset\mathbb{R}^N$ be a domain. For $T\in(0,\infty)$, denote $\Omega_T=\Omega\times(0,T)$. If $\Omega=\mathbb{R}^N$, then denote $Q_T=\mathbb{R}^N\times(0,T)$ instead of Ω_T.

As usual, the space $W^{1,p}(\Omega)$ $(p\geq 1)$ is the completion of $C^\infty(\Omega)$ with the norm

$$\|u\|_{W^{1,p}(\Omega)}=\|u\|_{p,\Omega}+\|\nabla u\|_{p,\Omega},$$

where $\|u\|_{p,\Omega}$ denotes the $L^p(\Omega)$ norm of u; the space $W_0^{1,p}(\Omega)$ $(p\geq 1)$ is the completion of $C_0^\infty(\Omega)$ with the norm

$$\|u\|_{W_0^{1,p}(\Omega)} = \|\nabla u\|_{p,\Omega}.$$

It is well-known that $W^{1,p}(\Omega)$ is equivalent to the space consisting of all functions which together with their first order weak derivatives, belong to $L^p(\Omega)$.

Let $q, r \geq 1$. We say that $u \in L^{q,r}(\Omega_T) \equiv L^r(0,T;L^q(\Omega))$, if u is measurable on Ω_T and $u(\cdot,t) \in L^q(\Omega)$ for almost all $t \in (0,T)$ and $\|u(\cdot,t)\|_{q,\Omega} \in L^r(0,T)$, that is,

$$\|u\|_{q,r,\Omega_T} = \left(\int_0^T \left(\int_\Omega |u|^q dx\right)^{r/q} dt\right)^{1/r} < \infty.$$

It is clear that $L^{q,q}(\Omega_T) = L^q(\Omega_T)$, $\|u\|_{q,q,\Omega_T} = \|u\|_{q,\Omega_T}$. We say that $u \in L^{q,r}_{loc}(\Omega_T)$, if for any compact subset of Ω and $[t_1,t_2] \subset (0,T)$, $u(\cdot,t) \in L^q(K)$ for almost all $t \in (t_1,t_2))$ and $\|u(\cdot,t)\|_{q,K} \in L^r(t_1,t_2)$.

$C(0,T;L^q(\Omega))(q \geq 1)$ is defined as the space of all measurable functions u on Ω_T such that for all $t \in [0,T]$, $u(\cdot,t) \in L^q(\Omega)$ and $u(\cdot,t)$ is a continuous function from $[0,T]$ to $L^q(\Omega)$, that is

$$\lim_{h\to 0} \|u(\cdot,t+h) - u(\cdot,t)\|_{q,\Omega} = 0.$$

Let $p, q \geq 1$. We say that $u \in L^q(0,T;W^{1,p}(\Omega))$, if u is measurable on Ω_T and for almost all $t \in (0,T)$, $u(\cdot,t) \in W^{1,p}(\Omega)$ and $\|u(\cdot,t)\|_{W^{1,p}(\Omega)} \in L^q(0,T)$. The spaces

$$L^q(0,T;W_0^{1,p}(\Omega)), \quad L^q_{loc}(0,T;W^{1,p}_{loc}(\Omega)), \quad C_{loc}(0,T;L^q(\Omega))$$

are defined in an obvious way.

Denote

$$V^{q,p}(\Omega_T) \equiv L^\infty(0,T;L^q(\Omega)) \cap L^p(0,T;W^{1,p}(\Omega)),$$

$$V_0^{q,p}(\Omega_T) \equiv L^\infty(0,T;L^q(\Omega)) \cap L^p(0,T;W_0^{1,p}(\Omega));$$

their norms are defined by

$$\|u\|_{V^{q,p}(\Omega_T)} = \text{ess} \sup_{0<t<T} \|u(\cdot,t)\|_{q,\Omega} + \|\nabla u(\cdot,t)\|_{p,\Omega_T}.$$

We have the following embedding inequality (see [LSU] p.62).

Lemma 2.1.1 *There exists a constant γ depending only on p, q, N such that for $v \in V_0^{q,p}(\Omega_T)$,*

$$\iint_{\Omega_T} |v|^h dxdt \le \gamma \left(\iint_{\Omega_T} |\nabla v|^p dxdt \right) \left(esssup \int_\Omega |v(x,t)|^q dx \right)^{p/N}, \quad (1.7)$$

where $h = \dfrac{p(q+N)}{N}$. Moreover,

$$\|v\|_{h,\Omega_T} \le \gamma \|v\|_{V^{q,p}(\Omega_T)}.$$

Let $v \in L^1(\Omega_T)$. For $0 < t < T$, define

$$v_h = v_h(x,t) = \begin{cases} \dfrac{1}{h} \displaystyle\int_t^{t+h} v(\cdot, \tau) d\tau, & \text{if } t \in (0, T-h), \\ 0, & \text{if } t > T-h, \end{cases}$$

$$v_{\overline{h}} = v_{\overline{h}}(x,t) = \begin{cases} \dfrac{1}{h} \displaystyle\int_{t-h}^{t} v(\cdot, \tau) d\tau, & \text{if } t \in (h, T), \\ 0, & \text{if } t < h, \end{cases}$$

which are called the Steklov mean value of v.

Lemma 2.1.2 *Let $v \in L^{q,r}(\Omega_T)$. Then for any $\varepsilon \in (0,T)$,*

$$v_h \to v, \quad \textit{in } L^{q,r}(\Omega_{T-\varepsilon}) \textit{as } h \to 0. \quad (1.8)$$

If $v \in C(0,T; L^q(\Omega))$, then for any $\varepsilon \in (0,T)$ and $t \in (0, T-\varepsilon)$,

$$v_h(\cdot, t) \to v(\cdot, t) \quad \text{in } L^q(\Omega) \text{as } h \to 0. \quad (1.9)$$

Proof. Since

$$v_h(\cdot, t) = \int_0^t v(\cdot, t + h\tau) d\tau,$$

using Minkowski's inequality, we have

$$\|v_h(x,t) - v(x,t)\|_{q,\Omega} \le \| \int_0^1 |v(\cdot, t+hs) - v(x,t)| ds\|_{q,\Omega}$$

$$\le \int_0^1 \|v(\cdot, t+hs) - v(x,t)\|_{q,\Omega} ds, \quad (1.10)$$

from which it follows that (1.9) holds for $v \in C(0,T;L^q(\Omega))$.

From (1.10), using Minkowski's inequality again yields

$$\begin{aligned}
&\left(\int_0^{T-\varepsilon}\left(\int_\Omega |v_h(x,t)-v(x,t)|^q dx\right)^{r/q} dt\right)^{1/r}\\
\le &\left(\int_0^{T-\varepsilon}\left(\int_0^1 \|v(x,t+sh)-v(x,t)\|_{q,\Omega} ds\right)^r dt\right)^{1/r}\\
\le &\int_0^1\left(\int_0^{T-\varepsilon} \|v(x,t+sh)-v(x,t)\|_{q,\Omega}^r dt\right)^{1/r} ds.
\end{aligned}$$

By the global continuity of functions in $L^{q,r}(Q_T)$ we see that the right hand side of the above inequality tends to zero as $h \to 0$ and thus (1.8) holds. □

Lemma 2.1.3 *Let $v \in W^{1,p}(\Omega)$. Then for any $k \in \mathbb{R}$, $(v-k)_\pm$, $|v-k| \in W^{1,p}(\Omega)$ and*

$$\begin{aligned}
\nabla(v-k)_+ &= \begin{cases} \nabla v, & \text{if } v > k,\\ 0, & \text{if } v \le k,\end{cases}\\
\nabla(v-k)_- &= \begin{cases} \nabla v, & \text{if } v < k,\\ 0, & \text{if } v \ge k,\end{cases} \qquad (1.11)\\
\nabla|v-k| &= \begin{cases} \nabla v, & \text{if } v > k,\\ -\nabla v, & \text{if } v < k.\end{cases}
\end{aligned}$$

In addition, if the trace of v on $\partial\Omega$ is essentially bounded and $\|v\|_{\infty,\partial\Omega} \le k_0$, then $(v-k)_+ \in W_0^{1,p}(\Omega)$ whenever $k > k_0$ and $(v-k)_- \in W_0^{1,p}(\Omega)$ whenever $k < -k_0$.

Proof. It is easily seen that if $f(s) \in C^1(R)$, $f'(s) \in L^\infty(R)$, $v \in W^{1,p}(\Omega)$, then $f(v) \in W^{1,p}(\Omega)$ and $\nabla f(v) = f'(v)\nabla v$.

For any $\varepsilon > 0$, choose

$$f_\varepsilon(s) = \begin{cases} [(s-k)^2+\varepsilon^2]^{1/2} - \varepsilon, & \text{if } s > k,\\ 0, & \text{if } s \le k.\end{cases}$$

It is easy to check that $f_\varepsilon(s) \in C^1(R)$, $f_\varepsilon'(s) \in L^\infty(R)$. Thus $f_\varepsilon(v) \in W^{1,p}(\Omega)$ and $\nabla f_\varepsilon(v) = f_\varepsilon'(v)\nabla v$. Hence for any $\varphi \in C_0^\infty(\Omega)$,

$$\int_\Omega f_\varepsilon(v)\nabla\varphi dx = -\int_\Omega f_\varepsilon'(v)\nabla v \cdot \varphi dx = -\int_{v>k} \frac{\varphi(v-k)\nabla v}{((v-k)^2+\varepsilon^2)^{1/2}} dx.$$

Let $\varepsilon \to 0$ and pass to the limit,

$$\int_\Omega (v-k)_+ \nabla\varphi dx = -\int_{v>k} \nabla v \cdot \varphi dx.$$

This shows that $\nabla(v-k)_+$ exists and

$$\nabla(v-k)_+ = \begin{cases} \nabla v, & \text{if } v > k, \\ 0, & \text{if } v \le k. \end{cases}$$

Since $(v-k)_- = (-v+k)_+$, $|v-k| = (v-k)_+ + (v-k)_-$, the assertion for $(v-k)_-$ and $|v-k|$ also holds. Noting that if $\|v\|_{\infty,\partial\Omega} \le k_0$, then the trace of $(v-k)_+$ on $\partial\Omega$ is zero whenever $k > k_0$ and the trace of $(v-k)_-$ is zero whenever $k < -k_0$, we conclude that $(v-k)_- \in W_0^{1,p}(\Omega)$ and $(v-k)_+ \in W_0^{1,p}(\Omega)$ whenever $k > k_0$ and $k < -k_0$ respectively. □

Lemma 2.1.4 *Let $f(t)$ be a nonnegative and bounded function on $[r_0, r_1]$ with $r_0 \ge 0$. If for $r_0 \le t < s \le r_1$,*

$$f(t) \le \theta f(s) + (A(s-t)^{-\alpha} + B), \tag{1.12}$$

where A, B, α, θ are nonnegative constants and $0 \le \theta < 1$, then for $r_0 \le \rho < R \le r_1$,

$$f(\rho) \le C(A(R-\rho)^{-\alpha} + B), \tag{1.13}$$

where C is a constant depending only on α, θ.

Proof. For fixed ρ, R, choose $\{t_j\}$ as follows

$$t_0 = \rho, \qquad t_{j+1} = t_j + (1-r)r^j(R-\rho).$$

Using (1.12) yields

$$f(t_0) \le \theta^n f(t_n) + (A(1-r)^{-\alpha}(R-\rho)^{-\alpha} + B)\sum_{j=0}^{n-1} \theta^j r^{-j\alpha}.$$

Choosing r such that $r^{-\alpha}\theta < 1$ and letting $n \to \infty$ we deduce (1.13), where

$$C = C(\alpha, \theta) = (1-r)^{-\alpha}(1-\theta r^{-\alpha})^{-1}.$$

□

Lemma 2.1.5 *Let $Q_n (n = 0, 1, 2, \cdots)$ be a sequence of bounded open subsets in Ω_T such that $Q_{n+1} \subset Q_n$. If for any $q \geq 1$, $v \in L^q(Q_0)$ and for some constants $\alpha_0 \geq 0$, λ, C_0, $C_1 > 0$, $K > 1$,*

$$\iint_{Q_{n+1}} |v|^{\alpha_0+\lambda K^{n+1}} dxdt \leq \left(C_0 C_1^n \iint_{Q_n} |v|^{\alpha_0+\lambda K^n} dxdt \right)^K, \tag{1.14}$$

then

$$\operatorname*{ess\,sup}_{Q_\infty} |v| \leq \left(C_0^{K/(K-1)} \overline{C}_1 \iint_{Q_{n_0}} |v|^{\alpha_0+\lambda K^{n_0}} dxdt \right)^{1/\lambda K^{n_0}}, \tag{1.15}$$

where $\overline{C}_1 = C_1^{K_1}$, $K_1 = \sum\limits_{n=n_0}^{\infty} nK^{-(n-n_0)}$ and n_0 is an arbitrary nonnegative integer.

Proof. From (1.14),

$$\begin{aligned} &\iint_{Q_{n+1}} |v|^{\alpha_0+\lambda K^{n+1}} dxdt \\ \leq\ & \left(C_0^{K_2} C_1^{K_3} \iint_{Q_{n+1}} |v|^{\alpha_0+\lambda K^{n_0}} dxdt \right)^{K^{n-n_0+1}}, \end{aligned}$$

where $K_2 = \sum\limits_{j=1}^{n+1-n_0} K^j$, $K_3 = \sum\limits_{j=n_0}^{n} jK^{n+1-j}$, that is,

$$\begin{aligned} &\left(\iint_{Q_{n+1}} |v|^{\alpha_0+\lambda K^{n+1}} dxdt \right)^{1/(\alpha_0+\lambda K^{n+1})} \\ \leq\ & \left(C_0^{K_4} \overline{C}_1 \iint_{Q_{n_0}} |v|^{\alpha_0+\lambda K^{n_0}} dxdt \right)^{(K^{n-n_0+1})/(\alpha_0+\lambda K^{n+1})}, \end{aligned}$$

where $K_4 = \sum_{j=n_0}^{n} K^{n_0-j}$. Letting $n \to \infty$ and noticing that

$$\lim_{n\to\infty}\left(\iint_{Q_{n+1}} |v|^{\alpha_0+\lambda K^{n+1}} dxdt\right)^{1/(\alpha_0+\lambda K^{n+1})} = \operatorname{esssup} Q_\infty |v|$$

yield (1.15). □

2.1.3 *Definitions of generalized solutions*

Definition 2.1.1 A function $u(x,t)$ is called a generalized solution of the equation (1.1) on $Q_T = \mathbb{R}^N \times (0,T)$, if

$$u \in C_{loc}(0,T; L^2_{loc}(\mathbb{R}^N)) \cap L^p_{loc}(0,T; W^{1,p}_{loc}(\mathbb{R}^N)),$$

$$\iint_{Q_T}\left(-u\frac{\partial\varphi}{\partial t} + |\nabla u|^{p-2}\nabla u\cdot\nabla\varphi\right) dxdt = 0, \quad \text{for } \varphi \in C_0^\infty(Q_T). \quad (1.16)$$

Definition 2.1.2 A function $u(x,t)$ is called a generalized solution of the Cauchy problem (1.1), (1.6) on $Q_T = \mathbb{R}^N \times (0,T)$, if $u(x,t)$ is a generalized solution of (1.1) on Q_T and

$$\lim_{t\to 0^+}\int_{\mathbb{R}^N} u(x,t)h(x)dx = \int_{\mathbb{R}^N} u_0(x)h(x)dx, \quad \text{for } h \in C_0^\infty(\mathbb{R}^N). \quad (1.17)$$

Similarly we can define generalized solutions of the boundary value problem. As an example, we consider the first boundary value problem on a bounded domain Ω; the boundary value condition and initial value condition are

$$u(x,t) = g(x,t), \qquad (x,t) \in \partial\Omega \times (0,T) \quad (1.18)$$

and

$$u(x,0) = u_0(x), \qquad x \in \overline{\Omega} \quad (1.19)$$

respectively, where $g \in L^\infty(\partial\Omega \times (0,T))$, $u_0 \in L^1(\Omega)$.

Definition 2.1.3 A function $u(x,t)$ is called a generalized solution of the first boundary value problem (1.1), (1.18), (1.19) on $\Omega_T = \Omega \times (0,T)$, if

$$
\begin{aligned}
&u \in C_{loc}(0,T;L^2(\Omega)) \cap L^p_{loc}(0,T;W^{1,p}(\Omega)),\\
&\iint_{\Omega_T} \left(-u\frac{\partial \varphi}{\partial t} + |\nabla u|^{p-2}\nabla u \cdot \nabla \varphi \right) dxdt = 0, \quad \text{for } \varphi \in C_0^\infty(\Omega_T)
\end{aligned}
\tag{1.20}
$$

$$
\lim_{t\to 0^+} \int_\Omega u(x,t)h(x)dx = \int_\Omega u_0(x)h(x)dx \quad \text{for } h \in C_0^\infty(\Omega) \tag{1.21}
$$

and

$$
u(x,t) = g(x,t) \quad \text{for } (x,t) \in \partial\Omega \times (0,T). \tag{1.22}
$$

Remark 2.1.1 Sometimes we need to use the concept of generalized super-solutions (sub-solutions). To define the generalized super-solutions (sub-solutions), it suffices to replace "=" in (1.16) by "$\geq$" ("$\leq$") and require φ to be nonnegative. To define the generalized super-solutions (sub-solutions) of the Cauchy problem (1.1), (1.16), it suffices to replace "=" in (1.16), (1.17) by "$\geq$" ("$\leq$") and require φ to be nonnegative. To define the generalized super-solutions (sub-solutions) of the first boundary value problem for (1.1), we need to replace (1.22) by $u(x,t) \geq g(x,t)$ ($u(x,t) \leq g(x,t)$) in addition to replacing "=" in (1.20), (1.21) by "$\geq$" ("$\leq$") and requiring φ, h to be nonnegative.

Remark 2.1.2 We may define generalized solutions for equation (1.2) similarly. The only difference is to replace $u \in L^p_{loc}(0,T;W^{1,p}_{loc}(\mathbb{R}^N))$ by $u^m \in L^p_{loc}(0,T;W^{1,p}_{loc}(\mathbb{R}^N))$ and require u to be nonnegative.

Remark 2.1.3 Let $\Omega \subset \mathbb{R}^N$ be a bounded domain, $0 < t_1 < t_2 < T$. By an approximate process, one may derive from (1.16) that for $\varphi \in L^p(t_1,t_2;W_0^{1,p}(\Omega))$ with $\varphi_t \in L^2(\Omega \times (t_1,t_2))$, the generalized solution u of (1.1) on Q_T satisfies

$$
\begin{aligned}
&\int_\Omega u(x,t_2)\varphi(x,t_2)dx - \int_\Omega u(x,t_1)\varphi(x,t_1)dx\\
&= -\int_{t_1}^{t_2}\int_\Omega \left(u\frac{\partial \varphi}{\partial t} - |\nabla u|^{p-2}\nabla u \cdot \nabla \varphi \right) dxdt.
\end{aligned}
\tag{1.23}
$$

In particular, for $\varphi \in W_0^{1,p}(\Omega)$,

$$\int_\Omega (u(x,t_2)-u(x,t_1))\varphi(x)dx + \int_{t_1}^{t_2}\int_\Omega |\nabla u|^{p-2}\nabla u\cdot\nabla\varphi dxdt = 0. \quad (1.24)$$

In fact, from $\varphi \in L^p(t_1,t_2;W_0^{1,p}(\Omega))$ and $\varphi_t \in L^2(\Omega\times(t_1,t_2))$ it follows that there exists a sequence of functions $\{\varphi_k\} \in C^\infty(\overline{\Omega}\times(t_1,t_2)$ such that for fixed $t\in(t_1,t_2)$, $\varphi_k(\cdot,t)\in C_0^\infty(\Omega)$ and

$$\|\varphi_{kt}-\varphi_t\|_{2,\Omega_{1,2}} \to 0, \quad \|\varphi_k-\varphi\|_{L^p(t_1,t_2;W^{1,p}(\Omega))}\to 0 \quad \text{as } k\to\infty,$$

where $\Omega_{1,2}=\Omega\times(t_1,t_2)$.

Choose $j(s)\in C_0^\infty(R)$ such that

$$j(s)\ge 0 \quad \text{for } s\in\mathbb{R};\quad j(s)=0 \qquad \text{for } |s|>1;\quad \int_R j(s)ds=1.$$

For $h>0$, define $j_h(s)=\dfrac{1}{h}j\left(\dfrac{s}{h}\right)$ and

$$\eta_h(t)=\int_{t-t_2-2h}^{t-t_1+2h} j_h(s)ds.$$

Then

$$\eta_h(t)\in C_0^\infty(t_1,t_2), \quad \lim_{h\to 0^+}\eta_h(t)=1, \quad \text{for } t\in(t_1,t_2).$$

Choosing $\varphi=\varphi_k(x,t)\eta_h(t)$ in (1.20), we obtain

$$\begin{aligned}&\int_{t_1}^{t_2}\int_\Omega u\varphi_k j_h(t-t_2+2h)dxdt - \int_{t_1}^{t_2}\int_\Omega u\varphi_k j_h(t-t_1-2h)dxdt\\&\quad -\int_{t_1}^{t_2}\int_\Omega u\varphi_{kt}\eta_h dxdt + \int_{t_1}^{t_2}\int_\Omega |\nabla u|^{p-2}\nabla u\cdot\nabla(\varphi_k\eta_h)dxdt = 0.\end{aligned} \quad (1.25)$$

Noting that

$$\begin{aligned}&\left|\int_{t_1}^{t_2}\int_\Omega u\varphi_k j_h(t-t_2+2h)dxdt - \int_\Omega (u\varphi_k)|_{t=t_2}dx\right|\\=&\left|\int_{t_2-3h}^{t_2-h}\int_\Omega u\varphi_k j_h(t-t_2+2h)dxdt\right.\\&\left.\quad -\int_{t_2-3h}^{t_2-h}\int_\Omega (u\varphi_k)|_{t=t_2} j_h(t-t_2+2h)dxdt\right|\end{aligned}$$

$$\leq \sup_{t_2-3h<t<t_2-h} \int_\Omega |(u\varphi_k)|_t - (u\varphi_k)|_{t_2}|\,dx$$

and $u \in C(0,T; L^2_{loc}(\mathbb{R}^N))$, we see that the right hand side tends to zero as $h \to 0$. Similarly, as $h \to 0$,

$$\left| \int_{t_1}^{t_2} \int_\Omega u\varphi_j j_h(t - t_1 - 2h)dxdt - \int_\Omega (u\varphi_j)|_{t=t_1} dx \right| \to 0.$$

Letting $h \to 0$ and then $k \to \infty$ yield (1.23).

Remark 2.1.4 Fix $\tau \in (0,T)$. Let h satisfy $0 < \tau < \tau + h < T$. Take $t_1 = \tau$, $t_2 = \tau + h$ in (1.24) and divide both sides by h. Then we have for $\varphi \in W_0^{1,p}(\Omega)$,

$$\begin{aligned} &\int_\Omega (u_h(x,\tau))_\tau \varphi(x)dx \\ &+ \int_\Omega (|\nabla u|^{p-2}\nabla u)_h(x,\tau) \cdot \nabla\varphi(x)dx = 0, \end{aligned} \tag{1.26}$$

where u_h is the Steklov mean value of u. Now we choose

$$\varphi(x) = \zeta(x,\tau) f(u_h(x,\tau))$$

in (1.26), where $\zeta \in C^1(\overline{\Omega} \times (0,T)$ is an arbitrary nonnegative function such that for $\tau \in (0,T)$, $\zeta(\cdot,\tau) \in C_0^1(\Omega)$ and $f(s) \in C^1(R)$ with $f'(s) \in L^\infty(\mathbb{R})$ and integrate with respect to $\tau \in (t_1, t_2)$ $(0 < t_1 < t_2 < T - h)$. Then

$$\begin{aligned} &\int_\Omega \int_0^{u_h(x,t_2)} f(s)ds \cdot \zeta(x,t_2)dx \\ &- \int_\Omega \int_0^{u_h(x,t_1)} f(s)ds \cdot \zeta(x,t_1)dx \\ &+ \int_{t_1}^{t_2} \int_\Omega (|\nabla u|^{p-2}\nabla u)_h f'(u_h)\nabla u_h \zeta dxdt \\ &+ \int_{t_1}^{t_2} \int_\Omega (|\nabla u|^{p-2}\nabla u))_h f(u_h)\nabla\zeta dxdt \\ = &\int_{t_1}^{t_2} \int_\Omega \int_0^{u_h} f(s)ds\zeta_\tau dxdt. \end{aligned}$$

Letting $h \to 0$ and using Lemma 2.1.2 we obtain

$$\begin{aligned}&\int_\Omega \int_0^{u(x,t_2)} f(s)ds \cdot \zeta(x,t_2)dx - \int_\Omega \int_0^{u(x,t_1)} f(s)ds \cdot \zeta(x,t_1)dx \\ &+ \int_{t_1}^{t_2} \int_\Omega |\nabla u|^p f'(u)\zeta dxdt + \int_{t_1}^{t_2} \int_\Omega |\nabla u|^{p-2}\nabla u \cdot f(u)\nabla\zeta dxdt \\ = &\int_{t_1}^{t_2} \int_\Omega \int_0^u f(s)ds\zeta_\tau dxdt. \end{aligned} \tag{1.27}$$

This integral identity will be applied very often in the sequel.

In particular, if we take $f(s) = s$, then (1.27) turns out to be

$$\begin{aligned}&\frac{1}{2}\int_\Omega u^2(x,t_2)\zeta(x,t_2)dx - \frac{1}{2}\int_\Omega u^2(x,t_1)\zeta(x,t_1)dx \\ &+ \int_{t_1}^{t_2} \int_\Omega |\nabla u|^p \zeta dxdt + \int_{t_1}^{t_2} \int_\Omega |\nabla u|^{p-2}\nabla u\, u\nabla\zeta dxdt \\ = &\frac{1}{2}\int_{t_1}^{t_2} \int_\Omega u^2\zeta_\tau dxdt. \end{aligned} \tag{1.28}$$

2.1.4 *Special solutions*

Denote

$$\begin{aligned}&E_{k,\rho}(x,t;\overline{x},\overline{t}) \\ = &k\rho^N S^{-N/\lambda_0}(t)\left(1 - \left(\frac{|x-\overline{x}|}{S^{1/\lambda_0}(t)}\right)^{p/(p-1)}\right)_+^{(p-1)/(m(p-1)-1)}, \end{aligned} \tag{1.29}$$

where $\lambda_0 = p + N[m(p-1)-1]$ and

$$S(t) = \lambda_0\left(\frac{mp}{m(p-1)-1}\right)^{p-1} k^{m(p-1)-1}\rho^{N[m(p-1)-1]}(t-\overline{t}) + \rho^{\lambda_0}, \quad t \ge \overline{t},$$

or

$$S(t) = \lambda_0\left(\frac{mp}{m(p-1)-1}\right)^{p-1} k^{m(p-1)-1}\rho^{N[m(p-1)-1]}(t-\overline{t}), \quad t \ge \overline{t}. \tag{1.30}$$

A direct calculation shows that for any $(\overline{x},\overline{t})$, $E_{k,\rho}(x,t;\overline{x},\overline{t})$ is a generalized solution of (1.2), a Barenblatt-type solution.

If $(\overline{x},\overline{t})=(0,0)$, then the generalized solution (1.29) satisfies the initial value condition

$$E_{k,\rho}(x,0;0,0)=M\delta(x),$$

where $M=\|E_{k,\rho}(\cdot,t;0,0)\|_{1,\mathbb{R}^N}$ and $\delta(x)$ is the Dirac measure. Thus (1.29) is also called the fundamental solution of (1.2).

In case $m=1$, $p>2$,

$$\begin{aligned} &E_{k,\rho}(x,t;\overline{x},\overline{t}) \\ &= k\rho^N S^{-N/\kappa}(t)\left(1-\left(\frac{|x-\overline{x}|}{S^{1/\kappa}(t)}\right)^{p/(p-1)}\right)_+^{(p-1)/(p-2)}, \end{aligned} \tag{1.31}$$

where $\kappa=N(p-2)+p$ and

$$S(t)=\kappa\left(\frac{p}{p-2}\right)^{p-1}k^{p-2}\rho^{N[p-2]}(t-\overline{t})+\rho^{\kappa},$$

or

$$S(t)=\kappa\left(\frac{p}{p-2}\right)^{p-1}k^{p-2}\rho^{N(p-2)}(t-\overline{t}).$$

Choosing k, ρ such that $k\rho^N\left(\kappa\left(\frac{p}{p-2}\right)^{p-1}k^{p-2}\rho^{N(p-2)}\right)^{-N/\kappa}=1$, in particular, we obtain a Barenblatt-type solution of (1.1) as follows

$$B(x,t)=t^{-N/\kappa}\left(1-\gamma_p\left(\frac{|x|}{t^{1/\kappa}}\right)^{p/(p-1)}\right)_+^{(p-1)/(p-2)},\quad t>0, \tag{1.32}$$

where

$$\gamma_p=\left(\frac{1}{\kappa}\right)^{1/(p-1)}\frac{p-2}{p},\quad p>2.$$

2.2 Existence of Solutions

In this section we discuss the existence of generalized solutions of the Cauchy problem (1.1), (1.6).

2.2.1 The case $u_0 \in C_0^\infty(\mathbb{R}^N)$ or $u_0 \in L^1(\mathbb{R}^N) \cap L^\infty(\mathbb{R}^N)$

Theorem 2.2.1 *Assume that $u_0 \in C_0^\infty(\mathbb{R}^N)$ and $p > 1$. Then the Cauchy problem (1.1), (1.6) admits a generalized solution u which possesses the following properties:*

(1) global properties:

$$u \in L^\infty(Q_T) \cap C(0,T; L^2(\mathbb{R}^N)) \cap L^p(0,T; W^{1,p}(\mathbb{R}^N)),$$

$$u_t \in L^\infty(Q_T),$$

$$\min_{\mathbb{R}^N} u_0 \le u \le \max_{\mathbb{R}^N} u_0 \quad a.e \ on \ \mathbb{R}^N, \tag{2.1}$$

$$\int_{\mathbb{R}^N} |u(x,t)| dx \le \int_{\mathbb{R}^N} |u_0(x)| dx \quad a.e \ on \ [0,T], \tag{2.2}$$

$$\int_0^T \int_{\mathbb{R}^N} |\nabla u|^p dx dt \le \frac{1}{2} \int_{\mathbb{R}^N} |u_0(x)|^2 dx, \quad a.e \ on \ [0,T]; \tag{2.3}$$

(2) local properties: for $\rho > 0$, $q \ge 1$, $t \in (0,T)$, there exists a constant C depending only on ρ, t, p, N, q and $||u||_{\infty, B_{4\rho} \times (t/4,T)}$ such that

$$\int_t^T \int_{B_\rho} |\nabla u|^q dx d\tau \le C, \qquad \int_t^T \int_{B_\rho} u_t^2 dx d\tau \le C, \tag{2.4}$$

where $B_\rho - \{x \subset \mathbb{R}^N; |x| < \rho\}$

To prove Theorem 2.2.1, let us consider the boundary value problem

$$\frac{\partial u}{\partial t} = \text{div}\left((|\nabla u|^2 + \frac{1}{n})^{(p-2)/2} \nabla u \right) \quad \text{on } B_n' \times (0,\infty), \tag{2.5}$$

$$u(x,t) = 0 \quad \text{on} \partial B_n \times (0,\infty), \tag{2.6}$$

$$u(x,0) = u_0(x) \quad \text{on } \overline{B}_n', \tag{2.7}$$

where $B_n' = \{x \in \mathbb{R}^N; |x| < n^{p/2(N+1)}\}$ with n large enough so that $\text{supp}\, u_0 \subset B_n'$. Here we use the ball of radius $n^{p/2(N+1)}$ for the convenience in the sequel.

According to the standard theory for parabolic equations [LSU], the problem (2.5)–(2.7) admits a classical solution $u_n \in C^\infty(\overline{B_n'} \times [0,\infty))$, which satisfies

$$\min_{\mathbb{R}^N} u_0 \le u_n \le \max_{\mathbb{R}^N} u_0. \tag{2.8}$$

Now we are ready to establish some uniform estimates on u_n.

Lemma 2.2.1 *The solution u_n problem (2.5)–(2.7) satisfies*

$$\int_{B_n'} |u_n(x,t)|dx \le \int_{\mathbb{R}^N} |u_0(x)|dx, \quad t \in (0,\infty), \tag{2.9}$$

$$\int_0^T \int_{B_n'} |\nabla u_n|^p dxdt \le \frac{1}{2}\int_{\mathbb{R}^N} u_0^2(x)dx. \tag{2.10}$$

Proof. For $\eta > 0$, let

$$\mathrm{sgn}_\eta s = \begin{cases} 1, & \text{for } s > \eta, \\ \dfrac{s}{\eta}, & \text{for } |s| \le \eta, \\ -1, & \text{for } s < -\eta. \end{cases} \tag{2.11}$$

Substitute $u = u_n$ into (2.5), multiply both sides by $\mathrm{sgn}_\eta u_n$ and integrate over $B_n' \times (0,t)$. Integrating by parts and using (2.6), (2.7), we derive

$$\int_{B_n'}\int_0^{u_n(x,t)} \mathrm{sgn}_\eta s ds dx - \int_{B_n'}\int_0^{u_0(x)} \mathrm{sgn}_\eta s ds dx$$
$$= \int_0^t \int_{B_n'} \left(|\nabla u_n|^2 + \frac{1}{n}\right)^{(p-2)/2} |\nabla u_n|^2 \mathrm{sgn}_\eta' u_n dxds.$$

Letting $\eta \to 0$ and noting that the first term of the right hand side tends to zero we obtain (2.9).

Multiplying (2.5) by $u = u_n$, integrating over $B_n' \times (0,T)$ and integrating by parts we may obtain (2.10). □

Lemma 2.2.2 *The solution u_n of (2.5)–(2.7) satisfies*

$$\max_{B_n' \times (0,T)} |u_{nt}| \le \max_{\mathbb{R}^N} \left| \mathrm{div}\left(\left(|\nabla u_0|^2 + \frac{1}{n}\right)^{(p-2)/2} \nabla u_0\right)\right|.$$

Proof. Differentiate (2.5) with respect to t,

$$\frac{\partial w}{\partial t} = a_{ij}\frac{\partial^2 w}{\partial x_i \partial x_j} + \frac{\partial a_{ij}}{\partial x_j}\frac{\partial w}{\partial x_i},$$

where $w = \frac{\partial u}{\partial t}$, $u = u_n$ and

$$a_{ij} = \left(|\nabla u|^2 + \frac{1}{n}\right)^{(p-2)/2} \cdot \left(\delta_{ij} + (p-2)\left(|\nabla u|^2 + \frac{1}{n}\right)^{-1} \frac{\partial u}{\partial x_i}\frac{\partial u}{\partial x_j}\right),$$

$$\delta_{ij} = \begin{cases} 1, & \text{for } i = j, \\ 0, & \text{for } i \neq j. \end{cases}$$

Clearly

$$w(x,t) = 0, \qquad (x,t) \in \partial B'_n \times [0,T],$$

$$w(x,0) = \operatorname{div}\left(\left(|\nabla u_0|^2 + \frac{1}{n}\right)^{(p-2)/2} \nabla u_0\right), \quad x \in B'_n.$$

Denote

$$a_0 = \left(|\nabla u|^2 + \frac{1}{n}\right)^{(p-2)/2}.$$

It is easy to see that for $\xi \in \mathbb{R}^N$,

$$\min\{p-1,1\}a_0|\xi|^2 < a_{ij}\xi_i\xi_j \leq \max\{p-1,1\}a_0|\xi|^2.$$

Thus, by the maximum principle, we obtain the conclusion of our lemma□

Lemma 2.2.3 *The solution u_n of (2.5)-(2.7) satisfies*

$$\int_t^T \int_{B_\rho} |\nabla u_n|^q dx d\tau \leq C, \tag{2.12}$$

$$\int_t^T \int_{B_\rho} u_{nt}^2 dx d\tau \leq C, \tag{2.13}$$

where $\rho > 0$, $t \in (0,T)$, $q \geq 1$ and the constant C depends only on ρ, p, q, t, N and $\|u_n\|_{\infty, B_{4\rho}\times(t/4,T)}$.

Proof. First we prove that $u = u_n$ satisfies

$$\begin{aligned}&\int_t^T \int_{B_{2\rho}} (|\nabla u|^2 + \frac{1}{n})^{(p-2)/2} |\nabla u|^2 dx d\tau \\ \leq\ & C(\rho, p, N) t^{-1} \left(\int_{t/4}^T \int_{B_{4\rho}} |u|^{\max\{p,2\}} dx d\tau + 1 \right),\end{aligned} \tag{2.14}$$

which offers a local L^p estimate on ∇u by means of the local $L^{\max\{p,2\}}$ norm of u.

Choose $\xi \in C^\infty(B_{4\rho} \times [0, T])$ such that

$$\begin{aligned}&0 \leq \xi \leq 1, && (x, \tau) \in B_{4\rho} \times [0, T], \\ &\xi = 1, && (x, \tau) \in B_{2\rho} \times (t/2, T), \\ &\xi = 0, && (x, \tau) \in \partial B_{4\rho} \times [0, T] \text{ or } \tau \leq t/4, \\ &|\nabla \xi| \leq C\rho^{-1}, && |\xi_\tau| \leq Ct^{-1}.\end{aligned}$$

Multiply (2.5) by $u\xi^p$ and integrate over $B_{4\rho} \times [0, T]$. After integrating by parts we obtain

$$\begin{aligned}&\frac{1}{2} \int_{B_{4\rho}} \xi^p u^2(x, T) dx \\ &\quad + \int_0^T \int_{B_{4\rho}} \xi^p (|\nabla u|^2 + \frac{1}{n})^{(p-2)/2} |\nabla u|^2 dx d\tau \\ &\leq p \int_0^T \int_{B_{4\rho}} \xi^{p-1} |\nabla \xi| |u| (|\nabla u|^2 + \frac{1}{n})^{(p-2)/2} |\nabla u| dx d\tau \\ &\quad + \frac{p}{2} \int_{B_{4\rho}} \xi^{p-1} |\xi_\tau| u^2 dx.\end{aligned} \tag{2.15}$$

Note that for $1 < p < 2$,

$$\left(|\nabla u|^2 + \frac{1}{n} \right)^{(p-2)/2} |\nabla u|^2 \geq \left(|\nabla u|^2 + \frac{1}{n} \right)^{p/2} - \left(\frac{1}{n} \right)^{p/2},$$

$$\left(|\nabla u|^2 + \frac{1}{n} \right)^{(p-2)/2} |\nabla u| \leq \left(|\nabla u|^2 + \frac{1}{n} \right)^{(p-1)/2}, \tag{2.16}$$

and for $p \geq 2$,

$$\left(|\nabla u|^2 + \frac{1}{n}\right)^{(p-2)/2} |\nabla u|^2 \geq |\nabla u|^p,$$

$$\left(|\nabla u|^2 + \frac{1}{n}\right)^{(p-2)/2} |\nabla u| \leq \left(|\nabla u|^{p-1} + 1\right). \tag{2.17}$$

Thus, if $1 < p < 2$, then we use (2.16) and Young's inequality to estimate the first term on the right hand side of (2.15) and obtain

$$\begin{aligned}
&\int_0^T \int_{B_{4\rho}} \xi^p(|\nabla u|^2 + \frac{1}{n})^{(p-2)/2}|\nabla u|^2 dx d\tau \\
\leq\ & C\int_0^T \int_{B_{4\rho}} |\nabla \xi|^p |u|^p dx d\tau + C\int_0^T \int_{B_{4\rho}} |\xi_\tau||u|^2 dx d\tau \\
&+C\left(\frac{1}{n}\right)^{p/2} \int_0^T \int_{B_{4\rho}} \xi^p dx d\tau;
\end{aligned}$$

if $p \geq 2$, then we use (2.17) and Young's inequality to estimate the same term and obtain

$$\begin{aligned}
&\int_0^T \int_{B_{4\rho}} \xi^p(|\nabla u|^2 + \frac{1}{n})^{(p-2)/2}|\nabla u|^2 dx d\tau \\
\leq\ & C\int_0^T \int_{B_{4\rho}} |\nabla \xi|^p |u|^p dx d\tau + C\int_0^T \int_{B_{4\rho}} |\xi_\tau||u|^2 dx d\tau \\
&+C\int_0^T \int_{B_{4\rho}} \xi^p dx d\tau.
\end{aligned}$$

In either case we can obtain (2.14).

Next we prove (2.12). Differentiate (2.5) with respect to x_j,

$$\frac{\partial u_{x_j}}{\partial t} = \left(\left(|\nabla u|^2 + \frac{1}{n}\right)^{(p-2)/2} u_{x_i}\right)_{x_i x_j}. \tag{2.18}$$

Choose $\xi \in C^\infty(\overline{B_{2\rho}} \times [0,T])$ such that

$$\begin{aligned}
0 \leq \xi \leq 1, &\quad \text{for } (x,\tau) \in B_{2\rho} \times [0,T]; \\
\xi = 1, &\quad \text{for } (x,\tau) \in B_\rho \times (t,T); \\
\xi = 0, &\quad \text{for } (x,\tau) \in \partial B_{2\rho} \times [0,T] \text{or } \tau \leq t/2; \\
|\nabla \xi| \leq C\rho^{-1}, &\quad |\xi_\tau| \leq Ct^{-1}.
\end{aligned}$$

Denote $v = |\nabla u|^2 + \frac{1}{n}$. Multiply (2.18) with $\xi^2 v^\alpha u_{x_j}$ $(\alpha \ge 0)$ and integrate over$B_{2\rho} \times [0, t_0]$. Then we obtain

$$\begin{aligned} &\frac{1}{2(\alpha+1)} \int_{B_{2\rho}} \xi^2 v^{\alpha+1}(x, t_0)dx \\ &\quad + \int_0^{t_0} \int_{B_{2\rho}} \xi^2 (v^\alpha u_{x_j})_{x_i} (v^{(p-2)/2} u_{x_i})_{x_j} dx d\tau \\ = &\frac{1}{\alpha+1} \int_0^{t_0} \int_{B_{2\rho}} \xi \xi_t v^{\alpha+1} dx d\tau \\ &\quad -2 \int_0^{t_0} \int_{B_{2\rho}} \xi v^\alpha u_{x_j} \left(v^{(p-2)/2} u_{x_i} \right)_{x_j} \xi_{x_i} dx d\tau. \end{aligned} \tag{2.19}$$

A simple calculation gives

$$\begin{aligned} &(v^\alpha u_{x_j})_{x_i} (v^{(p-2)/2} u_{x_i})_{x_j} \\ = \; & v^{(p+2\alpha-2)/2} u_{x_i x_j} u_{x_i x_j} + \frac{p+2\alpha-2}{4} v^{(p+2\alpha-4)/2} |\nabla v|^2 \\ &+ \frac{\alpha(p-2)}{2} v^{(p+2\alpha-6)/2} (\nabla u \cdot \nabla v)^2. \end{aligned}$$

Substituting this into (2.19) and using Young's inequality to the second term on the right hand side yield

$$\begin{aligned} &\frac{1}{2(\alpha+1)} \int_{B_{2\rho}} \xi^2 v^{\alpha+1}(x, t_0)dx \\ &\quad + (1-\varepsilon) \int_0^{t_0} \int_{B_{2\rho}} \xi^2 v^{(p+2\alpha-2)/2} \sum_{j=1}^N |\nabla u_{x_j}|^2 dx d\tau \\ &\quad + \frac{p+2\alpha-2}{4} \int_0^{t_0} \int_{B_{2\rho}} \xi^2 v^{(p+2\alpha-4)/2} |\nabla v|^2 dx d\tau \\ &\quad + \frac{\alpha(p-2)}{2} \int_0^{t_0} \int_{B_{2\rho}} \xi^2 v^{(p+2\alpha-6)/2} (\nabla u \cdot \nabla v)^2 dx d\tau \\ \le \; & C(\varepsilon) \int_0^{t_0} \int_{B_{2\rho}} v^{(p+2\alpha)/2} |\nabla \xi|^2 dx d\tau \\ &\quad + \frac{C}{\alpha+1} \int_0^{t_0} \int_{B_{2\rho}} \xi v^{\alpha+1} dx d\tau, \end{aligned} \tag{2.20}$$

where $0 < \varepsilon < 1$. Noting that

$$v^{(p+2\alpha-2)/2}\sum_{j=1}^{N}|\nabla u_{x_j}|^2 \geq \frac{1}{4}v^{(p+2\alpha-4)/2}|\nabla v|^2 \tag{2.21}$$

and the fact that all terms on the left hand side of (2.20) are nonnegative if $p \geq 2$, from (2.20) which holds for any $t_0 \in (0,T)$, we obtain

$$\begin{aligned}\sup_{0<\tau<T}\int_{B_{2\rho}} &\xi^2 v^{\alpha+1}(x,\tau)dx \\ &+\int_0^T\int_{B_{2\rho}}\xi^2 v^{(p+2\alpha-4)/2}|\nabla v|^2 dxd\tau \\ \leq\ C\int_0^T\int_{B_{2\rho}}&\xi v^{\alpha+1}dxd\tau + C\int_0^T\int_{B_{2\rho}} v^{(p+2\alpha)/2}|\nabla\xi|^2 dxd\tau.\end{aligned}$$

Let $w = v^{(p+2\alpha)/4}$, $\lambda = \dfrac{4(\alpha+1)}{p+2\alpha}$. Noting that since $p \geq 2$, we must have $\lambda \leq 2$ and hence $\xi^{4/\lambda} \leq \xi^2$, from the above inequality we derive

$$\begin{aligned}&\sup_{0<\tau<T}\int_{B_{2\rho}}\left(\xi^{2/\lambda}w\right)^{\lambda}(x,\tau)dx + \int_0^T\int_{B_{2\rho}}\left(\xi^{2/\lambda}|\nabla w|\right)^2 dxd\tau \\ \leq\ & C\int_0^T\int_{B_{2\rho}} w^2 dxd\tau + C\int_0^T\int_{B_{2\rho}} w^{\lambda}dxd\tau.\end{aligned} \tag{2.22}$$

First we take $\alpha = 0$. In this case, we can use (2.14) to estimate the right hand side of (2.22) and see that its upper bound depends only on ρ, p, t, N and $\|u\|_{\infty,B_{4\rho}\times(t/4,T)}$. Thus using the embedding inequality (1.7) to the function $\xi^{2/\lambda}w$ we can obtain

$$\int_0^T\int_{B_{2\rho}}\left(\xi^{2/\lambda}w\right)^r dxd\tau \leq C$$

and hence

$$\int_t^T\int_{B_\rho} w^r dxd\tau \leq C,$$

where $r = 2 + \dfrac{8}{Np}$. From this it follows that

$$\int_t^T\int_{B_{2\rho}}|\nabla u|^{s_1}dxdt \leq C, \tag{2.23}$$

where $s_1 = p + \frac{4}{N}$ and the constant C depends only on ρ, p, t, N and $||u||_{\infty, B_{4\rho}\times(t/4,T)}$.

Next we take $\alpha = \frac{2}{N}$. Using (2.23) and repeating the above argument, we can obtain $|\nabla u| \in L^{s_2}(B_\rho \times (t, T))$ and

$$\int_t^T \int_{B_{2\rho}} |\nabla u|^{s_2} dx d\tau \leq C,$$

where

$$s_2 = s_1 + \frac{4(N+2)}{N^2} = p + \frac{4}{N} + \frac{4(N+2)}{N^2}.$$

We can repeat this procedure up to some step, say, the k-th step so that we obtain $|\nabla u| \in L^{s_k}(B_\rho \times (t, T))$ and the estimate

$$\int_t^T \int_{B_{2\rho}} |\nabla u|^{s_k} dx d\tau \leq C$$

with s_k greater than the given $q \geq 1$. Thus (2.12) is proved in the case $p \geq 2$.

If $1 < p < 2$, then the third and fourth term of (2.20) are nonpositive. However we have

$$\begin{aligned} &\frac{\alpha(p-2)}{2} \int_0^{t_0} \int_{B_{2\rho}} \xi^2 v^{(p+2\alpha-6)/2} (\nabla u \cdot \nabla v)^2 dx d\tau \\ \geq\ &\frac{\alpha(p-2)}{2} \int_0^{t_0} \int_{B_{2\rho}} \xi^2 v^{(p+2\alpha-4)/2} |\nabla v|^2 dx d\tau. \end{aligned}$$

Using this inequality and (2.21), from (2.20) we see that there exist ε_1, $\varepsilon_2 > 0$ such that

$$\begin{aligned} &\sup_{0<\tau<T} \int_{B_{2\rho}} \xi^2 v^{\alpha+1}(x,\tau) dx + \varepsilon_1 \int_0^T \int_{B_{2\rho}} \xi^2 v^{\alpha_p} |\nabla v|^2 dx d\tau \\ &\quad + \varepsilon_2 \int_0^T \int_{B_{2\rho}} \xi^2 v^{\alpha_p+1} \sum_{j=1}^N |\nabla u_{x_j}|^2 dx d\tau \\ \leq\ & C \int_0^T \int_{B_{2\rho}} |\nabla \xi|^2 v^{\alpha_p+2} dx d\tau + C \int_0^T \int_{B_{2\rho}} \xi v^{\alpha+1} dx d\tau, \end{aligned} \tag{2.24}$$

where $\alpha_p = \dfrac{p+2\alpha-4}{2}$. It is easy to verify that

$$\begin{aligned}
&\int_0^T\int_{B_{2\rho}} \xi v^{\alpha+1} dx d\tau \\
\leq\ & C + C\int_0^T\int_{B_{2\rho}} \xi|\nabla u|^{2\alpha}\nabla u\cdot\nabla u dx d\tau \\
=\ & C - C\int_0^T\int_{B_{2\rho}} \xi|\nabla u|^{2\alpha} u\Delta u dx d\tau \\
& -C\int_0^T\int_{B_{2\rho}} u|\nabla u|^{2\alpha}\nabla u\cdot\nabla\xi dx d\tau \\
& -2C\alpha\int_0^T\int_{B_{2\rho}} u\xi u_{x_i}u_{x_j}u_{x_ix_j}|\nabla u|^{2\alpha-2} dx d\tau \\
\leq\ & C + \varepsilon_2\int_0^T\int_{B_{2\rho}} \xi^2 v^{(p+2\alpha-2)/2}\sum_{j=1}^N|\nabla u_{x_j}|^2 dx d\tau \\
& +C(\varepsilon_2)||u||^2_{\infty,B_{2\rho}\times(t/2,T)}\int_{t/2}^T\int_{B_{2\rho}} v^{(2-p+2\alpha)/2} dx d\tau \\
& +C||u||_{\infty,B_{2\rho}\times(t/2,T)}\int_0^T\int_{B_{2\rho}} |\nabla\xi| v^{1/2+\alpha} dx d\tau.
\end{aligned}$$

Substituting this into (2.24) yields

$$\begin{aligned}
&\sup_{0<\tau<T}\int_{B_{2\rho}} \xi^2 v^{\alpha+1}(x,\tau) dx \\
&\quad +\varepsilon_1\int_0^T\int_{B_{2\rho}} \xi^2 v^{(p+2\alpha-4)/2}|\nabla v|^2 dx d\tau \\
\leq\ & C + C\int_0^T\int_{B_{2\rho}} |\nabla\xi|^2 v^{(p+2\alpha)/2} dx d\tau \\
&\quad +C\int_{t/2}^T\int_{B_{2\rho}} v^{(2-p+2\alpha)/2} dx d\tau \\
&\quad +C\int_0^T\int_{B_{2\rho}} |\nabla\xi| v^{\alpha+1/2} dx d\tau,
\end{aligned} \tag{2.25}$$

where the constant C depends only on ρ, p, t, N and $||u||_{\infty,B_{4\rho\times(t/4,T)}}$. Noting that in the case $\alpha = 0$, the right hand side of (2.25) can be estimated by using (2.14), we can derive (2.12) similarly as we have done for the case $p \geq 2$.

Finally we prove (2.13). Multiply (2.5) by $\xi^{p+1}u_t$ and integrate over $B_{2\rho} \times (0,T)$. After integrating by parts we obtain

$$\begin{aligned}&\int_0^T \int_{B_{2\rho}} \xi^{p+1} u_\tau^2 dx d\tau \\ &\quad + \int_0^T \int_{B_{2\rho}} \left(|\nabla u|^2 + \frac{1}{n}\right)^{(p-2)/2} u_{x_i} u_{x_i \tau} \xi^{p+1} dx d\tau \\ &\quad + (p+1) \int_0^T \int_{B_{2\rho}} \xi^p u_\tau \left(|\nabla u|^2 + \frac{1}{n}\right)^{(p-2)/2} \\ &\quad \cdot \nabla u \cdot \nabla \xi dx d\tau = 0.\end{aligned} \tag{2.26}$$

Since

$$\begin{aligned}&\left|\xi^p u_\tau \left(|\nabla u|^2 + \frac{1}{n}\right)^{(p-2)/2} \nabla u \cdot \nabla \xi\right| \\ \leq\ & \frac{1}{2(p+1)} u_\tau^2 \xi^{p+1} + 8(p+1) \left(|\nabla u|^2 + \frac{1}{n}\right)^{p-2} |\nabla u|^2 |\nabla \xi|^2 \xi^{p-1}\end{aligned}$$

and

$$\begin{aligned}&\left(|\nabla u|^2 + \frac{1}{n}\right)^{(p-2)/2} u_{x_i} u_{x_i \tau} \\ =\ & \frac{1}{2} \frac{\partial}{\partial \tau} \int_0^{|\nabla u(x,\tau)|^2} \left(s + \frac{1}{n}\right)^{(p-2)/2} ds,\end{aligned}$$

from (2.26) we have

$$\begin{aligned}&\frac{1}{2} \int_0^T \int_{B_{2\rho}} \xi^{p+1} u_\tau^2 dx d\tau \\ \leq\ & \frac{p+1}{2} \int_0^T \int_{B_{2\rho}} \xi^p |\xi_\tau| \int_0^{|\nabla u(x,\tau)|^2} \left(s + \frac{1}{n}\right)^{(p-2)/2} ds dx d\tau \\ &\quad + C \int_0^T \int_{B_{2\rho}} \xi^{p-1} \left(|\nabla u|^2 + \frac{1}{n}\right)^{(p-2)/2} |\nabla u|^2 |\nabla \xi|^2 dx d\tau \\ \leq\ & C \left(1 + \int_0^T \int_{B_{2\rho}} \xi^p |\xi_\tau| |\nabla u|^2 \left(|\nabla u|^2 + \frac{1}{n}\right)^{(p-2)/2} dx d\tau\right)\end{aligned}$$

$$+C\int_0^T\int_{B_{2\rho}}|\nabla\xi|^2|\nabla u|^2\left(|\nabla u|^2+\frac{1}{n}\right)^{(p-2)/2}dxd\tau.$$

Therefore, (2.13) follows from (2.12). The proof of Lemma 2.2.3 is completed. □

Proof of Theorem 2.2.1. From (2.8), Lemma 2.2.1 and Lemma 2.2.2, we conclude that there exists a subsequence $\{u_n\}$, supposed to be $\{u_n\}$ itself, and a function $u\in L^\infty(Q_T)\cap L^\infty(0,T;L^1(\mathbb{R}^N))\cap L^p(0,T;W^{1,p}(\mathbb{R}^N))$, such that

$$u_t\in L^\infty(Q_T),$$

$$u_n\to u\quad \text{a.e in } Q_T,$$

$$\nabla u_n\to\nabla u\quad \text{weakly in } L^p(B_R\times(0,T)),$$

$$u_{nt}\to u_t\quad \text{weakly star in } L^\infty(B_R),$$

$$|\nabla u_n|^{p-2}u_{nx_i}\to\chi_i\quad \text{weakly in } L^{p/(p-1)}(B_R\times(0,T)),$$

where $R>0$ is an arbitrary constant, $i=1,2,\cdots,N$, $B_R=\{x\in\mathbb{R}^N;|x|<R\}$. Lemma 2.2.2 implies that $u_t\in L^\infty(Q_T)$. (2.1)–(2.4) follow from (2.8), (2.9), (2.10), (2.12), (2.13). Finally using Lemma 2.2.2 or the second inequality in (2.4) yields $u\in C(0,T;L^2_{loc}(R^N))$.

It is easy to verify that

$$\int_0^T\int_{\mathbb{R}^N}u\varphi_t dxd\tau-\int_0^T\int_{\mathbb{R}^N}\chi_i\varphi_{x_i}dxd\tau=0,\quad \text{for } \varphi\in C_0^\infty(Q_T).\tag{2.27}$$

Thus if we can prove that

$$\int_0^T\int_{\mathbb{R}^N}|\nabla u|^{p-2}\nabla u\cdot\nabla\varphi dxd\tau=\int_0^T\int_{\mathbb{R}^N}\chi_i\varphi_{x_i}dxd\tau,\quad \text{for } \varphi\in C_0^\infty(Q_T),\tag{2.28}$$

then u is a generalized solution of (1.1) on Q_T.

For any $v\in L^p_{loc}(0,T;W^{1,p}_{loc}(\mathbb{R}^N))$ and $\psi\in C_0^\infty(Q_T)$ with $0\le\psi\le 1$, $\operatorname{supp}\psi\subset B'_n$, we can easily obtain

$$\int_0^T\int_{\mathbb{R}^N}\psi\left(|\nabla u_n|^{p-2}\nabla u_n-|\nabla v|^{p-2}\nabla v\right)\nabla(u_n-v)dxdt\ge 0.\tag{2.29}$$

Noting

$$\begin{aligned}
&\int_0^T\int_{\mathbb{R}^N}\psi\left(|\nabla u_n|^2+\frac{1}{n}\right)^{(p-2)/2}|\nabla u_n|^2dxdt\\
=\ &\frac{1}{2}\int_0^T\int_{\mathbb{R}^N}\psi_t u_n^2dxdt\\
&-\int_0^T\int_{\mathbb{R}^N}u_n\left(|\nabla u_n|^2+\frac{1}{n}\right)^{(p-2)/2}\nabla u_n\cdot\nabla\psi dxdt
\end{aligned}$$

and using (2.16), (2.17), from (2.29) we derive

$$\begin{aligned}
&\frac{1}{2}\int_0^T\int_{\mathbb{R}^N}\psi_t u_n^2dxdt\\
&-\int_0^T\int_{\mathbb{R}^N}u_n\left(|\nabla u_n|^2+\frac{1}{n}\right)^{(p-2)/2}\nabla u_n\nabla\psi dxdt\\
&+\left(\frac{1}{n}\right)^{p/2}\text{mes}B_n'-\int_0^T\int_{\mathbb{R}^N}\psi|\nabla u_n|^{p-2}\nabla u_n\nabla vdxdt\\
&-\int_0^T\int_{\mathbb{R}^N}\psi|\nabla v|^{p-2}\nabla v\cdot\nabla(u_n-v)dxd\tau\geq 0.
\end{aligned}\tag{2.30}$$

Since

$$\begin{aligned}
&\left(|\nabla u_n|^2+\frac{1}{n}\right)^{(p-2)/2}\nabla u_n\\
=\ &|\nabla u_n|^{p-2}\nabla u_n+\frac{p-2}{2n}\int_0^1(|\nabla u_n|^2+s/n)^{(p-4)/2}ds\nabla u_n
\end{aligned}$$

and

$$\lim_{n\to\infty}\int_0^T\int_{\mathbb{R}^N}\frac{p-2}{2n}\int_0^1\left(|\nabla u_n|^2+\frac{s}{n}\right)^{(p-4)/2}ds\nabla u_n\nabla\psi u_ndxd\tau=0,$$

letting $n\to\infty$ in (2.30) and noting that $|B_n'|\leq Cn^{pN/2(N+1)}$ (the radius of B_n' is $n^{p/2(N+1)}$!) we obtain

$$\begin{aligned}
&\frac{1}{2}\int_0^T\int_{\mathbb{R}^N}\psi_t u^2dxd\tau-\int_0^T\int_{\mathbb{R}^N}\chi_i\psi_{x_i}udxd\tau-\int_0^T\int_{\mathbb{R}^N}\psi\chi_i v_{x_i}dxd\tau\\
&-\int_0^T\int_{\mathbb{R}^N}\psi|\nabla v|^{p-2}\nabla v\cdot\nabla(u-v)dxd\tau\geq 0.
\end{aligned}\tag{2.31}$$

Take $\varphi = \psi u$ in (2.27),

$$\frac{1}{2}\int_0^T\int_{\mathbb{R}^N}\psi_t u^2 dxdt - \int_0^T\int_{\mathbb{R}^N}\chi_i\psi_{x_i} udxdt = \int_0^T\int_{\mathbb{R}^N}\psi\chi_i u_{x_i} dxdt.$$

and substitute into (2.31). Then we deduce that

$$\int_0^T\int_{\mathbb{R}^N}\psi(\chi_i - |\nabla v|^{p-2}v_{x_i})(u_{x_i} - v_{x_i})dxd\tau \geq 0. \tag{2.32}$$

Take $v = u - \lambda\varphi$ with $\lambda \geq 0$, $\varphi \in C_0^\infty(Q_T)$ in (2.32),

$$\int_0^T\int_{\mathbb{R}^N}\psi(\chi_i - |\nabla(u-\lambda\varphi)|^{p-2}(u-\lambda\varphi)_{x_i})\varphi_{x_i} dxd\tau \geq 0$$

and let $\lambda \to 0$. Then we obtain

$$\int_0^T\int_{\mathbb{R}^N}\psi(\chi_i - |\nabla u|^{p-2}u_{x_i})\varphi_{x_i} dxd\tau \geq 0, \quad \text{for } \varphi \in C_0^\infty(Q_T).$$

If we take $\lambda \leq 0$, then we can obtain an opposite inequality. Therefore if we choose ψ such that $\operatorname{supp}\varphi \subset \operatorname{supp}\psi$ and $\psi = 1$ on $\operatorname{supp}\varphi$, then (2.28) follows.

Now we further prove that $u \in C(0,T;L^2(\mathbb{R}^N))$. Take $f(s) = s$, $\Omega = B_{2R}$, $\xi_R \in C_0^\infty(B_{2R})$ such that $0 \leq \xi_R \leq 1$ for $x \in B_{2R}$ and $\xi_R = 1$, for $x \in B_R$ in (1.28). Then as $R \to \infty$ we obtain

$$\int_{\mathbb{R}^N} u^2(x,t_2)dx - \int_{\mathbb{R}^N} u^2(x,t_1)dx = -2\int_{t_1}^{t_2}\int_{\mathbb{R}^N}|\nabla u|^p dxd\tau.$$

Taking $\Omega = B_{2R}$, $\varphi = \xi_R(x)u(x,t_1)$ in (1.24) and then letting $R \to \infty$, we obtain

$$\begin{aligned}&\int_{\mathbb{R}^N} u(x,t_2)u(x,t_1)dx - \int_{\mathbb{R}^N} u^2(x,t_1)dx \\ = \; & -\int_{t_1}^{t_2}\int_{\mathbb{R}^N}|\nabla u|^{p-2}\nabla u\cdot\nabla u(x,t_1)dxd\tau.\end{aligned}$$

Thus

$$\begin{aligned}
&\int_{\mathbb{R}^N} |u(x,t_2)-u(x,t_1)|^2 dx\\
=\;& \int_{\mathbb{R}^N} (u^2(x,t_2)-u^2(x,t_1))dx\\
&+2\int_{\mathbb{R}^N} (u^2(x,t_1)-u(x,t_1)u(x,t_2))dx\\
=\;& 2\int_{t_1}^{t_2}\int_{\mathbb{R}^N} |\nabla u|^{p-2}\nabla u\cdot\nabla u(x,t_1)dxd\tau.\\
&-2\int_{t_1}^{t_2}\int_{\mathbb{R}^N} |\nabla u|^p dxd\tau,
\end{aligned}$$

from which it follows that $u \in C(0,T;L^2(\mathbb{R}^N))$.

Finally since u_{nt} is uniformly bounded on $B'_n \times (0,T)$, we have

$$\int_{B_R} |u_n(x,t)-u_0(x)|dx \le \int_0^t \int_{B_R} |u_{nt}(x,\tau)|dxd\tau \le Ct.$$

Letting $n \to \infty$ gives

$$\int_{B_R} |u(x,t)-u_0(x)|dx \le Ct.$$

Hence

$$\lim_{t\to 0}\int_{B_R} |u(x,t)-u_0(x)|dx = 0. \tag{2.33}$$

This means that u satisfies the initial value condition (1.6) in a generalized sense. □

Theorem 2.2.2 *Assume that $u_0 \in L^1(\mathbb{R}^N)\cap L^\infty(\mathbb{R}^N)$ and $p>1$. Then the Cauchy problem (1.1), (1.6) admits a generalized solution u which possesses all properties in Theorem 2.2.1 except $u_t \in L^\infty(Q_T)$. Of course, $\min_{\mathbb{R}^N} u_0$ and $\max\limits_{\mathbb{R}^N} u_0$ should be replaced by $\operatorname{ess}\inf\limits_{\mathbb{R}^N}, u_0$ and $\operatorname{ess}\sup\limits_{\mathbb{R}^N}, u_0$ respectively.*

To prove this theorem it suffices to replace (2.7) by

$$u(x,0) = u_{0n}(x), \quad \text{on } \overline{B'}_n,$$

where $u_{0n}(x)$ is a smooth function approximating $u_0(x)$ with $\operatorname{supp} u_{0n} \subset B'_n$ and

$$\operatorname*{ess\,inf}_{\mathbb{R}^N} u_0(x) \le u_{0n}(x) \le \operatorname*{ess\,sup}_{\mathbb{R}^N} u_0(x).$$

In the present case, Lemma 2.2.1 and Lemma 2.2.3 still hold. We also have a result similar to Lemma 2.2.2 with u_0 on the right hand side of the inequality there replaced by u_{n0} and this is just why we can not obtain the conclusion $u_t \in L^\infty(Q_T)$.

2.2.2 *The case* $u_0 \in L^1_{loc}(\mathbb{R}^N)$

We are ready to discuss the case $u_0 \in L^1_{loc}(\mathbb{R}^N)$. It is well-known that for the Cauchy problem for the heat equation to have a solution, a most general condition on u_0 is given by

$$\int_{\mathbb{R}^N} \exp\left(-C|x|^2\right) |u_0(x)| dx < \infty$$

with some constant $C > 0$.

We have pointed out in §1.8 that the extension of this condition to the Newtonian filtration equation

$$\frac{\partial u}{\partial t} = \Delta u^m$$

with $m > 1$ is that for some constant $r > 0$,

$$\sup_{\rho \ge r} \rho^{-(N+2/(m-1))} \int_{B_\rho} u_0(x) dx < \infty.$$

We will prove that for the evolution p-Laplacian equation, we have a similar result in the case $p > 2$; the growth condition on u_0 is that for some constant $r > 0$,

$$|||u_0|||_r \equiv \sup_{\rho \ge r} \rho^{-\kappa/(p-2)} \int_{B_\rho} |u_0(x)| dx < \infty,$$

where $\kappa = N(p-2) + p$ and the notation $|||f|||_r$ is defined by

$$|||f|||_r \equiv \sup_{\rho \ge r} \rho^{-\kappa/(p-2)} \int_{B_\rho} |f(x)| dx.$$

Obviously, if $f \in L^1(\mathbb{R}^N)$, then for any $r > 0$, $|||f|||_r < \infty$. It is also easy to see that for any $r > 0$, $|||f|||_r < \infty$ if and only if for some $r_0 > 0$, $|||f|||_{r_0} < \infty$.

Theorem 2.2.3 *Assume that $p > 2$, $u_0 \in L^1_{loc}(\mathbb{R}^N)$ and there exists some constant $r > 0$ such that $|||u_0|||_r < \infty$. Then the Cauchy problem (1.1), (1.6) admits a generalized solution u on $Q_{T(u_0)}$, where*

$$T(u_0) = \begin{cases} C_0 \lim\limits_{r\to\infty} |||u_0|||_r^{2-p}, & \text{if } \lim\limits_{r\to\infty} |||u_0|||_r > 0, \\ +\infty, & \text{if } \lim\limits_{r\to\infty} |||u_0|||_r = 0 \end{cases}$$

with some constant C_0 depending only on p, N. Moreover, the generalized solution u possesses the following properties: for $0 < t < T_r(u_0) = C_0|||u_0|||_r^{2-p}$,

$$|||u(\cdot,t)|||_r \le C_1|||u_0|||_r, \tag{2.34}$$

$$|||u(\cdot,t)|||_{\infty,B_\rho} \le C_2 t^{-N/\kappa}\rho^{p/(p-2)}|||u_0|||_r^{p/\kappa}, \tag{2.35}$$

$$\int_0^t\int_{B_\rho} |\nabla u|^{p-1}dxd\tau \le C_2 t^{1/\kappa}\rho^{1+\kappa/(p-2)}|||u_0|||_r^{\mu}, \tag{2.36}$$

$$|\nabla u| \in L^q_{loc}(Q_T), \quad \text{for } q \ge 1, \quad u_t \in L^2_{loc}(Q_T), \tag{2.37}$$

where $\mu = 1 + \dfrac{(p-2)}{\kappa}$ and the constants C_1, C_2, C_3 depend only on p, N.

To prove Theorem 2.2.3, we first consider the Cauchy problem for (1.1) with initial value condition

$$u(x,0) = u_{0n}(x)$$

where

$$u_{0n} \in C_0^\infty(\mathbb{R}^N), \quad |||u_{0n}|||_r \le |||u_0|||_r$$

and

$$\lim_{n\to\infty}\int_{B_\rho} |u_{0n} - u_0|dx = 0 \qquad \forall\rho > 0.$$

According to Theorem 2.2.1, for any $T > 0$, this problem admits a generalized solution u_n satisfying the following conditions:

$$u_n \in \; L^\infty(Q_T) \cap C(0,T;L^2(\mathbb{R}^N)) \cap L^\infty(0,T;L^1(\mathbb{R}^N)) \cap L^p(0,T;W^{1,p}(\mathbb{R}^N)),$$

$$u_{nt} \in L^\infty(Q_T)$$

and

$$\int_t^T \int_{B_\rho} |\nabla u_n|^q dx d\tau \le C, \quad \int_t^T \int_{B_\rho} u_{nt}^2 dx d\tau \le C \tag{2.38}$$

for $\rho > 0$, $t > 0$, $q \ge 1$, where the constant C depends only on ρ, p, t, q, N and the upper bound of $||u_n||_{\infty, B_{4\rho}\times(t/4,T)}$.

If we can prove (2.34)–(2.36) for u_n uniformly in n, then (2.35) implies the uniform boundedness of $\{u_n\}$ on any compact subset of $Q_{T(u_0)}$ and thus, using (2.38) we can assert that there exists a subsequence of $\{u_n\}$ which converges to some function u almost everywhere on $Q_{T(u_0)}$. We can apply the argument similar to the proof of Theorem 2.2.1 to conclude that u is a generalized solution of (1.1) on $Q_{T(u_0)}$. From (2.34)–(2.36) and (2.38) which u_n satisfies, passing to the limit, we see that so does u itself.

Now we start to prove (2.34)–(2.36). For simplicity of notations, in what follows, we will omit the mark "n" of u_n and use γ to denote a constant depending only on p, T, N, which may take different values in different occasions. Without loss of generality we will suppose that $u \ge 0$; otherwise we may consider u_+ and u_- respectively and then obtain the same results. Here we notice that $u_\pm$ is a generalized sub-solution of (1.1).

Define

$$\phi(t) = \phi_r(t) = \sup_{\tau\in(0,t)} \tau^{N/\kappa} \sup_{\rho\ge r} \frac{||u(\cdot,\tau)||_{\infty,B_\rho}}{\rho^{p/(p-2)}}. \tag{2.39}$$

Lemma 2.2.4 *There exists a $\gamma = \gamma(N,p) > 0$ such that for any $t > 0$,*

$$||u(\cdot,t)||_{\infty,B_\rho} \le \gamma[K(t)]^{(N+p)/\lambda} \left(\int_{t/4}^t \int_{B_{2\rho}} u^p dx d\tau \right)^{p/\lambda}, \tag{2.40}$$

where

$$\lambda = N(p-2) + p^2, \qquad K(t) = t^{-1} + t^{-N(p-1)/\kappa} \phi^{p-2}(t).$$

Proof. For fixed $T>0, \rho>0, k>0$, set

$$
\begin{aligned}
T_n &= \frac{T}{2} - \frac{T}{2^{n+2}}, \quad \rho_n = \rho + \frac{\rho}{2^n}, \\
\overline{\rho_n} &= \frac{1}{2}(\rho_n + \rho_{n+1}) = \rho + \frac{3\rho}{2^{n+2}}, \\
k_n &= k - \frac{k}{2^{n+1}}, \quad B'_n = B_{\rho_n}, \quad B'_n = B_{\overline{\rho_n}}, \\
Q_n &= B'_n \times (T_n, T), \quad Q'_n = B'_n \times (T_{n+1}, T).
\end{aligned}
\tag{2.41}
$$

Let ξ_n be a cut-off smooth function on Q_n, satisfying

$$
\begin{aligned}
&0 \le \xi_n \le 1, \quad \forall (x,t) \in Q_n; \qquad \xi_n = 1, \quad \forall (x,t) \in Q'_n; \\
&|\nabla \xi_n| \le \frac{2^{n+2}}{\rho}, \quad 0 \le \xi_{nt} \le \frac{2^{n+2}}{T}.
\end{aligned}
\tag{2.42}
$$

Choose $\varphi = (u-k_n)_+^{p-1}\xi_n^p$ in (1.23) and observe that

$$
\begin{aligned}
&\int_{B'_n} u(x,t)\varphi(x,t)dx - \int_{T_n}^t \int_{B'_n} u\varphi_t dx d\tau \\
= \; & \int_{T_n}^t \int_{B'_n} u_t (u-k_n)_+^{p-1}\xi_n^p dx d\tau \\
\ge \; & \frac{1}{p}\int_{B'_n} (u-k_n)_+^p dx - \iint_{Q_n} (u-k_n)_+^p \xi_n^{p-1}\xi_{nt} dx d\tau, \\
& \qquad (T_{n+1} \le t \le T),
\end{aligned}
$$

$$
\begin{aligned}
&\iint_{Q_n} |\nabla u|^{p-2}\nabla u \cdot \nabla \varphi dx d\tau \\
= \; & (p-1)\int_{Q_n} \xi_n^p (u-k_n)_+^{p-2}|\nabla u|^{p-2}\nabla u \cdot \nabla (u-k_n)_+ dx d\tau \\
& + p\int_{Q_n} \xi_n^{p-1}(u-k_n)_+^{p-1}|\nabla u|^{p-2}\nabla u \cdot \nabla \xi_n dx d\tau \\
\ge \; & \frac{p-1}{2}\left(\frac{p}{2(p-1)}\right)^p \iint_{Q'_n} |\nabla (u-k_n)_+^{2(p-1)/p}|^p dx d\tau \\
& - \gamma \iint_{Q_n} (u-k_n)_+^{2(p-1)}|\nabla \xi_n|^p dx d\tau,
\end{aligned}
$$

and denote

$$w_n = (u - k_n)_+^{2(p-1)/p}, \qquad s = \frac{p^2}{2(p-1)}.$$

Then

$$\begin{aligned} &\sup_{T_{n+1}<t<T} \int_{B'_n} w_n^s(x,t)dx + \iint_{Q'_n} |\nabla w_n|^p dxd\tau \\ \leq\ & \frac{\gamma 2^{np}}{\rho^p} \iint_{Q_n} (u-k_n)_+^{p-2}(u-k_n)_+^p dxd\tau \\ &+ \frac{\gamma 2^n}{T} \iint_{Q_n} (u-k_n)_+^p dxd\tau. \end{aligned} \tag{2.43}$$

Since on Q_n,

$$(u-k_n)_+^{p-2}\rho^{-p} \leq \left(\frac{||u(\cdot,t)||_{\infty,B_\rho}}{\rho^{p/(p-2)}}\right)^{p-2} \leq t^{-N(p-2)/\kappa}\phi^{p-2}(t),$$

(2.43) can be written as

$$\begin{aligned} &\sup_{T_{n+1}<t<T} \int_{B'_n} w_n^s(x,t)dx + \iint_{Q'_n} |\nabla w_n|^p dxd\tau \\ \leq\ & \gamma 2^{np} K(T) \int_{Q_n} w_n^s dxd\tau. \end{aligned} \tag{2.44}$$

Choose another cut-off smooth function $\overline{\xi}_n$ on B'_n, such that $0 \leq \overline{\xi}_n \leq 1$, $|\nabla \overline{\xi}_n| \leq \dfrac{2^{n+1}}{\rho}$ and $\overline{\xi}_n = 1$ on B_{n+1}. Then

$$\begin{aligned} w_{n+1}\overline{\xi}_n &\in V_0^{s,p}(B'_n \times (T_{n+1},T)) \\ &= L^\infty(T_{n+1},T;L^s(B'_n)) \cap L^p(T_{n+1},T;W_0^{1,p}(B'_n)). \end{aligned}$$

By Lemma 2.1.1, we have

$$\begin{aligned} &\iint_{Q_{n+1}} w_{n+1}^q dxd\tau \\ \leq\ & \iint_{Q'_{n+1}} |w_{n+1}\overline{\xi}_n|^q dxd\tau \leq \iint_{Q'_n} |w_n\overline{\xi}_n|^q dxd\tau \\ \leq\ & \gamma\left(\iint_{Q'_n} |\nabla w_n|^p dxd\tau + \frac{2^{np}}{\rho^p}\iint_{Q'_n} w_n^p dxd\tau\right) \end{aligned}$$

$$\cdot \left(\sup_{T_{n+1}<t<T} \int_{B'_n} w_n^s(x,t)dx \right)^{p/N}. \tag{2.45}$$

Notice that the definition of w_n and $\phi(t)$ implies

$$\rho^{-p} \iint_{Q'_n} w_n^p dx d\tau \le \gamma K(T) \iint_{Q'_n} w_n^s dx d\tau.$$

Substituting this into (2.45) and using (2.44) to estimate its right hand side, we obtain

$$\begin{aligned} & \int_{Q_{n+1}} w_{n+1}^q dx d\tau \\ \le \;& \gamma (2^{np} K(T))^{(p+N)/N} \left(\iint_{Q_n} w_n^s dx d\tau \right)^{(p+N)/N}. \end{aligned} \tag{2.46}$$

Observe that from the definition of w_n and Hölder's inequality we have

$$\iint_{Q_{n+1}} (u-k_{n+1})_+^p dx d\tau \le \left(\iint_{Q_{n+1}} w_{n+1}^q dx d\tau \right)^{s/q} |A_{n+1}|^{1-s/q},$$

where

$$A_n = \{(x,t) \in Q_n : u(x,t) > k_n\},$$

$$|A_n| = \text{mes} A_n, \qquad n = 0, 1, 2, \cdots.$$

Noting

$$\begin{aligned} & \iint_{Q_n} (u-k_n)_+^p dx d\tau \\ \ge \;& \iint_{Q_n \cap \{u > k_{n+1}\}} (u-k_n)_+^p dx d\tau \\ \ge \;& |k_{n+1} - k_n|^p |A_{n+1}| = \frac{k^p}{2^{p(n+1)}} |A_{n+1}|, \end{aligned}$$

we derive from (2.46) that

$$\begin{aligned} & \iint_{Q_{n+1}} (u-k_{n+1})_+^p dx d\tau \\ \le \;& \gamma b^n (K(T))^{s(p+N)/qN} k^{-p(1-s/q)} \left(\iint_{Q_n} (u-k_n)_+^p dx d\tau \right)^{1+ps/qN}, \end{aligned}$$

where $b = 2^{p(1+ps/qN)}$. Thus we can apply the iteration lemma (Lemma 1.5.3) to conclude that

$$\iint_{Q_0} u^p dx d\tau \le \gamma(K(T))^{-(p+N)/p} k^{N(q/s-1)},$$

provided

$$\iint_{Q_n} (u-k_n)_+^p dx d\tau \to \iint_{Q_\infty} (u-k)_+^p dx d\tau = 0,$$

where

$$Q_\infty = B_\rho \times \left(\frac{T}{2}, T\right), \qquad Q_0 = B_{2\rho} \times \left(\frac{T}{4}, T\right).$$

Therefore if we take

$$k = \gamma(K(T))^{(p+N)/\lambda} \left(\iint_{Q_0} u^p dx d\tau\right)^{p/\lambda},$$

then $\sup\limits_{Q_\infty} u \le k$. Since $T > 0$ is arbitrary, Lemma 2.2.4 is proved. □

Define

$$\psi(t) = \sup_{\tau \in (0,t)} |||u(\cdot,\tau)|||_r, \quad r > 0. \tag{2.47}$$

Lemma 2.2.5 *For any* $t > 0$,

$$\phi(t) \le \gamma \int_0^t \tau^{-N(p-2)/\kappa} \phi^{p-1}(\tau) d\tau + \gamma \psi^{p/\kappa}(t).$$

Proof. Denote

$$\Phi(\rho, t) = t^{N/\kappa} \frac{||u(\cdot,t)||_{\infty, B_\rho}}{\rho^{p/(p-2)}},$$

where $\rho > 0$, $t > 0$. Multiplying (2.40) by $\tau^{N/\kappa} \rho^{-p/(p-2)}$ ($\tau \in (t/2, t)$) and noting the definition of $K(t)$, we see that

$$\begin{aligned}
&\Phi(\rho,\tau)\\
\leq\ & \gamma\phi^{(p-2)(p+N)/\lambda}(t)t^{2pN/\kappa\lambda}\left(\int_{t/4}^{t}\int_{B_{2\rho}}\rho^{-\lambda/(p-2)}u^p dxd\tau\right)^{p/\lambda}\\
&+\gamma t^{p(N(p-1)/\kappa-1)/\lambda}\left(\int_{t/4}^{t}\int_{B_{2\rho}}\rho^{-\lambda/(p-2)}u^p dxd\tau\right)^{p/\lambda}\\
=\ & H^{(1)}+H^{(2)}.
\end{aligned}\tag{2.48}$$

By the definition of $\phi(t), \psi(t), \Phi(\rho,t)$,

$$\begin{aligned}
H^{(1)} &\leq \gamma\phi^{1-2p/\lambda}(t)\left(\int_{t/4}^{t}\tau^{-N(p-2)/\kappa}\Phi^p(2\rho,\tau)d\tau\right)^{p/\lambda}\\
&\leq \gamma\phi^{1-p/\lambda}(t)\left(\int_{t/4}^{t}\tau^{-N(p-2)/\kappa}\phi^{p-1}(\tau)d\tau\right)^{p/\lambda}\\
&\leq \frac{1}{4}\phi(t)+\gamma\int_0^t\tau^{-N(p-2)/\kappa}\phi^{p-1}(\tau)d\tau;
\end{aligned}$$

$$\begin{aligned}
H^{(2)} &\leq \gamma\left(\frac{1}{t}\int_{t/4}^{t}\tau^{N(p-1)/\kappa}\left(\frac{\|u(\cdot,\tau)\|_{\infty,B_{2\rho}}}{(2\rho)^{p/(p-2)}}\right)^{p-1}f(\rho,\tau)d\tau\right)^{p/\lambda}\\
&\leq \gamma\phi^{p(p-1)/\lambda}(t)\left(\frac{1}{t}\int_0^t |||u(\cdot,\tau)|||_r d\tau\right)^{p/\lambda}\\
&\leq \gamma\phi^{p(p-1)/\lambda}(t)\psi^{p/\lambda}(t)\leq\frac{1}{4}\phi(t)+\gamma\psi^{p/\kappa}(t),
\end{aligned}$$

where

$$f(\rho,\tau)=(2\rho)^{-\kappa/(p-2)}\int_{B_{2\rho}}u(x,\tau)dx.$$

Substituting these estimates into (2.48) and taking the supreme with respect to $\rho\geq r$ and $\tau\in(0,t)$ yield the conclusion of our lemma. □

Lemma 2.2.6 *Assume that* $\rho\geq r>0$ *and* $\xi\in C_0^\infty(B_{2\rho})$ *such that* $\xi=1$

for $x \in B_\rho$, $0 \le \xi \le 1$, $|\nabla\xi| \le \rho^{-1}$ *for* $x \in B_{2\rho}$. *Then for any* $t > 0$,

$$\begin{aligned}
&\int_0^t \int_{B_{2\rho}} |\nabla u|^{p-1}\xi^{p-1} dx d\tau \\
&\le \gamma\rho^{1+\kappa/(p-2)} \left(\int_0^t \tau^{(p+1)/\kappa-1}\phi^{(p-2)(p+1)/p}(\tau)\psi(\tau)d\tau \right. \\
&\quad \left. + \int_0^t \tau^{1/\kappa-1}\phi^{(p-2)/p}(\tau)\psi(\tau)d\tau \right)^{(p-1)/p} \\
&\quad \cdot \left(\int_0^t \tau^{1/\kappa-1}\phi^{(p-2)/p}(\tau)\psi(\tau)d\tau \right)^{1/p}.
\end{aligned} \tag{2.49}$$

Proof. The following calculation is formal, since u is required to be positive. However we can replace u by $u+\varepsilon$ with $\varepsilon > 0$ and then let $\varepsilon \to 0$ in the resulting inequality. We simply suppose that u is positive.

By Hölder's inequality,

$$\begin{aligned}
&\int_0^t \int_{B_{2\rho}} |\nabla u|^{p-1}\xi^{p-1} dx d\tau \\
= \;& \int_0^t \int_{B_{2\rho}} \left(\tau^{(p-1)/p^2} |\nabla u|^{p-1} u^{-2(p-1)/p^2} \xi^{p-1} \right) \\
& \qquad \cdot \left(\tau^{-(p-1)/p^2} u^{2(p-1)/p^2} \right) dx d\tau \\
\le \;& \left(\int_0^t \int_{B_{2\rho}} \tau^{1/p} |\nabla u|^p u^{-2/p} \xi^p dx d\tau \right)^{(p-1)/p} \\
& \cdot \left(\int_0^t \int_{B_{2\rho}} \tau^{-(p-1)/p} u^{2(p-1)/p} dx d\tau \right)^{1/p} \\
\equiv \;& (J_1(t))^{(p-1)/p} (J_2(t))^{1/p}.
\end{aligned}$$

Using the integral identity (1.23) which u satisfies and choosing the test function $\varphi(x,\tau) = \tau^{1/p} u^{1-2/p}\, \xi^p$, we derive

$$\begin{aligned}
J_1(t) &= -\int_0^t \int_{B_{2\rho}} \tau^{1/p} |\nabla u|^p u^{-2/p} \xi^p dx d\tau \\
&\le \gamma\rho^{-p} \int_0^t \int_{B_{2\rho}} \tau^{1/p} u^{p-2/p} dx d\tau \\
&\quad + \gamma \int_0^t \int_{B_{2\rho}} \tau^{1/p-1} u^{2(p-1)/p} dx d\tau \\
&\equiv \gamma(L(t) + J_2(t)).
\end{aligned} \tag{2.50}$$

For $L(t)$ and $J_2(t)$, we have the following estimates

$$\begin{aligned} L(t) &\leq \rho^{1+\kappa/(p-2)} \int_0^t \tau^{(p+1)/\kappa-1} \Phi^{(p-2)(p+1)/p}(2\rho,\tau) \\ &\quad \cdot \left(\rho^{-\kappa/(p-2)} \int_{B_{2\rho}} u(x,\tau)dx \right) d\tau \\ &\leq \rho^{1+\kappa/(p-2)} \int_0^t \tau^{(p+1)/\kappa-1} \phi^{(p-2)(p+1)/p}(\tau)\psi(\tau)d\tau, \end{aligned} \tag{2.51a}$$

$$\begin{aligned} J_2(t) &\leq \rho^{1+\kappa/(p-2)} \int_0^t \tau^{1/\kappa-1} \Phi^{(p-2)/p}(2\rho,\tau) \\ &\quad \cdot \left(\rho^{-\kappa/(p-2)} \int_{B_{2\rho}} u(x,\tau)dx \right) d\tau \\ &\leq \rho^{1+\kappa/(p-2)} \int_0^t \tau^{1/\kappa-1} \phi^{(p-2)/p}(\tau)\psi(\tau)d\tau. \end{aligned} \tag{2.51b}$$

Combining (2.50) with (2.51a) and (2.50b) yields (2.49). □

Lemma 2.2.7 *$\psi(t)$ and $\phi(t)$ satisfy*

$$\begin{aligned} \psi(t) \leq \; & |||u_{0n}|||_r + \gamma \left(\int_0^t \tau^{(p+1)/\kappa-1} \phi^{(p-2)(p+1)/p}(\tau)\psi(\tau)d\tau \right. \\ & \left. + \int_0^t \tau^{1/\kappa-1} \phi^{(p-2)/p}(\tau)\psi(\tau)d\tau \right). \end{aligned} \tag{2.52}$$

Proof. From the integral identity (1.23) which u satisfies we can obtain

$$\begin{aligned} & \int_{B_\rho} u(x,t)dx \\ \leq \; & \int_{B_{2\rho}} u_{0n}(x)dx + \frac{\gamma}{\rho} \int_0^t \int_{B_{2\rho}} |\nabla u|^{p-1} \xi^{p-1} dx d\tau, \end{aligned}$$

where ξ is a cut-off smooth function on $B_{2\rho}$ such that $\xi = 1$ for $x \in B_\rho$ and $|\nabla \xi| \leq \rho^{-1}$ for $x \in B_{2\rho}$. Multiply both sides of the above inequality by $\rho^{-\kappa/(p-2)}$, use (2.49) and Young's inequality. Then (2.52) follows. □

Lemma 2.2.8 *Assume that $\phi(t)$ and $\psi(t)$ are nonnegative continuous function on $(0,\infty)$ satisfying*

$$\phi(t) \leq \int_0^t \tau^{-N(p-2)/\kappa} \phi^{p-1}(\tau)d\tau + \gamma\psi^{p/\kappa}(t), \tag{2.53}$$

$$\psi(t) \leq \quad |||u_{0n}|||_r + \gamma \int_0^t \tau^{(p+1)/\kappa - 1} \phi^{(p-2)(p+1)}(\tau)\psi(\tau)d\tau \\ + \gamma \int_0^t \tau^{1/\kappa - 1} \phi^{(p-2)/p}(\tau)\psi(\tau)d\tau. \tag{2.54}$$

Then there exist constants γ_0, γ_1, γ_2 *depending only on* γ, p, N *such that for* $0 < t < \gamma_0 |||u_{0n}|||_r^{2-p}$,

$$\phi(t) \leq \gamma_1 |||u_{0n}|||_r^{2-p}, \qquad \psi(t) \leq \gamma_2 |||u_{0n}|||_r.$$

Proof. Since $\psi(t)$ is increasing, from (2.53) it follows that for any fixed $t^* > 0$,

$$\phi(t) \leq \int_0^t \tau^{-N(p-2)/\kappa} \phi^{p-1}(\tau)d\tau + \gamma \psi^{p/\kappa}(t^*), \quad t \in (0, t^*].$$

Hence $\phi(t)$ is less than or equal to the solution of the problem

$$H'(t) = \gamma t^{-N(p-2)/\kappa} H^{p-1}(t), \quad H(0) = \gamma \psi^{p/\kappa}(t^*).$$

From this it follows that

$$\phi(t) \leq H(t) = \gamma \psi^{p/\kappa}(t^*) \left(1 - \gamma(p-2)\left(t\psi^{p-2}(t^*)\right)^{p/\kappa}\right)^{-1/(p-2)},$$

provided that the value in the brackets is positive. If we take t^* such that

$$1 - \gamma(p-2)\left(t^* \psi^{p-2}(t^*)\right)^{p/\kappa} > 0,$$

then the above inequality holds for $0 < t < t^*$. Thus there exist positive constants $\overline{\gamma}_0$, $\overline{\gamma}_1$ depending only on p, N, such that

$$\left(t\psi^{p-2}(t)\right)^{p/\kappa} \leq \overline{\gamma}_0 \tag{2.55}$$

implies

$$\phi(t) \leq \overline{\gamma}_1 \psi^{p/\kappa}(t). \tag{2.56}$$

Substituting (2.56) into (2.54) yields

$$\psi(t) \leq \gamma |||u_{0n}|||_r + \gamma \int_0^t \tau^{1/\kappa - 1} \psi^{1+(p-2)/\kappa}(\tau)d\tau$$

for all t satisfying (2.55). This means that $\psi(t)$ is less than or equal to the solution of the problem

$$M'(t) = \gamma t^{1/\kappa - 1} M^{1+(p-2)/\kappa}(t),$$

$$M(0) = \gamma |||u_{0n}|||_r.$$

From this it follows that

$$\psi(t) \le M(t) = \gamma |||u_{0n}|||_r \left(1 - \gamma \left(t|||u_{0n}|||_r^{p-2}\right)^{p/\kappa}\right)^{-\kappa/(p-2)},$$

provided that $\gamma \left(t|||u_{0n}|||_r^{p-2}\right)^{p/\kappa} < 1$. Thus we can choose a constant $\overline{\overline{\gamma}}_0$ such that if t satisfies (2.55) and

$$0 < t < \overline{\overline{\gamma}}_0 |||u_{0n}|||_r^{2-p}, \tag{2.57}$$

then

$$\psi(t) \le \gamma_2 |||u_{0n}|||_r. \tag{2.58}$$

Combining (2.55)with (2.57), (2.58) shows that there exists $\gamma_0 = \gamma_0(p, N)$ such that (2.55), (2.58) and hence (2.56) hold for u_n, provided that $0 < t < \gamma_0 |||u_{0n}|||_r^{2-p}$. This completes the proof of our lemma. □

Now we come back to the proof of Theorem 2.2.3. Combining Lemma 2.2.4 – Lemma 2.2.8 we derive (2.34) – (2.36) for u_n. Hence as we have pointed out, to complete the proof of Theorem 2.2.3, it remains to prove that u satisfies the initial value condition (1.6) in the generalized sense. In fact, we can prove that for any $R > 0$,

$$\lim_{t\to 0} \int_{B_R} |u(x,t) - u_0(x)| dx = 0. \tag{2.59}$$

However, this fact is an immediate corollary of the following lemma.

Lemma 2.2.9 *For any $R > 0$ and any integers m, n,*

$$\int_{B_R} |u_n(x,t) - u_m(x,t)| dx$$
$$\le \int_{B_{2R}} |u_{0n}(x) - u_{0m}(x)| dx + C_R(t),$$

where $C_R(t)$ is independent of m, n and

$$\lim_{t\to 0} C_R(t) = 0.$$

Proof. Since u_n is the generalized solution of (1.1) with $u_{nt} \in L^\infty(Q_T)$, from the definition of generalized solutions, it is easy to verify that

$$\int_0^t \int_{B_{2R}} \varphi u_{n\tau} dx d\tau + \int_0^t \int_{B_{2R}} |\nabla u_n|^{p-2} \nabla u_n \cdot \nabla \varphi dx d\tau = 0$$
$$\text{for } \varphi \in L^p(0,T; W_0^{1,p}(B_{2R})).$$

Let $v = u_n - u_m$. Then

$$\int_0^t \int_{B_{2R}} \varphi v_t dx d\tau + \int_0^t \int_{B_{2R}} \left(|\nabla u_n|^{p-2} \nabla u_n - |\nabla u_m|^{p-2} \nabla u_m\right) \cdot \nabla \varphi dx d\tau = 0$$

for $\varphi \in L^p(0,T; W_0^{1,p}(B_{2R}))$. Choose $\varphi = \xi(x)\mathrm{sgn}_\eta v$, where $\mathrm{sgn}_\eta s$ is the function in (2.11) and $\xi(x) \in C_0^1(B_{2R})$ such that $0 \le \xi \le 1$, $\xi = 1$ in B_R. Then we obtain

$$\begin{aligned}
&\int_0^t \int_{B_{2R}} \xi \left(\int_0^v \mathrm{sgn}_\eta s ds\right) dx d\tau \\
&+ \int_0^t \int_{B_{2R}} \left(|\nabla u_n|^{p-2} \nabla u_n - |\nabla u_m|^{p-2} \nabla u_m\right) \nabla \xi \mathrm{sgn}_\eta v dx d\tau \\
&+ \int_0^t \int_{B_{2R}} \left(|\nabla u_n|^{p-2} \nabla u_n - |\nabla u_m|^{p-2} \nabla u_m\right) \xi \nabla \xi \mathrm{sgn}'_\eta v dx d\tau = 0.
\end{aligned}$$

Noticing that the third term is nonnegative and

$$\int_0^t \int_{B_{2R}} \xi \left(\int_0^v \mathrm{sgn}_\eta s ds\right)_\tau dx d\tau = \int_{B_{2R}} \xi \int_0^v \mathrm{sgn}_\eta s ds \Big|_0^t dx,$$

we obtain

$$\begin{aligned}
&\int_{B_{2R}} \xi \int_0^v \mathrm{sgn}_\eta s ds \Big|_0^t dx \\
&+ \int_0^t \int_{B_{2R}} \left(|\nabla u_n|^{p-2} \nabla u_n - |\nabla u_m|^{p-2} \nabla u_m\right) \nabla \xi \mathrm{sgn}_\eta v dx d\tau \le 0.
\end{aligned}$$

Letting $\eta \to 0$ yields

$$\begin{aligned}
\int_{B_R} |u_n(x,t) - u_m(x,t)| dx \le & \int_{B_{2R}} |u_{0n} - u_{0m}| dx \\
& + C \int_0^t \int_{B_{2R}} \left(|\nabla u_n|^{p-1} + |\nabla u_m|^{p-1}\right) dx d\tau,
\end{aligned}$$

from which the conclusion of Lemma 2.2.9 follows by using (2.36). □

2.2.3 *Some remarks*

Remark 2.2.1 A result similar to Theorem 2.2.3 is valid for those initial data which are measures μ satisfying the following growth condition: for some $r > 0$,

$$\sup_{\rho \geq r} \rho^{-\kappa/(p-2)} \int_{B_\rho} |d\mu| < \infty.$$

In this case, to prove the result, one chooses the approximate initial function $u_{0n} \in C_0^\infty(\mathbb{R}^N)$ such that $|||u_{0n}|||_r \leq |||\mu|||_r$ and

$$\lim_{n \to \infty} \int_{\mathbb{R}^N} u_{0n} \varphi dx = \int_{\mathbb{R}^N} \varphi d\mu \quad \text{for } \varphi \in C_0(\mathbb{R}^N).$$

Here

$$|||\mu|||_r = \sup_{\rho \geq r} \rho^{-\kappa/(p-2)} \int_{B_\rho} |d\mu|.$$

The generalized solution u obtained satisfies the initial value condition in the following sense:

$$\lim_{t \to 0} \int_{\mathbb{R}^N} u(x,t) h(x) dx = \int_{\mathbb{R}^N} h(x) d\mu \quad \text{for } h \in C_0^1(\mathbb{R}^N).$$

Remark 2.2.2 In the next section, we will see that for the Cauchy problem for (1.1) with initial data $u_0 \geq 0$ to have a nonnegative generalized solution, the growth condition $|||u_0|||_r$ is not sufficient but also necessary.

Remark 2.2.3 [DH2] and [ZH7] are devoted to the case $p < 2$; the solvability of the Cauchy problem is proved for any Radon measure as its initial data and no growth condition is assumed provided $\dfrac{2N}{(N+1)} < p < 2$. In [DH2], the case $1 \leq p \leq \dfrac{2N}{N+1}$, $u_0 \in L^1_{loc}(\mathbb{R}^N)$ is discussed. In this case, since it is not able to obtain the estimate $|\nabla u| \in L^p_{loc}(Q_T)$, generalized solutions of some special form are defined and the existence is established.

Remark 2.2.4 In [ZY1], the authors have discussed the solvability of the Cauchy problem for the equation (1.2) with initial function $u_0 \in L^1(\mathbb{R}^N)$. In [ZX], it has been proved that the Cauchy problem for the equation (1.2) is solvable for $u_0 \in L^1_{loc}(\mathbb{R}^N)$ satisfying the following growth condition: for some $r > 0$,

$$\sup_{\rho \geq r} \rho^{-(N(m(p-1)-1)+p)/(m(p-1)-1)} \int_{B_\rho} |u_0| dx < \infty.$$

It is also pointed out that for (1.2) to have a nonnegative generalized solution, the growth condition is the best possibility. The Hölder continuity of generalized solutions are discussed in [I], [YU1] and [ZH].

2.3 Harnack Inequality and the Initial Trace of Solutions

We have discussed the existence of generalized solutions of (1.1) and proved that if u_0 satisfies

$$|||u_0|||_r < \infty \tag{3.1}$$

for some constant $r > 0$, then the corresponding Cauchy problem admits a generalized solution on $Q_{T(u_0)}$, where $T(u_0)$ is a certain positive constant.

From this section on, we study the properties of generalized solutions. In this section we first establish the Harnack inequality for nonnegative generalized solutions of (1.1) and then apply it to study the initial trace of generalized solutions. It will be indicated that for the Cauchy problem for (1.1) to have a generalized solution, the growth condition (3.1) is not only sufficient, but also necessary.

2.3.1 *Local Harnack inequality*

Theorem 2.3.1 *Assume that $p > 2$ and u is a nonnegative generalized solution of (1.1) on Q_T. Let $(x_0, t_0) \in Q_T$ and $u(x_0, t_0) > 0$. Then there exist positive constants C_0, $C_1 > 1$ depending only on p, N such that*

$$u(x_0, t_0) \leq C_0 \inf_{x \in B_\rho(x_0)} u(x, t_0 + \theta)$$

with

$$\theta = \frac{C_1 \rho^p}{u^{p-2}(x_0, t_0)}, \tag{3.2}$$

and ρ, θ such that $Q_{4\rho}(\theta) = \{|x - x_0| < 4\rho\} \times \{t_0 - 4\theta, t_0 + 4\theta\} \subset Q_T$.

Remark 2.3.1 In the discussion of regularity of solutions in §2.4, it will be proved that the generalized solution u of (1.1) is locally Hölder continuous. Hence we may regard u as defined everywhere on Q_T. It should be pointed out that in this section, we will use the Hölder continuity of solutions many times. Although this property is presented in §2.4, its proof and all discussions in §2.4 do not depend on any result of this section.

To prove Theorem 2.3.1, we need the following proposition and lemma.

Proposition 2.3.1 *(Local comparison theorem) Let u, v be generalized solutions of (1.1) on $\Omega_T = \Omega \times (0,T]$, where $\Omega \subset \mathbb{R}^N$ is a bounded and smooth domain, such that*

$$u, v \in C(0,T; L^2(\Omega)) \cap L^p(0,T; W^{1,p}(\Omega)),$$

$$u \le v \quad \text{a.e. on the parabolic boundary of } \Omega_T.$$

Then $u \le v$ a.e.on Ω_T.

Proof. Denote $w = u - v$. Then from (1.26) it is easy to see that for any fixed $\tau \in (0,T)$ and $\varphi \in W_0^{1,p}(\Omega)$, there holds

$$\int_\Omega w_{h\tau}\varphi dx + \int_\Omega \left(|\nabla u|^{p-2}\nabla u - |\nabla v|^{p-2}\nabla v\right)_h \cdot \nabla\varphi dx = 0,$$

where u_h is the Steklov mean value of u, $h > 0$ is sufficiently small so that $\tau + h < T$. Choose

$$\varphi(x,\tau) = (w_h)_+(x,\tau).$$

By the property of the Steklov mean value and noting that $w \le 0$ almost everywhere on the lateral boundary of Ω_T, we see that $\varphi \in L^p(0,T; W_0^{1,p}(\Omega))$. Substituting this function into the above integral equality and then integrating with respect to $\tau \in (0,t)$, we obtain

$$\begin{aligned} &\int_\Omega (w_h)_+^2 dx - \int_\Omega (w_h)_+^2(x,0)dx \\ = &-2\int_0^t \int_\Omega \left(|\nabla u|^{p-2}\nabla u - |\nabla v|^{p-2}\nabla v\right)_h \nabla((u-v)_h)_+ dxd\tau. \end{aligned} \tag{3.3}$$

Obviously

$$\int_\Omega (w_h)_+^2(x,0)dx \le \frac{1}{h}\int_0^h \int_\Omega w_+^2(x,\tau)d\tau.$$

By assumption, $u, v \in C(0,T;L^2(\Omega))$ and $w_+(x,0) = 0$ a.e on Ω. Therefore

$$\lim_{\tau \to 0} \int_\Omega (w)_+^2(x,\tau)dx = 0$$

and hence

$$\lim_{h \to 0} \int_\Omega (w_h)_+^2(x,0)dx = 0. \tag{3.4}$$

Letting $h \to 0$ in (3.3) and using (3.4) and Lemma 2.1.2, we further obtain

$$\int_\Omega w_+^2(x,t)dx = -2\int_0^t \int_\Omega \left(|\nabla u|^{p-2}\nabla u - |\nabla v|^{p-2}\nabla v\right) \cdot \nabla (u-v)_+ dxd\tau \le 0.$$

Hence $w_+ = 0$ or $u \le v$ a.e.on Ω_T. □

Corollary 2.3.1 *The first boundary value problem for (1.1) on $\Omega_T = \Omega \times (0,T)$ with Ω being a bounded and smooth domain, admits at most one generalized solution in the class of functions $C(0,T;L^2(\Omega) \cap L^p(0,T;w^{1,p}(\Omega))$.*

Lemma 2.3.1 *Let u be a nonnegative generalized solution of (1.1) on Q_T. If $|x - \overline{x}| < \rho$ implies $u(x,\overline{t}) \ge k$ where $(\overline{x},\overline{t}) \in Q_T$, then for $t \in [\overline{t},T]$,*

$$u(x,t) \ge E_{k,\rho}(x,t;\overline{x},\overline{t}),$$

where $E_{k,\rho}(x,t;\overline{x},\overline{t})$ is the fundamental solution of (1.1)(see (1.31)).

Proof. Given any $t^* \in (\overline{t},T)$. Denote

$$Q^* = B_{S^{1/k}(t^*)}(\overline{x}) \times (\overline{t},t^*).$$

Since $t \in (\overline{t},t^*)$, $\operatorname{supp} E_{k,\rho} \subset Q^*$ and $u \ge 0$, we have $u \ge E_{k,\rho} \equiv 0$ on the lateral boundary of Q^*. On the other hand, since we always have $E_{k,\rho} \le k$ and for $t = \overline{t}$, $\operatorname{supp} E_{k,\rho} = \{x; |x - \overline{x}| < S(\overline{t})^{1/\kappa} = \rho\}$, by assumption, we see that also $u \ge E_{k,\rho}$ on the bottom boundary of Q^*. Thus by Proposition 2.3.1, we have $u \ge E_{k,\rho}$ on Q^*. Since t^* is arbitrary and $u \ge 0$, it follows that $u \ge E_{k,\rho}$ holds for $x \in \mathbb{R}^N, t \ge \overline{t}$. □

Proof of Theorem 2.3.1. Let $(x_0,t_0) \in Q_T$ and $u(x_0,t_0) > 0$. Consider the cylinder

$$Q_{4\rho} = (|x - x_0| < 4\rho) \times \left(t_0 - \frac{4C\rho^p}{(u(x_0,t_0))^{p-2}}, t_0 + \frac{4C\rho^p}{(u(x_0,t_0))^{p-2}}\right),$$

where $C > 0$ is a constant to be determined later. By the transformation

$$x \to \frac{x - x_0}{\rho}, \qquad t \to \frac{(t - t_0)(u(x_0, t_0))^{p-2}}{\rho^p},$$

$Q_{4\rho}$ is changed to $Q \equiv Q^+ \cap Q^-$, where

$$Q^+ = B_4 \times [0, 4C), \qquad Q^- = B_4 \times (-4C, 0].$$

The new variables are still denoted by x, t. Set

$$v(x,t) = \frac{1}{u(x_0, t_0)} u\left(x_0 + \rho x, t_0 + \frac{t\rho^p}{(u(x_0, t_0))^{p-2}}\right).$$

We can verify that v is a bounded and nonnegative solution of the equation

$$v_t - \operatorname{div}(|\nabla v|^{p-2} \nabla v) = 0$$

on Q and $v(0,0) = 1$. To prove Theorem 2.3.1, it suffices to prove that there exist constants $\gamma_0 \in (0,1]$ and $C = C(p, N) > 1$ such that $\inf_{B_1} v(x, C) \ge \gamma_0$.

Denote

$$Q_\tau = (|x| < \tau) \times (-\tau^p, 0] \qquad \tau \in (0,1),$$

$$M_\tau = \sup_{Q_\tau} v, \qquad N_\tau = (1 - \tau)^{-\beta}, \qquad \tau \in [0, 1),$$

where the constant $\beta > 1$ is to be determined. Suppose that τ_0 is the maximal root of the equation $M_\tau = N_\tau$. Since $M_0 = N_0$, M_τ are bounded on $[0,1]$ and $\lim_{\tau \to 1} N_\tau = \infty$, we can assert the existence of τ_0, which satisfies $0 \le \tau_0 < 1$ and

$$\sup_{Q_\tau} v \le N_\tau \quad \text{for } \tau > \tau_0.$$

By the continuity of v (see §2.4), it follows that there exists $(\overline{x}, \overline{t}) \in Q_{\tau_0}$ such that

$$v(\overline{x}, \overline{t}) = N_{\tau_0} = (1 - \tau_0)^{-\beta}. \tag{3.5}$$

Now we try to prove that in a small neighborhood of $x = \overline{x}$, $v(x, \overline{t})$ is of order like $(1 - \tau_0)^{-\beta}$.

Let $R = \dfrac{1 - \tau_0}{2}$. Consider

$$D_R = (|x - \overline{x}| < R) \times (\overline{t} - R^p, \overline{t}).$$

Since $D_R \subset Q_{(1+\tau_0)/2}$, we have

$$\sup_{D_R} v \le N_{(1+\tau_0)/2} = 2^\beta(1-\tau_0)^{-\beta}$$

and

$$\text{osc}_{D_R} v \le 2^\beta(1-\tau_0)^{-\beta} \equiv \omega_0.$$

By the continuity of v (see §2.4),

$$\text{osc}_{B_\rho(\overline{x})} v \le A_0\omega_0 \left(\frac{\rho}{R}\right)^\sigma, \quad \text{for } 0<\rho<R, \tag{3.6}$$

where $\sigma \in (0,1)$, A_0 is a constant depending only on p, N and $||v||_\infty$.

Combining (3.5) with (3.6) gives

$$v(x,\overline{t}) \ge (1-\tau_0)^{-\beta} - A_0 2^\beta (1-\tau_0)^{-\beta}\left(\frac{\rho}{R}\right)^\sigma,$$

$$x \in B_\rho(\overline{x}), \qquad 0<\rho<R.$$

Taking $\rho = \epsilon R$ with small $\epsilon>0$ so that $A_0 2^\beta \epsilon^\sigma \le \frac{1}{2}$, we obtain

$$v(x,\overline{t}) \ge \frac{1}{2}(1-\tau_0)^{-\beta}, \quad \text{for } x \in B_{\epsilon R}(\overline{x}).$$

Taking $k=(1-\tau_0)^{-\beta}$, $\rho=\epsilon R$ and using Lemma 2.3.1 yield

$$v(x,t) \ge E_{k,\rho}(x,t;\overline{x},\overline{t}) \quad \text{for } t \ge \overline{t}.$$

From the definition of $E_{k,\rho}$,

$$\text{supp}\, E_{k,\rho}(\cdot,t;\overline{x},\overline{t})$$

$$= \left(|x-\overline{x}| < \left(\kappa(p/(p-2))^{p-1}\,\rho^{N(p-2)}k^{p-2}(t-\overline{t}) + \rho^\kappa\right)^{1/\kappa}\right).$$

Choosing

$$\beta = N, \quad C = \frac{3^\kappa(p-2)^{p-1}2^{(N+1)(p-2)}}{\kappa p^{p-1}\epsilon^{N(p-2)}},$$

and noting that $|\overline{x}| < 1, \overline{t} \in (-1, 0)$, we easily see that $\operatorname{supp} E_{k,\rho}(\cdot, C; \overline{x}, \overline{t}) \supset B_2$ and

$$\begin{aligned}\inf_{x\in B_1} v(x,C) &\geq \inf_{x\in B_1} E_{k,\rho}(x,C;\overline{x},\overline{t}) \\ &\geq \frac{1}{2\cdot 3^N}\left(\frac{\epsilon}{2}\right)^N\left(1-\left(\frac{2}{3}\right)^{p/(p-1)}\right)^{p/(p-2)} \\ &= \gamma_0.\end{aligned}$$

This completes the proof of Theorem 2.3.1. □

Theorem 2.3.2 *Assume that $p > 2$ and u is a nonnegative generalized solution of (1.1) on Q_T. Then for given $(x_0, t_0) \in Q_T$ and any ρ, θ such that $Q_{4\rho}(\theta) \subset Q_T$,*

$$\begin{aligned}& u(x_0, t_0) \\ \leq\ & C\left(\left(\frac{\rho^p}{\theta}\right)^{1/(p-2)} + \left(\frac{\theta}{\rho^p}\right)^{N/p}\left(\inf_{B_\rho(x_0)} u(\cdot, t_0+\theta)\right)\right)^{\kappa/p},\end{aligned} \tag{3.7}$$

where $\kappa = N(p-2) + p$, $C = C(p, N)$.

Noting that there exists $\widetilde{x} \in B_\rho(x_0)$ such that

$$u(\widetilde{x}, t_0) = \frac{1}{|B_\rho(x_0)|}\int_{B_\rho} u(x, t_0)dx,$$

from Theorem 2.3.2 we obtain, in particular,

Corollary 2.3.2 *Assume that $p > 2$ and u is a nonnegative generalized solution of (1.1) on Q_T. Then for given $(x_0, t_0) \in Q_T$ and any ρ, θ such that $Q_{4\rho}(\theta) \subset Q_T$,*

$$\begin{aligned}& \frac{1}{|B_\rho(x_0)|}\int_{B_\rho} u(x, t_0)dx \\ \leq\ & C\left(\left(\frac{\rho^p}{\theta}\right)^{1/(p-2)} + \left(\frac{\theta}{\rho^p}\right)^{N/p}(u(x_0, t_0+\theta))^{\kappa/p}\right),\end{aligned} \tag{3.8}$$

where $\kappa = N(p-2) + p, C = C(p, N)$.

Proof of Theorem 2.3.2. Without loss of generality, we suppose that $(x_0, t_0) = (0, 0)$. Denote $u_* = u(0, 0)$. We may also suppose that

$$t^* = \frac{C_1 \rho^p}{u_*^{p-2}} \le \frac{\theta}{2}, \tag{3.9}$$

where C_1 is the constant in Theorem 2.3.1; otherwise we have

$$u_* \le B \left(\frac{\rho^p}{\theta} \right)^{1/(p-2)}$$

with $B = (2C_1)^{1/(p-2)}$ and obviously (3.7) holds.

By Theorem 2.3.1,

$$u_* \le C_0 u(x, t^*) \quad \text{for } x \in B_\rho.$$

Consider the fundamental solution $E_{k,\rho}(x, t; 0, t^*)$ with $k = C_0^{-1} u_*$. By Lemma 2.3.1,

$$\begin{aligned} & u(x, \theta) \\ \ge \quad & \frac{u_* C_0^{-1} \rho^N}{S^{N/\kappa}(\theta)} \left(1 - \left(\frac{|x|}{S^{1/\kappa}(\theta)} \right)^{p/(p-1)} \right)_+^{(p-1)/(p-2)}, \end{aligned} \tag{3.10}$$

where

$$\begin{aligned} S(\theta) &= b(C_0^{-1} u_*)^{p-2} \rho^{N(p-2)} (\theta - t^*) + \rho^\kappa \\ &= \left(\frac{C_1 b}{C_0^{p-2}} \frac{u_*^{p-2}}{C_1 \rho^p} (\theta - t^*) + 1 \right) \rho^\kappa, \\ B &= \kappa \left(\frac{p}{p-2} \right)^{p-1}. \end{aligned}$$

Using (3.9), we have

$$\left(\frac{C_1 b}{C_0^{p-2}} + 1 \right) \rho^\kappa \le S(\theta) \le \left(\frac{b}{C_0^{p-2}} + \frac{1}{2C_1} \right) u_*^{p-2} \left(\frac{\theta}{\rho^p} \right) \rho^\kappa.$$

Therefore, it follows from (3.10) that

$$u(x, \theta) \ge C_2 u_*^{p/\kappa} \left(\frac{\rho^p}{\theta} \right)^{N/\kappa},$$

that is,

$$u_* \leq C_2^{-\kappa/p} \left(\frac{\theta}{\rho^p}\right)^{N/p} (u(x,\theta))^{\kappa/p}, \qquad \text{for } |x| < \rho,$$

where $C_2 = C_2(p,N)$. The proof of Theorem 2.3.2 is complete. □

From the proof of Theorem 2.3.2, we see that Theorem 2.3.1 implies Theorem 2.3.2. Conversely, if the conclusion of Theorem 2.3.2 is valid, then choosing

$$\theta = \frac{(2B)^{p-2}\rho^p}{u^{p-2}(x_0,t_0)}$$

in (3.7) gives

$$u(x_0,t_0) \leq 2B^{N(p-2)/\kappa} \inf_{x\in B_\rho(x_0)} u(\cdot, t_0+\theta).$$

This means that Theorem 2.3.2 implies Theorem 2.3.1. Summing up, we get

Proposition 2.3.2 *Theorem 2.3.1 and Theorem 2.3.2 are equivalent.*

2.3.2 *Global Harnack inequality*

Theorem 2.3.3 *Assume that $p > 2$ and u is a nonnegative generalized solution of (1.1) on Q_T. Then there exists a constant $C = C(p,N) > 1$ such that for any $(x_0,t_0) \in Q_T$ and $\rho > 0, \theta > 0$,*

$$\frac{1}{|B_\rho(x_0)|}\int_{B_\rho} u(x,t_0)dx \leq C\left(\left(\frac{\rho^p}{\theta}\right)^{1/(p-2)} + \left(\frac{\theta}{\rho^p}\right)^{N/p}(u(x_0,t_0+\theta))^{\kappa/p}\right) \tag{3.11}$$

provided $t_0+\theta < T$, where $\kappa = N(p-2)+p$.

To prove this theorem, we need some propositions and lemmas.

Proposition 2.3.3 *(Global comparison theorem) Let u and v be generalized solutions of (1.1) such that*

$$u, v \in C(0,T; L^2(\mathbb{R}^N)) \cap L^p(0,T; W^{1,p}(\mathbb{R}^N)),$$

$$u(x,0) \leq v(x,0) \qquad a.e \text{ on } \mathbb{R}^N.$$

Then $u(x,t) \le v(x,t)$ a.e. on Q_T.

Proof. Choose $\xi_R \in C_0^\infty(B_{2R})$ such that $0 \le \xi_R \le 1$, $\xi_R = 1$ for $x \in B_R$ and $|\nabla \xi_R| \le R^{-1}$. Similar to the proof Proposition 2.3.1, we can derive

$$\begin{aligned}
&\int_{B_{2R}} \xi_R w_+^2(x,t)dx \\
= & -2\int_0^t \int_{B_{2R}} \xi_R \left(|\nabla u|^{p-2}\nabla u - |\nabla v|^{p-2}\nabla v\right) \cdot \nabla (u-v)_+ dxd\tau \\
& -2\int_0^t \int_{B_{2R}} \left(|\nabla u|^{p-2}\nabla u - |\nabla v|^{p-2}\nabla v\right) \cdot \nabla \xi_R w_+ dxd\tau \\
\le & \int_0^t \int_{B_{2R}} \left(|\nabla u|^{p-2}\nabla u - |\nabla v|^{p-2}\nabla v\right) \cdot \nabla \xi_R w_+ dxd\tau,
\end{aligned}$$

where $w = u - v$. Using Hölder's inequality to estimate the integral on the right hand side of the above formula, we further obtain

$$\begin{aligned}
&\int_{B_R} w_+^2(x,t)dx \\
\le & \; CR^{-1}||w_+||_p(||\nabla u||_p^{p-1} + ||\nabla v||_p^{p-1}) \to 0, \quad \text{as } R \to \infty.
\end{aligned}$$

From this it follows that $u \le v$ a.e.on Q_T. □

Corollary 2.3.3 *The Cauchy problem for (1.1) admits at most one generalized solution on Q_T in the class $C(0,T;L^2(\mathbb{R}^N)) \cap L^p(0,T;W^{1,p}(\mathbb{R}^N))$.*

Consider the following Cauchy problem

$$v_t - \operatorname{div}(|\nabla v|^{p-2}\nabla v) = 0, \tag{3.12}$$

$$v(x,0) = \overline{v}_0(x), \qquad x \in \mathbb{R}^N, \tag{3.13}$$

where

$$\overline{v}_0(x) = v_0(x), \quad \text{for } x \in \overline{B}_r; \quad \overline{v}_0(x) = 0, \quad \text{for } x \in \mathbb{R}^N \backslash \overline{B}_r$$

and $v_0 \ge 0$, $v_0 \in C(\overline{B}_r)$ is a given nonnegative function.

By Theorem 2.2.2, this problem admits a generalized solution v on $Q = \mathbb{R}^N \times (0,\infty)$ which possesses all properties in Theorem 2.2.1 except $u_t \in L^\alpha(Q_T)$. In addition, v possesses the following property.

Proposition 2.3.4 *For any $T > 0$, there exists $R(T)$ such that*

$$supp\, v(\cdot,t) \subset B_{R(T)}, \quad \text{for } t \in (0,T), \tag{3.14}$$

$$\int_{\mathbb{R}^N} v(x,t)dx = \int_{B_r} v_0(x)dx, \quad \text{for } t \in (0,\infty). \tag{3.15}$$

Proof. We will use the fundamental solution of (3.12)$E_{k,2r}(x,t;0,0)$ (see (1.31)). For sufficiently large k,

$$\begin{aligned} E_{k,2r}(x,0;0,0) &\geq k\left(1-\left(\frac{1}{2}\right)^{p/(p-1)}\right)^{(p-1)/(p-2)} \\ &\geq \sup_{B_r} v_0(x) = \sup_{\mathbb{R}^N} v_0. \end{aligned}$$

Thus, by Proposition 2.3.3,

$$v(x,t) \leq E_{k,2r}(x,t;0,0) \quad \text{for } (x,t) \in Q_T.$$

From this (3.14) follows. Noting that for $\varphi(x) \in C_0^\infty(\mathbb{R}^N)$,

$$\begin{aligned} &\int_{\mathbb{R}^N} v(x,t)\varphi(x)dx - \int_{\mathbb{R}^N} v_0(x)\varphi(x)dx \\ = &\int_0^t \int_{\mathbb{R}^N} |\nabla v|^{p-2}\nabla v \cdot \nabla\varphi(x)dxd\tau, \end{aligned} \tag{3.16}$$

and choosing $\varphi \in C_0^\infty(B_{2\rho})$ such that $\varphi = 1$ for B_ρ, we obtain (3.15) by letting $\rho \to \infty$ and using (3.14). The proof of our proposition is complete. ∎

Denote

$$E_0 = \fint_{B_r} v_0 dx \equiv \frac{1}{|B_r|}\int_{B_r} v_0 dx.$$

Then using (3.15) we obtain

$$\begin{aligned} |||v|||_r = &\sup_{t\in(0,\infty)} \sup_{\rho\geq r} \rho^{-\kappa/(p-2)} \int_{B_\rho} v(x,t)dx \\ &\leq \sup_{t\in(0,\infty)} \sup_{\rho\geq r} \rho^{-\kappa/(p-2)} \int_{\mathbb{R}^N} v(x,t)dx \\ &= |B_1| r^{-p/(p-2)} E_0. \end{aligned}$$

This combining with (2.36) in which u is changed to v, yields the existence of a constant $C_3 = C_3(p,N)$ such that for $0 < t < C_0 r^p E_0^{2-p}$,

$$\frac{1}{\rho}\int_0^t \fint_{B_\rho} |\nabla v|^{p-1} dxd\tau \leq C_3\left(\frac{t}{r^p}\right)^{1/\kappa}\left(\frac{\rho}{r}\right)^{p/(p-2)} E_0^{1+(p-2)/\kappa}. \tag{3.17}$$

Lemma 2.3.2 *There exists a constant $C_* = C_*(p, N) > 0$ such that*

$$\fint_{B_{2r}} v(x, t_0)dx \geq 2^{-(N+1)} E_0,$$

where $t_0 = C_* \dfrac{r^p}{E_0^{p-2}}$.

Proof. Let

$$C_* = \left(2^{2(p-1)/(p-2)+N+1} pC_3\right)^{-\kappa},$$

where C_3 is the constant in (3.17). In (3.16) take $t = t_0$ and choose $\varphi \in C_0^\infty(B_{2r})$ such that

$$0 \leq \varphi \leq 1, \quad |\nabla\varphi| \leq pr^{-1}, \quad \text{for } x \in B_{2r}, \quad \varphi = 1 \quad \text{for } x \in B_r.$$

Then using (3.17) gives

$$\begin{aligned} \fint_{B_{2r}} v(x, t_0)dx &\geq 2^{-N} \fint_{B_r} v_0(x)dx - \frac{p}{r} \int_0^{t_0} \fint_{B_{2r}} |\nabla v|^{p-1} dxdt \\ &\geq 2^{-N} E_0 - 2^{2(p-1)/(p-2)} pC_3 C_*^{1/\kappa} E_0. \\ &= 2^{-(N+1)} E_0. \end{aligned}$$

□

Lemma 2.3.3 *There exists a constant $B(p, N) > 1$ such that for any $\theta > 0$,*

$$\fint_{B_r} v_0(x)dx \leq B\left(\left(\frac{r^p}{\theta}\right)^{1/(p-2)} + \left(\frac{\theta}{r^p}\right)^{N/p} \left(\inf_{B_r} v(\cdot, \theta)\right)^{\kappa/p}\right).$$

Proof. Since $v \in C(Q_T)$ (see §2.4), Lemma 2.3.2 implies that for some $\overline{x} \in B_{2r}$,

$$v(\overline{x}, t_0) \geq 2^{-(N+1)} E_0. \tag{3.18}$$

Let $\delta = \left(\dfrac{C_*}{2^\kappa C_1}\right)^{1/p}$, where C_* and C_1 are the constants in Lemma 2.3.2 and Theorem 2.3.1. From $t_0 = C_* \dfrac{r^p}{E_0^{p-2}}$ and (3.18), we have

$$t_0 \geq \frac{4C_1(\delta r)^p}{(v(\overline{x}, t_0))^{p-2}}.$$

Now we can use Theorem 2.3.1 on the cylinder

$$Q = B_{4\delta r}(\overline{x}) \times \left\{ t_0 - \frac{4C_1(\delta r)^p}{(v(\overline{x},t_0))^{p-2}}, t_0 + \frac{4C_1(\delta r)^p}{(v(\overline{x},t_0))^{p-2}} \right\}$$

to obtain

$$v(x,\overline{t}) \geq C_0^{-1} v(\overline{x},t_0) \geq 2^{-(N+1)} C_0^{-1} E_0 \equiv aE_0, \tag{3.19}$$

where $\overline{t} = t_0 + 4C_1 \dfrac{(\delta r)^p}{v(\overline{x},t_0)^{p-2}}$, $x \in B_{\delta r}(\overline{x})$. By virtue of the choice of t_0 and (3.18),

$$\overline{t} \leq \left(C_* + C_1 \delta^p 2^{(N+1)(p-2)} \right) r^p E_0^{2-p} \equiv B_1 \frac{r^p}{E_0^{p-2}}. \tag{3.20}$$

Fix $\theta > 0$. We may suppose that

$$\theta \geq 2B_1 \frac{r^p}{E_0^{p-2}}; \tag{3.21}$$

otherwise $E_0 \leq B \left(\dfrac{r^p}{\theta} \right)^{1/(p-2)}$ with $B \equiv (2B_1)^{1/(p-2)}$, obviously our lemma holds. □

Now we estimate the lower bound of $v(x,\theta)$ on B_r. By Lemma 2.3.1, we obtain from (3.19),

$$v(x,t) \geq E_{aE_0,\delta r}(x,t;\overline{x},\overline{t}), \quad \text{for } x \in \mathbb{R}^N, \quad t \geq \overline{t}. \tag{3.22}$$

Since from (3.20) and (3.21), $\theta - \overline{t} \geq \theta/2$, we can conclude that if

$$\begin{aligned} S(\theta) &= b(aE_0)^{p-2}(\delta r)^{N(p-2)}(\theta - \overline{t}) + (\delta r)^\kappa \\ &\geq \frac{b}{2}(aE_0)^{p-2}(\delta r)^{N(p-2)}\theta \geq (8r)^\kappa, \end{aligned} \tag{3.23}$$

that is,

$$E_0^{p-2} > B_2 \left(\frac{r^p}{\theta} \right) \tag{3.24}$$

with $B_2 = \dfrac{2^{3\kappa+1}}{b(a\delta^N)^{p-2}}$, then $\operatorname{supp} E_{aE_0,\delta r}(x,\theta;\overline{x},\overline{t}) \subset B_{4r}$. We may suppose that (3.24) holds; otherwise the conclusion of our lemma is valid obviously.

Moreover,

$$\begin{aligned} S(\theta) &\le b(a\delta^N)^{p-2}E_0^{p-2}r^{N(p-2)}\theta + \delta^\kappa r^{N(p-2)+p} \\ &\le \left(b(a\delta^N)^{p-2} + \frac{\delta^\kappa}{B_2}\right)E_0^{p-2}r^{N(p-2)}\theta \\ &\equiv \delta_1 E_0^{p-2}r^{N(p-2)}\theta. \end{aligned} \tag{3.25}$$

Substituting (3.25), (3.23) into (3.22) we conclude that for $x \in B_r$,

$$\begin{aligned} v(x,\theta) &\ge \frac{aE_0(\delta r)^N}{\delta_1^{N/\kappa}(E_0^{p-2}r^{N(p-2)}\theta)^{N/\kappa}}\left(1-\left(\frac{1}{2}\right)^{p/(p-1)}\right)^{(p-1)/(p-2)} \\ &\equiv \delta_0\left(\frac{r^p}{\theta}\right)^{N/\kappa}E_0^{p/\kappa}, \end{aligned}$$

from which it follows that

$$E_0 \equiv \fint_{B_r} v_0(x)dx \le \delta_0^{-\kappa/p}\left(\frac{\theta}{r^p}\right)^{N/p}\left(\inf_{B_r} v(\cdot,\theta)\right)^{\kappa/p}.$$

Choosing $B = \max\left\{(2B_1)^{1/(p-2)}, B_2^{1/(p-2)}, \delta_0^{-\kappa/p}\right\}$, we then obtain the conclusion of Lemma 2.3.3. Now we come back to the

Proof of Theorem 2.3.3. For $(x_0,t_0) \in Q_T, \theta \in (0, T-t_0), \rho > 0$, consider the Cauchy problem for (3.12) with initial value

$$v(x,t_0) = \begin{cases} u(x,t_0), & \text{for } x \in \overline{B}_\rho, \\ 0, & \text{for } x \in \mathbb{R}^N \setminus \overline{B}_\rho. \end{cases}$$

Let v be its generalized solution. Then by Proposition 2.3.4, there exists $B_{R(T)}$ such that $\operatorname{supp} v(\cdot,t) \subset B_{R(T)}$. Applying the comparison theorem (Proposition 2.3.1) on $B_{R(T)} \times (t_0,T)$ we obtain

$$v(x,t) \le u(x,t), \qquad (x,t) \in B_{R(T)} \times (t_0,T),$$

which combining with Lemma 2.3.3 implies the conclusion of our theorem. □

2.3.3 *Initial trace of solutions*

Theorem 2.3.4 *Assume that $p > 2$ and u is a nonnegative generalized solution of (1.1) on Q_T. Then there exists a unique Radon measure μ such* *that*

$$\lim_{t\to 0}\int_{\mathbb{R}^N} u(x,t)\varphi(x)dx = \int_{\mathbb{R}^N}\varphi(x)d\mu, \quad \text{for } \varphi\in C_0^\infty(\mathbb{R}^N) \tag{3.26}$$

and

$$\sup_{\rho>r}\rho^{-(N+p/(p-2))}\int_{B_\rho} d\mu < \infty, \quad \text{for } r>0. \tag{3.27}$$

Theorem 2.3.4 shows that for the Cauchy problem for (1.1) to have a nonnegative generalized solution, the growth condition (3.1) on the initial data is not only sufficient but also necessary.

To prove Theorem 2.3.4, we first prove the following lemma.

Lemma 2.3.4 *Assume that u is a nonnegative generalized solution of (1.1) on Q_T. Then for any $\rho>0$, $\sigma\in(0,1)$, $0<\tau<t<\frac{T}{2}$,*

$$\int_{B_{(1+\sigma)\rho}} u(x,t)dx \ge \int_{B_\rho} u(x,\tau)dx - \frac{\gamma}{\sigma}(t-\tau)^{1/\kappa}\rho^{\kappa/(p-2)}H_r(\tau,t), \tag{3.28}$$

where $H_r(\tau,t)=\sup_{s\in(\tau,t)}|||u(\cdot,s)|||_r^{1+(p-2)/\kappa}$.

Proof. Let $\xi(x)$ be a smooth cut-off function defined on $B_{(1+\sigma)\rho}$ such that $|\nabla\xi|\le\frac{2}{\sigma}$, $\xi=1$ on B_ρ. Choosing

$$\varphi=\xi,\qquad \Omega=B_{(1+\sigma)\rho},\qquad t_1=\tau,\qquad t_2=t,$$

in (1.23), we obtain

$$\int_{B_{(1+\sigma)\rho}} u(x,t)dx \ge \int_{B_\rho} u(x,\tau)dx - \frac{\gamma}{\sigma\rho}\int_\tau^t\int_{B_{(1+\sigma)\rho}}|\nabla u|^{p-1}dxds. \tag{3.29}$$

An argument similar to the proof of (2.36) in Theorem 2.2.3 gives

$$\int_\tau^t\int_{B_{(1+\sigma)\rho}}|\nabla u|^{p-1}dxds \le C(t-\tau)^{1/\kappa}\rho^{1+\kappa/(p-2)}H_r(\tau,t), \tag{3.30}$$

which combining with (3.29) yields (3.28). □

Proof of Theorem 2.3.4. It suffices to prove the existence, since the uniqueness is obvious.

By Theorem 2.3.3, for $\varphi \in C_0^\infty(B_\rho)$, $0 < t < T - \epsilon$,

$$\left| \int_{B_\rho} u(x,t)\varphi(x)dx \right| \le \gamma(p, N, T, u(0, T-\epsilon), \rho) ||\varphi||_{\infty, B_\rho},$$

which implies that the linear operator $\displaystyle\int_{\mathbb{R}^N} u(x,t)\varphi(x)dx$ on $C_0^\infty(\mathbb{R}^N)$ is bounded uniformly in $0 < t < T - \epsilon$. Hence there exists $t_n \downarrow 0$ such that

$$\lim_{t_n \to 0} \int_{\mathbb{R}^N} u(x, t_n)\varphi(x)dx = \int_{\mathbb{R}^N} \varphi(x)d\mu, \quad \text{for } \varphi \in C_0^\infty(\mathbb{R}^N),$$

where μ is a nonnegative finite Borel measure. From the uniqueness of generalized solutions (which will be proved in §2.5) and Remark 2.2.1, we may conclude that μ satisfies

$$|||\mu|||_r = \sup_{\rho \ge r} \rho^{-\kappa/(p-2)} \int_{B_\rho} d\mu < +\infty.$$

Suppose that there exists $\tau_n \downarrow 0$ and a measure ν possessing all properties as μ. Then

$$\lim_{\tau_n \to 0} \int_{\mathbb{R}^N} u(x, \tau_n)\varphi(x)dx = \int_{\mathbb{R}^N} \varphi(x)d\nu, \quad \text{for } \varphi \in C_0^\infty(\mathbb{R}^N). \tag{3.32}$$

Now choose a subsequence $\{t_n'\}$ of $\{t_n\}$ and a subsequence $\{\tau_n'\}$ of $\{\tau_n\}$ such that $t_n' \ge \tau_n'$. Set $t = t_n'$, $\tau = \tau_n'$ in (3.28), recall Remark 2.2.1 and let $n \to \infty$. Then we obtain by using (3.31) and (3.32) that

$$\int_{B_{(1+\sigma)\rho}} d\mu \ge \int_{B_\rho} d\nu.$$

Choose another subsequence $\{t_n''\}$ of $\{t_n\}$ and another subsequence $\{\tau_n''\}$ of $\{\tau_n\}$ such that $t_n'' \le \tau_n''$. Then we obtain

$$\int_{B_{(1+\sigma)\rho}} d\mu \le \int_{B_\rho} d\nu.$$

Since $\sigma \in (0,1)$ is arbitrary, there must hold $\mu = \nu$. □

2.4 Regularity of Solutions

2.4.1 *Boundedness of solutions*

For $(x_0, t_0) \in Q_T$, denote

$$Q(\rho, t_0) = B_\rho(x_0) \times (t_0, T).$$

Theorem 2.4.1 *Assume that* $p > \max\left\{1, \dfrac{2N}{N+2}\right\}$ *and* u *is a generalized solution of (1.1) on* Q_T*. If* $p > 2$*, then*

$$\sup_{Q(\rho,t_0/2)} |u| \leq \max\left\{C\rho^{-(N+p)}I(\rho, t_0), J(\rho, t_0)\right\}, \tag{4.1}$$

and if $p < 2$*, then*

$$\sup_{Q(\rho,t_0/2)} |u| \leq \max\left\{Ct_0^{-(N+p)/p}I(\rho, t_0), J(\rho, t_0)\right\}, \tag{4.2}$$

where $q = \dfrac{p(N+2)}{N}$, $\lambda = N(p-2) + qp$, $C = C(N, p)$, $\rho > 0$, $t_0 \in (0, T)$ *and*

$$I(\rho, t_0) = \left(\iint_{Q(2\rho,t_0/4)} |u|^q dxdt\right)^{p/\lambda},$$

$$J(\rho, t_0) = \left(\frac{t_0}{\rho^p}\right)^{1/(2-p)}.$$

Since from the definition of generalized solutions,

$$u \in C_{loc}(0, T; L^2_{loc}(\mathbb{R}^N)) \cap L^p_{loc}(0, T; W^{1,p}_{loc}(\mathbb{R}^N)),$$

we have, by the embedding inequality (1.7),

$$u \in L^q_{loc}(Q_T)$$

with $q = p\dfrac{N+2}{N}$. Hence Theorem 2.4.1 implies that $u \in L^\infty_{loc}(Q_T)$.

Proof of Theorem 2.4.1. Set

$$T_n = \frac{t_0}{2} - \frac{t_0}{2^{n+2}}, \quad \rho_n = \rho + \frac{\rho}{2^n},$$

$$\overline{\rho}_n = \frac{1}{2}(\rho_n + \rho_{n+1}) = \rho + \frac{3\rho}{2^{n+2}},$$

$$B_n = B_{\rho_n}(x_0), \quad B'_n = B_{\overline{\rho}_n}(x_0),$$

$$Q_n = B_n \times (T_n, T),$$
$$Q'_n = B'_n \times (T_{n+1}, T), \quad k_n = k - \frac{k}{2^n},$$
$$k > 0, \quad n = 0, 1, 2, \cdots.$$

Let ξ_n be a smooth cut-off function defined on $B_n \times (0, T]$ such that $\operatorname{supp} \xi_n(\cdot, t) \subset B_n$ for $t \in (T_n, T)$, $\xi_n = 0$ for $t \le T_n$, $\xi_n = 1$ for $(x,t) \in Q'_n$, $|\nabla \xi_n| \le \dfrac{2^{n+2}}{\rho}$, $0 \le \xi_{nt} \le \dfrac{2^{n+3}}{t_0}$.

In what follows, we suppose that $u \ge 0$, otherwise we may consider $u_+ = \max\{u, 0\}$ and $u_- = \max\{-u, 0\}$ separately and obtain the same conclusion. Here we notice that both u_+ and u_- are generalized sub-solutions of (1.1).

Choosing

$$\Omega = B_n, \qquad t_1 = 0, \qquad t_2 = t,$$
$$f(s) = (s - k_n)_+, \qquad \zeta(x, \tau) = \xi^p(x, \tau)$$

in (1.27), we obtain

$$\begin{aligned}
&\frac{1}{2}\int_{B_n} (u(x,t) - k_n)_+^2 \xi_n^p(x,t)dx \\
&+ \int_0^t \int_{B_n} |\nabla u|^{p-2} \nabla u \cdot \nabla (u - k_n)_+ \xi_n^p dx d\tau \\
&+ p\int_0^t \int_{B_n} |\nabla u|^{p-2} \nabla u \cdot (u - k_n)_+ \nabla \xi_n \, \xi_n^{p-1} dx d\tau \\
= \; &\frac{p}{2}\int_0^t \int_{B_n} (u - k_n)_+^2 \xi_n^{p-1} \xi_{n\tau} dx d\tau.
\end{aligned} \tag{4.3}$$

Denoting $w_n = (u - k_n)_+$ and applying Hölder's inequality to the third term on the left hand side of (4.3), we further have

$$\begin{aligned}
&\sup_{T_{n+1} < t < T} \int_{B_n} \omega_n^2 \xi_n^p dx + \iint_{Q_n} |\nabla \omega_n|^p \xi_n^p dx dt \\
\le \; &\frac{\gamma 2^{np}}{\rho^p} \iint_{Q_n} \omega_n^p dx dt + \frac{\gamma 2^n}{t_0} \iint_{Q_n} \omega_n^2 dx dt.
\end{aligned} \tag{4.4}$$

Here and below, γ denotes a constant depending only on N, p, T.

Let $\widetilde{\xi}_n(x) \in C_0^1(B'_n)$ be such that $0 \le \widetilde{\xi}_n \le 1$, $|\nabla \widetilde{\xi}_n| \le \dfrac{2^{n+2}}{\rho}$, $\widetilde{\xi}_n = 1$ for $x \in B_{n+1}$. Clearly

$$\omega_n \widetilde{\xi}_n \in L^\infty(T_{n+1}, T; L^2(B'_n)) \cap L^p(T_{n+1}, T; W_0^{1,p}(B'_n)).$$

Using the embedding inequality (1.7) and (4.4) we obtain

$$\begin{aligned}
&\iint_{Q_{n+1}} \omega_{n+1}^q dxdt \le \iint_{Q'_n} (\omega_{n+1}\widetilde{\xi}_n)^q dxdt \\
\le\ & \gamma\left(\iint_{Q'_n} |\nabla \omega_{n+1}|^p dxdt + \frac{2^{np}}{\rho^p}\iint_{Q'_n} \omega_{n+1}^p dxdt\right) \\
&\cdot \left(\sup_{T_{n+1}<t<T} \int_{B'_n} \omega_{n+1}^2(x,t)dx\right)^{p/N} \\
\le\ & \gamma\left(\sup_{T_{n+1}<t<T} \int_{B'_n} \omega_{n+1}^2(x,t)dx\right. \\
&\left. + \iint_{Q'_n} |\nabla \omega_{n+1}|^p dxdt + \frac{2^{np}}{\rho^p}\iint_{Q'_n} \omega_{n+1}^p dxdt\right)^{(N+p)/N} \\
\le\ & \gamma\left(\frac{2^{np}}{\rho^p}\iint_{Q_n} \omega_{n+1}^p dxdt + \frac{2^n}{t_0}\iint_{Q_n} \omega_{n+1}^2 dxdt\right)^{(N+p)/N}. \qquad (4.5)
\end{aligned}$$

Since $q = \dfrac{p(N+2)}{N} > p$ and $p \ge \dfrac{2N}{N+2}$, we have $q \ge 2$. Thus

$$\left\{\begin{aligned}
&\iint_{Q_n} \omega_{n+1}^p dxdt \le \left(\iint_{Q_n} \omega_{n+1}^q dxdt\right)^{p/q} |A_{n+1}|^{1-p/q}, \\
&\iint_{Q_n} \omega_{n+1}^2 dxdt \le \left(\iint_{Q_n} \omega_{n+1}^q dxdt\right)^{2/q} |A_{n+1}|^{1-2/q},
\end{aligned}\right. \qquad (4.6)$$

where $A_{n+1} = \{(x,t) \in Q_n; u(x,t) > k_{n+1}\}$, $|A_{n+1}| = \operatorname{mes} A_{n+1}$. Since $\omega_{n+1} \le \omega_n$ and

$$\iint_{Q_n} \omega_n^q dxdt \ge \iint_{Q_n \cap \{u > k_{n+1}\}} (u - k_n)_+^q dxdt \ge \left(\frac{k}{2^{n+1}}\right)^q |A_{n+1}|,$$

from (4.5), (4.6) we derive

$$\iint_{Q_{n+1}} \omega_{n+1}^q dxdt \le \gamma b^n((\rho^{-p}k^{p-q} + t_0^{-1}k^{2-q})I_q)^{(N+p)/N}, \qquad (4.7)$$

where $b = 2^q$, $I_q = \iint_{Q_n} \omega_n^q dxdt$.

If $p > 2$, then we require $k \geq \left(\frac{\rho^p}{t_0}\right)^{1/(p-2)}$. In this case, from (4.7),

$$\iint_{Q_{n+1}} \omega_{n+1}^q dxdt \leq \gamma b^n (2\rho^{-p} k^{p-q} I_q)^{(N+p)/N}.$$

It follows that, if

$$\iint_{Q_0} |u|^q dxdt \leq \gamma^{-N/p} (2\rho^{-p} k^{p-q})^{-(N+p)/p} b^{-(N/p)^2},$$

then using the iteration lemma (Lemma 1.5.3 in §1.1.5) yields

$$\iint_{Q_\infty} (u-k)_+^q dxdt = 0,$$

where $Q_0 = B_{2\rho}(x_0) \times (t_0/4, T)$, $Q_\infty = B_\rho(x_0) \times (t_0/2, T)$. Therefore there exists a constant $C = C(N, p)$ such that if we choose

$$k = \max\left\{ \left(C\rho^{-(N+p)} \iint_{Q_0} |u|^q dxdt \right)^{N/2(p+N)}, \left(\frac{\rho^p}{t_0}\right)^{1/(p-2)} \right\},$$

then $\sup_{Q_\infty} |u| \leq k$ and (4.1) follows.

If $p < 2$, we require $k \geq \left(\frac{t_0}{\rho^p}\right)^{1/(2-p)}$, In this case, from (4.7),

$$\iint_{Q_{n+1}} \omega_{n+1}^q dxdt \leq \gamma b^n \left((2t_0^{-1} k^{2-q}) \iint_{Q_n} \omega_n^q dxdt \right)^{(N+p)/N}.$$

Similar to the case $p > 2$, we can conclude that there exists a constant $C = C(N, p)$ such that if we choose

$$k = \max\left\{ \left(Ct_0^{-(N+p)/p} \iint_{Q_0} |u|^q dxdt \right)^{N/\lambda}, \left(\frac{t_0}{\rho^p}\right)^{1/(2-p)} \right\},$$

then $\sup_{Q_\infty} |u| \leq k$ and (4.2) follows. □

2.4.2 Boundedness of the gradient of solutions

Theorem 2.4.2 *Assume that* $p > \max\left\{1, \dfrac{2N}{N+2}\right\}$ *and* u *is a generalized solution of (1.1) on* Q_T. *Then* $\dfrac{\partial u}{\partial x_i} \in L^\infty_{loc}(Q_T)$ $(i = 1, 2, \cdots, N)$.

For any $R > 0$, $t_1 \in (0, T)$, $u \in C(t_1, T; L^2(B_R)) \cap L^p(t_1, T; W^{1,p}(B_R))$ can be regarded as a generalized solution of the first boundary value problem for (1.1)on $B_R \times (t_1, T)$. By virtue of the uniqueness of solutions (Proposition 2.3.1) and the discussions in §2.2.2, u can be obtained as the limit of the smooth solution u_n of the following boundary value problem:

$$\frac{\partial u_n}{\partial t} = \operatorname{div}\left(\left(|\nabla u_n|^2 + \frac{1}{n}\right)^{(p-2)/2} \nabla u_n\right), \quad \text{in } B_R \times (t_1, T), \tag{4.8}$$

$$u_n(x,t) = u(x,t) \quad \text{on } \partial B_R \times (t_1, T), \tag{4.9}$$

$$u_n(x,0) = u(x,t_1) \quad \text{on } \overline{B}_R. \tag{4.10}$$

Therefore, Theorem 2.4.2 is an immediate corollary of the following proposition.

Proposition 2.4.1 *Assume that* $p > \max\left\{1, \dfrac{2N}{N+2}\right\}$ *and* u_n *is a solution of the boundary value problem (4.8)–(4.10). Then for any compact subset* K *of* $B_R \times (t_1, T)$,

$$\left\|\frac{\partial u_n}{\partial x_i}\right\|_{L^\infty(K)} \le C \quad (i = 1, 2, \cdots, N),$$

where C *is a constant independent of* n.

To prove Proposition 2.4.1, we need the following lemma, which can be proved similar to Lemma 2.2.3.

Lemma 2.4.1 *Assume that* $p > \max\left\{1, \dfrac{2N}{N+2}\right\}$ *and* u_n *is a solution of the boundary value problem (4.8)–(4.10). Then for any compact subset* K *of* $B_R \times (t_1, T)$,

$$\iint_K |\nabla u_n|^q dxdt \le C(q, K, pN) \quad \text{for } q \ge 1,$$

where $C(q, K, p, N)$ *is a constant independent of* n.

Proof of Proposition 2.4.1. Differentiate (4.8) with respect to x_j,

$$\frac{\partial u_{x_j}}{\partial t} = \left(\left(|\nabla u|^2 + \frac{1}{n}\right)^{(p-2)/2} u_{x_i}\right)_{x_i x_j}. \tag{4.11}$$

Suppose that $t_0 \in (t_1, T)$, $t_0 > 4t_1$. Notations $Q(\rho, t_0)$, T_n, ρ_n, $\overline{\rho}_n$, B_n, B'_n, Q_n, Q'_n and function ξ_n are the same as in the proof of Theorem 2.4.1. Take $x_0 = 0$ and suppose $B_{2\rho} \subset B_R$. Denote $v = |\nabla u|^2 + \dfrac{1}{n}$. Multiplying (4.11) with $\xi_n^2 v^\alpha u_{x_j}$ ($\alpha \geq 1$) and integrating over $B_{2\rho} \times (t_0/4, t)$ yield

$$\begin{aligned}
&\frac{1}{2(\alpha+1)} \int_{B_{2\rho}} \xi_n^2 v^{\alpha+1}(x,t)dx \\
&+ \int_0^t \int_{B_{2\rho}} \xi_n^2 (v^\alpha u_{x_j})_{x_i} (v^{(p-2)/2} u_{x_i})_{x_j} dx d\tau \\
= \;& \frac{1}{\alpha+1} \int_0^t \int_{B_{2\rho}} \xi_n \xi_{n\tau} v^{\alpha+1} dx d\tau \\
&-2 \int_0^t \int_{B_{2\rho}} \xi_n v^\alpha u_{x_j} (v^{(p-2)/2} u_{x_i})_{x_j} \xi_{n x_i} dx d\tau.
\end{aligned}$$

If $p > 2$, then similar to the derivation of (2.22) we can obtain

$$\begin{aligned}
&\sup_{T_n < t < T} \int_{B_n} (\xi_n^{2/\lambda} \omega)^\lambda (x, \tau) dx + \iint_{Q_n} (\xi_n^{2/\lambda} |\nabla \omega|)^2 dx d\tau \\
\leq \;& C\left(\frac{2^{2n}}{\rho^2} \iint_{Q_n} \omega^2 dx d\tau + \frac{2^n}{t_0} \iint_{Q_n} \omega^\lambda dx d\tau\right),
\end{aligned} \tag{4.12}$$

where $\omega = v^{(p+2\alpha)/4}$, $\lambda = \dfrac{4(\alpha+1)}{p+2\alpha}$. Using the embedding inequality (1.7) and (4.12) we further obtain

$$\iint_{Q_{n+1}} \omega^{2+2\lambda/N} dx d\tau \leq C\left(\frac{2^{2n}}{\rho^2} \iint_{Q_n} \omega^2 dx d\tau + \frac{2^n}{t_0} \iint_{Q_n} \omega^\lambda dx d\tau\right)^{1+2/N},$$

that is,

$$\begin{aligned}
&\iint_{Q_{n+1}} v^{(p+2\alpha)/2 + 2(\alpha+1)/N} dx d\tau \\
\leq \;& C\left(\frac{2^{2n}}{\rho^2} \iint_{Q_n} v^{(p+2\alpha)/2} dx d\tau + \frac{2^n}{t_0} \iint_{Q_n} v^{\alpha+1} dx d\tau\right)^{1+2/N}.
\end{aligned}$$

Denote $k = 1 + \frac{2}{N}$ and choose $2\alpha = \frac{k^n}{2} - 1$. Then the above inequality turns out to be

$$\iint_{Q_{n+1}} v^{(p-2)/2+k^{n+1}/2} dx d\tau \\ \leq C\left(\frac{2^{2n}}{\rho^2}\iint_{Q_n} v^{(p-2)/2+k^n/2} dx d\tau + \frac{2^n}{t_0}\iint_{Q_n} v^{k^n/2} dx d\tau\right)^k. \tag{4.13}$$

If there exists a sequence $\{n_i\}$ of positive integers with $n_i \to \infty$ $(i \to \infty)$ such that

$$\frac{2^{n_i}}{t_0}\iint_{Q_{n_i}} v^{k_i^n/2} dx d\tau \geq \frac{2^{2n_i}}{\rho^2}\iint_{Q_{n_i}} v^{(p-2)/2+k^{n_i}/2} dx d\tau, \tag{4.14}$$

then by Hölder's inequality, we have

$$\iint_{Q_{n_i}} v^{(p-2)/2+k^{n_i}/2} dx d\tau \leq \left(\frac{\rho^2}{t_0}\right)^{(p-2+k^{n_i})/(p-2)} \operatorname{mes} Q_{n_i}$$

and hence

$$\sup_{B_\rho \times (t_0/2,T)} v \leq \left(\frac{\rho^2}{t_0}\right)^{2/(p-2)}. \tag{4.15}$$

If there is not any sequence $\{n_i\}$ of positive integers such that (4.14) holds, then there exists a positive integer n_0 such that for $n \geq n_0$,

$$\iint_{Q_{n+1}} v^{(p-2)/2+k^{n+1}/2} dx d\tau \leq C\left(\frac{2^{2n}}{\rho^2}\iint_{Q_n} v^{(p-2)/2+k^n/2} dx d\tau\right)^k.$$

Thus, using Lemma 2.1.5 we obtain

$$\sup_{Q_\infty} v \leq C\left(\rho^{-(N+2)}\iint_{Q_{n_0}} v^{(p-2)/2+k^{n_0}/2} dx d\tau\right), \tag{4.16}$$

where n_0 is an integer such that $k^{n_0} > 2$. The conclusion of Proposition 2.4.1 then follows from Lemma 2.4.1.

Now we consider the case $p < 2$. Similar to (2.24), we can obtain

$$\sup_{t_0/4<t<T}\int_{B_{2\rho}} \xi_n^2 v^{\alpha+1}(x,\tau) dx + \iint_{Q(2\rho,t_0/4)} \xi_n^2 v^{\alpha_p} |\nabla v|^2 dx d\tau \\ \leq C\iint_{Q(2\rho,t_0/4)} v^{\alpha_p+2} |\nabla \xi_n|^2 dx d\tau + C\iint_{Q(2\rho,t_0/4)} \xi_n \xi_{nt} v^{\alpha+1} dx d\tau,$$

where $\alpha_p = \dfrac{p+2\alpha-4}{2}$. Similar to the discussion for the case $p > 2$, using the embedding inequality we can derive

$$\begin{aligned}
&\left(\iint_{Q_{n+1}} v^{N(2-p)/4_k^{n+1}/2} dx d\tau\right)^{1/k} \\
\leq\ & C\left(\frac{2^{2n}}{t_0}\iint_{Q_n} v^{N(2-p)/4+k^n/2} dx d\tau\right. \qquad (4.17)\\
&\left. + \frac{2^{2n}}{\rho^2}\iint_{Q_n} v^{(N-2)(2-p)/4+k^n/2} dx d\tau\right).
\end{aligned}$$

If there exists $\{n_i\}$ with $n_i \to \infty$ $(i \to \infty)$ such that

$$\begin{aligned}
&\frac{2^{2n_i}}{\rho^2}\iint_{Q_{n_i}} v^{(N-2)(2-p)/4+k^{n_i}/2} dx d\tau \\
\geq\ & \frac{2^{2n_i}}{t_0}\iint_{Q_{n_i}} v^{N(2-p)/4+k^{n_i}/2} dx d\tau,
\end{aligned} \qquad (4.18)$$

then using Hölder's inequality we derive

$$\iint_{Q_{n_i}} v^{N(2-p)/2+k^{n_i}/2} dx d\tau \leq \left(\frac{t_0}{\rho^2}\right)^{(N(2-p)+k^{n_i})/(2-p)} \operatorname{mes} Q_{n_i},$$

from which it follows that

$$\sup_{B_\rho \times (t_0/2.T)} v \leq \left(\frac{t_0}{\rho^2}\right)^{2/(2-p)}.$$

Otherwise there exists an integer n_0 such that for $n \geq n_0$,

$$\begin{aligned}
&\left(\iint_{Q_{n+1}} v^{N(2-p)/4+k^{n+1}/2} dx d\tau\right)^{1/k} \\
\leq\ & \gamma\left(\frac{2^{2n}}{\rho^2}\iint_{Q_n} v^{N(2-p)/4+k^n/2} dx d\tau\right).
\end{aligned}$$

Using Lemma 2.1.5, we then obtain

$$\sup_{B_\rho \times (t_0/2,T)} v \leq C\left(\rho^{-(N+2)}\iint_{Q_{n_0}} v^{N(2-p)/4+k^{n_0}/2} dx d\tau\right)^{2/K^{n_0}}$$

and the conclusion of Proposition 2.4.1 also follows from Lemma 2.4.1. □

2.4.3 *Hölder continuity of solutions*

Theorem 2.4.3 *Assume that* $p > \max\left\{1, \dfrac{2N}{N+2}\right\}$ *and* u *is a generalized solution of (1.1) on* Q_T. *Then for any compact subset* $K \subset Q_T$ *and* $(x_1,t_1),(x_2,t_2) \in K$,

$$|u(x_1,t_1) - u(x_2,t_2)| \le C(|x_1 - x_2| + |t_1 - t_2|^{1/2}), \tag{4.19}$$

where C *is a constant depending only on* N, p *and* $\|u\|_{\infty,K}$.

Proof. It suffices to prove that for any $R > 0$ and $t_0 \in (0,T)$, u satisfies (4.19) on $B_R \times (t_0,T)$.

Consider the mollifier

$$u_\varepsilon(x,t) = J_\varepsilon u(x,t) = \int_0^T \int_{\mathbb{R}^N} j_\varepsilon(x-y,t-\tau)u(y,\tau)dyd\tau,$$
$$0 < \varepsilon < t_0 < t < T - \varepsilon.$$

For any $x_1, x_2 \in B_R$,

$$\begin{aligned}
&u_\varepsilon(x_1,t) - u_\varepsilon(x_2,t)\\
= {}& \int_0^T \int_{\mathbb{R}^N} \int_0^1 \frac{d}{ds} j_e(sx_1 + (1-s)x_2 - y, t-\tau)u(y,\tau)dsdyd\tau\\
= {}& \int_0^T \int_{\mathbb{R}^N} \int_0^1 \nabla_x j_\varepsilon(sx_1 + (1-s)x_2 - y, t-\tau)\\
&\cdot u(y,\tau)dsdyd\tau \cdot (x_1 - x_2)\\
= {}& -\int_0^1 \int_{\mathbb{R}^N} \int_0^T \nabla_y j_\varepsilon(sx_1 + (1-s)x_2 - y, t-\tau)\\
&\cdot u(y,\tau)dyd\tau ds \cdot (x_1 - x_2)\\
= {}& \int_0^1 \int_{\mathbb{R}^N} \int_0^T j_\varepsilon(sx_1 + (1-s)x_2 - y, t-\tau)\\
&\cdot \nabla_y u(y,\tau)dyd\tau ds \cdot (x_1 - x_2).
\end{aligned}$$

Hence, by Theorem 2.4.2,

$$\begin{aligned}
&|u_\varepsilon(x_1,t)-u_\varepsilon(x_2,t)| \\
\le\ & \int_0^1\int_{\mathbb{R}^N}\int_0^T |j_\varepsilon(sx_1+(1-s)x_2-y,t-\tau)| \\
&\cdot|\nabla_y u(y,\tau)|dyd\tau ds\cdot|x_1-x_2| \\
\le\ & C|x_1-x_2|.
\end{aligned} \tag{4.20}$$

Here and below, C denotes a constant independent of ε.

Let $0<\varepsilon<t_0<t_1<t_2<T$, $B(\Delta t)=B_{(\Delta t)^{1/2}}(x_0)$, $\varphi\in C_0^1(B(\Delta t))$, $x_0\in B_R$, $\Delta t=t_2-t_1$. Then

$$\begin{aligned}
&\int_{B(\Delta t)}\varphi(x)(u_\varepsilon(x,t_2)-u_\varepsilon(x,t_1))dx \\
=\ & \int_{B(\Delta t)}\varphi(x)\int_0^1\frac{d}{ds}u_\varepsilon(x,st_2+(1-s)t_1)dsdx \\
=\ & \Delta t\int_{B(\Delta t)}\varphi(x)\int_0^1\int_0^T\int_{\mathbb{R}^N} j_{\varepsilon t}(x-y,st_2+(1-s)t_1-\tau) \\
&\cdot u(y,\tau)dyd\tau dsdx
\end{aligned} \tag{4.21}$$

$$\begin{aligned}
=\ & -\Delta t\int_{B(\Delta t)}\varphi(x)\int_0^1\int_0^T\int_{\mathbb{R}^N} j_{\varepsilon\tau}(x-y,st_2+(1-s)t_1-\tau) \\
&\cdot u(y,\tau)dyd\tau dsdx.
\end{aligned}$$

Noting that for any fixed $(x,t)\in Q_T$ with $0<\varepsilon<t_0<t<T-\varepsilon$, $J_\varepsilon(x-y,t-\tau)\in C_0^1(Q_T)$, we obtain from the definition of generalized solutions,

$$\begin{aligned}
&\int_0^T\int_{\mathbb{R}^N} j_{\varepsilon\tau}(x-y,st_2+(1-s)t_1-\tau)u(y,\tau)dyd\tau \\
=\ & \int_0^T\int_{\mathbb{R}^N}|\nabla_y u|^{p-2}\nabla_y u\cdot\nabla_y j_\varepsilon(x-y,st_2+(1-s)t_1-\tau)dyd\tau.
\end{aligned}$$

Substituting this into (4.21) gives

$$\begin{aligned}
&\int_{B(\Delta t)}\varphi(x)(u_\varepsilon(x,t_2)-u_\varepsilon(x,t_1))dx \\
=\ & -\Delta t\int_{B(\Delta t)}\varphi(x)\int_0^1\int_0^T\int_{\mathbb{R}^N}|\nabla_y u|^{p-2}\nabla_y u
\end{aligned}$$

$$\cdot\nabla_y j_\varepsilon(x-y, st_2+(1-s)t_1-\tau)dyd\tau dsdx$$

$$\begin{aligned}
= & -\Delta t\int_0^1\int_0^T\int_{\mathbb{R}^N}|\nabla_y u|^{p-2}\nabla_y u \\
& \left(\int_{B(\Delta t)}\nabla_x\varphi\cdot j_\varepsilon(x-y, st_2+(1-s)t_1-\tau)dx\right)dyd\tau ds \\
= & -\Delta t\int_0^1\int_{B(\Delta t)}\nabla_x\varphi\cdot\left(\int_0^T\int_{\mathbb{R}^N}j_\varepsilon(x-y, st_2+(1-s)t_1-\tau)\right. \\
& \left.\cdot|\nabla_y u|^{p-2}\nabla_y u dyd\tau\right)dxds \\
= & -\Delta t\int_0^1\int_{B(\Delta t)}\nabla_x\varphi \\
& \cdot J_\varepsilon(|\nabla u|^{p-2}\nabla u)(x, st_2+(1-s)t_1)dxds. \qquad (4.22)
\end{aligned}$$

Now choose $\delta(s)\in C_0^1(\mathbb{R})$ such that $\delta(s)\geq 0$, $\delta(s)=0$ for $|s|\geq 1$ and $\int_{\mathbb{R}}\delta(s)ds=1$. For $h>0$, define $\delta_h(s)=\frac{1}{h}\delta\left(\frac{s}{h}\right)$. By approximation, we see that (4.22) holds for $\varphi\in W_0^{1,1}(B(\Delta t))$. Thus, if we choose

$$\varphi=\varphi_h(x)=\int_{-h}^{(\Delta t)^{1/2}-|x-x_0|-2h}\delta_h(s)ds$$

in (4.22), then we have

$$\begin{aligned}
& \int_{B(\Delta t)}\varphi_h(x)(u_\varepsilon(x,t_2)-u_\varepsilon(x,t_1))dx \\
= & -\Delta t\int_0^1\int_{B(\Delta t)}\delta_h((\Delta t)^{1/2}-|x-x_0|-2h)\cdot\frac{x_{0i}-x_i}{|x-x_0|} \\
& \cdot J_\varepsilon(|\nabla u|^{p-2}u_{x_i})(x, st_2+(1-s)t_1)dxds. \qquad (4.23)
\end{aligned}$$

Noting that for $x\in B(\Delta t)$, $\lim_{h\to 0}\varphi_h(x)=1$ and if $|x-x_0|<(\Delta t)^{1/2}-h$, then $\delta_h((\Delta t)^{1/2}-|x-x_0|-2h)=0$, $\delta_h\leq\frac{C}{h}$ and

$$\text{mes}\,(B(\Delta t)\backslash B_{(\Delta t)^{1/2}-h}(x_0))\leq Ch(\Delta t)^{(N-1)/2},$$

from (4.23), we can use Theorem 2.4.2 to obtain

$$\left|\int_{B(\Delta t)} \varphi_h(x)(u_\varepsilon(x,t_2) - u_\varepsilon(x,t_1))dx\right| \le C(\Delta t)^{(N+1)/2}.$$

Letting $h \to 0$ yields

$$\left|\int_{B(\Delta t)} (u_\varepsilon(x,t_2) - u_\varepsilon(x,t_1))dx\right| \le C(\Delta t)^{(N+1)/2},$$

from which it follows by the mean value theorem that there exists $x^* \in B(\Delta t)$ such that

$$|u_\varepsilon(x^*,t_2) - u_\varepsilon(x^*,t_1)| \le C(\Delta t)^{1/2}.$$

Using this inequality and (4,20, we derive

$$\begin{aligned} |u_\varepsilon(x_0,t_2) - u_\varepsilon(x_0,t_1)| &\le |u_\varepsilon(x_0,t_2) - u_\varepsilon(x^*,t_2)| \\ &\quad + |u_\varepsilon(x^*,t_2) - u_\varepsilon(x^*,t_1)| + |u_\varepsilon(x^*,t_1) - u_\varepsilon(x_0,t_1)| \qquad (4.24) \\ &\le C(\Delta t)^{1/2}. \end{aligned}$$

Combining (4.20) with (4.24) and letting $\varepsilon \to 0$ to pass to the limit show that u satisfies (4.19). This completes the proof of our theorem. □

2.4.4 *Hölder continuity of the gradient of solutions*

Theorem 2.4.4 *Assume that* $p > \max\left\{1, \dfrac{2N}{N+2}\right\}$ *and* u *is a generalized solution of (1.1) on* Q_T. *Then* u_{x_j} $j = 1, 2, \cdots, N$ *is locally Hölder continuous.*

We will give a proof of the theorem only for the case $p > 2$. The proof is based on three basic propositions stated below, which describe the local properties of ∇u.

Similar to §2.4.2, to prove Theorem 2.4.4, it suffices to establish the uniform Hölder estimate for the gradient of the classical solution u_n of the boundary value problem (4.8)–(4.10). For simplicity of expression and notation, we simply suppose that u is the classical solution of (1.1) and wish to establish the Hölder estimate for ∇u with Hölder coefficient and exponent depending only on p, N, $\|u\|_{L^\infty_{loc}(Q_T)}$.

Let $P_0=(x_0,t_0)\in Q_T$. For $0<R\le 1$, $\mu>0$, denote

$$Q_\mu(P_0,R)=\left\{(x,t);|x-x_0|<R,t_0-\frac{R^2}{\mu^{p-2}}<t\le t_0\right\},$$
$$M_{i\mu}^{\pm}(R)=\operatorname{ess}\sup_{Q_\mu(P_0,R)}(\pm u_{x_i}),\ i=1,2,\cdots,N,$$
$$M_\mu(R)=\max_{1\le i\le N}\operatorname{ess}\sup_{Q_\mu(P_0,R)}|u_{x_i}|.$$

We always assume that $Q_\mu(P_0,R)\subset Q_T$. We will state and prove our propositions for u_{x_1} as an example.

Proposition 2.4.2 *Assume that* $2M_{1\mu}^+(R)\ge M_\mu(R)$ *and* μ *satisfies*

$$2M_{1\mu}^+(R)\ge\mu\ge M_\mu(R). \tag{4.25}$$

Then there exists $\varepsilon_0=\varepsilon_0(p,N)$ *such that*

$$⨌_{Q_\mu(P_0,R)}\left(M_{1\mu}^+(R)-u_{x_1}\right)^2dxdt\le\varepsilon_0(M_{1\mu}^+(R))^2 \tag{4.26}$$

implies

$$ess\sup_{Q_\mu(P_0,R/2)}u_{x_1}\ge\frac{M_{1\mu}^+(R)}{2}, \tag{4.27}$$

where

$$⨌_{Q_\mu(P_0,R)}fdxdt=\frac{1}{mes\,Q_\mu(P_0,R)}\iint_{Q_\mu(P_0,R)}fdxdt.$$

Proposition 2.4.3 *Assume that* $2M_{1\mu}^+(R)\ge M_\mu(R)$ *and* μ *satisfies (4.25). Then for any* $\varepsilon_0>0$*, there exist constants* λ, $\beta\in(0,1)$ *depending only on* p, N, ε_0*, such that*

$$⨌_{Q(P_0,R)}\left(M_{1\mu}^+-u_{x_1}\right)^2dxdt>\varepsilon_0(M_{1\mu}^+)^2 \tag{4.28}$$

implies

$$\begin{aligned}&mes\{(x,t)\in Q_\mu(P_0,R);\ u_{x_1}(x,t)\ge(1-\beta)M_{1\mu}^+(R)\}\\>\ &\lambda mes\,Q_\mu(P_0,R).\end{aligned} \tag{4.29}$$

Proposition 2.4.4 *Assume that* $2M_{1\mu}^{+}(R) \geq M_{\mu}(R)$ *and* μ *satisfies (4.25). If for some constants* λ, $\beta \in (0,1)$,

$$\begin{aligned} & mes\{(x,t) \in Q_{\mu}(P_0,R);\ u_{x_1}(x,t) \geq (1-\beta)M_{1\mu}^{+}(R)\} \\ \geq\ & \lambda mes\, Q_{\mu}(P_0,R), \end{aligned} \tag{4.30}$$

then there exist constants δ, $\gamma \in (0,1)$ *depending only on* N, p, λ, β, *such that*

$$M_{1\mu}^{+}(\delta R) \leq \gamma M_{1\mu}^{+}(R). \tag{4.31}$$

Remark 2.4.1 In view of the special form of (1.1), it is easy to see that if we replace $M_{1\mu}^{+}$, u_{x_1} by $M_{1\mu}^{-}$, $-u_{x_1}$, the corresponding conclusions are also valid.

We will first apply these propositions to prove Theorem 2.4.4 and then come back to the proof of these propositions.

First we choose ε_0 in Proposition 2.4.2. Then we determine λ, β in Proposition 2.4.3 by ε_0. Finally we determine δ, γ in Proposition 2.4.4 by λ, β. Of course, constants γ, δ depend only on N, p.

Choose $s \in (1,2)$ so close to 2 that

$$\delta^{2(2-s)/s(p-2)} > \max\left\{\frac{1}{2}, \gamma\right\}. \tag{4.32}$$

Let $0 < \eta_0 < T$, $\Omega_{\eta_0,T} = \Omega \times (\eta_0, T) \subset\subset Q_T$ be a bounded open set. By Theorem 2.4.2, ∇u is bounded on $\Omega_{\eta_0 T}$. Suppose that

$$\|\nabla u\|_{\infty,\Omega_{\eta_0 T}} \leq \overline{M}_0 \tag{4.33}$$

and denote

$$M_0 = \overline{M}_0 \delta^{-2(2-s)/s(p-2)}.$$

Now we choose $R_0 \in (0,1]$ such that $Q_{2M_0}(P_0,R_0) \subset \Omega_{\eta_0,T}$ and denote

$$t_R = R^s R_0^{2-s}(2M_0)^{2-p},$$

$$\hat{Q}(P_0.R) = \{(x,t); |x-x_0| < R, t_0 - t_R < t < t_0\},$$

$$M_i^{\pm}(R) = \operatorname{ess}\sup_{\hat{Q}(P_0,R)} (\pm u_{x_i}), \quad i = 1,2,\cdots,N,$$

$$M(R) = \max_{1\le i\le N} \operatorname{ess}\sup_{\hat{Q}(P_0,R)} |u_{x_i}|,$$

$$\operatorname*{osc}_{\hat{Q}(P_0,R)} u_{x_i} = \operatorname{ess}\sup_{\hat{Q}(P_0,R)} u_{x_i} - \operatorname{ess}\inf_{\hat{Q}(P_0,R)} u_{x_i} = M_i^+(R) + M_i^-(R).$$

We are ready to prove that there exist constants $\rho \in (0,1)$, $C > 0$ depending only on N, p, such that

$$\operatorname*{osc}^{\hat{Q}(P_0,R)} u_{x_i} \le CM_0 \left(\frac{R}{R_0}\right)^\rho, \quad \text{for } 0 < R \le R_0, i = 1,2,\cdots,N, \tag{4.34}$$

which implies the Hölder continuity of u_{x_i} $(i = 1,2,\cdots,N)$ on $\Omega_{\eta_0,T}$ and completes the proof of our theorem.

Define

$$R_1 = \sup\{R \in [0,R_0];\text{ there exist } j, 1 \le j \le N, \theta \in \{+,-\} \text{ such that}$$
$$|M_j^\theta(R)| \ge 2M_0(R/R_0)^{(2-s)/(p-2)}\}. \tag{4.35}$$

We may suppose that $R_1 > 0$; otherwise (4.34) obviously holds. By the definition of M_0, $\overline{M}_0$, from (4.35) we can derive $R_1 \le \delta^{2/s}R_0 < R_0$. Hence there exists R_2 such that $\delta^{2/s}R_2 < R_1 < R_2 < R_0$ and

$$|M_j^{\pm}(R_2)| \le 2M_0 \left(\frac{R_2}{R_0}\right)^{(2-s)/(p-2)}, j = 1,2,\cdots,N \tag{4.36}$$

and there exist i_0, θ, without loss of generality, suppose $i_0 = 1$, $\theta = +$ such that

$$M_1^+(\delta^{2/s}R_2) > 2M_0 \left(\frac{\delta^{2/s}R_2}{R_0}\right)^{(2-s)/(p-2)}. \tag{4.37}$$

Let

$$\mu = 2M_0 \left(\frac{R_2}{R_0}\right)^{(2-s)/(p-2)}. \tag{4.38}$$

We first prove that

$$\iint_{Q_\mu(P_0,R_2)} (M_1^+(R_2) - u_{x_1})^2 dxdt \le \varepsilon_0 (M_1^+(R_2))^2. \tag{4.39}$$

By the choice of μ, we have

$$Q_\mu(P_0,R_2) = \hat{Q}(P_0,R_2), \quad M_{1\mu}^+(R_2) = M_1^+(R_2).$$

Hence, from (4.38), (4.37) and (4.32) we derive

$$M_{1\mu}^+(R_2) \ge M_1^+(\delta^{2/s}R_2) > \max\{\gamma, 1/2\}\mu, \tag{4.40}$$

which together with (4.36), (4.38) implies

$$2M_{1\mu}^+(R_2) > \mu \ge M_\mu(R_2). \tag{4.41}$$

If (4.39) is not true, then according to Proposition 2.4.3 and Proposition 2.4.4, we must have

$$M_{1\mu}^+(\delta R_2) \le \gamma M_{1\mu}^+(R_2)$$

and then noting that $Q_\mu(P_0,\delta R_2) = \hat{Q}(P_0,\delta^{2/s}R_2)$,

$$M_1^+(\delta^{2/s}R_2) = M_{1\mu}^+(\delta R_2) \le \gamma M_{1\mu}^+(R_2) \le \gamma\mu,$$

which contradicts (4.40). Therefore (4.39) holds.

Now from (4.40) and Proposition 2.4.2, we can derive

$$\operatorname*{ess\,inf}_{Q_\mu(P_0,R_2/2)} u_{x_1} \ge \frac{M_{1\mu}^+(R_2)}{2} \ge \frac{\mu}{4}. \tag{4.42}$$

We will use (4.42) to prove that u_{x_m} satisfies (4.34). For this purpose, we differentiate (1.1) with respect to x_m to obtain

$$\frac{\partial u_{x_m}}{\partial t} - (a_{ij}|\nabla u|^{p-2}u_{x_m x_j})_{x_i} = 0, \tag{4.43}$$

where

$$a_{ij} = \delta_{ij} + \frac{(p-2)u_{x_i}u_{x_j}}{|\nabla u|^2}, \ \delta_{ij} = \begin{cases} 1, & \text{if } i = j, \\ 0, & \text{if } i \ne j. \end{cases}$$

Let

$$\xi = x - x^0, \tau = \mu^{p-2}(t-t^0), v(\xi,\tau) = u_{x_m}(x,t),$$
$$Q'(R) = \{(\xi,\tau); |\xi| < R, -R^2 < \tau \le 0\}.$$

Then v satisfies

$$\frac{\partial v}{\partial \tau} - \left(a_{ij}\frac{|\nabla u|^{p-2}}{\mu^{p-2}}v_{\xi_j}\right)_{\xi_i} = 0. \tag{4.44}$$

From (4.33) and (4.42),

$$\frac{1}{C}|\eta|^2 \le a_{ij}\left(\frac{|\nabla u|}{\mu}\right)^{p-2}\eta_i\eta_j \le c|\eta|^2,$$
$$\text{for } \eta \in \mathbb{R}^N,\ (\xi,\tau) \in Q'(R_2/2),$$

which shows that (4.44) is uniformly parabolic on $Q'(R_0/2)$. Thus applying Hölder's estimate for such kind of equations, we have

$$\operatorname*{osc}_{Q'(R)} v \le C\left(\frac{R}{R_2}\right)^{\overline{\beta}} \operatorname{osc} Q'(R_2/4)v, \quad \text{for } 0 < R < \frac{R_2}{4},$$

where $C > 0$, $\overline{\beta} \in (0,1)$ depend only on N, p. Coming back to the original variables (x,t), we obtain

$$\operatorname*{osc}_{Q_\mu(P_0,R)} u_{x_m} \le C\left(\frac{R}{R_2}\right)^{\overline{\beta}} \operatorname*{osc}_{Q_\mu(P_0,R_2/4)} u_{x_m}. \tag{4.45}$$
$$\text{for } 0 < R < R_2/4,\ m = 1,2,\cdots,N.$$

If $R \ge R_2$, then by the definition of R_2, we have

$$\operatorname*{osc}_{\hat{Q}(P_0,R)} u_{x_m} \le |M_m^+(R)| + |M_m^-(R)| \le 4M_0\left(\frac{R}{R_0}\right)^{(2-s)/(p-2)}. \tag{4.46}$$

If $\dfrac{R_2}{4} \le R \le R_2$, then

$$\operatorname*{osc}_{\hat{Q}(P_0,R)} u_{x_m} \le \operatorname*{osc}_{\hat{Q}(P_0,4R)} u_{x_m} \le 4M_0\left(\frac{4R}{R_0}\right)^{(2-s)/(p-2)}. \tag{4.47}$$

If $0 < R < \dfrac{R_2}{4}$, then from (4.45) and (4.47),

$$\operatorname*{osc}_{\hat{Q}(P_0,4R)} u_{x_m} \le c\left(\frac{R}{R_2}\right)^{\overline{\beta}} 4M_0\left(\frac{4R_2}{R_0}\right)^{(2-s)/(p-2)}.$$

Thus, if we set $\rho = \min\left\{\overline{\beta}, \dfrac{2-s}{p-2}\right\}$, then

$$\operatorname*{osc}_{Q_\mu(P_0,R)} u_{x_m} \le cM_0\left(\frac{R}{R_2}\right)^{\rho}\left(\frac{R_2}{R_0}\right)^{\rho} = CM_0\left(\frac{R}{R_0}\right)^{\rho},$$

which implies (4.34), since $\hat{Q}(P_0,R) \subset Q_\mu(P_0,R)$, The proof of Theorem 2.4.4 is completed.

Proof of Proposition 2.4.2. Without loss of generality, we suppose that $(x_0,t_0) = (0,0)$. Denote

$$Q_\mu(R) = Q_\mu(P_0,R),\ M_1 = M_{1\mu}^+(R),\ v = u_{x_1}.$$

Let $\xi(x,t)$ be a smooth cut-off function on $Q_\mu(R)$, which vanishes near the parabolic boundary of $Q_\mu(R)$. In the integral form of (4.43), take $m = 1$, $\varphi = \xi^2(k-v)_+$, $\dfrac{M_1}{2} \le k < M_1$. Then we have

$$\begin{aligned}
&\iint_{Q_\mu(R)} \frac{1}{2}\xi^2\frac{\partial(k-v)_+^2}{\partial t}dxdt \\
&+ \iint_{Q_\mu(R)\cap\{v<k\}} \xi^2 a_{ij}|\nabla u|^{p-2}v_{x_i}v_{x_j}dxdt \\
=\ & 2\iint_{Q_\mu(R)} \xi a_{ij}|\nabla u|^{p-2}(k-v)_+\xi_{x_i}v_{x_j}dxdt.
\end{aligned}$$

A simple calculation shows that

$$\begin{aligned}
&\operatorname*{ess\ sup}_{-R^2/\mu^{p-2}<t<0}\int_{B_R}(k-v(x,t))_+^2\xi^2dx \\
&+ \iint_{Q_\mu(R)} \xi^2|\nabla u|^{p-2}|\nabla(k-v)_+|^2dxdt \\
\le\ & C\left(\iint_{Q_\mu(R)} |\nabla u|^{p-2}(k-v)_+^2|\nabla\xi|^2dxdt\right. \\
&\left.+ \iint_{Q_\mu(R)} |\xi_t|(k-v)_+^2dxdt\right).
\end{aligned} \tag{4.48}$$

Define

$$\varphi_k(s) = \begin{cases} 0, & \text{for } s > k, \\ k - s, & \text{for } k \geq s \geq k - \dfrac{M_1}{4}, \\ \dfrac{M_1}{4}, & \text{for } s < k - \dfrac{M_1}{4}. \end{cases}$$

From $\frac{M_1}{2} \leq k \leq M_1$ and the boundedness of v, we have

$$\varphi_k(v) \leq (k - v)_+ \leq C\varphi_k(v).$$

Thus from (4.48) we obtain

$$\begin{aligned} &\operatorname*{ess\,sup}_{-R^2/\mu^{p-2}<t<0} \int_{B_R} \xi^2 \varphi_k^2(v) dx \\ &\quad + \iint_{Q_\mu(R)} \xi^2 |\nabla u|^{p-2} |\nabla \varphi_k(v)|^2 dx dt \\ &\leq C_1 \Big(\iint_{Q_\mu(R)} |\nabla u|^{p-2} \varphi_k^2(v) |\nabla \xi|^2 dx dt \iint_{Q_\mu(R)} |\xi_t| \varphi_k^2(v) dx dt \Big). \end{aligned} \tag{4.49}$$

Since (4.25) implies

$$|\nabla u|^{p-2} \leq N^{(p-2)/2} \mu^{p-2}$$

and for $k \geq v \geq k - \frac{M_1}{4}$,

$$|\nabla u|^{p-2} \geq 2^{-3(p-2)} \mu^{p-2},$$

from (4.49) we further obtain

$$\begin{aligned} &\operatorname*{ess\,sup}_{-R^2/\mu^{p-2}<t<0} \int_{B_R} \xi^2 \varphi_k^2(v) dx + \mu^{p-2} \iint_{Q_\mu(R)} |\nabla(\xi \varphi_k(v))|^2 dx dt \\ &\leq C \iint_{Q_\mu(R)} \varphi_k^2(v) (\mu^{p-2} |\nabla \xi|^2 + |\xi_t|) dx dt. \end{aligned}$$

Let $\tau = t\mu^{p-2}$, $\widetilde{Q}(R) = \{(x,\tau); |x| < R, -R^2 < \tau < 0\}$. Then the above inequality turns out to be

$$\begin{aligned}
&\operatorname*{ess\,sup}_{-R^2<\tau<0} \int_{B_R} \xi^2 \varphi_k^2(v)dx + \iint_{\widetilde{Q}(R)} |\nabla(\xi\varphi_k(v))|^2 dxd\tau \\
\le\ & C \iint_{\widetilde{Q}(R)} \varphi_k^2(v)(|\nabla\xi|^2 + \mu^{2-p}|\xi_t|)dxd\tau,
\end{aligned}$$

from which, using Lemma 2.1.1, we derive

$$\begin{aligned}
&\left(\iint_{\widetilde{Q}(R)} (\xi\varphi_k(v))^{2(N+2)/N} dxd\tau\right)^{N/(N+2)} \\
\le\ & C \iint_{\widetilde{Q}(R)} \varphi_k^2(v)(|\nabla\xi|^2 + \mu^{2-p}|\xi_t|)dxd\tau.
\end{aligned}$$

Coming back to the original variable t, we obtain

$$\begin{aligned}
&\left(\iint_{Q_\mu(R)} (\xi\varphi_k(v))^{2(N+2)/N} dxdt\right)^{N/(N+2)} \\
\le\ & C\mu^{2(p-2)/(N+2)} \iint_{Q_\mu(R)} \varphi_k^2(v)(|\nabla\xi|^2 + \mu^{2-p}|\xi_t|)dxdt.
\end{aligned} \tag{4.50}$$

Let

$$k_l = M_1\left(\frac{1}{2} + \frac{1}{2^{l+1}}\right), \quad R_l = R\left(\frac{1}{2} + \frac{1}{2^{l+1}}\right), \quad l = 1, 2, \cdots$$

and ξ_l be a smooth cut-off function on $Q_\mu(R_l)$ with $\xi_l = 1$ on $Q_\mu(R_{l+1})$, $0 \le \xi_l \le 1$, $|\nabla\xi_l|^2 \le \dfrac{C4^l}{R^2}$, $|\xi_{lt}| \le \dfrac{C4^l\mu^{p-2}}{R^2}$. Taking $\xi = \xi_l$, $k = k_l$, $R = R_l$ in (4.50) gives

$$\begin{aligned}
&\left(\iint_{Q_\mu(R_l)} (\xi_l\varphi_{k_l}(v))^{2(N+2)/N} dxdt\right)^{N/(N+2)} \\
\le\ & \frac{C4^l\mu^{2(p-2)/(N+2)}}{R^2} \iint_{Q_\mu(R_l)} \varphi_{k_l}^2(v)dxdt.
\end{aligned} \tag{4.51}$$

Denote

$$J_l = \iint_{Q_\mu(R_l)} \varphi_{k_l}^2(v)dxdt.$$

Using Hölder's inequality yields

$$\begin{aligned}
&J_{l+1}\\
&= \iint_{Q_\mu(R_{l+1})} \varphi_{k_{l+1}}^2(v)dxdt\\
&\le \left(\iint_{Q_\mu(R_{l+1})} (\varphi_{k_{l+1}}(v))^{2(N+2)/N} dxdt\right)^{N/(N+2)}\\
&\quad \cdot \mathrm{mes}\,(Q_\mu(R_{l+1}) \cap \{\varphi_{k_{l+1}}(v) > 0\})^{2/(N+2)}.
\end{aligned} \tag{4.52}$$

On the other hand,

$$\begin{aligned}
J_l &\ge \frac{1}{C}\iint_{Q_\mu(R_l)} (k_l - v)_+^2 dxdt\\
&\ge \frac{1}{C}(k_l - k_{l+1})^2 \mathrm{mes}\,(Q_\mu(R_l) \cap \{v < k_{l+1}\})\\
&\ge \frac{M_1^2}{C4^l} \mathrm{mes}\,(Q_\mu(R_{l+1}) \cap \{\varphi_{k_{l+1}} > 0\}),
\end{aligned}$$

or

$$\mathrm{mes}\,(Q_\mu(R_{l+1}) \cap \{\varphi_{k_{l+1}} > 0\}) \le \frac{C4^l}{M_1^2} J_l. \tag{4.53}$$

Combining (4.51) with (4.52), (4.53), we obtain

$$J_{l+1} \le \frac{C16^l \mu^{2(p-2)/(N+2)}}{M_1^{4/(N+2)} R^2} J_l^{1+2/(N+2)},$$

which can be written as

$$Y_{l+1} \le C16^l Y_l^{1+2/(N+2)},$$

where

$$Y_l = J_l \mu^{p-2}/M_1^2 R^{N+2}.$$

Therefore, by the iteration lemma (Lemma 1.5.3), there exists $\hat{\varepsilon}_0 > 0$ such that if $Y_0 < \hat{\varepsilon}_0$, then $\lim_{l\to\infty} Y_l = 0$. From this it follows that there exists ε_0 such that if

$$⨏\!\!\!\!\!⨏_{Q_\mu(P_0,R)} (M_{1\mu}^+(R) - u_{x_1})^2 dxdt \le \varepsilon_0 (M_{1\mu}^+(R))^2,$$

then

$$\text{ess} \inf_{Q_\mu(P_0,R/2)} u_{x_1} \geq \frac{M_{1\mu}^+(R)}{2}.$$

□

Proof of Proposition 2.4.3. If (4.29) does not hold, then

$$\begin{aligned}
&\iint_{Q_\mu(R)} (M_1 - u_{x_1})^2 dxdt \\
= &\iint_{Q_\mu(R) \cap \{u_{x_1} \leq (1-\beta)M_1\}} (M_1 - u_{x_1})^2 dxdt \\
&+ \iint_{Q_\mu(R) \cap \{u_{x_1} > (1-\beta)M_1\}} (M_1 - u_{x_1})^2 dxdt \\
\leq &(M_1 + M_1(R))^2 \lambda \text{mes}\, Q_\mu(R) + (\beta M_1)^2 \text{mes}\, Q_\mu(R) \\
\leq &(9\lambda + \beta^2) M_1^2 \text{mes}\, Q_\mu(R),
\end{aligned} \tag{4.54}$$

where$M_1 = M_{1\mu}^+(R)$. If we choose λ, β such that

$$(9\lambda + \beta^2)\text{mes}\, Q_\mu(R) \leq \varepsilon_0, \quad \beta \leq \frac{1}{2},$$

then from (4.54) we obtain an inequality, which is opposite to (4.28). This shows that for such λ, β, (4.29) should be valid, provided that (4.28) holds. Thus the proof of Proposition 2.4.3 is completed. □

To prove Proposition 2.4.3, we need the following lemma.

Lemma 2.4.2 *Assume that $v \in C(-R^2, 0; W^{1,2}(B_R))$ is a nonnegative generalized super-solution of the equation*

$$v_t - (\hat{a}_{ij}(x,t)v_{x_j})_{x_i} = 0, \tag{4.55}$$

on $\widetilde{Q}(R) = \{(x,t); |x| < R, -R^2 < t \leq 0\}$, where $\hat{a}_{ij} \in L^\infty(\widetilde{Q}(R))$ and

$$\hat{a}_{ij}\xi_i\xi_j \geq a_0|\xi|^2, \quad \textit{for } \xi \in \mathbb{R}^N, (x,t) \in \widetilde{Q}(R).a_0 > 0.$$

If there exists a constant $\lambda \in (0,1)$ such that

$$mes\{(x,\tau) \in \widetilde{Q}(R); v(x,t) \geq 1\} \geq \lambda mes\, \widetilde{Q}(R), \tag{4.56}$$

then there exist constants δ, $\gamma \in (0,1)$ depending only on λ, N, a_0, such that

$$ess \inf_{\widetilde{Q}(\delta_R)} v \geq 1 - \gamma.$$

Proof. First we prove that there exist constants $0<\beta<1$, $0<\alpha<\lambda$, $h<\dfrac{1}{2}$, such that for $t\in[-\alpha R^2,0]$,

$$\operatorname{mes}\{x\in B_{\beta R}; v(x,t)\geq h\}\geq\frac{\lambda}{2}\operatorname{mes}B_{\beta R}. \tag{4.57}$$

Denote

$$\mu(t)=\operatorname{mes}\{x\in B_R; v(x,t)\geq 1\}.$$

From the assumption (4.56), it is easily seen that for any $\alpha\in(0,\lambda)$, there exists a subset $E\subset[-R^2,-\alpha R^2]$ of measure zero, such that for $\tau\in E$,

$$\mu(\tau)\geq(\lambda-\alpha)(1-\alpha^{-1})\operatorname{mes}B_R. \tag{4.58}$$

Set

$$w=f(v)=\ln^+\frac{1}{v+h}=\max\left\{\ln\frac{1}{v+h},0\right\},$$

where $h\in(0,1/2)$ is to be determined. Let $\xi(x)$ be a smooth cut-off function on B_R with $\xi=1$ on $B_{\beta R}$, $0\leq\xi\leq 1$, $|\nabla\xi|\leq C(1-\beta)^{-1}$. By assumption, v is a generalized super-solution of (4.55). Choose $\varphi=f'_\varepsilon(v)\xi^2$ in the integral inequality which v satisfies, where $f_\varepsilon(\sigma)$ is a smooth approximation of $f(\sigma)$ possessing the following properties: $f'_\varepsilon(\sigma)\leq 0$, $f''_\varepsilon(\sigma)\geq 0$ and uniformly

$$\begin{aligned}
&\lim_{\varepsilon\to 0}f_e(\sigma)=f(\sigma), && \text{for } \sigma\geq 0,\\
&\lim_{\varepsilon\to 0}f'_e(\sigma)=f'(\sigma)=-\frac{1}{\sigma+h}, && \text{for } \sigma\leq 1-h,\\
&\lim_{\varepsilon\to 0}f''_e(\sigma)=f''(\sigma)=\frac{1}{(\sigma+sh)^2}, && \text{for } \sigma\leq 1-h.
\end{aligned} \tag{4.59}$$

Then using an argument similar to the one in Remark 2.4.1 gives

$$\begin{aligned}
&\int_{B_R}\xi^2(x)f_\varepsilon(v(x,s))dx+\int_\tau^s\int_{B_R}\xi^2 f''_\varepsilon(v)\hat{a}^{ij}v_{x_i}v_{x_j}dxdt\\
\leq\;& -2\int_\tau^s\int_{B_R}\hat{a}^{ij}f'_\varepsilon(v)v_{x_j}\xi\xi_{x_i}dxdt\\
&\quad+\int_{B_R}\xi^2(x)f_e(v(x,\tau))dx,
\end{aligned}$$

where $\tau \in E$, $s \in [-\alpha R^2, 0]$. Letting $\varepsilon \to 0$ and using (4.59) further give

$$\begin{aligned}&\int_{B_R} \xi^2(x) w(x,s)dx + a_0 \int_\tau^s \int_{B_R} \xi^2 |\nabla w|^2 dxdt \\ \leq\ & -2\int_\tau^s \int_{B_R} \hat{a}^{ij} w_{x_j} \xi \xi_{x_i} dxdt + \int_{B_R} \xi^2(x) w(x,\tau)dx.\end{aligned}$$

Noting that $\hat{a}^{ij} \in L^\infty(\widetilde{Q}(R))$, we may use Young's inequality to obtain

$$\begin{aligned}&\int_{B_{\beta R}} w(x,s)dx + a_0 \frac{1}{2} \int_\tau^s \int_{B_{\beta R}} |\nabla w|^2 dxdt \\ \leq\ & C\int_\tau^s \int_{B_R} |\nabla \xi|^2 dxdt + \int_{B_R} w(x,s)dx,\end{aligned} \tag{4.60}$$

from which it follows by using (4.58) that

$$\begin{aligned}&\ln\frac{1}{2h} \mathrm{mes}\,(B_{\beta R}\backslash N_s) \\ \leq\ & C\alpha(1-\beta)^{-2}\mathrm{mes}\, B_R + \ln\frac{1}{h}(\mathrm{mes}\, B_R - \mu(\tau)) \\ \leq\ & (C\beta^{-N}(1-\beta)^{-2} \\ & + \ln\frac{1}{h}(1-\lambda)\beta^{-N}(1-\alpha)^{-1})\mathrm{mes}\, B_{\beta R},\end{aligned} \tag{4.61}$$

where

$$N_s = \{x \in B_{\beta R}; v(x,s) \geq h\}.$$

Take $\alpha \in (0,\lambda)$, $\beta \in (0,1)$ such that

$$(1-\lambda)\beta^{-N}(1-\alpha)^{-1} < (1-\lambda/2)(1-\delta)$$

with $\delta \in (0,1)$. Since

$$\lim_{h\to 0} \ln\frac{1}{2h}\Big/\ln\frac{1}{h} = 1,$$

from (4.61) we can obtain

$$\mathrm{mes}\,(B_{\beta R}\backslash N_s) \leq (1-\lambda/2)\mathrm{mes}\, B_{\beta R},$$

and (4.57) follows, provided that $h \in \left(0, \dfrac{1}{2}\right)$ is small enough.

Now let

$$g = g(v) = \ln^+ \frac{h}{v+\varepsilon}, \quad 0 < \varepsilon < \frac{h}{2}.$$

Similar to the derivation of (4.60), we can obtain

$$\begin{aligned}
&\int_{-\alpha R^2}^{0} \int_{B_{\beta R}} |\nabla g|^2 dx dt \\
\le\ & C \operatorname{mes} B_R + C \int_{B_R} g(x, -\alpha R^2) dx \\
\le\ & C \mathbb{R}^N \ln \frac{3h}{\varepsilon}.
\end{aligned}$$

Since $g = 0$ on N_t with $t \in [-\alpha R^2, 0]$ and (4.57) implies that $\operatorname{mes} N_t \ge \frac{\lambda}{2} \operatorname{mes} B_{\beta R}$, we may use Poincaré's inequality to obtain

$$\begin{aligned}
&\int_{-\alpha R^2}^{0} \int_{B_{\beta R}} g^2 dx dt \\
\le\ & C R^2 \int_{-\alpha R^2}^{0} \int_{B_{\beta R}} |\nabla g|^2 dx dt \\
\le\ & C R^{N+2} \ln \frac{3h}{\varepsilon}.
\end{aligned} \tag{4.62}$$

It is easy to check that g is a generalized sub-solution of (4.55). According to the local estimate on the upper bound of generalized sub-solutions of uniformly parabolic equations, we have

$$\operatorname*{ess\ sup}_{\widetilde{Q}(\beta R/2)} g^2 \le C R^{-(N+2)} \iint_{\widetilde{Q}(\beta R)} g^2 dx dt,$$

which combining with (4.62) yields

$$\operatorname*{ess\ sup}_{\widetilde{Q}(\beta R/2)} g^2 \le C \ln \frac{3h}{\varepsilon}.$$

Hence for small $\varepsilon > 0$,

$$\ln^+ \frac{h}{v+\varepsilon} \le C \left(\ln \frac{3h}{\varepsilon} \right)^{1/2} \le \ln \frac{h}{\sqrt{\varepsilon}} s. \tag{4.63}$$

If $v \geq h-\varepsilon$, then $v \geq \frac{h}{2}$. If $v < h-\varepsilon$, then (4.63) implies that $v \geq \sqrt{\varepsilon}-\varepsilon$. Hence if we choose $\delta = \frac{\beta}{2}$, $\gamma = \max\{1-h/2, 1+\varepsilon-\sqrt{\varepsilon}\}$, then

$$\operatorname{essinf}_{\widetilde{Q}(\delta R)} v \geq 1-\gamma$$

and the lemma is proved. □

Proof of Proposition 2.4.4. Set

$$k = (1-\beta)M_{1\mu}^{+}(R), \quad \omega = (u_{x_1}-k)_+,$$

$$\operatorname{sgn}_\eta(u_{x_1}-k)_+ = \begin{cases} 1 & \text{for } u_{x_1}-k > \eta, \\ (u_{x_1}-k)_+ & \text{for } (u_{x_1}-k) \leq \eta \end{cases}$$

Choose $\varphi \operatorname{sgns}_\eta(u_{x_1}-k)_+$ with $\varphi \geq 0$ and $\varphi \in C_0^1(Q_\mu(R))$ as the test function in the integral form of (4.43) and let $\eta \to 0$. Then we obtain

$$\iint_{Q_\mu} (\omega\varphi_t - \widetilde{a}_{ij}\omega_{x_j}\varphi_{x_i})dxdt \leq 0,$$

where

$$\widetilde{a}_{ij} = \begin{cases} a_{ij}|\nabla u|^{p-2}, & \text{for } u_{x_1} \geq k, \\ \mu^{p-2}\delta_{ij}, & \text{for } u_{x_1} < k. \end{cases}$$

Let $\tau = t\mu^{p-2}$. Then $\omega(x,\tau)$ satisfies

$$\omega_\tau - (\widetilde{a}_{ij}\mu^{2-p}\omega_{x_j})_{x_i} \leq 0$$

on $\widetilde{Q}(R)$ in the sense of distributions. Denote

$$\hat{a}_{ij} = \widetilde{a}_{ij}\mu^{2-p}, \; v = \frac{\operatorname*{ess\,sup}_{\widetilde{Q}(R)} \omega - \omega}{\operatorname*{ess\,sup}_{\widetilde{Q}(R)} \omega}.$$

Then v satisfies

$$v_\tau - (\hat{a}_{ij}v_{x_j})_{x_i} \geq 0$$

in the sense of distributions and (4.30) implies that

$$\operatorname{mes}\{(x,\tau) \in \widetilde{Q}(R); \; v(x,\tau) \geq 1\} \geq \lambda \operatorname{mes}\widetilde{Q}(R).$$

Notice that there exists a constant $C > 0$ depending only on p, N, β, such that

$$\frac{1}{C}|\xi|^2 \leq \hat{a}_{ij}\xi_i\xi_j \leq C|\xi|^2$$

$$\text{for } \xi \in \mathbb{R}^N, \quad (x,t) \in \widetilde{Q}(R).$$

Thus we can use Lemma 2.4.2 to conclude that there exist constants δ, $\gamma \in (0,1)$, such that

$$\operatorname{essinf}_{\widetilde{Q}(\delta R)} v \geq 1 - \gamma > 0,$$

or

$$\operatorname*{ess\,sup}_{\widetilde{Q}(\delta R)} \omega \leq \gamma \operatorname*{ess\,sup}_{\widetilde{Q}(R)} \omega.$$

Hence

$$M_{1\mu}^+(\delta R) \leq (1 - (1-\gamma)\beta) M_{1\mu}^+(R)$$

which is just what we desire to prove. □

2.5 Uniqueness of Solutions

By Corollary 2.3.3, if $p > 2$, then the uniqueness of generalized solutions of the Cauchy problem for equation (1.1) is valid in the class of functions

$$C(0,T; L^2(\mathbb{R}^N)) \cap L^p(0,T; W^{1,p}(\mathbb{R}^N))$$

and the proof is quite easy.

In this section, we will prove the uniqueness in a class of functions, which probably is the broadest class for the uniqueness to be valid. The proof is rather difficult. To do this, we need to prove a series of auxiliary propositions, which have their own independent significance.

2.5.1 *Auxiliary propositions*

As in §2.2.2, we will use the notation

$$|||f|||_r = \sup_{\rho \geq r} \rho^{-\kappa/(p-2)} \int_{B_\rho} |f(x)| dx$$

with $r > 0$, $\kappa = N(p-2) + p$.

Proposition 2.5.1 *Assume that $p > 2$ and u is a generalized solution of (1.1) on Q_T and for some $r > 0$,*

$$\sup_{0<t<T} |||u(\cdot,t)|||_r = \Lambda < \infty. \tag{5.1}$$

Then there exist constants $C_i = C_i(N,p)(i = 0,1,2,3)$ such that the following estimates hold for $t \in (0,T_0)$ with $T_0 = \min\{R, C_0\Lambda^{2-p}\}$:

$$\|u(\cdot,t)\|_{\infty,B_\rho} \le C_1 t^{-N/\kappa} \rho^{p/(p-2)} \Lambda^{p/\kappa}, \tag{5.2}$$

$$\|\nabla u(\cdot,t)\|_{\infty,B_\rho} \le C_2 t^{-(N+1)/\kappa} \rho^{2/(p-2)} \Lambda^{2/\kappa}, \tag{5.3}$$

$$\int_0^t \int_{B_\rho} |\nabla u|^{p-1} dx d\tau \le C_3 t^{1/\kappa} \rho^{1+(p-2)/\kappa} \Lambda^{1+(p-2)/\kappa}. \tag{5.4}$$

Proof. First we prove (5.2). For any $t_0 \in (0,T)$, if we choose

$$\rho_n = \rho + \frac{\sigma}{2^n}\rho, T_n = \frac{t_0}{2} - \frac{\sigma}{2^{n+1}} t_0, \sigma \in (0,1/2], n = 1,2,\cdots$$

in the proof of Theorem 2.4.1, then we can obtain

$$\sup_{B_\rho \times (t_0/2,t_0)} |u| \le \max\left\{ I_\sigma, \left(\frac{\rho^p}{t_0}\right)^{1/(p-2)} \right\}, \tag{5.5}$$

where $q = p\dfrac{N+2}{N}$ and

$$I_\sigma = \gamma(\rho\sigma)^{-N/2} \left(\int_{t_0(1-\sigma)/2}^{t_0} \int_{B_{\rho(1+\sigma)}} |u|^q dx d\tau \right)^{N/2(p+N)}$$

with some positive constant $\gamma = \gamma(N,p)$.

If

$$\sup_{B_\rho \times (t_0/2,t_0)} |u| \le I_\sigma,$$

then using Young's inequality derives

$$\begin{aligned}
&\sup_{B_\rho\times(t_0/2,t_0)}|u|\\
\le\ &\gamma\sup_{B_{\rho(1+\sigma)}\times(t_0(1-\sigma)/2,t_0)}|u|^{p/2(N+p)}\\
&\cdot(\rho\sigma)^{-N/2}\left(\int_{t_0(1-\sigma)/2}^{t_0}\int_{B_{\rho(1+\sigma)}}|u|^{p(N+1)/N}dxd\tau\right)^{N/2(p+N)}\\
\le\ &\frac{1}{2}\sup_{B_{\rho(1+\sigma)}\times(t_0(1-\sigma)/2,t_0)}|u|+\gamma(\rho\sigma)^{-N(N+p)/(2N+p)}\\
&\cdot\left(\int_{t_0(1-\sigma)/2}^{t_0}\int_{B_{\rho(1+\sigma)}}|u|^{p(N+1)/N}dxd\tau\right)^{N/(2N+p)},
\end{aligned}$$

from which, taking $\sigma=\frac{1}{2}$, we obtain by Lemma 2.1.4,

$$\begin{aligned}
&\sup_{B_\rho\times(t_0/2,t_0)}|u|\\
\le\ &\gamma\rho^{-N(N+p)/(2N+p)}\left(\int_{t_0/4}^{t_0}\int_{B_{3\rho/2}}|u|^{p(N+1)/N}dxd\tau\right)^{N/(2N+p)}.
\end{aligned}\tag{5.6}$$

Similarly we can derive

$$\sup_{B_\rho\times(t_0/2,t_0)}|u|\le\gamma\rho^{-(N+p)/2}\left(\int_{t_0/4}^{t_0}\int_{B_{3\rho/2}}|u|^p dxd\tau\right)^{1/2}.\tag{5.7}$$

Choose $\xi\in C_0^1(B_{2\rho(1+\sigma)})$ such that $0\le\xi\le1$, $\xi=1$ for $x\in B_{2\rho}$, $|\nabla\xi|\le\frac{\gamma}{\sigma\rho}$ and $\eta(t)\in C_0^1(0,T+1)$ such that $0\le\eta\le1$, $\eta(t)=1$ for $t\in\left(\frac{t_0}{4},t_0\right)$, $\eta(t)=0$ for $t\le\frac{t_0}{4}(1-\sigma^p)$, $|\eta'(t)|\le\frac{\gamma}{\sigma^p t_0}$. Using the embedding inequality, we obtain

$$\begin{aligned}
&\int_{t_0/4}^{t_0}\int_{B_{2\rho}}|u|^{p(N+1)/N}dxd\tau\\
\le\ &\int_{t_0/4}^{t_0}\int_{B_{2\rho(1+\sigma)}}|\xi^p\eta u|^{p(N+1)/N}dxd\tau\\
\le\ &\int_{t_0/4}^{t_0}\int_{B_{2\rho(1+\sigma)}}|\nabla(\xi^p\eta u)|^p dxd\tau
\end{aligned}$$

$$\cdot\left(\sup_{t_0/4<t<t_0}\int_{B_{2\rho(1+\sigma)}}|u|dx\right)^{p/N}. \tag{5.8}$$

Taking $t_1=\frac{t_0}{4}(1-\sigma^p)$, $t_2=t_0$, $\Omega=B_{2\rho(1+\sigma)}$, $\zeta=\xi^p\eta$ in (1.28), we deduce

$$\begin{aligned}
&\int_{t_0(1-\sigma^p)/4}^{t_0}\int_{B_{2\rho(1+\sigma)}}|\nabla u|^p\xi^p\eta dxd\tau\\
&+p\int_{t_0(1-\sigma^p)/4}^{t_0}\int_{B_{2\rho(1+\sigma)}}u|\nabla u|^{p-2}\nabla u\cdot\nabla\xi\xi^{p-1}\eta dxd\tau\\
\leq\;&\gamma t^{-1}\sigma^{-p}\int_{t_0(1-\sigma^p)/4}^{t_0}\int_{B_{2\rho(1+\sigma)}}u^2dxd\tau.
\end{aligned}$$

Using Young's inequality to the inner integral of the second term on the left hand side, we further obtain

$$\begin{aligned}
&\int_{t_0(1-\sigma^p)/4}^{t_0}\int_{B_{2\rho(1+\sigma)}}|\nabla u|^p\xi^p\eta dxd\tau\\
\leq\;&\gamma(\rho\sigma)^{-p}\int_{t_0(1-\sigma^p)/4}^{t_0}\int_{B_{2\rho(1+\sigma)}}|u|^pdxd\tau\\
&+\gamma t_0^{-1}\sigma^{-p}\int_{t_0(1-\sigma^p)/4}^{t_0}\int_{B_{2\rho(1+\sigma)}}u^2dxd\tau.
\end{aligned}$$

Hence, in view of

$$|\nabla(\xi^p\eta u)|^p\leq\gamma(|\nabla u|^p\xi^p\eta+|u|^p|\nabla\xi|^p\eta),$$

we derive

$$\begin{aligned}
&\int_{t_0(1-\sigma^p)/4}^{t_0}\int_{B_{2\rho(1+\sigma)}}|\nabla(\xi^p\eta u)|^pdxd\tau\\
\leq\;&\gamma(\rho\sigma)^{-p}\int_{t_0(1-\sigma^p)/4}^{t_0}\int_{B_{2\rho(1+\sigma)}}|u|^pdxd\tau\\
&+\gamma t_0^{-1}\sigma^{-p}\int_{t_0(1-\sigma^p)/4}^{t_0}\int_{B_{2\rho(1+\sigma)}}u^2dxd\tau.
\end{aligned} \tag{5.9}$$

If

$$(\rho\sigma)^{-p}\int_{t_0(1-\sigma^p)/4}^{t_0}\int_{B_{2\rho(1+\sigma)}}|u|^p dxd\tau \\ \leq t_0^{-1}\sigma^{-p}\int_{t_0(1-\sigma^p)/4}^{t_0}\int_{B_{2\rho(1+\sigma)}}u^2 dxd\tau, \tag{5.10}$$

then using Hölder's inequality gives

$$\int_{t_0(1-\sigma^p)/4}^{t_0}\int_{B_{2\rho(1+\sigma)}}|u|^p dxd\tau \leq \gamma\left(\frac{\rho^p}{t_0}\right)^{p/(p-2)}\rho^N t_0.$$

Substituting this into (5.7), we obtain

$$\sup_{B_\rho\times(t_0/2,t_0)}|u| \leq \gamma\rho^{p/(p-2)}t_0^{-1/(p-2)}. \tag{5.11}$$

If (5.10) does not hold, then (5.9) implies

$$\int_{t_0/4}^{t_0}\int_{B_{2\rho(1+\sigma)}}|\nabla(\xi^p\eta u)|^p dxd\tau \\ \leq \gamma(\rho\sigma)^{-p}\int_{t_0(1-\sigma^p)/4}^{t_0}\int_{B_{2\rho(1+\sigma)}}|u|^p dxd\tau.$$

Substituting this into (5.8) and using Hölder's inequality, we obtain

$$\begin{aligned}
&\int_{t_0/4}^{t_0}\int_{B_{2\rho}}|u|^{p(N+1)/N}dxd\tau \\
\leq\ &\gamma(\rho\sigma)^{-p}\left(\int_{t_0(1-\sigma^p)/4}^{t_0}\int_{B_{2\rho(1+\sigma)}}|u|^{p(N+1)/N}dxd\tau\right)^{N/(N+1)} \\
&\quad\cdot(\mathrm{mes}(B_{2\rho}\times(t_0(1-\sigma^p)/4,t_0)))^{1/(N+1)} \\
&\quad\cdot\left(\sup_{t_0/4<t<t_0}\int_{B_{2\rho(1+\sigma)}}|u|dx\right)^{p/N} \\
\leq\ &\frac{1}{2}\int_{t_0(1-\sigma^p)/4}^{t_0}\int_{B_{2\rho(1+\sigma)}}|u|^{p(N+1)/N}dxd\tau \\
&\quad+\gamma\sigma^{-p(N+1)}\rho^{N-\rho N-p}\left(\sup_{t_0/4<t<t_0}\int_{B_{2\rho(1+\sigma)}}|u|dx\right)^{p(N+1)/N},
\end{aligned}$$

$$\int_{t_0/4}^{t_0}\int_{B_{2\rho}} |u|^{p(N+1)/N}dxd\tau$$
$$\leq \gamma\rho^{N-pN-p}\left(\sup_{t_0/4<t<t_0}\int_{B_{3\rho}}|u|dx\right)^{p(N+1)/N}.$$

Thus, taking $\sigma = \frac{1}{2}$, we obtain by Lemma 1.1.5,

$$\sup_{B_\rho\times(t_0/2,t_0)} |u| \leq \gamma\rho^{p/(p-2)}\Lambda^{p(N+1)/(2N+p)}. \tag{5.12}$$

Combining (5.5) with (5.11), (5.12) we see that for any $t \in (0,T)$,

$$\frac{\|u(\cdot,t)\|_{\infty,B_\rho}}{\rho^{p/(p-2)}} \leq \gamma t^{-1/(p-2)}.$$

Thus for any $\varepsilon_0 \in (0,T)$ we may define the function

$$\phi(t) = \sup_{\tau\in(0,t)} \tau^{N/K}\sup_{\rho\geq r}\frac{\|u(\cdot,\tau+\varepsilon_0)\|_{\infty,B_\rho}}{\rho^{p/(p-2)}}$$

on $t \in (0, T-\varepsilon_0)$ for $u(x,t+\varepsilon_0)$ and use an argument similar to the proof of Theorem 2.2.3 to conclude that there exists $\gamma_0 > 0$, such that $u(x,t+\varepsilon_0)$ satisfies the estimate (5.2) for $0 < t < \gamma_0\Lambda^{2-p}$. The proof of (5.2) is completed by letting $\varepsilon_0 \to 0$.

The proof of (5.3) is more difficult. For this purpose we first prove that for any $0 < \varepsilon_0 < T$,

$$\sup_{t\in(\varepsilon_0,T)}\sup_{\rho\geq r}\frac{\|\nabla u(\cdot,t)\|_{\infty,B_\rho}}{\rho^{2/(p-2)}} < \infty. \tag{5.13}$$

For any given $t_0 \in (0,T)$, we may take

$$\rho_n = \rho + \frac{\sigma}{2^n}\rho, \quad T_n = \frac{t_0}{2} - \frac{\sigma}{2^{n+1}}t_0,$$
$$\sigma \in \left(0, \frac{1}{2}\right], \quad n = 0,1,2,\cdots,$$

in the proof of Theorem 2.4.2 to derive

$$\sup_{B_\rho\times(t_0/2,t_0)} |\nabla u|$$
$$\leq \max\left\{\gamma((\rho\sigma)^{-(N+2)}I_0)^{1/k_0}, \left(\frac{\rho^2}{t_0}\right)^{1/(p-2)}\right\}, \tag{5.14}$$

where $k = 1 + \frac{2}{N}$, the integer n_0 is so large that $k^{n_0} > 3$ and

$$I_0 = \iint_{Q_{n_0}} |\nabla u|^{p-2+k^{n_0}} dx d\tau.$$

Obviously, if

$$\sup_{B_\rho \times (t_0/2, t_0)} |\nabla u| \le \left(\frac{\rho^2}{t_0} \right)^{1/(p-2)}, \tag{5.15}$$

then (5.13) holds. Now we suppose that

$$\sup_{B_\rho \times (t_0/2, t_0)} \nabla u \le \gamma \left((\rho\sigma)^{-(N+2)} \iint_{Q_{n_0}} |\nabla u|^{p-2+k^{n_0}} dx d\tau \right)^{1/k_0}.$$

Then, by Young's inequality,

$$\begin{aligned} &\sup_{B_\rho \times (t_0/2, t_0)} |\nabla u| \\ \le\ & \gamma \sup_{Q_{n_0}} |\nabla u|^{(k^{n_0}-2)/k^{n_0}} \left((\rho\sigma)^{-(N+2)} \iint_{Q_{n_0}} |\nabla u|^p dx d\tau \right)^{1/k_{n_0}} \\ \le\ & \frac{1}{2} \sup_{Q_{n_0}} |\nabla u| + \gamma \left((\rho\sigma)^{-(N+2)} \iint_{Q_{n_0}} |\nabla u|^p dx d\tau \right)^{1/2}. \end{aligned}$$

Thus, taking $\sigma = \frac{1}{2}$, we obtain by Lemma 1.1.4,

$$\begin{aligned} &\sup_{B_\rho \times (t_0/2, t_0)} |\nabla u| \\ \le\ & \gamma \left(\rho^{-(N+2)} \iint_{Q_{n_0}} |\nabla u|^p dx d\tau \right)^{1/2} \\ \le\ & \gamma \left(\rho^{-(N+2)} \int_{t_0/4}^{t_0} \int_{B_{3\rho/2}} |\nabla u|^p dx d\tau \right)^{1/2}, \end{aligned} \tag{5.16}$$

from which it follows by using (5.9) with $\sigma = \frac{1}{2}$, that

$$\begin{aligned}
&\sup_{B_\rho\times(t_0/2,t_0)} |\nabla u| \\
\le\ & \gamma\rho^{-(N+2)/2}\Big(\rho^p\int_{t_0/8}^{t_0}\int_{B_{3\rho}}|u|^p dxd\tau \\
&+ t_0^{-1}\int_{t_0/8}^{t_0}\int_{B_{3\rho}} u^2 dxd\tau\Big)^{1/2} \\
\le\ & \gamma\rho^{-p(p-3)/(2p-4)-1}\sup_{t\in(t_0/8,t_0)}\|u\|_{\infty,B_{3\rho}}^{(p-1)/2} M t_0^{1/2} \\
&+\gamma\rho^{(4-p)/(2p-4)}\sup_{t\in(t_0/8,t_0)}\|u\|_{\infty,B_{3\rho}}^{1/2} M t_0^{1/2},
\end{aligned}$$

where $M = \left(\sup_{0<t<T}|||u(\cdot,t)|||_r\right)^{1/2}$. Substituting (5.2) into the above inequality we further obtain

$$\sup_{B_\rho\times(t_0/2,t_0)} |\nabla u| \le \gamma\rho^{2/(p-2)} t_0^{-N/(2k)}$$

which shows that (5.13) holds.

Now we use (5.13) to derive the estimate (5.3) which is sharper than (5.13). It is clear that if we can prove (5.3) on $(0, T_0)$ for $u(x, t + \varepsilon_0)$ for any small $\varepsilon_0 > 0$, then we can conclude by letting $\varepsilon_0 \to 0$ that $u(x, t)$ also satisfies (5.3). The discussion in what follows is aimed at $u(x, t + \varepsilon_0)$. Similar to the argument in §2.4.2 and §2.4.4, we may simply suppose that u is a classical solution of (1.1).

Multiplying (4.43) by $\xi_n^2(v-k)_+^\alpha u_{x_m}$ $(\alpha > 0, k > 0)$, integrating over Q_n and summing up for n from 1 to N, we can obtain

$$\begin{aligned}
&\sup_{T_{n+1}\le t\le t_0}\int_{B'_n}(v(x,t)-k)_+^{\alpha+1}dx \\
&+\iint_{Q'_n}|\nabla u|^{p-2}|\nabla(v-k)_+^{(\alpha+1)/2}|^2 dxd\tau \\
\le\ & \frac{\gamma 2^{2n}}{\rho^2}\int_{Q_n}|\nabla u|^p(v-k)_+^\alpha dxd\tau \\
&+\frac{\gamma 2^{2n}}{t_0}\iint_{Q_n}(v-k)_+^{\alpha+1}dxd\tau,
\end{aligned} \tag{5.17}$$

where $v = |\nabla u|^2$ and B_n, B_n', Q_n, Q_n', ξ_n are the same as in the proof of Theorem 2.4.1.

Denote

$$k_n = k - \frac{k}{2^{n+1}}, \quad A_n = \{(x,t) \in Q_n; v(x,t) > k_n\}.$$

It is clear that

$$|\nabla u|^{p-2} \geq \left(\frac{1}{2}k\right)^{(p-2)/2}.$$

Set

$$F(t) = \sup_{0<\tau<t} \tau^{(N+1)/\kappa} \sup_{\rho>r>0} \frac{\|\nabla u\|_{\infty,B_\rho}}{\rho^{2/(p-2)}}.$$

Using (5.13) to $u = u(x, t + \varepsilon_0)$ shows that $F(t)$ is well-defined for $t \in [0, T_0 - \varepsilon_0]$. Thus for $t \in (t_0/2, t_0)$, $\rho > r$,

$$\frac{|\nabla u|^{p-2}}{\rho^2} \leq \gamma t_0^{-(N+1)(p-2)/\kappa} F(t)^{p-2}.$$

Since

$$(v - k_n)_+^{\alpha+1} \geq \frac{v}{2}(v - k_n)_+^\alpha \geq Cv(v - k_{n+1})_+^\alpha,$$

whenever $v > 2k_n$ and

$$(v - k_n)_+^{\alpha+1} \geq (v - k_n)_+^\alpha (k_{n+1} - k_n) \geq C2^{-n} v(v - k_{n+1})_+^\alpha,$$

whenever $k_{n+1} \leq v \leq 2k_n$, we have

$$v(v - k_{n+1})_+^\alpha \leq C2^n (v - k_n)_+^{\alpha+1}.$$

Taking $k = k_{n+1}$ and denoting $w_n = (v - k_n)_+^{(\alpha+1)/2}$ in (5.17) yield

$$\begin{aligned} &\sup_{T_{n+1}<t<t_0} \int_{B_n'} w_{n+1}^2(x,t)dx + k^{(p-2)/2} \iint_{Q_n'} |\nabla w_{n+1}|^2 dx d\tau \\ \leq\ & \gamma 2^{2n} H(t_0) \iint_{Q_n} w_n^2 dx d\tau, \end{aligned} \tag{5.18}$$

where

$$H(t_0) = t_0^{-(N+1)(p-2)/k} F(t)^{p-2} + t_0^{-1}.$$

Let $\tilde{\xi}_n(x)$ be a smooth cut-off function on B'_n such that $\tilde{\xi}_n(x) = 1$ on B_{n+1} and $|\nabla\tilde{\xi}_n| \le \frac{2^n}{\rho}$. Applying the embedding inequality (1.7) to $w_{n+1}\tilde{\xi}_n$ derives

$$\iint_{Q_{n+1}} w_{n+1}^{2(N+2)/N} dx d\tau \le \iint_{Q'_n} (w_{n+1}\tilde{\xi}_n)^{2(N+2)/N} dx d\tau$$

$$\le \gamma \left(\sup_{\tau_{n+1}<t<t_0} \int_{B'_n} w_{n+1}^2 dx \right)^{2/N} \tag{5.19}$$

$$\cdot \left(\iint_{Q'_n} |\nabla w_{n+1}|^2 dx d\tau + \frac{2^{2n}}{\rho^2} \iint_{Q'_n} w_{n+1}^2 dx d\tau \right).$$

Suppose that

$$0 < k < \rho^{4/(p-2)} t_0^{-2(N+1)/\kappa} F^2(t_0). \tag{5.20}$$

Then $\frac{1}{\rho^2} < \frac{H(t_0)}{k^{(p-2)/2}}$ and hence

$$\frac{2^{2n}}{\rho^2} \iint_{Q'_n} w_{n+1}^2 dx d\tau \le \gamma 2^{2n} \frac{H(t_0)}{k^{(p-2)/2}} \iint_{Q_n} w_{n+1}^2 dx d\tau.$$

Substituting this into (5.19) and using (5.18) to estimate the right hand side, we obtain

$$\iint_{Q_{n+1}} w_{n+1}^{2(N+2)/N} dx d\tau \le \gamma (2^{2n} H(t_0))^{(N+2)/N} k^{-(p-2)/2}$$

$$\cdot \left(\iint_{Q_n} w_n^2 dx d\tau \right)^{(N+2)/N}. \tag{5.21}$$

Notice that by Hölder's inequality, we have

$$\iint_{Q_{n+1}} w_{n+1}^2 dx d\tau \le \left(\int_{Q_{n+1}} w_{n+1}^{2(N+2)/N} dx d\tau \right)^{N/(N+2)} \tag{5.22}$$

$$\cdot |A_{n+1}|^{2/(N+2)},$$

and clearly,

$$|A_{n+1}| \le 2^{(\alpha+1)(n+1)} k^{-(\alpha+1)} \iint_{Q_n} (v - k_n)_+^{\alpha+1} dx d\tau.$$

Thus from (5.21), (5.22), we arrive at

$$\begin{aligned}&\iint_{Q_{n+1}} (v-k_n)_+^{\alpha+1}dxd\tau\\ \le\ &\gamma b^n H(t_0)k^{-\sigma/2(N+2)}\left(\iint_{Q_n}(v-k_n)_+^{\alpha+1}dxd\tau\right)^{(N+4)/(N+2)},\end{aligned}$$

where $\sigma = N(p-2)+4(\alpha+1)$, $b = 2^{2+2(\alpha+1)/(N+2)}$, $\gamma = \gamma(N,p,\alpha)$. From this it follows by the iteration lemma (Lemma 1.5.3) that if

$$\iint_{Q_0}(v-k/2)_+^{\alpha+1}dxd\tau \le \gamma(H(t_0))^{-(N+2)/2}k^{\sigma/4}, \tag{5.23}$$

then

$$\iint_{Q_\infty}(v-k)_+^{\alpha+1}dxd\tau = \lim_{n\to\infty}\iint_{Q_n}(v-k_n)_+^{\alpha+1}dxd\tau = 0,$$

that is, $v = |\nabla u|^2 \le k$ on $B_\rho \times (t_0/2, t_0)$. Obviously, if we choose

$$k = \gamma(H(t_0))^{2(N+2)/\sigma}\left(\int_{t_0/4}^{t_0}\int_{B_{2\rho}}|\nabla u|^{2(\alpha+1)}dxd\tau\right)^{4/\sigma}, \tag{5.24}$$

then (5.23) holds. Therefore if (5.24) and (5.20) are compatible, then for $t \in (t_0/2, t_0)$,

$$\|\nabla u\|_{\infty,B_\rho}(t) \le \gamma(H(t_0))^{(N+2)/\sigma}\left(\int_{t_0/4}^{t_0}\int_{B_{2\rho}}|\nabla u|^{2(\alpha+1)}dxd\tau\right)^{2/\sigma}; \tag{5.25}$$

if (5.24) and (5.20) are incompatible, then

$$\rho^{2/(p-2)}t_0^{-(N+1)/k}F(t_0) \le \gamma H(t_0)^{(N+2)/\sigma}\left(\int_{t_0/4}^{t_0}\int_{B_{2\rho}}|\nabla u|^{2(\alpha+1)}dxd\tau\right)^{2/\sigma}.$$

In the latter case, by the definition of $F(t)$, we also obtain an inequality like (5.25). Thus if we enlarge the constant γ appropriately, then (5.25) holds in both cases.

Now we choose $\alpha = \dfrac{p-2}{2}$ and multiply (5.25) by $\dfrac{t^{(N+1)/\kappa}}{\rho^{2/(p-2)}}$. Then

$$\begin{aligned} & t^{(N+1)/\kappa}\frac{\|\nabla u\|_{\infty,B_\rho(t)}}{\rho^{2/(p-2)}} \\ \le\ & \gamma F(t)^{(N+2)(p-2)/\sigma} t^{4(N+1)/\kappa\sigma} \\ & \cdot\left(\int_{t/4}^{t}\rho^{-\sigma/(p-2)}\int_{B_{2\rho}}|\nabla u|^p dxd\tau\right)^{2/\sigma} \\ & +\gamma\left(t^{N/\kappa}\int_{t/4}^{t}\rho^{-\sigma/(p-2)}\int_{B_{2\rho}}|\nabla u|^p dxd\tau\right)^{2/\sigma} \\ \equiv\ & G^{(1)}+G^{(2)}. \end{aligned} \tag{5.26}$$

We need to estimate $G^{(1)}$ and $G^{(2)}$. We have

$$\begin{aligned} G^{(1)} \le\ & \gamma F(t)^{(N+2)(p-2)/\sigma}\left(\int_{t/4}^{t}\tau^{-(N+1)(p-2)/\kappa}\tau^{(N+1)p/\kappa}A_r(\tau)d\tau\right)^{2/\sigma} \\ \le\ & \gamma F(t)^{(N+2)(p-2)/\sigma+2/\sigma}\left(\int_0^t \tau^{-(N+1)(p-2)/\kappa}F(\tau)^{p-1}d\tau\right)^{2/\sigma} \\ \le\ & \frac{1}{4}F(t)+\gamma\int_0^t \tau^{-(N+1)(p-2)/\kappa}F(\tau)^{p-1}d\tau, \end{aligned} \tag{5.27}$$

where$A_r(\tau) = \left(\sup\limits_{\rho\ge r}\dfrac{\|\nabla u\|_{\infty,B_\rho(\tau)}}{\rho^{2/(p-2)}}\right)^p$. Taking $\sigma = \dfrac{1}{2}$ in (5.9) yields

$$\begin{aligned} & \int_{t/4}^{t}\int_{B_{2\rho}}|\nabla u|^p dxd\tau \\ & \le\ \frac{\gamma}{\rho^p}\int_{t/8}^{t}\int_{B_{3\rho}}|u|^p dxd\tau+\frac{\gamma}{t}\int_{t/8}^{t}\int_{B_{3\rho}}u^2 dxd\tau. \end{aligned}$$

We may use this inequality to obtain

$$\begin{aligned} G^{(2)} \ \le\ & \gamma\left(t^{N/\kappa}\sup_{\rho\ge r}\int_{t/4}^{t}\rho^{-\sigma/(p-2)}\int_{B_{2\rho}}|\nabla u|^p dxd\tau\right)^{2/\sigma} \\ \le\ & \gamma\Bigg(\int_{t/8}^{t}\tau^{N/\kappa}\left(\sup_{\rho\ge r}\frac{\|u(\cdot,t)\|_{\infty,B_\rho}}{\rho^{2/(p-2)}}\right)^{p-1} \end{aligned}$$

$$
\begin{aligned}
&\cdot(\sup_{\rho\geq r}\rho^{-\kappa/(p-2)}\int_{B_\rho}|u(x,\tau)dx)d\tau\Big)^{2/\sigma}\\
&+\gamma\Big(\int_{t/8}^{t}\tau^{N/\kappa-1}\left(\sup_{\rho\geq r}\frac{\|u(\cdot,t)\|_{\infty,B_\rho}}{\rho^{2/(p-2)}}\right)\\
&\cdot\left(\sup_{\rho\geq r}\rho^{-\kappa/(p-2)}\int_{B_\rho}|u(x,\tau)dx\right)d\tau\Big)^{2/\sigma}\\
\leq\ &\gamma\Big(\int_{t/8}^{t}\tau^{-N(p-2)/\kappa}B(\tau)|||u(\cdot,\tau)|||d\tau\Big)^{2/\sigma}\\
&+\gamma\Big(\int_{t/8}^{t}\tau^{-1}B(\tau)|||u(\cdot,\tau)|||d\tau\Big)^{2/\sigma}\\
\leq\ &\gamma\left(t^{p/\kappa}\phi(t)^{p-2}\right)^{2/\sigma}\left(\psi(t)^{1+p/\kappa}\right)^{2/\sigma}+\gamma\left(\psi(t)^{1+p/\kappa}\right)^{2/\sigma}
\end{aligned}
\tag{5.28}
$$

where $\phi(t)$ and $\psi(t)$ are defined by (2.39) and (2.47), $0<t<\gamma_0\Lambda^{2-p}$,

$$B(\tau)=\left(\tau^{N/\kappa}\sup_{\rho\geq r}\frac{\|u(\cdot,t)\|_{\infty,B_\rho}}{\rho^{2/(p-2)}}\right)^{p-1}.$$

Since (5.2) implies that for $0<t<\gamma_0\Lambda^{2-p}$,

$$t^{p/\kappa}\phi(t)^{p-2}\leq\gamma=\gamma(N,p),$$

from (5.28) we have

$$G^{(2)}\leq\gamma\Lambda^{2/\kappa},\qquad\text{for } 0<t<\gamma_0\Lambda^{2-p}.$$

Substituting $G^{(1)}$ and $G^{(2)}$ into (5.26) derives

$$F(t)\leq\int_0^t\tau^{-(N+1)(p-2)/\kappa}F(\tau)^{p-1}d\tau+\gamma\Lambda^{2/\kappa}$$
$$\text{for } 0<t<\gamma_0\Lambda^{2-p}.$$

This implies that $F(t)$ is less than or equal to the solution of the following problem

$$V'(t)=\gamma t^{-(N+1)(p-2)/\kappa}V^{p-1}(t)$$
$$V(0)=\gamma\Lambda^{2/\kappa},\quad 0<t<\gamma_0\Lambda^{2-p},$$

that is,

$$F(t) \le \gamma\Lambda^{2/\kappa}\left(1-\gamma(t\Lambda^{p-2})^{2/\kappa}\right)^{-1/(p-2)},$$

provided that the value in the brackets is positive. Therefore if we require t to satisfy

$$\left(1-\gamma(t\Lambda^{p-2})^{2/\kappa}\right)^{-1/(p-2)} le2, \quad 0 < t < \gamma_0\Lambda^{2-p},$$

that is,

$$0 < t < \min\{\gamma_0\Lambda^{2-p}, (1-2^{2-p})^{\kappa/2}\gamma^{-1}\Lambda^{2-p}\},$$

then

$$t^{(N+1)/\kappa}\|\nabla u\|_{\infty,B_\rho}(t)\rho^{2/(p-2)} \le C_2\Lambda^{2/\kappa}.$$

From this it follows that if we choose $C_0 = \min\{\gamma_0, (1-2^{2-p})^{\kappa/2}\gamma^{-1}\}$, then (5.3) holds.

Similar to the proof of Theorem 2.2.3, we can derive (5.4) from (5.2) and (5.3). Thus the proof of our proposition is completed. □

Denote

$$A_\alpha(x) = (1+|x|^p)^{-\alpha},$$

$$h_\alpha(t) = \sup_{0<\tau<t}\int_{\mathbb{R}^N}|u(x,\tau)|A_\alpha(x)dx, \ t\in(0,T_0).$$

Proposition 2.5.2 *Under the assumptions of Proposition 2.5.1, for* $\alpha > \dfrac{\kappa}{p(p-2)}$*, there exists a constant* $C(\alpha) = C(\alpha, N, p)$*, such that*

$$h_\alpha(t) \le C(\alpha)\Lambda, \quad \textit{for } t\in(0,T_0).$$

Proof. Clearly

$$\int_{\mathbb{R}^N}|u(x,\tau)|A_\alpha(x)dx$$
$$\le \int_{|x|\le r}|u(x,\tau)|dx + \int_{|x|>r}\frac{|u(x,\tau)|}{|x|^\alpha p}dx.$$

Since

$$\frac{1}{|x|^{p\alpha}} = p\alpha\int_{|x|}^\infty\frac{dR}{R^{p\alpha+1}},$$

using Fubini's theorem and (5.1), noting that $\alpha > \dfrac{\kappa}{p(p-2)}$, we obtain

$$\begin{aligned}
&\int_{|x|>r} \frac{|u(x,\tau)|}{|x|^{p\alpha}} dx \\
=\ & p\alpha \int_r^\infty \frac{1}{R^{p\alpha+1}} \left(\int_{r<|x|<R} |u(x,\tau)| dx \right) dR \\
\le\ & \sup_{R\ge r} R^{-\kappa/(p-2)} \int_{B_R} |u(x,\tau)| dx \\
& \cdot p\alpha \int_r^\infty R^{\kappa/(p-2)-p\alpha-1} dR \\
=\ & -\frac{p\alpha(p-2)}{\kappa - p\alpha(p-2)} r^{\kappa/(p-2)-p\alpha} \\
& \cdot \sup_{R\ge r} R^{-\kappa/(p-2)} \int_{B_R} |u(x,\tau)| dx \\
<\ & C\Lambda.
\end{aligned}$$

□

Proposition 2.5.3 *Under the assumptions of Proposition 2.5.1, for* $\alpha > \dfrac{\kappa}{p(p-2)}$, *there exists a constant* $\gamma = \gamma(\Lambda, N, p)$, *such that*

$$\int_0^t \int_{\mathbb{R}^N} |\nabla u|^{p-1} A_{\alpha+1/p} dx d\tau \le \gamma t^{1/\kappa}, \quad \textit{for } t \in (0, T_0).$$

Proof. Without loss of generality, we may suppose that $u \ge 0$; otherwise we may treat u_+ and u_- separately. Also we may suppose that $u > 0$; otherwise we may replace u by $u + \varepsilon$ ($\varepsilon > 0$) and then let $\varepsilon \to 0$.

Take

$$f(s) = s^{1-2/p}, \quad \zeta(x,t) = (t-\varepsilon)_+^{1/p} (A_{\alpha+1/p}^{1/p} \xi)^p,$$

in (1.27), where $\varepsilon \in (0, T_0)$, $\xi(x)$ is a smooth cut-off function on B_ρ such that $0 \le \xi \le 1$, $|\nabla \xi| \le \dfrac{2}{\rho}$, $\xi(x) = 1$ on $B_{\rho/2}$. Then we obtain

$$\begin{aligned}
&\int_\varepsilon^t \int_{B_\rho} (\tau-\varepsilon)^{1/p} |\nabla u|^p u^{-2/p} A_{\alpha+1/p} \xi^p dx d\tau \\
\le\ & \gamma \int_\varepsilon^t \int_{B_\rho} (\tau-\varepsilon)^{1/p} u^{p-2/p} |\nabla (A_{\alpha+1/p}^{1/p} \xi)|^p dx d\tau
\end{aligned}$$

$$+\gamma\int_{\varepsilon}^{t}\int_{B_\rho}(\tau-\varepsilon)^{1/p-1}u^{(p-2)/p}A_{1/p}uA_\alpha dxd\tau$$

$$\equiv J_\rho^{(1)}+J_\rho^{(2)}. \tag{5.29}$$

For $J_\rho^{(2)}$, we have

$$J_\rho^{(2)} \le \gamma\int_{\varepsilon}^{t}(\tau-\varepsilon)^{1/\kappa-1}\int_{B_\rho}\tau^{N(p-2)/\kappa p}\frac{|u(x,\tau)|^{(p-2)/p}}{(1+|x|^p)^{1/p}} \cdot|u(x,\tau)|A_\alpha(x)dxd\tau.$$

Using (5.2) and Proposition 2.5.2 yields

$$J_\rho^{(2)} \le \gamma(t-\varepsilon)^{1/\kappa} \quad \forall\rho\ge r. \tag{5.30}$$

For $J_\rho^{(1)}$, we have

$$\begin{aligned}J_\rho^{(1)} &\le \gamma\int_{\varepsilon}^{t}\int_{B_\rho}(\tau-\varepsilon)^{1/p}u^{p-2/p}A_{\alpha+1/p}|\nabla\xi|^p dxd\tau\\ &\quad+\gamma\int_{\varepsilon}^{t}\int_{B_\rho}(\tau-\varepsilon)^{1/p}u^{p-2/p}|\nabla A_{\alpha+1/p}^{1/p}|^p dxd\tau\\ &\equiv J_\rho^{(1,1)}+J_\rho^{(1,2)}.\end{aligned}$$

Since

$$|\nabla A_{\alpha+1/p}^{1/p}|^p \le \gamma A_{\alpha/p+1/p+1/p^2}^p \le \gamma A_{1+1/p}A_\alpha,$$

we may use (5.2) and Proposition 2.5.2, similar to the derivation of (5.30), to obtain

$$\begin{aligned}J_\rho^{(1,1)} &\le \gamma h_\alpha(t)\int_{\varepsilon}^{t}(\tau-\varepsilon)^{1/\kappa-1}d\tau \le \gamma(t-\varepsilon)^{1/\kappa}.\\ J_\rho^{(1,2)} &\le \gamma\int_{\varepsilon}^{t}\int_{B_\rho}(\tau-\varepsilon)^{1/p}\frac{u(x,\tau)^{p-1-2/p}}{(1+|x|^p)^{1+1/p}}uA_\alpha dxd\tau\\ &\le \gamma h_\alpha(t)\int_{\varepsilon}^{t}(\tau-\varepsilon)^{1/p-N(p-2)/\kappa p-N(p-2)/\kappa}d\tau\\ &\le \gamma(t-\varepsilon)^{1/\kappa}.\end{aligned}$$

Thus, for $J_\rho^{(1)}$, we have

$$J_\rho^{(1)} \le \gamma(t-\varepsilon)^{1/\kappa} \quad \text{for } \rho\ge r.$$

Substituting the estimates on $J_\rho^{(1)}$ and $J_\rho^{(2)}$ into (5.29) yields

$$\int_\varepsilon^t \int_{B_\rho} (\tau-\varepsilon)^{1/p} |\nabla u|^p u^{-2/p} A_{\alpha+1/p} dx d\tau \leq \gamma (t-\varepsilon)^{1/\kappa}. \text{ for } \rho \geq r.$$

For any $\varepsilon \in (0,t)$, $t \in (0,T_0)$, using this estimate we derive

$$\begin{aligned}
&\int_\varepsilon^t \int_{B_{\rho/2}} |\nabla u|^{p-1} A_{\alpha+1/p} dx d\tau \\
= &\int_\varepsilon^t \int_{B_{\rho/2}} (\tau-\varepsilon)^{(p-1)/p^2} \frac{|\nabla u|^{p-1}}{u^{2(p-1)/p^2}} A_{\alpha+1/p}^{(p-1)/p} \\
&\quad \cdot (\tau-\varepsilon)^{-(p-1)/p^2} u^{2(p-1)/p^2} A_{\alpha+1/p}^{1/p} dx d\tau \\
\leq &\left(\int_\varepsilon^t \int_{B_{\rho/2}} (\tau-\varepsilon)^{1/p} \frac{|\nabla u|^p}{u^{2/p}} A_{\alpha+1/p} dx d\tau \right)^{(p-1)/p} \\
&\quad \cdot \left(\int_\varepsilon^t \int_{B_{\rho/2}} (\tau-\varepsilon)^{-(p-1)/p} u^{(p-2)/p} A_{1/p} u A_\alpha dx d\tau \right)^{1/p} \\
\leq &\ \gamma (t-\varepsilon)^{(p-1)/pk} \left(\int_\varepsilon^t (\tau-\varepsilon)^{1/k-1} h_\alpha(\tau) d\tau \right)^{1/p} \\
\leq &\ \gamma (t-\varepsilon)^{1/k} \qquad \text{for } \gamma > r,\ t \in (0,T_0).
\end{aligned}$$

The conclusion of our proposition follows by letting $\rho \to \infty$, $\varepsilon \to 0$. □

2.5.2 *Uniqueness theorem and its proof*

Theorem 2.5.1 *Assume that $p > 2$ and u, v are generalized solutions of the equation (1.1) on Q_T such that for some $r > 0$*

$$\sup_{t\in(0,T)} |||u(\cdot,t)|||_r, \ \sup_{t\in(0,T)} |||v(\cdot,t)|||_r < \infty. \tag{5.31}$$

If

$$\lim_{t\to 0} \int_{B_R} |u(x,t) - v(x,t)| dx = 0, \quad \textit{for } R > 0, \tag{5.32}$$

then $u \equiv v$ a.e. on Q_T.

Remark 2.5.1 Since by Theorem 2.3.3, any nonnegative generalized solution of the equation (1.1) possesses the property (5.31), Theorem 2.5.1

implies the uniqueness of nonnegative generalized solutions of (1.1) with initial data $u_0(x)$ in the following sense:

$$\lim_{t\to 0}\int_{B_R}|u(x,t)-u_0(x)|dx=0 \quad \text{for } R>0.$$

Proof. Let $w = u - v$. It is easy to see that from the definition of generalized solutions and Remark 2.1.3 that for any $t_1, t_2 \in (0,T)$ and any function $\varphi \in W^{1,p}(\mathbb{R}^N)$ with compact support, we have

$$\begin{aligned}&\int_{\mathbb{R}^N} w(x,t_2)\varphi(x)dx-\int_{\mathbb{R}^N} w(x,t_1)\varphi(x)dx\\&\quad+\int_{t_1}^{t_2}\int_{\mathbb{R}^N} a^{ij}w_{x_i}\varphi_{x_j}dxdt=0,\end{aligned}\tag{5.33}$$

where

$$\begin{aligned}a^{ij}(x,t)&=\int_0^1|s\nabla u+(1-s)\nabla v|^{p-2}ds\cdot\delta_{ij}\\&\quad+(p-2)\int_0^1|s\nabla u+(1-s)\nabla v|^{p-4}(su+(1-s)v)_{x_j}\\&\quad\cdot(su+(1-s)v)_{x_i}ds,\\ \delta_{ij}&=1\quad(i=j),\qquad \delta_{ij}=0\quad(i\neq j).\end{aligned}$$

Obviously (a^{ij}) is nonnegative definite and

$$a_0(x,t)|\xi|^2\le a^{ij}(x,t)\xi_i\xi_j\le(p-1)a_0(x,t)|\xi|^2$$

$$\text{for } \xi\in\mathbb{R}^N,\ (x,t)\in Q_T,$$

where

$$a_0(x,t)=\int_0^1|s\nabla u+(1-s)\nabla v|^{p-2}ds.$$

Using the Steklov mean value and processing as in Remark 2.1.4, we obtain from (5.33),

$$\begin{aligned}&\int_{\mathbb{R}^N} w_{h\tau}(x,\tau)\phi(x)dx\\&\quad+\int_{\mathbb{R}^N}(a^{ij}w_{x_i})_h(x,\tau)\nabla\phi(x)dx=0.\end{aligned}\tag{5.34}$$

Similar to the derivation from (1.26) to (1.27), we may derive from (5.34),

$$\begin{aligned}\int_{\mathbb{R}^N}\int_0^{w(x,t_2)} & f(s)ds\cdot\zeta(x,t_2)dx \\ & -\int_{\mathbb{R}^N}\int_0^{w(x,t_1)} f(s)ds\cdot\zeta(x,t_1)dx \\ & +\int_{t_1}^{t_2}\int_{\mathbb{R}^N} a^{ij}w_{x_i}w_{x_j}f'(w)\zeta dxd\tau \qquad (5.35) \\ & +\int_{t_1}^{t_2}\int_{\mathbb{R}^N} a^{ij}w_{x_i}f(w)\nabla\zeta dxd\tau \\ = & \int_{t_1}^{t_2}\int_{\mathbb{R}^N}\int_0^{w} f(s)ds\cdot\zeta_\tau dxd\tau,\end{aligned}$$

where $f(s)$ is an arbitrary increasing piecewise smooth function and $\zeta\in C^1(\overline{Q}_T)$ is a nonnegative function such that $\zeta(\cdot,\tau)\in C_0^1(\mathbb{R}^N)$.

We first prove

$$\int_{\mathbb{R}^N}|w(x,t)|A_\alpha(x,t)dx\le\gamma t^{1/k},\ \text{for } t\in(0,T_0)$$

$$\alpha>\frac{\kappa}{p(p-2)}, \tag{5.36}$$

$$\lim_{t\to 0}\int_{B_\rho}|w(x,t)|^{1+\varepsilon}=0,\ \text{for } \rho>0,\varepsilon\in(0,1/N), \tag{5.37}$$

where T_0 is the constant in Proposition 2.5.1.

To prove (5.36), we take

$$f(s)=\operatorname{sgn}_\eta s,\eta>0,\zeta(x,\tau)=A_\alpha(x)\xi(x),t_1=0,t_2=t \tag{5.38}$$

in (5.35), where $\xi(x)$ is a smooth cut-off function on B_ρ such that $0\le\xi\le 1$, $|\nabla\xi|\le\gamma\rho^{-1}$, $\xi(x)=1$ on $B_{\rho/2}$. Using the condition (5.32), we obtain

$$\begin{aligned}\int_{B_\rho}\int_0^{w(x,t)} & \operatorname{sgn}_\eta s ds A_\alpha(x)\xi dx \\ & +\int_0^t\int_{B_\rho} a^{ij}w_{x_i}w_{x_j}\operatorname{sgn}'_\eta w A_\alpha(x)\xi(x)w_{x_i}dxd\tau \\ = & -\int_0^t\int_{B_\rho} a^{ij}w_{x_i}\operatorname{sgn}_\eta w(A_\alpha(x)\xi(x))_{x_j}dxd\tau.\end{aligned}$$

Giving up the second term on the left hand side, which is nonnegative and then letting $\eta \longrightarrow 0$, we derive

$$\begin{aligned}
&\int_{B_\rho} |w(x,t)| A_\alpha(x)\xi(x)dx \\
\le\ & \int_0^t \int_{B_\rho} (|\nabla u| + |\nabla v|)^{p-1} |\nabla(A_\alpha \xi)| dx d\tau \\
\le\ & \gamma \int_0^t \int_{B_\rho} (|\nabla u| + |\nabla v|)^{p-1} A_\alpha |\nabla \xi| dx d\tau \\
& + \gamma \int_0^t \int_{B_\rho} (|\nabla u| + |\nabla v|)^{p-1} |\nabla A_\alpha| dx d\tau.
\end{aligned}$$

Clearly $|\nabla A_\alpha| \le \gamma A_{\alpha+1/p}$. Since $|\nabla \xi| = 0$ for $|x| \le \frac{\rho}{2}$, we have $A_\alpha |\nabla \xi| \le \gamma A_{\alpha+1/p}$ for $x \in B_\rho$. Substituting these into the above formula and letting $\rho \to \infty$, we further obtain

$$\begin{aligned}
&\int_{\mathbb{R}^N} |w(x,t)| A_\alpha(x) dx \\
\le\ & \gamma \int_0^t \int_{\mathbb{R}^N} (|\nabla u| + |\nabla v|)^{p-1} A_{\alpha+1/p} dx d\tau,
\end{aligned}$$

from which and Proposition 2.5.3, (5.36) follows.

Fixed $\varepsilon \in (0, 1/N)$. For $t \in (0, T_0)$, we have

$$\begin{aligned}
&\int_{B_\rho} |w(x,t)|^{1+\varepsilon} dx \\
\le\ & \gamma \rho^{(\alpha+\varepsilon/(p-2))p} \int_{\mathbb{R}^N} |w(x,t)|^{1+\varepsilon} A_{\alpha+\varepsilon/(p-2)} dx \\
\le\ & \gamma \rho^{(\alpha+\varepsilon/(p-2))p} \int_{\mathbb{R}^N} |w(x,t)|^{\varepsilon} A_{\varepsilon/(p-2)}(x) |w(x,t)| A_\alpha(x) dx.
\end{aligned}$$

Since (5.2) implies that

$$|w(x,t)|^\varepsilon A_{\varepsilon/(p-2)} \le \gamma t^{-N\varepsilon/\kappa},$$

using (5.36) we obtain

$$\int_{B_\rho} |w(x,t)|^{1+\varepsilon} dx \le \gamma(\Lambda, N, p, \rho) t^{(1-N\varepsilon)/\kappa}.$$

Thus (5.37) holds.

Now we choose

$$f(s)=(|s|+\delta)^{\varepsilon}\operatorname{sgn}_{\eta}s,\ \varepsilon\in(0,1/N),\ \delta\in(0,T_0),\ \eta>0,$$
$$\zeta(x,\tau)=(A_{\alpha}^{1/2}\xi)^2,\ t_1=0,\ t_2=t. \tag{5.39}$$

Then

$$\begin{aligned}&\int_{B_\rho}\int_0^{w(x,t)}(|s|+\delta)^{\varepsilon}\operatorname{sgn}_{\eta}s ds\, A_{\alpha}\xi^2dx\\
&\quad+\varepsilon\int_\delta^t\int_{B_\rho}a_0(x,\tau)\frac{|\nabla w|^2}{(|w|+\delta)^{1-\varepsilon}}\frac{w\operatorname{sgn}_{\eta}w}{|w|}(A_{\alpha}^{1/2}\xi)^2dxd\tau\\
&\quad+\int_\delta^t\int_{B_\rho}a_0(x,\tau)|\nabla w|^2(|w|+\delta)^{\varepsilon}\operatorname{sgn}'_{\eta}w(A_{\alpha}^{1/2}\xi)^2dxd\tau\\
\le&\int_{B_\rho}\int_0^{w(x,\delta)}(|s|+\delta)^{\varepsilon}\operatorname{sgn}_{\eta}s ds\, A_{\alpha}\xi^2dx\\
&\quad+\gamma\int_\delta^t\int_{B_\rho}a_0(x,\tau)|\nabla w|(|w|+\delta)^{\varepsilon}A_{\alpha}^{1/2}\xi|\nabla(A_{\alpha}^{1/2}\xi)|dxd\tau.\end{aligned}$$

Giving up the third term on the left hand side, which is nonnegative, and letting $\eta\to 0$, we further obtain

$$\begin{aligned}&\frac{1}{1+\varepsilon}\int_{B_\rho}(|w(x,t)|+\delta)^{1+\varepsilon}A_{\alpha}\xi^2dx\\
&\quad+\varepsilon\int_\delta^t\int_{B_\rho}a_0(x,\tau)\frac{|\nabla w|^2}{(|w|+\delta)^{1-\varepsilon}}(A_{\alpha}^{1/2}\xi)^2dxd\tau\\
\le&\ \frac{1}{1+\varepsilon}\int_{B_\rho}(|w(x,\delta)+\delta)^{1+\varepsilon}A_{\alpha}\xi^2dx\\
&\quad+\gamma\int_\delta^t\int_{B_\rho}a_0(x,\tau)|\nabla w|(|w|+\delta)^{\varepsilon}\\
&\quad\cdot A_{\alpha}^{1/2}\xi|\nabla(A_{\alpha}^{1/2}\xi)|dxd\tau.\end{aligned} \tag{5.40}$$

By Young's inequality we see that the last term in (5.40) is less than

$$\begin{aligned}&\frac{\varepsilon}{2}\int_\delta^t\int_{B_\rho}a_0(x,\tau)\frac{|\nabla w|^2}{(|w|+\delta)^{1-\varepsilon}}(A_{\alpha}^{1/2}\xi)^2dxd\tau\\
&\quad+\gamma(\varepsilon)\int_\delta^t\int_{B_\rho}a_0(x,\tau)(|w|+\delta)^{1+\varepsilon}\cdot\\
&\quad\cdot(A_{\alpha}|\nabla\xi|^2+|\nabla A_{\alpha}^{1/2}|^2)dxd\tau.\end{aligned}$$

Again noting that on B_ρ,

$$A_\alpha|\nabla\xi|^2 + |\nabla A_\alpha^{1/2}|^2 \le \gamma A_\alpha(x)A_{2/p}(x),$$

from (5.40) we obtain

$$\begin{aligned}
&\int_{B_\rho}(|w(x,t)+\delta)^{1+\varepsilon}A_\alpha(x)\xi^2(x)dx \\
\le\ &\int_{B_\rho}(|w(x,\delta)|+\delta)^{1+\varepsilon}A_\alpha(x)dx \\
&+\gamma\int_\delta^t\int_{B_\rho}a_0(x,\tau)A_{2/p}(x) \\
&\cdot(|w(x,\tau)|+\delta)^{1+\varepsilon}A_\alpha(x)dxd\tau.
\end{aligned} \tag{5.41}$$

By the definition of $a_0(x,t)$ and (5.3),

$$\begin{aligned}
&a_0(x,\tau)A_{2/p}(x) \\
\le\ &\gamma\frac{\max(r^2,|x|^2)}{(1+|x|^p)^{2/p}}\Lambda^{2(p-2)/\kappa}\tau^{-(N+1)(p-2)/\kappa} \\
\le\ &\gamma\tau^{-(N+1)(p-2)/\kappa}.
\end{aligned}$$

Substituting this into (5.41), letting $\delta\to 0$, $\rho\to\infty$, and using (5.37) yield

$$\begin{aligned}
&\int_{\mathbb{R}^N}|w(x,t)|^{1+\varepsilon}A_\alpha(x)dx \\
\le\ &\gamma\int_0^t\tau^{(N+1)(p-2)/\kappa}\int_{\mathbb{R}^N}|w(x,t)|^{1+\varepsilon}A_\alpha(x)dxd\tau.
\end{aligned} \tag{5.42}$$

Express α as $\alpha=\alpha_1+\dfrac{\varepsilon}{p-2}$. Then for sufficiently small $\varepsilon>0$, $\alpha_1>\dfrac{\kappa}{p(p-2)}$. Since $|w(x,t)|^\varepsilon A_{\varepsilon/(p-2)}\le Ct^{-\varepsilon N/\kappa}$, (5.36) implies

$$\begin{aligned}
&\int_{\mathbb{R}^N}|w(x,t)|^{1+\varepsilon}A_{\alpha_1}(x)dx \\
\le\ &Ct^{-\varepsilon N/\kappa}\int_{\mathbb{R}^N}|w(x,t)|A_\alpha(x)dx \\
\le\ &Ct^{(1-\varepsilon N)/\kappa},
\end{aligned}$$

which shows that for sufficiently small $\varepsilon > 0$, the right hand side of (5.42) is integrable. Thus we can use Gronwall's lemma to (5.42) to obtain

$$\int_{\mathbb{R}^N} |w(x,t)|^{1+\varepsilon} A_\alpha(x)dx = 0, \quad \text{for } t \in (0, T_0).$$

Hence on $\mathbb{R}^N \times (0, T_0)$, we have $u = v$. If $T_0 < T$, then we repeat the above argument on $\mathbb{R}^N \times (T_0, T_1)$ where $T_1 \leq T$ with $T_1 - T_0 < C_0 \Lambda^{2-p}$. To do this, we need to replace $\mathbb{R}^N \times (0, T_0)$ in Proposition 2.5.1 by $\mathbb{R}^N \times (T_0, T_1)$. In the present case, on the right hand side of (5.2)–(5.4), t should be replaced by $t - T_0$. Thus we can further prove that on $\mathbb{R}^N \times (T_0, T_1)$, $u = v$. Going on this way, we finally arrive at the conclusion that on $\mathbb{R}^N \times (0, T)$, $u = v$ and complete the proof of Theorem 2.5.1. □

Similarly we can prove the following comparison theorem.

Theorem 2.5.2 *Assume that $p > 2$ and u, v are generalized solutions of the equation (1.1) on Q_T such that for some $r > 0$,*

$$\sup_{t \in (0,T)} |||u(\cdot, t)|||_r, \ \sup_{t \in (0,T)} |||v(\cdot, t)|||_r < \infty.$$

If

$$\lim_{t \to 0} \int_{B_R} (u(x,t) - v(x,t))_+ dx = 0, \quad \textit{for } R > 0,$$

then $u \leq v$ on Q_T.

Proof. First we replace $f(s)$ in (5.38) by

$$f(s) = \operatorname{sgn}_\eta s_+,$$

and prove that

$$\int_{\mathbb{R}^N} w(x,t)_+ A_\alpha(x)dx \leq \gamma t^{1/\kappa},$$

$$\alpha > \frac{k}{p(p-2)}, \forall t \in (0, T_0),$$

$$\lim_{t \to 0} \int_{B_\rho} w(x,t)_+^{1+\varepsilon} dx = 0, \forall \rho > 0, \varepsilon \in (0, 1/N).$$

Next we replace $f(s)$ in (5.39) by

$$f(s) = (s_+ + \delta)^\varepsilon \operatorname{sgn}_\eta s_+,$$

and prove that

$$\int_{\mathbb{R}^N} w(x,t)_+^{1+\varepsilon} A_\alpha(x)dx = 0 \quad \forall t \in (0, T_0).$$

□

2.6 Properties of the Free Boundary

In this section, we always assume that $p > 2$, u_0 is a nonnegative and continuous function on $\mathbb{R}^N$ with compact support and u is the nonnegative generalized solution on $Q = \mathbb{R}^N \times (0, \infty)$ with initial value u_0. As shown in §2.2 and §2.4, such solution exists and is Hölder continuous on Q. Denote

$$\Omega = \{(x,t); u(x,t) > 0, t > 0\},$$

$$\Omega(t) = \{x \in \mathbb{R}^N; u(x,t) > 0\},$$

$$\Gamma = \partial\Omega \cap \{t > 0\},$$

$$\Gamma(t) = \partial\Omega(t).$$

Proposition 2.3.4 shows that for each $t > 0$, $\Omega(t)$ is bounded, that is, the generalized solution u has compact support. By the continuity of u, Ω and $\Omega(t)$ are open sets in Q and $\mathbb{R}^N$ respectively. We call Γ the **free boundary** or **interface** of the generalized solution u. This section is devoted to a study of properties of the free boundary. Since here the argument is parallel to §1.7 at many points, it is not our intention to give all proofs in great detail. In fact, all those similar proofs are omitted or described briefly.

2.6.1 *Monotonicity and Hölder continuity of the free boundary*

Proposition 2.6.1 *There holds*

$$u_t \geq -\frac{u}{(p-2)t}$$

in the sense of distributions.

Proof. Denote

$$u_r(x,t) = ru(x, t_0 + r^{p-2}t),$$

where $r > 0$, $t_0 > 0$. It is easy to check that u_r is also a generalized solution of the equation (1.1), whose initial value is $u_r(x,0) = ru(x,t_0)$.

If $r < 1$, then $u_r(x,0) \le u(x,t_0)$. Thus, by the comparison theorem (Proposition 2.3.3), for $t > 0$, we have $u_r(x,t) \le u(x,t_0+t)$. Hence

$$u(x,t_0+r^{p-2}t) - u(x,t_0+t) \le (1-r)u(x,t_0+r^{p-2}t).$$

Similarly, if $r > 1$, then

$$u(x,t_0+r^{p-2}t) - u(x,t_0+t) \ge (1-r)u(x,t_0+r^{p-2}t).$$

In both cases,

$$\begin{aligned}&\frac{u(x,t_0+r^{p-2}t) - u(x,t_0+t)}{(r^{p-2}-1)t}\\ \ge\ & -\frac{1-r}{(1-r^{p-2})t}u(x,t_0+r^{p-2}t).\end{aligned}$$

Hence for $\varphi \in C_0^\infty(Q)$ with $\varphi \ge 0$,

$$\begin{aligned}&-r^{2-p}\iint_Q u(x,t_0+t)\frac{\varphi(x,r^{2-p}t)-\varphi(x,t)}{(r^{2-p}-1)t}dxdt\\ \ge\ & -\frac{1-r}{1-r^{p-2}}\iint_Q \frac{u(x,t_0+r^{p-2}t)}{t}\varphi(x,t)dxdt.\end{aligned}$$

First let $r \to 1$ and then let $t_0 \to 0$ to pass to the limit. Then we obtain

$$-\iint_Q u\varphi_t dxdt \ge -\iint_Q \frac{u}{(p-2)t}\varphi dxdt,$$

which is just what we want to prove. □

Theorem 2.6.1 *For any $t_2 > t_1 > 0$,*

$$\Omega(t_1) \subset \Omega(t_2)$$

and

$$\inf\{|x|; x \in \Gamma(t)\} \ge Ct^{1/\kappa},$$

where $C > 0$ is a certain constant and $\kappa = N(p-2)+p$.

Proof. The first conclusion follows from Proposition 2.6.1. To prove the second conclusion, we may use the comparison theorem to u and the Barenblatt-type solution (see (1.34))

$$B(x,t,c) = t^{-N/\kappa}\left(c - \gamma_p\left(\frac{|x|}{t^{1/k}}\right)^{p/(p-1)}\right)_+^{(p-1)/(p-2)},$$

where $\gamma_p = \left(\frac{1}{\kappa}\right)^{1/(p-2)} \frac{p-2}{p}$ and c is an appropriate constant. □

Theorem 2.6.2 *Let $D = \Omega(0) \subset \mathbb{R}^N$ be a bounded C^1 domain. If there exist constants $c_0 > 0, \delta > 0$ and $\gamma \in (0, (p-1)/(p-2))$ such that for $x \in D$ with $dist\,(x, \partial D) < \delta$, there holds*

$$u_0(x) \geq c_0(dist\,(x, \partial D))^\gamma,$$

then

$$\overline{\Omega(0)} = \overline{D} \subset \Omega(t) \quad \textit{for } t > 0.$$

Proof. Similar to Theorem 1.7.2 in §1.2.7, we can prove the conclusion by means of the comparison theorem and the fundamental solution of (1.1) (see (1.33)):

$$\begin{aligned} & E_{k,\rho}(,t;\overline{x},0) \\ = \; & k\rho^N(S(t))^{-N/\kappa}\left(1 - \left(\frac{|x-\overline{x}|}{S(t)^{1-1/\kappa}}\right)^{p/(p-1)}\right)_+^{(p-1)/(p-2)}, \end{aligned}$$

where

$$S(t) = b(N,p)k^{p-2}\rho^{N(p-2)}t + \rho^\kappa, \; b(N,p) = \kappa\left(\frac{p}{p-2}\right)^{p-1}.$$

□

To discuss the Hölder continuity of the free boundary, we need some auxiliary propositions.

Proposition 2.6.2 *For R, $\beta > 0$, $t_0 \geq 0$, $x_0 \in \mathbb{R}^N$, if*

$$u(x, t_0) = 0, \quad \textit{for } x \in B_R(x_0),$$

then there exists $C = C(p,N) > 0$ *such that*

$$\sup_{Q^\beta_{R/2}(x_0)} u \le CR^{-p}\int_{t_0}^{t_0+\beta} \fint_{B_R(x_0)} u^{p-1}dxdt, \tag{6.1}$$

where

$$Q^\beta_R = B_R(x_0)\times(t_0, t_0+\beta).$$

Here and below, we use the notation

$$\fint_\Omega f(x)dx = \frac{1}{|\Omega|}\int_\Omega f(x)dx.$$

Proof. Denote $R_n = \tau R + (1-\tau)R/2^n$, $\overline{R}_n = \frac{1}{2}(R_n + R_{n+1})$, $\tau \in [1/2, 1]$. Let $\xi_n(x)$ and $\overline{\xi}_n(x)$ be smooth cut-off functions on $B_{R_n}(x_0)$ and $B_{\overline{R}_n}(x_0)$ respectively such that $0 \le \xi_n, \overline{\xi}_n \le 1$, $|\nabla\xi_n|, |\nabla\overline{\xi}_n| \le \frac{C2^{n+1}}{(1-\tau)R}$, $\xi_n(x) = 1$ on $B_{R_n}(x_0)$, $\overline{\xi}_n(x) = 1$ on $B_{R_{n+1}}(x_0)$.

Choosing $\zeta = \xi_n^p(x)$, $f(s) = s^\alpha$ in (1.27), we have

$$\begin{aligned}&\frac{1}{\alpha+1}\int_{B_{R_n}(x_0)} \xi_n^p(x)u^{\alpha+1}(x,t)dx\\&-\frac{1}{\alpha+1}\int_{B_{R_n}(x_0)} \xi_n^p(x)u^{\alpha+1}(x,t_0)dx\\&+\alpha\int_{t_0}^t\int_{B_{R_n}(x_0)} \xi_n^p|\nabla u|^p u^{\alpha-1}dxd\tau\\&+p\int_{t_0}^t\int_{B_{R_n}(x_0)} \xi_n^{p-1}|\nabla u|^{p-2}\nabla u\nabla\xi_n u^\alpha dxd\tau = 0.\end{aligned}$$

Noting that the second term on the left hand side is equal to zero and using Young's inequality to the fourth term, we further obtain

$$\begin{aligned}&\sup_{t_0\le t\le t_0+\beta}\int_{B_{R_n}(x_0)} \xi_n^p(x)u^{\alpha+1}(x,t)dx\\&\quad+\iint_{Q^\beta_{R_n}} \xi_n^p|\nabla u^\gamma|^p dxd\tau\\&\le C\iint_{Q^\beta_{R_n}} |\nabla\xi_n|^p u^{\gamma p}dxd\tau,\end{aligned}\tag{6.2}$$

where $\gamma = 1 + \frac{\alpha-1}{p}$, $Q^\beta_R = Q^\beta_R(x_0)$.

Using the embedding inequality (1.7) to $v = \overline{\xi}_n u^\gamma$ yields

$$\begin{aligned}
&\iint_{Q^\beta_{R_n}} (\overline{\xi}_n u^\gamma)^{p(\alpha+1)+N_\gamma/N\gamma} dxdt \\
\leq\ & C\Big(\iint_{Q^\beta_{R_n}} |\nabla\overline{\xi}_n|^p u^{\gamma p} dxdt + \iint_{Q^\beta_{R_n}} \overline{\xi}^p_n |\nabla u^\gamma|^p dxdt\Big) \\
&\cdot\left(\sup_{t_0\leq t\leq t_0+\beta}\int_{B_{R_n}(x_0)} \overline{\xi}_n^{(\alpha+1)/\gamma}(x) u^{\alpha+1}(x,t)dx\right)^{p/N}.
\end{aligned}$$

Hence

$$\begin{aligned}
&\iint_{Q^\beta_{R_{n+1}}} u^{\alpha(1+p/N)+p+p/N-1} dxdt \\
\leq\ & C\Big(\iint_{Q^\beta_{R_n}} |\nabla\overline{\xi}_n|^p u^{\gamma p} dxdt + \iint_{Q^\beta_{R_n}} \overline{\xi}^p_n |\nabla u^\gamma|^p dxdt\Big) \\
&\cdot\left(\sup_{t_0\leq t\leq t_0+\beta}\int_{B_{R_n}(x_0)} \overline{\xi}_n^{(\alpha+1)/\gamma}(x) u^{\alpha+1}(x,t)dx\right)^{p/N}.
\end{aligned}$$

Using (6.2) we derive

$$\begin{aligned}
&\iint_{Q^\beta_{R_{n+1}}} u^{\alpha(1+p/N)+p+p/N-1} dxdt \\
\leq\ & \left(\frac{C2^{np}}{(1-\tau)^p R^p}\iint_{Q^\beta_{R_n}} u^{p+\alpha-1} dxdt\right)^{1+p/N},
\end{aligned}$$

which can be rewritten as

$$\iint_{Q^\beta_{R_n}} u^{p-2+\lambda^{n+1}} dxdt \leq \left(\frac{C2^{np}}{(1-\tau)^p R^p}\iint_{Q^\beta_{R_n}} u^{p-2+\lambda^n} dxdt\right)^\lambda,$$

where $\lambda = 1 + \dfrac{p}{N}$, $\alpha = \lambda^n - 1$ $(n = 1, 2, \cdot)$. This is a standard iteration inequality. Thus we may use Lemma 2.1.5 to obtain

$$\sup_{Q^\beta_{\tau R}} u \leq \left(\frac{C}{((1-\tau)R)^{N+p}}\iint_{Q^\beta_R} u^{p-2+\lambda} dxdt\right)^{1/\lambda}.$$

Again using Young's inequality we easily see that

$$\sup_{Q^\beta_{\tau R}} u \leq \frac{1}{2}\sup_{Q^\beta_R} u + \frac{C}{(1-\tau)R)^{N+p}}\iint_{Q^\beta_R} u^{p-1} dxdt.$$

Finally, we use Lemma 2.1.4 to derive (6.1) and complete the proof of our proposition. □

Proposition 2.6.3 *For $R > 0$, $x_0 \in \mathbb{R}^N$, $t_0 > 0$, there exists $\alpha_0 = \alpha_0(p, N)$ such that if*

$$u(x, t_0) = 0 \quad \textit{for } x \in B_R(x_0)$$

and

$$\fint_{B_R(x_0)} u^{p-1}(x, t_0 + \beta)dx \leq \alpha_0 \left(\frac{R^p}{\beta}\right)^{(p-1)/(p-2)} \tag{6.3}$$

where $\beta \in (0, t_0/4)$, then

$$u(x, t_0 + \beta) = 0 \quad \textit{for } x \in B_{R/4}(x_0).$$

Proof. From Theorem 2.3.1, for any $\theta \in (0, t_0/4)$,

$$u(x, t) \leq Cu(x, t + \theta) \quad \text{for } (x, t) \in \mathbb{R}^N \times [t_0, \infty).$$

In particular,

$$u(x, t) \leq C_0 u(x, t_0 + \beta) \quad \text{for } t \in [t_0, t_0 + \beta].$$

From this and (6.3), using Proposition 2.6.2, we obtain

$$\sup_{Q^{\beta}_{R/2}(x_0)} u \leq CC_0^{p-1}\alpha_0 \left(\frac{R^p}{\beta}\right)^{1/(p-2)}. \tag{6.4}$$

Let $R_n = \dfrac{R}{2^n}$, $n = 1, 2, \cdots$. For any $\overline{x} \in B_{R/4}(x_0)$, since $B_{R/2}(\overline{x}) \subset B_R(x_0)$, by Proposition 2.6.2, we have

$$\sup_{Q^{\beta}_{R_{n+1}}(\overline{x})} u \leq \frac{C\beta}{R^p}(2^p)^n (\sup_{Q^{\beta}_{R_n}(\overline{x})} u)^{p-1}.$$

Thus by the iteration lemma (Lemma 1.5.4 in §1.5) we conclude that there exists $\delta_0 = \delta_0(p, N) > 0$ such that if

$$\sup_{Q^{\beta}_{R/2}(\overline{x})} u \leq \delta_0 \left(\frac{R^p}{\beta}\right)^{1/(p-2)}, \tag{6.5}$$

then

$$\lim_{n\to\infty} \sup_{Q^{\beta}_{R_n}(\overline{x})} u = 0,$$

from which, in particular,

$$u(\overline{x}, t_0 + \beta) = 0.$$

For (6.5) to hold, it suffices to choose α_0 such that $CC_0^{p-1}\alpha_0 < \delta_0$. Thus the proposition is proved. □

Denote

$$\sigma(x,t) = \{(x,\tau); 0 < \tau < t\},$$

$$\Gamma_m(x,t) = \{(x,t) \in \Gamma; \sigma(x,t) \cap \Gamma = \emptyset\},$$

$$\Gamma_s(x,t) = \{(x,t) \in \Gamma; \sigma(x,t) \subset \Gamma\}.$$

Clearly $\Gamma_m \cap \Gamma_s = \emptyset$.

Theorem 2.6.3 $\Gamma = \Gamma_m \cup \Gamma_s$, *that is, for any* $(x_0, t_0) \in \Gamma$, *either (i)* $\sigma(x_0, t_0) \subset \Gamma$ *or (ii)* $\sigma(x_0, t_0) \cap \Gamma = \emptyset$.

Proof. Suppose that the conclusion is false. Then there exist $0 < t_1 < t_2 < t_0$, such that

$$(x_0, t_2) \in \Gamma, \quad (x_0, t_1) \notin \Gamma.$$

Without loss of generality, we may suppose that $t_0 - t_2$ is sufficiently small and $\lambda = \dfrac{t_0 - t_2}{t_2 - t_1}$ is sufficiently large. Theorem 2.6.1 implies that $u(x_0, t_1) = 0$ and there exists $R > 0$ such that $u(x, t_1) = 0$ for $x \in B_R(x_0)$. Thus, by Theorem 2.6.3 we have

$$\begin{aligned} &\int_{B_{R/2}(x_0)} u^{p-1}(x, t_2)dx \\ \geq\ & \alpha_0 \lambda^{(p-1)/(p-2)} \left(\frac{2^{-p}R^p}{t_0 - t_2}\right)^{(p-1)/(p-2)}. \end{aligned} \tag{6.6}$$

Using Theorem 2.3.2 gives

$$\begin{aligned} &u(x_0, t_2) \\ \leq\ & C\left(\left(\frac{R^p}{\theta}\right)^{1/(p-2)} + \left(\frac{\theta}{R^p}\right)^{N/p}\left(\inf_{x\in B_R(x_0)} u(x, t_2 + \theta)\right)^{\kappa/p}\right), \end{aligned}$$

$$\theta = t_0 - t_2.$$

Hence

$$\begin{aligned}
&\fint_{B_{R/2}(x_0)} u^{p-1}(x,t_2)dx \\
\leq\ & 2^{p-1}C^{p-1}\left(\left(\frac{R^p}{\theta}\right)^{(p-1)/(p-2)}\right. \\
&\left. + \left(\frac{\theta}{R^p}\right)^{N(p-1)/p} (u(x_0,t_0))^{\kappa(p-1)/p}\right).
\end{aligned} \tag{6.7}$$

Now we choose t_1, t_2 such thatλ is sufficiently large and

$$\alpha_0(\lambda 2^p)^{(p-1)/(p-2)} > 2^{p-1}C^{p-1}.$$

Combining (6.6) with (6.7) we conclude that $u(x_0,t_0) > 0$, which contradicts the assumption $(x_0,t_0) \in \Gamma$. Thus the theorem is proved. □

Remark 2.6.1 Under the assumption of Theorem 2.6.2, $\Gamma = \Gamma_m$.

Theorem 2.6.4 *For any $\eta_0 > 0$, there exist constants C, h, $\gamma > 0$, such that for any $(x_0,t_0) \in \Gamma_m$ with $t_0 > \eta_0$,*

$$u(x,t) = 0, \ \textit{for}\ |x - x_0| \leq C(t_0 - t)^\gamma, t_0 - h < t < t_0, \tag{6.8}$$

$$u(x,t) > 0, \ \textit{for}\ |x - x_0| \leq C(t - t_0)^\gamma, t_0 < t < t_0 + h. \tag{6.9}$$

Proof. For fixed $\tau \in (7t_0/8, t_0)$, let $h = t_0 - \tau$. Since $(x_0,t_0) \in \Gamma_m$, there must be a constant $R > 0$, such that

$$u(x,\tau) = 0 \quad \text{for } x \in B_R(x_0).$$

Let $t_1 = \tau + (1-\lambda)h = t_0 - \lambda h$ with $\lambda \in (0,1)$ to be determined. Suppose

$$\operatorname{dist}(x_0, \Gamma(t_1)) < \alpha R, \tag{6.10}$$

with $\alpha \in (0,1)$ to be determined. Then, by Proposition 2.6.3, for $x_1 \in \Gamma(t_1)$ such that $|x_1 - x_0| \leq \alpha R$, we have

$$\fint_{B_{(1-\alpha)R}(x_1)} u^{p-1}(x,t_1)dx \geq \alpha_0 \left(\frac{(1-\alpha)^p R^p}{(1-\lambda)h}\right)^{(p-1)/(p-2)}.$$

From this it follows that

$$\fint_{B_R(x_0)} u^{p-1}(x,t_1)dx \geq \alpha_0(1-\alpha)^N \left(\frac{(1-\alpha)^p R^p}{(1-\lambda)h}\right)^{(p-1)/(p-2)}.$$

Taking $\lambda = 1 - \dfrac{1}{k}$, $\alpha = \left(1 - \dfrac{1}{k}\right)^k$ with k large enough, such that

$$\alpha_0(1-\alpha)^N \left(\frac{(1-\alpha)^p R^p}{(1-\lambda)h}\right)^{(p-1)/(p-2)} > C\left(\frac{R^p}{\lambda h}\right)^{(p-1)/(p-2)},$$

where C is the constant in Theorem 2.3.2. Similar to the proof of Theorem 2.6.3, we may obtain $u(x_0,t_0) > 0$, which contradicts the assumption $(x_0,t_0) \in \Gamma_m$. This shows that (6.10) must be false, that is

$$\text{dist}\,(x_0, \Gamma(t_1)) \geq \alpha R, \quad t_1 = t_0 - \lambda h,$$

in particular,

$$u(x,t_1) = 0, \quad \text{for } x \in B_{\alpha R}(x_0).$$

Replacing R by αR, τ by t_1 and h by λh and repeating the above argument, we further obtain

$$\text{dist}\,(x_0, \Gamma(t_2)) \geq \alpha^2 R, \quad t_2 = t_0 - \lambda^2 h.$$

Going on this way we finally arrive at

$$\text{dist}\,(x_0, \Gamma(t)) \geq \alpha^n R, \quad t = t_0 - \lambda^n h, n = 1, 2, \cdots.$$

The remainder of the proof is just the same as Theorem 1.7.4 in §1.7. □

As a corollary of Theorem 2.6.7, we have

Theorem 2.6.5 *For any $(x_0,t_0) \in \Gamma_m$, there exists a neighborhood of (x_0,t_0), such that the portion $U(x_0,t_0)$ of Γ which lies in this neighborhood can be expressed by a Hölder continuous function*

$$t = S(x) \qquad x \in U_x,$$

where U_x is the projection of $U(x_0,t_0)$ on the plane $t = 0$.

Furthermore we have

Theorem 2.6.6 *Under the assumptions of Theorem 2.6.2, the free boundary Γ can be expressed by a function*

$$t = S(x) \qquad x \in (\mathbb{R}^N \backslash D)$$

and $S(x)$ is uniformly Hölder continuous on any compact subset which does not intersect with $\overline{D}$.

2.6.2 *Lipschitz continuity of the free boundary*

Similar to the Newtonian filtration equation, we can prove the Lipschitz continuity of the free boundary for large time and estimate the time T_0 such that for $t \geq T_0$ Γ can be expressed by a Lipschitz continuous function, which is just $\inf\{t > 0; B_{R_0} \subset \Omega(t)\}$, where $B_{R_0}(0)$ is the smallest ball containing $D = \Omega(0)$. The proof of this result can be founded in [ZY3], where the following result on the global Lipschitz continuity of the free boundary is also presented.

Theorem 2.6.7 *Assume that $u_0 \in C^*$. Then the free boundary Γ can be expressed by a function*

$$t = S(x) \qquad x \in \mathbb{R}^N \backslash D$$

and $S(x)$ is Lipschitz continuous on $\mathbb{R}^N \backslash D$.

Here C^* denotes the set of all functions u_0 satisfying the following conditions:

(i) $u_0 \geq 0$, $u_0 \in C^{1+\alpha}(\mathbb{R}^N)$, $\alpha \in (0,1)$, and

$$D = \{x \in \mathbb{R}^N; u_0(x) > 0\}$$

is a bounded C^1 domain;

(ii) $div(|\nabla u_0|^{p-2}\nabla u_0) \in L^\infty(\mathbb{R}^N)$ and there exists an open set $S \subset D$ such that

$$\overline{S} \backslash D = \partial D,$$

$$div(|\nabla u_0|^{p-2}\nabla u_0) \geq K_0 |\nabla u_0| \quad on\ S,$$

where $K_0 > 0$ is a constant.

Remark 2.6.2 The $C^{1,\alpha}$-regularity of the free boundary for (1.1) and (1.2) was proved by KO,Youngsang in [KY1], [KY2]. Daskalopoulos, P.

and Hamilton, R. further studied the C^∞-regularity of the free boundary in [DH3].

2.7 Other Problems

In this section we briefly introduce some other problems and the related results.

2.7.1 *p-Laplacian equation with strongly nonlinear sources*

Consider the equation

$$u_t = \operatorname{div}(|\nabla u|^{p-2}\nabla u) + \lambda u^q, \tag{7.1}$$

where $p > 1$, $q > 0$ and λ are some constants, in which the nonlinear term λu^q describes the nonlinear source in the diffusion process, called "heat source" if $\lambda > 0$ and "cold source" if $\lambda < 0$. Just as the Newtonian equation, the appearance of nonlinear sources will exert a great influence to the properties of solutions and the influence of "heat source" and "cold source" is completely different. For instance, for the equation (7.1) to have a generalized solution (the definition is similar to the case $\lambda = 0$), the condition on the growth of initial value

$$u(x,0) = u_0(x) \geq 0 \qquad \text{for } x \in \mathbb{R}^N \tag{7.2}$$

is completely contrary.

In the case $\lambda > 0$, we have the following result on the local existence of solutions: if $u_0 \in L^h_{loc}(\mathbb{R}^N)$ and $[u_0]_h < \infty$ with $h = 1$ whenever $q < p - 1 + \frac{p}{N}$, $h > \frac{N}{p}(q - p + 1)$ whenever $q \geq p - 1$, then there exists $\gamma_0 = \gamma_0(N,p,q,h)$ such that the Cauchy problem (7.1), (7.2) admits a generalized solution u on Q_{T_0} possessing the following properties:

$$[u(\cdot,t)]_h \leq \gamma[u_0]_h,$$

$$u(x,t) \leq \gamma t^{-N/\kappa_h}[u_0]_h^{ph/\kappa_h},$$

$$|\nabla u(x,t)| \leq \gamma t^{-(N+1)/\kappa} \max\{1, [u_0]_1^{1+(p-2)/\kappa}\},$$

$$\int_0^t \int_{B_\rho(x_0)} |\nabla u|^\sigma dx d\tau$$

$$\leq \gamma t^{1-\sigma/p-N(2\sigma-p)/\kappa p} \left(\sup_{0<\tau<t} \int_{B_\rho(x_0)} u(x,\tau)dx \right)^{1+(2\sigma-p)/\kappa},$$

provided that T_0 satisfies

$$T_0[u_0]_h^{p-2} + T_0^{1-N(q-p+1)/ph}[u_0]_h^{q-1} = \gamma_0^{-1},$$

where $\kappa_h = N(p-2)+hp$, $\kappa = \kappa_1$, $\gamma = \gamma(N,p,q,h,\sigma)$, and $[f]_h$ is defined in §1.1.9. The basic idea of the proof is to consider the approximate problem

$$u_t = \text{div}\,(|\nabla u|^{p-2}\nabla u) + \lambda \min\{n, u^q\},$$

$$u(x,0) = u_{0n}(x),$$

and establish the locally uniform estimate on the bound of the approximate solution u_n, where $u_{0n} \in C_0^\infty(\mathbb{R}^N)$ with $[u_{0n}]_h \leq \gamma[u_0]_h$ and

$$\lim_{n\to\infty} \int_{B_\rho} |u_{0n} - u_0|^h dx = 0 \quad \text{for } \rho > 0.$$

It can be seen in the proof that this result is still valid for Radom measure initial μ with $[\mu_0]_1 < \infty$, provided that $1 < q < p - 1 + \dfrac{p}{N}$, $p > \dfrac{2N}{N+1}$.

Similar method can be applied to prove that if $q > p-1+\dfrac{p}{N}$, then the Cauchy problem (7.1), (7.2) admits a global solution for any "small" initial value.

Based on the a prior estimate on the solutions of (7.1), one can also prove that for the Cauchy problem (7.1), (7.2) to have a local solution, the condition $[u_0]_h < \infty$ is not only sufficient, but also necessary and if $p-1 < q < p-1+\dfrac{p}{N}$, then the equation (7.1) does not admit any nontrivial global solution.

The uniqueness of generalized solutions of the Cauchy problem (7.1), (7.2) is established in the class $\mathcal{R}$ of functions. By $u \in \mathcal{R}$, we mean that u satisfies

$$[u(\cdot,t)]_1 \leq C \qquad \text{for } t \in (0,T),$$

$$\sup_{x\in\mathbb{R}^N} u(x,t) \leq Ct^{-\delta}, \quad \text{for } t \in (0,T),$$

$$|\nabla u(x,t)| \leq Ct^{-\delta_1} \qquad \text{for } t \in (0,T),$$

where δ, δ_1 and C are positive constants with $\delta < (a-1)^{-1}$, $\delta_1 < \dfrac{1}{p-2}$ and $a = \max\{q, p-1\}$.

For the proofs of the above results we refer to [ZH6].

In the case $\lambda < 0$, we have the following results (see [ZH2]).

First, the existence of generalized solutions of the Cauchy problem (7.1), (7.2) on $Q = \mathbb{R}^N \times (0, \infty)$ is established in the following two cases:

(i) $p - 1 < q < p - 1 + \dfrac{p}{N}$ and the initial value u_0 is a Radom measure;

(ii) $q > p - 1$ and $u_0 \in L^1_{loc}(\mathbb{R}^N)$.

It can also be proved that if $q \geq p - 1 + \dfrac{p}{N}$ and the initial value is $\delta(x)$, then the Cauchy problem (7.1), (7.2) does not admit any generalized solution. This means that if the initial value is a measure, then for the Cauchy problem)7.1), (7.2) to have a solution, the condition $q < p-1+\dfrac{p}{N}$ is also necessary. If $q > p - 1$, then the Cauchy problem (7.1), (7.2) admits a global solution for any $u_0 \in L^1_{loc}(\mathbb{R}^N)$, completely different to the case $\lambda = 0$.

For the Cauchy problem (7.1), (7.2) to have a solution, one requires that $u_0 \in L^{1+\alpha}_{loc}(\mathbb{R}^N)$ $(\alpha > 0)$ and

$$\int_{\mathbb{R}^N} \exp\{-c\sqrt{1+|x|^2}\} u_0^{1+\alpha}(x)dx < \infty,$$

whenever $q = p - 1$, where $c < p(p-1)^{(1-p)/p}\alpha^{(p-1)/p}$; and requires $u_0 \in L^\infty_{loc}(\mathbb{R}^N)$ and

$$u_0(x) \leq c_1(c_2 + |x|^2)^{p/2(p-1-q)} \quad \text{a.e. on } \mathbb{R}^N, \tag{7.3}$$

whenever $1 < q < p - 1$, where c_1, c_2 are positive constants with

$$c_1 < \left(\frac{(p-1-q)^p}{p^{p-1}(pq + N(p-1-q))} \right)^{1/(p-1-q)}.$$

In both cases, the generalized solution exists on $Q = \mathbb{R}^N \times (0, \infty)$.

The condition (7.3) for the growth of u_0 does not exceed $|x|^{p/(p-1-q)}$ is less restrictive compared with the case $\lambda = 0$. For the Cauchy problem (7.1), (7.2) to have a solution, the condition (7.3) is almost the best possible, since one can prove that in the case $1 < q < p - 1$, if $u_0 \in L^\infty_{loc}(\mathbb{R}^N)$ and for some constants $\alpha > \dfrac{p}{p-1-q}$ and $B > 0$ such that

$$\lim_{|x|\to\infty} \frac{u_0(x)}{|x|^\alpha} = B,$$

then the Cauchy problem (7.1), (7.2) does not admits any generalized solution.

Again one can prove that if $\lambda < 0$, $p > \dfrac{2N}{N+1}$ and $\max\{1, p-1\} < q < p - 1 + \dfrac{p}{N}$, the equation (7.1) admits a very singular solution $U(x,t)$. The proof is similar to that for the Newtonian equation with "cold source" (see [ZH3]).

For other related works, see [LAU1], [QW].

2.7.2 *Asymptotic properties of solutions*

Assume that

$$\lim_{|x|\to\infty} |x|^\alpha u_0(x) = A,$$

where α and A are positive constants. Denote

$$\gamma = \alpha(p-2) + p, \ \mu = p - 2 + \frac{p}{N}, \ \beta = \frac{p(q-1)}{q-p+1}.$$

Again assume that $p > \dfrac{2N}{N+1}$.

In the case $\lambda = 0$, we have the following results:

(1) If $0 \le \alpha < N$, then the solution u of (7.1), (7.2) satisfies

$$t^{\alpha/\gamma}|u(x,t) - w_A(x,t)| \to 0 \ \text{ as } t \to \infty,$$

uniformly on $\{x \in \mathbb{R}^N; |x| \le bt^{1/\gamma}\}$ with $b > 0$, where $w_A(x,t)$ is the solution of (7.1) with initial value

$$w_A(x,0) = A|x|^{-\alpha}. \tag{7.4}$$

$w_A(x,t)$ can be expressed as

$$w_A = t^{-\alpha/\gamma} f(\eta, A),$$

where $\eta = |x|t^{-1/\gamma}$ and f is the solution of the problem

$$(|f'|^{p-2}f')' + \frac{N-1}{\eta}|f'|^{p-2}f' + \frac{1}{\gamma}\eta f' + \frac{\alpha}{\gamma}f = 0, \eta > 0,$$

$$f'(0) = 0, f \geq 0, \lim_{\eta\to\infty} \eta^\alpha f(\eta) = A.$$

(2) If $\alpha > N$, then the solution u of (7.1), (7.2) satisfies

$$t^{1/\mu}|u(x,t) - E_c(x,t)| \to 0, \quad \text{as } t \to \infty,$$

uniformly on $\{x \in \mathbb{R}^N; |x| \leq bt^{1/\gamma}\}$ with $b > 0$, where E_c is the Barenblatt-type solution of (7.1):

$$E_c = (N\mu)^{-1/(p-2)}t^{-1/\mu} \cdot \left(b - \frac{p-2}{p}|x|^{p/(p-1)}t^{-p/(N\mu(p-1))}\right)_+^{(p-1)/(p-2)}, \tag{7.5}$$

and $c > 0$ is a constant such that $\int_{\mathbb{R}^N} E_c(x,t)dx = c$.

In the case $\lambda < 0$, the asymptotic behavior of solutions of (7.1), (7.2) depends not only on the behavior of u_0 as $|x| \to \infty$, but also on the "competition" of diffusion and absorption. We have the following results:

(1) If $q > \max\{p-1, 1\}$, $\alpha < \dfrac{p}{q-p+1}$, then the solution u of (7.1), (7.2) satisfies

$$t^{1/(q-1)}u(x,t) \to \left(\frac{1}{q-1}\right)^{1/(q-1)}, \quad \text{as } t \to \infty,$$

uniformly on $\{x \in \mathbb{R}^N; |x| \leq bt^{1/\beta}\}$ with $b > 0$.

(2) If $q > p - 1 + \dfrac{p}{N}$, $\dfrac{p}{q-p+1} < \alpha < N$, then the solution u of (7.1), (7.2) satisfies

$$t^{\alpha/\gamma}|u(x,t) - w_A(x,t)| \to 0, \quad \text{as } t \to \infty,$$

uniformly on $\{x \in \mathbb{R}^N; |x| \leq bt^{1/\gamma}\}$ with $b > 0$, where w_A is the solution of (7.1), (7.4) with $\lambda = 0$.

(3) If $\alpha > N$, $q > p - 1 + \dfrac{p}{N}$, then the solution u of (7.1), (7.2) satisfies

$$t^{\alpha/\gamma}|u(x,t) - E_c(x,t)| \to 0, \quad \text{as } t \to \infty,$$

uniformly on $\{x \in \mathbb{R}^N; |x| \le b^{1/N\mu}\}$ with $b > 0$, where

$$c = \int_{\mathbb{R}^N} u_0(x)dx + \lambda \int_0^\infty \int_{\mathbb{R}^N} u^q(x,t)dxdt,$$

and $E_c(x,t)$ is given by (7.5) with constant c such that $\int_{\mathbb{R}^N} E_c(x,t)dx = c$.

(4) If $\max\{1, p-1\} < q < p - 1 + \frac{p}{N}$, $\alpha > \frac{p}{q-p+1}$, then the solution u of (7.1), (7.2) satisfies

$$t^{1/(q-1)}|u(x,t) - U(x,t)| \to 0, \quad \text{as } t \to \infty,$$

uniformly on $\{x \in \mathbb{R}^N; |x| \le b^{1/\beta}\}$ with $b > 0$, where $U(x,t)$ is the very singular solution of (7.1).

For the proofs of these results we refer to [ZH4].

Remark 2.7.1 Various kinds of singular limits have been studied by many authors, see for example, [AV1], [AV2], [EFG], [SA1], [SA2], [YZ].

Chapter 3

General Quasilinear Equations of Second Order

3.1 Introduction

We have studied the Newtonian filtration equation

$$\frac{\partial u}{\partial t} = \Delta(|u|^{m-1}u) \quad (m > 1) \tag{1.1}$$

in Chapter 1 and the non-Newtonian filtration equation

$$\frac{\partial u}{\partial t} = \operatorname{div}(|\nabla u|^{p-2}\nabla u) \quad (p > 2) \tag{1.2}$$

in Chapter 2. Sometimes we also talked about the polytropic filtration equation

$$\begin{gathered}\frac{\partial u}{\partial t} = \operatorname{div}(|\nabla(|u|^{m-1}u)|^{p-2}\nabla(|u|^{m-1}u)) \\ (p > 1, m > 0, m(p-1) > 1).\end{gathered} \tag{1.3}$$

The obvious character of these equations is their simplicity in form and the essential point is that they have only one "point of degeneracy", namely, (1.1) and (1.2) degenerate only if $u = 0$ and $\nabla u = 0$ respectively and (1.3) degenerates only if $u = 0$ or $\nabla u = 0$.

This chapter is devoted to general quasilinear degenerate parabolic equations. Here by the word "general", we mean the following two aspects. The first is the generality in form of the equations, for example, instead of (1.1), we consider

$$\frac{\partial u}{\partial t} = \frac{\partial}{\partial x_i}\left(a^{ij}(x,t,u)\frac{\partial u}{\partial x_j}\right) + \frac{\partial}{\partial x_i}b^i(x,t,u) + c(x,t,u), \tag{1.4}$$

where $a^{ij} = a^{ji}$ and

$$a^{ij}(x,t,u)\xi_i\xi_j \geq 0, \quad \forall \xi = (\xi_1, \cdots, \xi_N) \in R^N.$$

As before, repeated indices imply a summation from 1 up to N. Also instead of (1.2) and (1.3), we consider

$$\frac{\partial u}{\partial t} = \text{div}\vec{A}(\nabla B(u)), \tag{1.5}$$

where $\vec{A}(v) = \big(A^1(v), \cdots, A^N(v)\big)$, and the more general equations, for example, in one dimensional case,

$$\frac{\partial u}{\partial t} = \frac{\partial}{\partial x}F\left(\frac{\partial}{\partial x}A(u)\right) + \frac{\partial B(u)}{\partial x} \tag{1.6}$$

with $F'(s) \geq 0$, $A'(s) \geq 0$. The second and the more essential aspect is that the equations under consideration are allowed to have many "points of degeneracy" or even have arbitrary degeneracy. We will treat two kinds of degeneracy, namely, weak degeneracy and strong degeneracy. For example, (1.6) is called weakly degenerate, if neither $E_F = \{s; F'(s) = 0\}$ nor $E_A = \{s; A'(s) = 0\}$ has interior point. Otherwise, (1.6) is called strongly degenerate.

In some literature (see [BP]), (1.6) is also called strongly degenerate, if one of the conditions

$$\lim_{s\to+\infty} F(s) < +\infty, \qquad \lim_{s\to-\infty} F(s) > -\infty \tag{1.7}$$

is satisfied. Notice that in this case, E_F even might be empty and degeneracy happens at infinity, since we have

$$\underline{\lim}_{s\to+\infty} F'(s) = 0 \quad or \quad \underline{\lim}_{s\to-\infty} F'(s) = 0.$$

Also notice that for weakly degenerate equation (1.6), E_F or E_A might be a set containing points of infinite numbers, even be a set of positive measure.

In this Chapter, for simplicity, we will present our arguments basically for equations of the form

$$\frac{\partial u}{\partial t} = \Delta A(u) + \text{div}\vec{B}(u)$$

with $A(0) = 0, A'(s) \geq 0$, $\vec{B}(s) = (B^1(s), \cdots, B^N(s))$. Main results and typical methods will be introduced for these equations. Extension of our arguments to equations of more general form will be mentioned in the related part of each section, readers who are interested, may refer to the literature quoted there.

3.2 Weakly Degenerate Equations in One Dimension

This section is devoted to weakly degenerate equations in one dimension. As indicated in §3.3.1, we will basically discuss equations of the form

$$\frac{\partial u}{\partial t} = \frac{\partial^2 A(u)}{\partial x^2} + \frac{\partial B(u)}{\partial x}, \tag{2.1}$$

where $A(s)$, $B(s) \in C^1(\mathbb{R})$, $A(0) = 0$, $A'(s) \geq 0$, however the set $E = \{s; A'(s) = 0\}$ does not contain any interior point. Only the first boundary value problem for (2.1) will be discussed in detail, the corresponding initial and boundary conditions are

$$u(0,t) = u(1,t) = 0, \tag{2.2}$$

$$u(x,0) = u_0(x). \tag{2.3}$$

Denote $Q_T = (0,1) \times (0,T)$. Similar to Definition 1.1.4, we introduce the following definition.

Definition 3.2.1 A function $u \in L^1(Q_T)$ is called a **generalized solution** of the boundary value problem (2.1)–(2.3), if $A(u)$, $B(u) \in L^1(Q_T)$ and the integral equality

$$\iint_{Q_T} \left(u\frac{\partial \varphi}{\partial t} + A(u)\frac{\partial^2 \varphi}{\partial x^2} - B(u)\frac{\partial \varphi}{\partial x} \right) dxdt + \int_0^1 u_0(x)\varphi(x,0)dx = 0$$

is fulfilled for any function $\varphi \in C^\infty(\overline{Q_T})$ with $\varphi(0,t) = \varphi(1,t) = \varphi(x,T) = 0$.

It is easy to check that if $u \in C^1(\overline{Q_T}) \bigcap C^2(Q_T)$ is a generalized solution of the boundary value problem (2.1)–(2.3), then u is a classical solution of the problem. Here we notice that $A(u(x,t))|_{x=0,1} = 0$ implies $u(x,t)|_{x=0,1} = 0$ because of the strict monotonicity of $A(s)$.

3.2.1 *Uniqueness of bounded and measurable solutions*

Theorem 3.2.1 *Assume that $u_0 \in L^\infty(0,1)$, $A(s)$, $B(s) \in C^1(\mathbb{R})$, $A'(s) \geq 0$ and $E = \{s; A'(s) = 0\}$ has no interior point. Then the first boundary value problem (2.1)–(2.3) has at most one bounded and measurable solution.*

We will prove the theorem by means of Holmgren's approach. The crucial step is to establish the L^1 estimate for the derivatives of the solutions of the adjoint equation. Our proof will be completed by using this estimate, together with some L^2-type estimates for the solutions.

Let $u_1, u_2 \in L^\infty(Q_T)$ be solutions of the boundary value problem (2.1)–(2.3). By Definition 3.2.1, we have

$$\iint_{Q_T} (u_1 - u_2)\left(\frac{\partial \varphi}{\partial t} + \widetilde{A}\frac{\partial^2 \varphi}{\partial x^2} - \widetilde{B}\frac{\partial \varphi}{\partial x}\right) dxdt = 0$$

for any $\varphi \in C^\infty(\overline{Q_T})$ with $\varphi(0,t) = \varphi(1,t) = \varphi(x,T) = 0$, where

$$\widetilde{A} = \widetilde{A}(u_1, u_2) = \int_0^1 A'(\theta u_1 + (1-\theta)u_2)d\theta,$$

$$\widetilde{B} = \widetilde{B}(u_1, u_2) = \int_0^1 B'(\theta u_1 + (1-\theta)u_2)d\theta.$$

If for any $f \in C_0^\infty(Q_T)$, the adjoint problem

$$\frac{\partial \varphi}{\partial t} + \widetilde{A}\frac{\partial^2 \varphi}{\partial x^2} - \widetilde{B}\frac{\partial \varphi}{\partial x} = f,$$
$$\varphi(0,t) = \varphi(1,t) = 0,$$
$$\varphi(x,T) = 0$$

had a solution in $C^\infty(\overline{Q_T})$, then we would have

$$\iint_{Q_T} (u_1 - u_2) f dxdt = 0$$

and the uniqueness would follow from the arbitrariness of f. However, since the coefficients $\widetilde{A}$ and $\widetilde{B}$ are merely bounded and measurable, it is difficult to discuss the solvability of the adjoint problem directly. Even if we have established the existence of solutions, the solutions is not smooth in general. This situation competes us to consider some approximation of the adjoint equation.

For sufficiently small $\eta > 0$ and $\delta > 0$, let

$$\lambda_\eta^\delta = \begin{cases} (\eta + \widetilde{A})^{-1/2}\widetilde{B}, & \text{if } |u_1 - u_2| \geq \delta, \\ 0, & \text{if } |u_1 - u_2| < \delta. \end{cases}$$

Since $A(s)$ is strictly increasing and $u_1,\ u_2 \in L^\infty(Q_T)$, there must be constants $L(\delta) > 0$, $K(\delta) > 0$ depending on δ, but independent of η, such that

$$\widetilde{A} = \frac{A(u_1) - A(u_2)}{u_1 - u_2} \geq L(\delta), \quad \text{whenever } |u_1 - u_2| \geq \delta,$$
$$|\lambda_\eta^\delta| \leq K(\delta).$$

Let $\widetilde{A}_\varepsilon$ and $\lambda_{\eta,\varepsilon}^\delta$ be a C^∞ approximation of $\widetilde{A}$ and λ_η^δ respectively, such that

$$\lim_{\varepsilon \to 0} \widetilde{A}_\varepsilon = \widetilde{A}, \quad \text{a.e. in } Q_T,$$
$$\lim_{\varepsilon \to 0} \lambda_{\eta,\varepsilon}^\delta = \lambda_\eta^\delta, \quad \text{a.e. in } Q_T,$$
$$\widetilde{A}_\varepsilon \leq C,$$
$$|\lambda_{\eta,\varepsilon}^\delta| \leq K(\delta).$$

Denote

$$\widetilde{B}_{\eta,\varepsilon}^\delta = \lambda_{\eta,\varepsilon}^\delta (\eta + \widetilde{A}_\varepsilon)^{1/2}.$$

For given $f \in C_0^\infty(Q_T)$, consider the approximate adjoint problem

$$\frac{\partial \varphi}{\partial t} + (\eta + \widetilde{A}_\varepsilon)\frac{\partial^2 \varphi}{\partial x^2} - \widetilde{B}_{\eta,\varepsilon}^\delta \frac{\partial \varphi}{\partial x} = f, \tag{2.4}$$

$$\varphi(0,t) = \varphi(1,t) = 0, \tag{2.5}$$

$$\varphi(x,T) = 0. \tag{2.6}$$

Lemma 3.2.1 *The solution φ of (2.4)–(2.6) satisfies*

$$\sup_{Q_T} |\varphi(x,t)| \leq C, \tag{2.7}$$

$$\iint_{Q_T} (\eta + \widetilde{A}_\varepsilon)\left(\frac{\partial^2 \varphi}{\partial x^2}\right)^2 dxdt \leq K(\delta)\eta^{-1}, \tag{2.8}$$

$$\iint_{Q_T} \left(\frac{\partial \varphi}{\partial x}\right)^2 dxdt \le K(\delta)\eta^{-1}. \tag{2.9}$$

Here and in the sequel, we use C to denote a universal constant, independent of δ, η and ε, and $K(\delta)$ a constant, depending only on δ, which may take different values on different occasions.

Proof. (2.7) follows from the maximum principle. To prove (2.8) and (2.9), we multiply (2.4) by $\dfrac{\partial^2 \varphi}{\partial x^2}$ and integrate over Q_T. Integrating by parts and using (2.5), (2.6) yield

$$\frac{1}{2}\int_0^1 \left(\frac{\partial \varphi(x,0)}{\partial x}\right)^2 dx + \iint_{Q_T} \left(\eta + \widetilde{A}_\varepsilon\right)\left(\frac{\partial^2 \varphi}{\partial x^2}\right)^2 dxdt$$
$$- \iint_{Q_T} \widetilde{B}^\delta_{\eta,\varepsilon} \frac{\partial \varphi}{\partial x}\frac{\partial^2 \varphi}{\partial x^2} dxdt = \iint_{Q_T} f \frac{\partial^2 \varphi}{\partial x^2} dxdt.$$

Using Young's inequality and noticing that $|\lambda^\delta_{\eta,\varepsilon}| \le K(\delta)$, we obtain

$$\begin{aligned} &\iint_{Q_T} (\eta + \widetilde{A}_\varepsilon)\left(\frac{\partial^2 \varphi}{\partial x^2}\right)^2 dxdt \\ \le\; &\iint_{Q_T} \lambda^\delta_{\eta,\varepsilon}(\eta + \widetilde{A}_\varepsilon)^{1/2} \frac{\partial \varphi}{\partial x}\frac{\partial^2 \varphi}{\partial x^2} dxdt + \iint_{Q_T} f \frac{\partial^2 \varphi}{\partial x^2} dxdt \\ \le\; &\frac{1}{4}\iint_{Q_T} (\eta + \widetilde{A}_\varepsilon)\left(\frac{\partial^2 \varphi}{\partial x^2}\right)^2 dxdt + K(\delta)\iint_{Q_T} (\frac{\partial \varphi}{\partial x})^2 dxdt + C\eta^{-1}. \end{aligned} \tag{2.10}$$

Using (2.5) and Young's inequality again gives

$$\begin{aligned} &\iint_{Q_T} (\frac{\partial \varphi}{\partial x})^2 dxdt = -\iint_{Q_T} \varphi \frac{\partial^2 \varphi}{\partial x^2} dxdt \\ \le\; &\alpha \iint_{Q_T} (\eta + \widetilde{A}_\varepsilon)\left(\frac{\partial^2 \varphi}{\partial x^2}\right)^2 dxdt + C\alpha^{-1}\eta^{-1} \end{aligned} \tag{2.11}$$

for any $\alpha > 0$. Substituting this into (2.10) and choosing $\alpha > 0$ small enough, we derive (2.8). (2.9) follows from (2.8) and (2.11). The proof is complete. □

Lemma 3.2.2 *The solution φ of (2.4)–(2.6) satisfies*

$$\sup_{0<t<T} \int_0^1 |\frac{\partial \varphi(x,t)}{\partial x}| dx \le C, \tag{2.12}$$

where the constant C is independent of δ, η and ε.

Proof. For small $\beta > 0$, let

$$\mathrm{sgn}_\beta s = \begin{cases} 1, & \text{if } s \geq \beta, \\ \dfrac{s}{\beta}, & \text{if } |s| < \beta, \\ -1, & \text{if } s \leq -\beta, \end{cases} \qquad I_\beta(s) = \int_0^s \mathrm{sgn}_\beta \theta d\theta.$$

Differentiate (2.4) with respect to x, multiply the resulting equality by $\mathrm{sgn}_\beta \dfrac{\partial \varphi}{\partial x}$ and integrate over $S_t \equiv (0,1) \times (t,T)$. Then we obtain

$$\begin{aligned} &\iint_{S_t} \frac{\partial}{\partial t} I_\beta(\frac{\partial \varphi}{\partial x}) dx d\tau \\ &+ \iint_{S_t} \frac{\partial}{\partial x}[(\eta + \widetilde{A}_\varepsilon)\frac{\partial^2 \varphi}{\partial x^2}]\mathrm{sgn}_\beta(\frac{\partial \varphi}{\partial x}) dx d\tau \\ &- \iint_{S_t} \frac{\partial}{\partial x}[\widetilde{B}^\delta_{\eta,\varepsilon}\frac{\partial \varphi}{\partial x}]\mathrm{sgn}_\beta(\frac{\partial \varphi}{\partial x}) dx d\tau \\ &= \iint_{S_t} \mathrm{sgn}_\beta(\frac{\partial \varphi}{\partial x})\frac{\partial f}{\partial x} dx d\tau. \end{aligned}$$

Hence, integrating by parts and using (2.6) yield

$$\begin{aligned} &\int_0^1 I_\beta(\frac{\partial \varphi(x,t)}{\partial x}) dx \\ = &- \iint_{S_t} (\eta + \widetilde{A}_\varepsilon)(\frac{\partial^2 \varphi}{\partial x^2})^2 \mathrm{sgn}'_\beta(\frac{\partial \varphi}{\partial x}) dx d\tau \\ &+ \iint_{S_t} \widetilde{B}^\delta_{\eta,\varepsilon}\frac{\partial \varphi}{\partial x}\frac{\partial^2 \varphi}{\partial x^2}\mathrm{sgn}'_\beta(\frac{\partial \varphi}{\partial x}) dx d\tau \\ &+ \int_t^T \left((\eta + \widetilde{A}_\varepsilon)\frac{\partial^2 \varphi}{\partial x^2} - \widetilde{B}^\delta_{\eta,\varepsilon}\frac{\partial \varphi}{\partial x}\right)\mathrm{sgn}_\beta(\frac{\partial \varphi}{\partial x})\Big|_{x=0}^{x=1} d\tau \\ &- \iint_{S_t} \mathrm{sgn}_\beta(\frac{\partial \varphi}{\partial x})\frac{\partial f}{\partial x} dx d\tau. \end{aligned} \tag{2.13}$$

The first term on the right side is nonpositive. The last term is bounded.

Using (2.4), (2.5) and the fact $f \in C_0^\infty(Q_T)$ we see that

$$\left((\eta + \widetilde{A}_\varepsilon)\frac{\partial^2 \varphi}{\partial x^2} - \widetilde{B}^\delta_{\eta,\varepsilon}\frac{\partial \varphi}{\partial x}\right) \operatorname{sgn}_\beta(\frac{\partial \varphi}{\partial x})\Big|_{x=0}^{x=1}$$
$$= \left(f - \frac{\partial \varphi}{\partial t}\right) \operatorname{sgn}_\beta(\frac{\partial \varphi}{\partial x})\Big|_{x=0}^{x=1} = 0$$

which shows that the third term on the right side of (2.13) vanishes. Therefore we have

$$\int_0^1 I_\beta\left(\frac{\partial \varphi(x,t)}{\partial x}\right) dx \le C(\eta, \varepsilon, \delta)\int_t^T d\tau \int_{[0,1]\bigcap\{x;|\partial\varphi/\partial x|\le\beta\}} |\frac{\partial^2 \varphi}{\partial x^2}|dx + C$$

from which (2.12) follows by letting $\beta \to 0$ and using a known result (see S. Sakes, Theory of Integration, 131–133) to conclude that

$$\int_{[0,1]\bigcap\{x;|\partial\varphi/\partial x|\le\beta\}} |\frac{\partial^2 \varphi}{\partial x^2}|dx \to 0 \quad \text{as} \quad \beta \to 0.$$

□

Proof of Theorem 3.2.1. Given $f \in C_0^\infty(Q_T)$. Let φ be a solution of (2.4)–(2.6). Then

$$\iint_{Q_T} (u_1 - u_2) f dx dt$$
$$= \iint_{Q_T} (u_1 - u_2)\left(\frac{\partial \varphi}{\partial t} + (\eta + \widetilde{A}_\varepsilon)\frac{\partial^2 \varphi}{\partial x^2} - \widetilde{B}^\delta_{\eta,\varepsilon}\frac{\partial \varphi}{\partial x}\right) dx dt.$$

As indicated above, from the definition of generalized solutions, we have

$$\iint_{Q_T} (u_1 - u_2)\left(\frac{\partial \varphi}{\partial t} + \widetilde{A}\frac{\partial^2 \varphi}{\partial x^2} - \widetilde{B}\frac{\partial \varphi}{\partial x}\right) dx dt = 0.$$

Thus

$$\iint_{Q_T} (u_1 - u_2) f dx dt$$
$$= \iint_{Q_T} (u_1 - u_2)\eta\frac{\partial^2 \varphi}{\partial x^2} dx dt + \iint_{Q_T} (u_1 - u_2)(\widetilde{A}_\varepsilon - \widetilde{A})\frac{\partial^2 \varphi}{\partial x^2} dx dt$$
$$- \iint_{Q_T} (u_1 - u_2)(\widetilde{B}^\delta_{\eta,\varepsilon} - \widetilde{B})\frac{\partial \varphi}{\partial x} dx dt. \tag{2.14}$$

Now we are ready to estimate all terms on the right side of (2.14).

First, from Lemma 3.2.1,

$$\begin{aligned}
&\left|\iint_{Q_T}(u_1-u_2)(\widetilde{A}_\varepsilon-\widetilde{A})\frac{\partial^2\varphi}{\partial x^2}dxdt\right| \\
\leq\ & C\left(\iint_{Q_T}(\widetilde{A}_\varepsilon-\widetilde{A})^2dxdt\right)^{1/2}\left(\iint_{Q_T}(\frac{\partial^2\varphi}{\partial x^2})^2dxdt\right)^{1/2} \\
\leq\ & K(\delta)\eta^{-1}\left(\iint_{Q_T}(\widetilde{A}_\varepsilon-\widetilde{A})^2dxdt\right)^{1/2}.
\end{aligned}$$

Hence

$$\lim_{\varepsilon\to 0}\iint_{Q_T}(u_1-u_2)(\widetilde{A}_\varepsilon-\widetilde{A})\frac{\partial^2\varphi}{\partial x^2}dxdt=0. \tag{2.15}$$

Denote

$$\begin{aligned}
G_\delta &= \{(x,t)\in Q_T; |u_1-u_2|<\delta\}, \\
F_\delta &= \{(x,t)\in Q_T; |u_1-u_2|\geq\delta\}.
\end{aligned}$$

Using Lemma 3.2.2, we have

$$\begin{aligned}
&\left|\iint_{G_\delta}(u_1-u_2)(\widetilde{B}^\delta_{\eta,\varepsilon}-\widetilde{B})\frac{\partial\varphi}{\partial x}dxdt\right| \\
\leq\ & \delta K(\delta)\iint_{G_\delta}|\frac{\partial\varphi}{\partial x}|dxdt\leq\delta K(\delta).
\end{aligned}$$

By Lemma 3.2.1,

$$\begin{aligned}
&\left|\iint_{F_\delta}(u_1-u_2)(\widetilde{B}^\delta_{\eta,\varepsilon}-\widetilde{B})\frac{\partial\varphi}{\partial x}dxdt\right| \\
\leq\ & C\left(\iint_{F_\delta}(\widetilde{B}^\delta_{\eta,\varepsilon}-\widetilde{B})^2dxdt\right)^{1/2}\left(\iint_{F_\delta}(\frac{\partial\varphi}{\partial x})^2dxdt\right)^{1/2} \\
\leq\ & K(\delta)\eta^{-1/2}\left(\iint_{F_\delta}(\widetilde{B}^\delta_{\eta,\varepsilon}-\widetilde{B})^2dxdt\right)^{1/2}.
\end{aligned}$$

Since $\lim\limits_{\varepsilon\to 0}\lambda^\delta_{\eta,\varepsilon}=\lambda^\delta_\eta=(\eta+\widetilde{A})^{-1/2}\widetilde{B}$, a.e. on F_δ, we have $\lim\limits_{\varepsilon\to 0}\widetilde{B}^\delta_{\eta,\varepsilon}=\widetilde{B}$ a.e. on F_δ. Thus

$$\lim_{\varepsilon\to 0}\iint_{F_\delta}(u_1-u_2)(\widetilde{B}^\delta_{\eta,\varepsilon}-\widetilde{B})\frac{\partial\varphi}{\partial x}dxdt=0$$

and hence

$$\overline{\lim_{\varepsilon\to 0}}\left|\iint_{Q_T}(u_1-u_2)(\widetilde{B}^{\delta}_{\eta,\varepsilon}-\widetilde{B})\frac{\partial\varphi}{\partial x}dxdt\right|\le \delta K(\delta). \qquad (2.16)$$

Using Lemma 3.2.1 again, we derive

$$\begin{aligned}
&\left|\iint_{Q_T}(u_1-u_2)\frac{\partial^2\varphi}{\partial x^2}dxdt\right|\\
\le\ & C\left(\sup_{F_\delta}\widetilde{A}_\varepsilon^{-1}\right)^{1/2}\left(\iint_{F_\delta}\widetilde{A}_\varepsilon(\frac{\partial^2\varphi}{\partial x^2})^2dxdt\right)^{1/2}\\
&+C\left(\iint_{G_\delta}(u_1-u_2)^2(\frac{\partial^2\varphi}{\partial x^2})^2dxdt\right)^{1/2}\\
\le\ & K(\delta)\eta^{-1/2}\left(\sup_{F_\delta}\widetilde{A}_\varepsilon^{-1}\right)^{1/2}+C\delta\left(\iint_{G_\delta}\left(\frac{\partial^2\varphi}{\partial x^2}\right)^2dxdt\right)^{1/2}\\
\le\ & K(\delta)\eta^{-1/2}\left(\sup_{F_\delta}\widetilde{A}_\varepsilon^{-1}\right)^{1/2}+K(\delta)\delta\eta^{-1}
\end{aligned}$$

Since $\lim_{\varepsilon\to 0}\widetilde{A}_\varepsilon=\widetilde{A}$ a.e on Q_T and $\widetilde{A}\ge L(\delta)$ on F_δ, letting $\varepsilon\to 0$ in the above inequality yields

$$\overline{\lim_{\varepsilon\to 0}}\left|\iint_{Q_T}(u_1-u_2)\frac{\partial^2\varphi}{\partial x^2}dxdt\right|\le K(\delta)L(\delta)^{-1/2}\eta^{-1/2}+K(\delta)\delta\eta^{-1}$$

and hence

$$\overline{\lim_{\varepsilon\to 0}}\left|\iint_{Q_T}\eta(u_1-u_2)\frac{\partial^2\varphi}{\partial x^2}dxdt\right|\le K(\delta)L(\delta)^{-1/2}\eta^{1/2}+K(\delta)\delta. \qquad (2.17)$$

Combining (2.14)–(2.17) we finally obtain

$$\left|\iint_{Q_T}(u_1-u_2)fdxdt\right|\le K(\delta)\delta+K(\delta)L(\delta)^{-1/2}\eta^{1/2}$$

which implies that

$$\iint_{Q_T}(u_1-u_2)fdxdt=0$$

by letting $\eta\to 0$ and then $\delta\to 0$. The proof of our theorem is complete. □

3.2.2 Existence of continuous solutions

Theorem 3.2.2 *Assume that $u_0 \in Lip[0,1]$ with $u_0(0) = u_0(1) = 0$, $A(s)$, $B(s)$ are appropriately smooth, $A'(s) \geq 0$, $\lim\limits_{s\to\pm\infty} A(s) = \pm\infty$ and the set $E = \{s; A'(s) = 0\}$ has no interior point. Then the first boundary value problem (2.1)–(2.3) admits a continuous solution.*

To prove the theorem, we consider the following regularized problem

$$\frac{\partial u_\varepsilon}{\partial t} = \frac{\partial^2 A_\varepsilon(u_\varepsilon)}{\partial x^2} + \frac{\partial B(u_\varepsilon)}{\partial x}, \tag{2.18}$$

$$u_\varepsilon(0,t) = u_\varepsilon(1,t) = 0, \tag{2.19}$$

$$u_\varepsilon(x,0) = u_{0\varepsilon}(x), \tag{2.20}$$

where $A_\varepsilon(s) = \varepsilon s + A(s)(\varepsilon > 0)$ and $u_{0\varepsilon}$ is a smooth function approximating u_0 uniformly with $u_{0\varepsilon}^{(k)}(0) = u_{0\varepsilon}^{(k)}(1) = 0(k = 0,1,2)$ and $|u'_{0\varepsilon}|$ uniformly bounded.

Let u_ε be a smooth solution of the problem, whose existence follows from the classical theory. We need some estimates for u_ε to ensure the compactness of $\{u_\varepsilon\}$.

First, the maximum principle implies that

$$\sup_{Q_T} |u_\varepsilon(x,t)| \leq M \tag{2.21}$$

with constant M independent of ε.

Next, we have

Lemma 3.2.3 *Let u_ε be a solution of the problem (2.18)–(2.20). Then*

$$\left| \frac{\partial A_\varepsilon(u_\varepsilon)}{\partial x}\bigg|_{x=0,1} \right| \leq C \tag{2.22}$$

with constant C independent of ε.

Proof. Let

$$\lambda_\varepsilon(s) = \frac{A'_\varepsilon(s)}{\theta(s)}, \quad w_\varepsilon = \int_0^{u_\varepsilon} \lambda_\varepsilon(s) ds,$$

where $\theta(s)$ is an auxiliary function of the form $\theta(s) = \alpha + s$ with an arbitrary constant α greater than M, the constant in (2.21). For example, we may choose $\alpha = M + 1$. Then

$$\frac{\partial w_\varepsilon}{\partial x} = \frac{\partial A_\varepsilon(u_\varepsilon)}{\partial x}/\theta(u_\varepsilon) = A'_\varepsilon(u_\varepsilon)\frac{\partial u_\varepsilon}{\partial x}/\theta(u_\varepsilon),$$

$$\frac{\partial w_\varepsilon}{\partial t} = \frac{\partial A_\varepsilon(u_\varepsilon)}{\partial t}/\theta(u_\varepsilon) = A'_\varepsilon(u_\varepsilon)\frac{\partial u_\varepsilon}{\partial t}/\theta(u_\varepsilon).$$

Using (2.18), one can easily check that w_ε satisfies

$$\frac{\partial w_\varepsilon}{\partial t} - A'_\varepsilon(u_\varepsilon)\frac{\partial^2 w_\varepsilon}{\partial x^2} - (\frac{\partial w_\varepsilon}{\partial x} + B'(u_\varepsilon))\frac{\partial w_\varepsilon}{\partial x} = 0,$$

in which we have a term $(\frac{\partial w_\varepsilon}{\partial x})^2$; as will be seen below, this term plays an important role in our proof. If we set $w_\varepsilon = \frac{\partial A_\varepsilon(u_\varepsilon)}{\partial x}$, then in the equation which w_ε satisfies, this term disappears. This is just why we introduce the auxiliary function $\theta(s)$.

Define an operator H as follows:

$$H[w] \equiv \frac{\partial w}{\partial t} - A'_\varepsilon(u_\varepsilon)\frac{\partial^2 w}{\partial x^2} - (\frac{\partial w}{\partial x} + B'(u_\varepsilon)\frac{\partial w}{\partial x}.$$

Then $H[w_\varepsilon] = 0$. Let

$$v_\varepsilon = K(x-1) - w_\varepsilon$$

where the constant K is to be determined. By calculation we see that, for sufficient large $K > 0$,

$$\begin{aligned} H[v_\varepsilon] &= -2(\frac{\partial w_\varepsilon}{\partial x} - \frac{K}{2})^2 - \frac{K^2}{2} - KB'(u_\varepsilon) - H[w_\varepsilon] \\ &\le -\frac{K^2}{2} - KB'(u_\varepsilon) < 0 \quad \text{in } Q_T. \end{aligned}$$

From this it follows that v_ε can not achieve its maximum at any point inside Q_T. In addition, since from (2.19) (2.20) and the uniform boundedness of $u'_{0\varepsilon}$ we have

$$v_\varepsilon(0,t) = -K < 0, \quad v_\varepsilon(1,t) = 0,$$

$$\left.\frac{\partial v_\varepsilon}{\partial x}\right|_{t=0} = K - w'_\varepsilon(x,0) > 0$$

for large K, we can assert that the maximum of v_ε must be zero and must be achieved at $x = 1$.Then

$$\left.\frac{\partial v_\varepsilon}{\partial x}\right|_{x=1} \geq 0$$

and hence

$$\left.\frac{\partial A_\varepsilon(u_\varepsilon)}{\partial x}\right|_{x=1} = (\alpha + u_\varepsilon)\left.\frac{\partial w_\varepsilon}{\partial x}\right|_{x=1} \leq (\alpha + u_\varepsilon)\Big|_{x=1} K \leq C.$$

Similarly, we can prove that

$$\left.\frac{\partial A_\varepsilon(u_\varepsilon)}{\partial x}\right|_{x=1} \geq -C.$$

Therefore the conclusion (2.22) for $x = 1$ is proved. Similarly, we can prove another part of the conclusion (2.22). □

Lemma 3.2.4 *Let u_ε be a solution of the problem (2.18)–(2.20). Then*

$$\left|\frac{\partial A_\varepsilon(u_\varepsilon)}{\partial x}\right| \leq C \tag{2.23}$$

with constant C independent of ε.

Proof. Let

$$\lambda_\varepsilon(s) = \frac{A'_\varepsilon(s)}{\theta(s)}, \quad w_\varepsilon = \int_0^{u_\varepsilon} \lambda_\varepsilon(s)ds,$$

where $\theta(s)$ is an auxiliary function to be determined, the first requirement is that it has positive upper bound and lower bound on $|s| \leq M$ (M is the constant in (2.21)). Then from (2.18) we see that w_ε satisfies

$$\frac{\partial w_\varepsilon}{\partial t} - A'_\varepsilon(u_\varepsilon)\frac{\partial^2 w_\varepsilon}{\partial x^2} - \left(\theta'(u_\varepsilon)\frac{\partial w_\varepsilon}{\partial x} + B'(u_\varepsilon)\right)\frac{\partial w_\varepsilon}{\partial x} = 0$$

and $v_\varepsilon = \dfrac{\partial w_\varepsilon}{\partial x}$ satisfies

$$\frac{\partial v_\varepsilon}{\partial t} - A'_\varepsilon(u_\varepsilon)\frac{\partial^2 v_\varepsilon}{\partial x^2} - \left(2\theta'(u_\varepsilon)v_\varepsilon + B'(u_\varepsilon) + A''_\varepsilon(u_\varepsilon)\frac{\partial u_\varepsilon}{\partial x}\right)\frac{\partial v_\varepsilon}{\partial x}$$
$$-\frac{\theta''(u_\varepsilon)}{\lambda_\varepsilon(u_\varepsilon)}v_\varepsilon^3 - \frac{B''(u_\varepsilon)}{\lambda_\varepsilon(u_\varepsilon)}v_\varepsilon^2 = 0.$$

Multiplying this equality by v_ε gives

$$\begin{aligned}\frac{1}{2}\frac{\partial v_\varepsilon^2}{\partial t} &- A'_\varepsilon(u_\varepsilon)v_\varepsilon\frac{\partial^2 v_\varepsilon}{\partial x^2}\\ &-\frac{1}{2}\left(2\theta'(u_\varepsilon)v_\varepsilon + B'(u_\varepsilon) + A''_\varepsilon(u_\varepsilon)\frac{\partial u_\varepsilon}{\partial x}\right)\frac{\partial v_\varepsilon^2}{\partial x}\\ &-\frac{\theta''(u_\varepsilon)}{\lambda_\varepsilon(u_\varepsilon)}v_\varepsilon^4 - \frac{B''(u_\varepsilon)}{\lambda_\varepsilon(u_\varepsilon)}v_\varepsilon^3 = 0.\end{aligned} \tag{2.24}$$

If v_ε^2 achieves its maximum at some point of the parabolic boundary, then, by Lemma 3.2.3, (2.23) holds clearly. Suppose that v_ε^2 achieves the maximum at some point (x_0, t_0) not on the parabolic boundary. Then at (x_0, t_0) the sum of the first three terms on the left side of (2.24) is nonnegative and hence

$$-\frac{\theta''(u_\varepsilon)}{\lambda_\varepsilon(u_\varepsilon)}v_\varepsilon^4 - \frac{B''(u_\varepsilon)}{\lambda_\varepsilon(u_\varepsilon)}v_\varepsilon^3 \le 0,$$

namely

$$-\theta''(u_\varepsilon)v_\varepsilon^2 - B''(u_\varepsilon)v_\varepsilon \le 0$$

from which it follows by using Young's inequality that for $\delta > 0$,

$$-\theta''(u_\varepsilon)v_\varepsilon^2 \le \delta v_\varepsilon^2 + \frac{(B''(u_\varepsilon))^2}{4\delta},$$

namely

$$(-\theta''(u_\varepsilon) - \delta)v_\varepsilon^2 \le \frac{(B''(u_\varepsilon))^2}{4\delta}.$$

If $\theta(s)$ is chosen such that $\theta''(s)$ has negative upper bound on $|s| \le M$, then we can choose $\delta > 0$ so small that

$$v_\varepsilon^2 \le C$$

with constant C independent of ε. The last inequality implies (2.23), if $\theta(s)$ is required to have positive upper bound and lower bound on $|s| \le M$. The choice of such functions $\theta(s)$ is quite free, for example, we can choose $\theta(s) = 1 + (M - s)(M + s)$. This complete the proof of our lemma. □

Lemma 3.2.5 *Let u_ε be a solution of the problem (2.18)-(2.20). Then for any $(x_1,t_1),(x_2,t_2)\in Q_T$,*

$$|A_\varepsilon(u_\varepsilon(x_1,t_1))-A_\varepsilon(u_\varepsilon(x_2,t_2))|\le C\left(|x_1-x_2|+|t_1-t_2|^{1/2}\right), \quad (2.25)$$

where the constant C is independent of ε.

Proof. Since Lemma 3.2.4 implies that

$$|A_\varepsilon(u_\varepsilon(x_1,t))-A_\varepsilon(u_\varepsilon(x_2,t))|\le C\,|x_1-x_2|, \quad \forall (x_1,t),(x_2,t)\in Q_T, \quad (2.26)$$

it remains to further prove

$$|A_\varepsilon(u_\varepsilon(x,t_1))-A_\varepsilon(u_\varepsilon(x,t_2))|\le C\,|t_1-t_2|^{1/2}, \quad \forall (x,t_1),(x,t_2)\in Q_T. \quad (2.27)$$

Suppose, for example, $\Delta t=t_2-t_1>0$. Given $\alpha\in(0,1)$ arbitrarily and denote $d=\Delta t^\alpha$. We may suppose that $d\le 1/2$; otherwise, (2.27) follows immediately from the uniform boundedness of $\{u_\varepsilon\}$.

In case $x+d\le 1$, we integrate (2.18) over $(x,x+d)\times(t_1,t_2)$. Integrating by parts gives

$$\int_x^{x+d}(u_\varepsilon(\xi,t_2)-u_\varepsilon(\xi,t_1))\,d\xi=\int_{t_1}^{t_2}\frac{\partial A_\varepsilon(u_\varepsilon)}{\partial x}\Big|_x^{x+d}dt+\int_{t_1}^{t_2}B(u_\varepsilon)\Big|_x^{x+d}dt.$$

Using the mean value theorem for integrals, we see that

$$\int_x^{x+d}(u_\varepsilon(\xi,t_2)-u_\varepsilon(\xi,t_1))\,d\xi=d\,(u_\varepsilon(x^*,t_2)-u_\varepsilon(x^*,t_1))$$

for some $x^*\in[x,x+d]$. Combining this with the above equality and using (2.21) and Lemma 3.2.4, we obtain

$$|u_\varepsilon(x^*,t_2)-u_\varepsilon(x^*,t_1)|\le C\Delta t^{1-\alpha}.$$

This, together with (2.26) gives

$$\begin{aligned}
&|A_\varepsilon(u_\varepsilon(x,t_2))-A_\varepsilon(u_\varepsilon(x,t_1))|\\
\le\;& |A_\varepsilon(u_\varepsilon(x,t_2))-A_\varepsilon(u_\varepsilon(x^*,t_2))|\\
&+|A_\varepsilon(u_\varepsilon(x^*,t_2))-A_\varepsilon(u_\varepsilon(x^*,t_1))|\\
&+|A_\varepsilon(u_\varepsilon(x^*,t_1))-A_\varepsilon(u_\varepsilon(x,t_1))|\\
\le\;& C\Delta t^\alpha+C\Delta t^{1-\alpha}+C\Delta t^\alpha=C(2\Delta t^\alpha+\Delta t^{1-\alpha})
\end{aligned}$$

which implies (2.27), if we take $\alpha = \dfrac{1}{2}$.

In case $x + d > 1$, since $d \le \dfrac{1}{2}$, we have $x > 1 - d \ge \dfrac{1}{2}$ and can obtain the same conclusion by integrating (2.18) over $(x - d, x) \times (t_1, t_2)$. The proof is complete. □

Proof of Theorem 3.2.2. Denote $w_\varepsilon = A_\varepsilon(u_\varepsilon)$. Lemma 3.2.5 and (2.21) imply the uniform boundedness and equicontinuity of $\{w_\varepsilon\}$ on Q_T. Hence there exists a subsequence, still denoted by $\{w_\varepsilon\}$, and a function $w \in C^{1,1/2}(Q_T)$, such that

$$\lim_{\varepsilon\to 0} w_\varepsilon(x,t) = w(x,t), \quad \text{uniformly in } Q_T.$$

Let $\psi(s)$ be the inverse function of $A(s)$, whose existence for $s \in R$ follows from the assumption $\lim\limits_{s\to\pm\infty} A(s) = \pm\infty$ and the strict monotonicity of $A(s)$. Then

$$u(x,t) = \lim_{\varepsilon\to 0} u_\varepsilon = \lim_{\varepsilon\to 0} \psi(w_\varepsilon - \varepsilon u_\varepsilon)$$

exists and $u \in C(\overline{Q_T})$. To prove that u is a generalized solution of the problem (2.1)–(2.3), notice that from (2.18)–(2.21), for any $\varphi \in C^\infty(\overline{Q_T})$ with $\varphi(0,t) = \varphi(1,t) = \varphi(x,T) = 0$, one has

$$\iint_{Q_T} \left(u_\varepsilon \frac{\partial\varphi}{\partial t} + w_\varepsilon \frac{\partial^2\varphi}{\partial x^2} - B(u_\varepsilon)\frac{\partial\varphi}{\partial x} \right) dxdt + \int_0^1 u_{0\varepsilon}(x)\varphi(x,0)dx = 0$$

and hence

$$\iint_{Q_T} \left(u \frac{\partial\varphi}{\partial t} + w \frac{\partial^2\varphi}{\partial x^2} - B(u)\frac{\partial\varphi}{\partial x} \right) dxdt + \int_0^1 u_0(x)\varphi(x,0)dx = 0$$

by letting $\varepsilon \to 0$. Since $w = A(u)$, by definition, u is a generalized solution of (2.1)–(2.3). The proof of Theorem 3.2.2 is complete. □

3.2.3 *Hölder continuity of solutions*

Theorem 3.2.3 *If in addition to the conditions of Theorem 3.2.2, assume that*

$$|A(s_1) - A(s_2)| \ge \lambda |s_1 - s_2|^m \tag{2.28}$$

for some constants $m > 1$, $\lambda > 0$, then the generalized solution u of the problem (2.1)–(2.3) given in Theorem 3.2.2, is Hölder continuous, precisely, $u \in C^{1/m,1/(m+1)}(Q_T)$.

Proof. In the proof of Theorem 3.2.2, in fact, we have reached $A(u(x,t)) \in C^{1,1/2}(Q_T)$ which follows from (2.25) by letting $\varepsilon \to 0$. Thus, using the assumption (2.28), we obtain

$$\begin{aligned}
&|u(x_1,t_1) - u(x_2,t_2)| \\
\le\ & \lambda^{-1/m} \left|A(u(x_1,t_1)) - A(u(x_2,t_2))\right|^{1/m} \\
\le\ & C\left(|x_1 - x_2| + |t_1 - t_2|^{1/2}\right)^{1/m} \\
\le\ & C\left(|x_1 - x_2|^{1/m} + |t_1 - t_2|^{1/2m}\right) \quad \forall (x_1,t_1),(x_2,t_2) \in Q_T,
\end{aligned}$$

namely $u \in C^{1/m,1/2m}(Q_T)$. We further prove that $u \in C^{1/m,1/m+1}(Q_T)$.

First, using (2.28) and Lemma 3.2.5 gives

$$\begin{aligned}
&|u_\varepsilon(x_1,t) - u_\varepsilon(x_2,t)| \\
\le\ & \lambda^{-1/m} \left|A(u_\varepsilon(x_1,t)) - A(u_\varepsilon(x_2,t))\right|^{1/m} \\
\le\ & C\left|A_\varepsilon(u_\varepsilon(x_1,t)) - A_\varepsilon(u_\varepsilon(x_2,t))\right|^{1/m} \\
& + C\varepsilon^{1/m}\left|u_\varepsilon(x_1,t) - u_\varepsilon(x_2,t)\right|^{1/m} \\
\le\ & C\left|x_1 - x_2\right|^{1/m} + C\varepsilon^{1/m}.
\end{aligned} \tag{2.29}$$

Next, for any given $\alpha \in (0,1)$, by an argument similar to the proof of Lemma 3.2.5, we can assert that for any $x \in (0,1)$, there exists $x^* \in (0,1)$ with $|x - x^*| \le d = \Delta t^\alpha$, such that

$$|u_\varepsilon(x^*,t_1) - u_\varepsilon(x^*,t_2)| \le C\Delta t^{1-\alpha}.$$

Combining this with (2.29) gives

$$\begin{aligned}
&|u_\varepsilon(x,t_1) - u_\varepsilon(x,t_2)| \\
\le\ & |u_\varepsilon(x,t_1) - u_\varepsilon(x^*,t_1)| + |u_\varepsilon(x^*,t_1) - u_\varepsilon(x^*,t_2)| \\
& + |u_\varepsilon(x^*,t_2) - u_\varepsilon(x,t_2)| \\
\le\ & C\left(\Delta t^{\alpha/m} + \varepsilon^{\alpha/m} + \Delta t^{1-\alpha}\right).
\end{aligned} \tag{2.30}$$

Let $\varepsilon \to 0$ in (2.29), (2.30) and choosing $\alpha = \dfrac{m}{m+1}$ yield

$$|u(x_1,t) - u(x_2,t)| \le C\,|x_1 - x_2|^{1/m}, \quad \forall (x_1,t),(x_2,t) \in Q_T,$$

$$|u(x,t_1) - u(x,t_2)| \le C\,|t_1 - t_2|^{1/(m+1)}, \quad \forall (x,t_1),(x,t_2) \in Q_T$$

which imply $u \in C^{1/m,1/m+1}(Q_T)$. □

3.2.4 *Some extensions*

The results presented in the above theorems were obtained by Zhao [ZH1]. The equations he considered are more general in form, namely,

$$\frac{\partial u}{\partial t} = \frac{\partial}{\partial x}\left(a(x,t,u)\frac{\partial u}{\partial x}\right) + \frac{\partial}{\partial x} b(x,t,u) + c(x,t,u),$$

where $a(x,t,u) \ge 0$ and for any (x,t), $E = \{s; a(x,t,s) = 0\}$ does not contain any interior point. In addition, the boundary value is not necessarily required to be homogenious. Also the argument can be applied to the Cauchy problem.

In the introduction of this chapter, we have mentioned another kind of weakly degenerate parabolic equation

$$\frac{\partial u}{\partial t} = \frac{\partial}{\partial x} F\left(\frac{\partial A(u)}{\partial x}\right) + \frac{\partial B(u)}{\partial x} \tag{2.31}$$

with $F'(s) \ge 0$, $A'(s) \ge 0$. Since degeneracy happens either when $A'(u) = 0$ or when $F'\left(\dfrac{\partial A(u)}{\partial x}\right) = 0$,such kind of equations are called double degenerate. (2.31) is just (2.1) when $F(s) = s$. In particular, it turns out to be the Newtonian equation when $F(s) = s$, $A(s) = s^m (m > 1)$, $B(s) = 0$ and the non-Newtonian equation when $F(s) = |s|^{p-2}s$, $A(s) = s^m (m > 1)$, $B(s) = 0$. It was Kalashnikov[KA4] who first studied

$$\frac{\partial u}{\partial t} = \frac{\partial}{\partial x} F\left(\frac{\partial A(u)}{\partial x}\right), \tag{2.32}$$

a special case of (2.31). To treat such kind of equations, one has to overcome more technical difficulty, which comes from the strong nonlinearity in the equations. For example, if we study the existence by means of parabolic regularity, namely, to obtain a solution as the limit of solutions of the regularized equations

$$\frac{\partial u_\varepsilon}{\partial t} = \frac{\partial}{\partial x} F_\varepsilon \left(\frac{\partial A_\varepsilon(u_\varepsilon)}{\partial x} \right),$$

similar to the proof of Theorem 3.2.2, one can obtain uniform estimates for the maximum norm of u_ε and $w_\varepsilon = \frac{\partial A_\varepsilon(u_\varepsilon)}{\partial x}$ and the Hölder norm of $A_\varepsilon(u_\varepsilon)$. However it seems almost impossible to obtain the equi-continuity norm for $F_\varepsilon(w_\varepsilon)$. Kalashnikov accomplished the limit process for approximate solutions under the convexity condition on the nonlinear functions $A(s)$ and $F(s)$ and thus proved the existence. However the uniqueness of solutions is remained.

The convexity condition which Kalashnikov assumed implies that both $E_A = \{s; A'(s) = 0\}$ and $E_F = \{s; F'(s) = 0\}$ contain at most one point, namely, the equation considered has at most one point of degeneracy. Yin[YI1] removed the convexity condition and accomplished a limit process for approximate solutions to obtain a continuous solution by means of parabolic regularization, using the technique of BV estimates. What he considered is the boundary value problem for equations (2.31) with convection term. To prove the existence, the only condition needed is

$$\lim_{s \to \pm\infty} F(s) = \pm\infty \tag{2.33}$$

which is necessary in order to obtain continuous solutions. An investigation of Bertsch and Dal Passo[BP] shows that even for a special case of (2.32), namely, for the equation

$$u_t = F(u_x)_x,$$

the solution might be discontinuous if the condition (2.33) is removed. In addition, Yin [YI1] proved the uniqueness of continuous BV solutions of the first boundary value problem for (2.31) without any structure condition.

3.3 Weakly Degenerate Equations in Higher Dimension

In this section, we consider equations of the form

$$\frac{\partial u}{\partial t} = \Delta A(u) + \mathrm{div}\vec{B}(u), \tag{3.1}$$

where

$$A(u) = \int_0^u a(s)ds, \quad \vec{B}(u) = \int_0^u \vec{b}(s)ds$$

with $a(s) \geq 0$, $a(s)$ and $\vec{b}(s)$ being continuous and $E = \{s; a(s) = 0\}$ containing no interior point. Without loss of generality, we may assume that $A(0) = 0$. We will discuss the Cauchy problem for (3.1) with initial value condition

$$u(x,0) = u_0(x). \tag{3.2}$$

Denote $Q_T = \mathbb{R}^N \times (0,T)$.

As shown in §3.3.2, the existence and uniqueness for weakly degenerate equations in one dimension has been solved satisfactorily: uniqueness is proved in $L^\infty(Q_T)$, existence is established in $C(Q_T)$ and the solution obtained is Hölder continuous under certain structure condition.

However, the problem for such equations in higher dimension is more difficult. So far the existence of continuous solutions for such equations is remained as an open problem. The results on uniqueness are not as good as those for equations in one dimension. In this section we first prove the existence of continuous solutions for equations with two points of degeneracy [ZH9] and then establish the uniqueness of BV solutions of the Cauchy problem for general weakly degenerate equations [YI2].

Similar to Definition 1.1.2, we define the generalized solutions for (3.1) as follows:

Definition 3.3.1 A function $u \in L^1_{loc}(Q_T)$ is called a **generalized solution** of the Cauchy problem (3.1)–(3.2), if $A(u) \in L^1_{loc}(Q_T)$, $\vec{B}(u) \in L^1_{loc}(Q_T)$, and for any $\varphi \in C_0^\infty(Q_T)$,

$$\iint_{Q_T} u\frac{\partial \varphi}{\partial t} dxdt + \iint_{Q_T} A(u)\Delta\varphi dxdt - \iint_{Q_T} \vec{B}(u)\nabla\varphi dxdt = 0;$$

furthermore, for any $h \in C_0^\infty(\mathbb{R}^N)$,

$$\lim_{t\to 0+} \int_{\mathbb{R}^N} u(x,t)h(x)dx = \int_{\mathbb{R}^N} u_0(x)h(x)dx.$$

If in addition, $u \in L^\infty(Q_T) \cap BV(Q_T)$, with $\nabla A(u) \in L^2_{loc}(Q_T)$, then we will simply say that u is a BV **solution**.

Here and below, we denote by $BV(Q_T)$ the set of all functions of locally bounded variation on Q_T. In the appendix of this chapter, we will list all results on the class $BV(Q_T)$ and more general class $BV_x(Q_T)$ needed in this chapter without proofs.

3.3.1 *Existence of continuous solutions for equations with two points of degeneracy*

In this subsection, we consider equations of the form (3.1) with two points of degeneracy, namely, with $E = \{0, 1\}$, or $a(0) = a(1) = 0$ and $a(s) > 0$ for any $s \neq 0, 1$. Assume that $a(s)$, $\vec{b}(s)$ are appropriately smooth and satisfy the condition

$$a^{-1}(s)\left|\vec{b}(s)\right| \leq b_0, \qquad \forall s \in (0,1) \tag{3.3}$$

for some constant b_0.

Theorem 3.3.1 *Assume that $E = \{0, 1\}$, and (3.3) holds, $u_0 \in C^\alpha(\mathbb{R}^N)$ for some $\alpha \in (0, 1)$, $0 \leq u_0(x) \leq 1$ and for some $\eta \in (0, 1)$, $m \geq 1$,*

$$1 \leq \frac{a(s)s}{A(s)} \leq m, \qquad \text{if} \quad 0 < s \leq \eta, \tag{3.4}$$

$$1 \leq \frac{a(s)(1-s)}{A(1) - A(s)} \leq m, \qquad \text{if} \quad 1 - \eta \leq s \leq 1. \tag{3.5}$$

Then the Cauchy problem (3.1)–(3.2) admits a locally Hölder continuous solution.

In order to prove Theorem 3.3.1, for sufficient small $\varepsilon > 0$, choose smooth approximate function $u_{0\varepsilon}$, such that

$$\varepsilon \leq u_{0\varepsilon}(x) \leq 1 - \varepsilon$$

and consider the regularized problem for (3.1) with initial data

$$u_\varepsilon(x, 0) = u_{0\varepsilon}(x). \tag{3.6}$$

From the classical theory, this problem admits a smooth solution u_ε. We need to establish a series of estimates for u_ε.

First we have

$$\varepsilon \leq u_\varepsilon \leq 1 - \varepsilon.$$

For simplicity, we will drop the subscript "ε" and simply denote u_ε by u temporarily.

Let $u = \Phi(w)$ be the inverse function of $w = A(u)$. As in §1.3.5, denote $B_\rho = B_\rho(x_0) = \{x \in \mathbb{R}^N; |x - x_0| < \rho\}$, $G_\rho = B_\rho \times (0,T)$, and

$$A_{k,\rho}(t) = \{x \in B_\rho; w(x,t) > k\},$$
$$B_{k,\rho}(t) = \{x \in B_\rho; w(x,t) < k\}$$

where $w(x,t) = A(u(x,t))$.

Applying the method in the proof of Proposition 3.5.1, we can obtain

Lemma 3.3.1 *For any cut-off function $\zeta(x)$, there holds*

$$\frac{\partial}{\partial t}\int_{A_{k,\rho}(t)} \zeta^2 \chi_k(w-k)dx + \frac{1}{2}\int_{A_{k,\rho}(t)} \zeta^2|\nabla w|^2 dx$$
$$\leq \gamma \int_{A_{k,\rho}(t)} |\nabla\zeta|^2(w-k)^2 dx, \tag{3.7}$$

$$\frac{\partial}{\partial t}\int_{B_{k,\rho}(t)} \zeta^2 \widetilde{\chi}_k(k-w)dx + \frac{1}{2}\int_{B_{k,\rho}(t)} \zeta^2|\nabla w|^2 dx$$
$$\leq \gamma \int_{B_{k,\rho}(t)} |\nabla\zeta|^2(w-k)^2 dx, \tag{3.8}$$

where γ depends only on b_0, the constant in (3.3) and

$$\chi_k(s) = \int_0^s \Phi'(k+\theta)\theta d\theta, \quad \widetilde{\chi}_k(s) = \int_0^s \Phi'(k-\theta)\theta d\theta.$$

Lemma 3.3.2 *Assume that $G_\rho \subset Q_T$. Then there exists a constant $\delta > 0$ depending only on b_0, N, T, ρ, such that*

$$h = A(1) - k > 0, k > 0, mes A_{k,\rho}(0) = 0, \quad \max_{s \in (\Phi(k),1)} a(s) < \delta$$

imply

$$mes A_{k+h/2,\rho/2}(t) = 0, \qquad \forall \quad 0 < t < T.$$

Proof. Denote

$$k_n = k + \frac{h}{2} - \frac{h}{2^{n+1}}, \quad \rho_n = \frac{\rho}{2} + \frac{\rho}{2^{n+1}},$$

and choose a cut-off function $\xi_n(x)$ on B_{ρ_n} such that $0 \le \xi_n \le 1$, $\xi_n = 1$ on $B_{\rho_{n+1}}$. Take $\zeta = \xi_n$, $k = k_n$, in (3.7) and notice that

$$\int_0^{(w-k_n)_+} \Phi'(k_n + s) s ds \ge \frac{1}{2\delta}(w - k_n)_+^2.$$

Then we can derive

$$\begin{aligned} &\frac{1}{\delta} \max_{0\le t\le T} \int_{A_{k_n,\rho_n}(t)} (w - k_n)^2 \xi_n^2 dx \\ &\quad + \frac{1}{4}\int_0^T \int_{A_{k_n,\rho_n}(t)} |\nabla((w - k_n)\xi_n)|^2 dxdt \\ \le\ & (\gamma + \frac{1}{2}) \int_0^T \int_{A_{k_n,\rho_n}(t)} (w - k_n)^2 |\nabla \xi_n|^2 dxdt. \end{aligned}$$

Using Sobolev's inequality (see Lemma 1.1.1), we obtain

$$\begin{aligned} &\int_0^T \int_{A_{k_{n+1},\rho_{n+1}}(t)} (w - k_n)^{2(N+2)/N} dxdt \\ \le & \int_0^T \int_{A_{k_n,\rho_n}(t)} (\xi_n(w - k_n))^{2(N+2)/N} dxdt \\ \le & C\delta^{2/N} \left(\frac{1}{\delta} \max_{0\le t\le T} \int_{A_{k_n,\rho_n}(t)} (w - k_n)^2 \xi_n^2 dx \right)^{2/N} \cdot \\ & \int_0^T \int_{A_{k_n,\rho_n}(t)} |\nabla((w - k_n)\xi_n)|^2 dxdt \\ \le & C\delta^{2/N} \Bigg(\frac{1}{\delta} \max_{0\le t\le T} \int_{A_{k_n,\rho_n}(t)} (w - k_n)^2 \xi_n^2 dx \\ & + \int_0^T \int_{A_{k_n,\rho_n}(t)} |\nabla((w - k_n)\xi_n)|^2 dxdt \Bigg)^{(N+2)/N} \\ \le & C \left(4^n \rho^{-2} \delta^{2/(N+2)} \int_0^T \int_{A_{k_n,\rho_n}(t)} (w - k_n)^2 dxdt \right)^{(N+2)/N}. \qquad (3.9) \end{aligned}$$

Denote

$$A_n = \{(x,t) \in G_{\rho_n}; w(x,t) > k_n\}.$$

From (3.9) we obtain

$$(k_{n+1}-k_n)^{2(N+2)/N}\text{mes}A_n \le C(\rho)\left(4^n\delta^{2/(N+2)}\text{mes}A_{kn}\right)^{(N+2)/N},$$

or

$$\text{mes}A_{n+1} \le C(\rho)\delta^{2/N}(\text{mes}A_n)^{(N+2)/N}(4^{2(N+2)/N})^n.$$

Using Lemma 1.5.4, from this it follows

$$\lim_{n\to\infty}\text{mes}A_n = 0$$

provided $\delta > 0$ is chosen so small that

$$\text{mes}A_0 \le C(\rho)^{-N/2}\delta^{-1}4^{2(N+2)/N}.$$

Lemma 3.3.2 is proved. □

Similarly, we can prove

Lemma 3.3.3 *Assume that $G_\rho \subset Q_T$. Then there exists a constant $\delta > 0$ depending only on b_0, N, T, ρ, such that*

$$k > 0, mesB_{k,\rho}(0) = 0, \quad \max_{s\in(0,\Phi(k))} a(s) < \delta$$

imply

$$mesB_{k/2,\rho/2}(t) = 0, \qquad \forall \quad 0 < t < T.$$

Proof of Theorem 3.3.1. First we prove that for any $x_0 \in \mathbb{R}^N$, there exists a constant $\rho > 0$, such that $\{u_\varepsilon\}$ is Hölder equi-continuous on $B_\rho(x_0)\times(0,T)$.

If $u(x_0) \le 1/2$, then there exists a constant $\rho > 0$, such that $u_{0\varepsilon}(x) \le 3/4$ for $x \in B_{2\rho}(x_0)$. Choose $k \ge A(3/4)$ such that $\max\limits_{s\in(\Phi(k),1)} a(s) < \delta$, where δ is the constant determined in Lemma 3.3.2.

Since

$$A_{k,2\rho}(0) = \text{mes}\{x \in B_{2\rho}; A(u_{0\varepsilon}) > k\} = 0,$$

by Lemma 3.3.2, we have $\text{mes}A_{k+h/2,\rho}(t) = 0$ for any $t \in (0,T)$, namely,

$$u_\varepsilon \le \Phi\left(\frac{A(1)+k}{2}\right).$$

This and the facts $u_\varepsilon \ge \varepsilon$ and $\Phi(\frac{A(1)+k}{2}) < 1$ which follows from $k < A(1)$ and the strict monotonicity of $A(s)$ show that, actually, for $(x,t) \in B_\rho \times (0,T)$, the equation (3.1) has $u = 0$ as its only possible point of degeneracy. Thus we can apply the method used in §1.1.5 (see [CH2]) to prove the Hölder equi-continuity of u_ε on $B_\rho \times (0,T)$.

If $u_0(x_0) > 1/2$, then we can find a constant $\rho > 0$, such that $u_{0\varepsilon}(x) > 1/4$ for $x \in B_{2\rho}(x_0)$. Now choose $0 < k < A(1/4)$ such that $\max\limits_{s\in(0,\Phi(k))} a(s) < \delta$, with δ determined in Lemma 3.3.2.

Since

$$B_{k,2\rho}(0) = \text{mes}\,\{x \in B_{2\rho}; A(u_{0\varepsilon}) < k\} = 0,$$

by Lemma 3.3.3, we have $\text{mes}A_{k/2,\rho}(t) = 0$ for any $t \in (0,T)$, namely,

$$\Phi\,(k/2) \le u_\varepsilon.$$

This and the facts $u_\varepsilon \le 1-\varepsilon$ and $\Phi(k/2) > 0$ which follows from $A(0) = 0$ and the strict monotonicity of $A(s)$, show that, for $(x,t) \in B_\rho \times (0,T)$, the equation (3.1) has $u = 1$ as its only possible point of degeneracy. Thus the method used in §1.1.5 can be applied to prove the Hölder equi-continuity of u_ε on $B_\rho \times (0,T)$.

Summing up, we may conclude that u_ε is uniformly bounded on Q_T and equi-continuous on any bounded domain of Q_T. Hence a uniformly convergent subsequence of $\{u_\varepsilon\}$ can be found, whose limit is locally Hölder continuous and is a generalized solution of the Cauchy problem (3.1), (3.2). The proof of Theorem 3.3.1 is thus completed. □

3.3.2 *Uniqueness of BV solutions*

Theorem 3.3.2 *Assume that $E = \{s; a(s) = 0\}$ has no interior point. Let u_1, and u_2 be BV solutions of (3.1) with initial value u_1^0, and u_2^0 respectively. Then for almost all $t \in (0,T)$,*

$$\int_{\mathbb{R}^N} |u_1(x,t) - u_2(x,t)|\,\omega_\lambda(x)dx \le e^{K_\lambda t}\int_{\mathbb{R}^N} |u_1^0(x) - u_2^0(x)|\,\omega_\lambda(x)dx,$$

where $\lambda > 0$, K_λ *is a constant depending only on* λ *and the bound of* u_1, *and* u_2, *and*

$$\omega_\lambda(x) = exp\left(-\lambda\sqrt{1+|x|^2}\right).$$

We first prove the following lemma which plays an important role in the proof of Theorem 3.3.2.

Lemma 3.3.4 *Assume that* $E = \{s; a(s) = 0\}$ *has no interior point. Let* u *be a BV solution of (3.1)–(3.2). Then* $H(\Gamma_u) = 0$, *where* $H(S)$ *denotes the Hausdorff measure of* S.

Here and below, for the notation Γ_u and many other notations, we refer to §3.3.7 of this chapter.

Proof. Denote

$$\Gamma_1 = \{(x,t) \in \Gamma_u; \gamma_1(x,t) = \cdots = \gamma_N(x,t) = 0\},$$

$$\Gamma_2 = \left\{(x,t) \in \Gamma_u; \gamma_1^2(x,t) + \cdots + \gamma_N^2(x,t) > 0\right\}.$$

Clearly, $\Gamma_u = \Gamma_1 \cup \Gamma_2$.

First prove $H(\Gamma_1) = 0$. Since any measurable subset of Γ_1 can be expressed as the union of a Borel set and a set of measure zero, it suffices to prove $H(S) = 0$ for any Borel subset $S \subset \Gamma_1$. We may suppose that $\overline{S}$ is compact. By Lemma 3.7.8 of the appendix of this chapter, for any bounded function $f(x,t)$ which is measurable with respect to the measure $\dfrac{\partial u}{\partial x_i}$, we have

$$\iint_S f(x,t)\frac{\partial u}{\partial x_i} = \int_0^T dt \int_{S^t} f(x,t)\frac{\partial u(\cdot,t)}{\partial x_i}, \tag{3.10}$$

where $S^t = \{x; (x,t) \in S\}$. Since by Lemma 3.7.4, for any Borel subset $S_1 \subset S$,

$$\frac{\partial u}{\partial x_i}(S_1) = \int_{S_1} (u^+(x,t) - u^-(x,t))\gamma_i dH,$$

$$\frac{\partial u(\cdot,t)}{\partial x_i}(S_1^t) = \int_{S_1^t} (u_+^t(x,t) - u_-^t(x,t))\gamma_i^t dH^t,$$

(3.10) is equivalent to

$$\iint_S f(x,t)(u^+(x,t) - u^-(x,t))\gamma_i dH$$
$$= \int_0^T dt \int_{S^t} f(x,t)(u_+^t(x,t) - u_-^t(x,t))\gamma_i^t dH^t.$$

The definition of Γ_1 implies that the left hand side vanishes, so we have

$$\int_0^T dt \int_{S^t} f(x,t)(u_+^t(x,t) - u_-^t(x,t))\gamma_i^t dH^t = 0.$$

Choose

$$f(x,t) = \chi_S(x,t)\text{sgn}(u_+^t(x,t) - u_-^t(x,t))\text{sgn}\gamma_i^t$$

in the above equality, where $\chi_S(x,t)$ denotes the characteristic function of S, and sum up for i from 1 up to N. Then we obtain

$$\int_G dt \int_{S^t} \left|u_+^t(x,t) - u_-^t(x,t)\right| \left(|\gamma_1^t| + \cdots + |\gamma_N^t|\right) dH^t = 0, \tag{3.11}$$

where G is the projection of S on the t-axis. (3.11) implies that for almost all $t \in G$,

$$\int_{S^t} \left|u_+^t(x,t) - u_-^t(x,t)\right| \left(|\gamma_1^t| + \cdots + |\gamma_N^t|\right) dH^t = 0,$$

and hence for almost all $t \in G$,

$$\gamma_1^t = \cdots = \gamma_N^t = 0$$

H^t-almost everywhere on S^t, which is impossible unless $\text{mes}G = 0$.

For any α, β with $0 < \alpha < \beta < T$, we can choose $\psi_j(t) \in C_0^\infty(0,T)$ such that

$$0 \le \psi_j(t) \le 1, \quad \lim_{j\to\infty} \psi_j(t) = \chi_{(\alpha,\beta]}(t) \qquad \forall \ \ t \in (0,T).$$

By Lemma 3.7.6, we can choose $\varphi_k(x,t) \in C_0^\infty(Q_T)$ such that

$$|\varphi_k(x,t)| \le 1, \quad \lim_{k\to\infty} \varphi_k = \chi_S \quad \text{in } L^1\left(Q_T, \left|\frac{\partial u}{\partial t}\right|\right).$$

Now from the definition of BV solutions, we have

$$\begin{aligned}&\iint_{Q_T}\varphi_k(x,t)\psi_j(t)\frac{\partial u}{\partial t}\\=&\iint_{Q_T}A(u)\psi_j(t)\Delta\varphi_k dxdt-\iint_{Q_T}\vec{B}(u)\psi_j(t)\nabla\varphi_k dxdt.\end{aligned}$$

Letting $j\to\infty$ leads to

$$\begin{aligned}&\iint_{Q_T}\varphi_k(x,t)\chi_{(\alpha,\beta]}(t)\frac{\partial u}{\partial t}\\=&\iint_{Q_T}A(u)\chi_{(\alpha,\beta]}(t)\Delta\varphi_k dxdt-\iint_{Q_T}\vec{B}(u)\chi_{(\alpha,\beta]}(t)\nabla\varphi_k dxdt.\end{aligned}$$

Clearly, this equality also holds if $(\alpha,\beta]$ is replaced by (α,β) and hence it holds even if $(\alpha,\beta]$ is replaced by any open set I with $\overline{I}\subset(0,T)$. Since G is a Borel set, by approximation, we may conclude that

$$\begin{aligned}&\iint_{Q_T}\varphi_k(x,t)\chi_G(t)\frac{\partial u}{\partial t}\\=&\iint_{Q_T}A(u)\chi_G(t)\Delta\varphi_k dxdt-\iint_{Q_T}\vec{B}(u)\chi_G(t)\nabla\varphi_k dxdt.\end{aligned}$$

Since $\text{mes}G=0$, the two terms on the right hand side vanish and

$$\iint_{Q_T}\varphi_k(x,t)\chi_G(t)\frac{\partial u}{\partial t}=0.$$

Letting $k\to\infty$ gives

$$\iint_S\frac{\partial u}{\partial t}=\iint_{Q_T}\chi_S(x,t)\chi_G(t)\frac{\partial u}{\partial t}=0.$$

Hence

$$\int_S(u^+(x,t)-u^-(x,t))\gamma_t dH=0 \tag{3.12}$$

which implies $H(S)=0$ and $H(\Gamma_1)=0$ by the arbitrariness of S.

Next we prove $H(\Gamma_2)=0$. Let S be any Borel subset of Γ_2, which is compact in Q_T. Since S is a set of $N+1$-dimensional measure zero and $\nabla A(u)\in L^2_{loc}(Q_T)$, we have

$$\iint_S\frac{\partial}{\partial x_i}A(u)dxdt=0,\quad i=1,\cdots,N,$$

and hence

$$\int_S (A(u^+(x,t)) - A(u^-(x,t)))\gamma_i dH = 0, \quad i = 1, \cdots, N.$$

From this it follows by the definition of Γ_2 and the strict monotonicity of $A(S)$ that $H(S) = 0$ and hence $H(\Gamma_2) = 0$ by the arbitrariness of S. Thus the lemma is proved. □

Proof of Theorem 3.3.2. By the definition of generalized solutions, we have

$$\iint_{Q_T} \{(\bar{u}_1 - \bar{u}_2)\frac{\partial \varphi}{\partial t} + (A(\bar{u}_1) - A(\bar{u}_2)) \triangle \varphi - (\vec{B}(\bar{u}_1) - \vec{B}(\bar{u}_2))\nabla\varphi\} dxdt = 0$$

or

$$\iint_{Q_T} (\overline{z}\frac{\partial \varphi}{\partial t} + \alpha\overline{z}\Delta\varphi - \vec{\beta}\overline{z}\nabla\varphi) dxdt = 0 \tag{3.13}$$

for any $\varphi \in C_0^\infty(Q_T)$, where

$$z = u_1 - u_2, \quad \alpha = \int_0^1 a(\lambda\overline{u}_1 + (1-\lambda)\overline{u}_2)d\lambda, \quad \vec{\beta} = \int_0^1 \vec{b}(\lambda\overline{u}_1 + (1-\lambda)\overline{u}_2)d\lambda.$$

Here for convenience of the following discussion we have replaced u_i by $\bar{u}_i$, the symmetric mean value of u_i. Note that doing so does not change the value of the related integrals. Since u_i $(i = 1, 2)$ are BV solutions, from the properties of BV functions we see that

$$\alpha\overline{z} \in BV(Q_T), \qquad \vec{\beta}\overline{z} \in BV(Q_T)$$

and (3.13) can be written as

$$\iint_{Q_T} (-\varphi\frac{\partial \overline{z}}{\partial t} - \nabla(\alpha\overline{z})\nabla\varphi + \varphi \text{div}(\vec{\beta}\overline{z})) dxdt = 0 \tag{3.14}$$

for any $\varphi \in C_0^\infty(Q_T)$.

The crucial step in proving Theorem 3.3.2 is to establish the following inequality

$$J(u_1, u_2, \varphi) \equiv \iint_{Q_T} \text{sgn}\overline{z}\left\{\overline{z}\frac{\partial \varphi}{\partial t} - \nabla(\alpha\overline{z})\nabla\varphi - \vec{\beta}\overline{z}\nabla\varphi\right\} dxdt \geq 0 \tag{3.15}$$

for any $0 \leq \varphi \in C_0^\infty(Q_T)$.

To this purpose, we define

$$H_\varepsilon(s) = \int_0^s h_\varepsilon(\rho)d\rho$$

for small $\varepsilon > 0$, where

$$h_\varepsilon(s) = \frac{2}{\varepsilon}\left(1 - \frac{|s|}{\varepsilon}\right)_+.$$

Obviously $h_\varepsilon \in C(R)$ and for all $s \in R$,

$$h_\varepsilon(s) \geq 0, \quad |sh_\varepsilon(s)| \leq 1, \quad |H_\varepsilon(s)| \leq 1,$$

$$\lim_{\varepsilon\to 0} H_\varepsilon(s) = \text{sgn}s, \quad \lim_{\varepsilon\to 0} sh_\varepsilon(s) = 0.$$

From the properties of BV solutions, one has

$$H_\varepsilon(\alpha\bar{z}) \in BV(Q_T), \quad \nabla H_\varepsilon(\alpha\bar{z}) \in L^2_{loc}(Q_T).$$

Instead of $J(u_1, u_2, \varphi)$, we first consider

$$J_\varepsilon(u_1, u_2, \varphi) \equiv \iint_{Q_T} H_\varepsilon(\alpha\bar{z})\left\{\bar{z}\frac{\partial\varphi}{\partial t} - \nabla(\alpha\bar{z})\nabla\varphi - \vec{\beta}\bar{z}\nabla\varphi\right\}dxdt.$$

By the strict monotonicity of $A(s)$, $\alpha\bar{z} = A(\bar{u}_1) - A(\bar{u}_2)$ and $\bar{z} = \bar{u}_1 - \bar{u}_2$ have the same sign, so $\text{sgn}\bar{z} = \text{sgn}(\alpha\bar{z})$ and hence

$$\lim_{\varepsilon\to 0} J_\varepsilon(u_1, u_2, \varphi) = J(u_1, u_2, \varphi).$$

Replace φ by $H_\varepsilon(\alpha\bar{z})\varphi$ in (3.14), (This is possible by approximation)

$$\iint_{Q_T}(-H_\varepsilon(\alpha\bar{z})\varphi\frac{\partial\bar{z}}{\partial t} - \nabla(\alpha\bar{z})\nabla(H_\varepsilon(\alpha\bar{z})\varphi) + H_\varepsilon(\alpha\bar{z})\varphi\text{div}(\vec{\beta}\bar{z}))dxdt = 0.$$

Using this we obtain

$$\begin{aligned}
&J_\varepsilon(u_1, u_2, \varphi)\\
= &\iint_{Q_T} H_\varepsilon(\alpha\bar{z})\frac{\partial}{\partial t}(\varphi z) - \iint_{Q_T} H_\varepsilon(\alpha\bar{z})\text{div}(\varphi\vec{\beta}\bar{z})\\
&\quad + \iint_{Q_T} \hat{h}_\varepsilon(\alpha\bar{z})\left|\nabla(\alpha\bar{z})\right|^2\varphi dxdt\\
\geq &\iint_{Q_T} H_\varepsilon(\alpha\bar{z})\frac{\partial}{\partial t}(\varphi z) - \iint_{Q_T} H_\varepsilon(\alpha\bar{z})\text{div}(\varphi\vec{\beta}\bar{z})
\end{aligned}$$

for $0 \le \varphi \in C_0^\infty(Q_T)$, where $\hat{h}_\varepsilon(w)$ is the functional superposition of $h_\varepsilon(s)$ and $w(x,t)$, namely,

$$\hat{h}_\varepsilon(w) = \int_0^1 h_\varepsilon(\lambda w^+ + (1-\lambda)w^-)d\lambda.$$

Hence

$$\begin{aligned} J(u_1,u_2,\varphi) &= \lim_{\varepsilon\to 0} J_\varepsilon(u_1,u_2,\varphi) \\ \ge \iint_{Q_T} &\operatorname{sgn}(\alpha\overline{z})\frac{\partial}{\partial t}(\varphi z) - \iint_{Q_T} \operatorname{sgn}(\alpha\overline{z})\operatorname{div}(\varphi\vec{\beta}\overline{z}) \\ = \iint_{Q_T} &\operatorname{sgn}(\overline{z})\frac{\partial}{\partial t}(\varphi z) - \iint_{Q_T} \operatorname{sgn}(\overline{z})\operatorname{div}(\varphi\vec{\beta}\overline{z}). \end{aligned} \tag{3.16}$$

By Lemma 3.3.4, we have

$$H(\Gamma_{u_1}) = H(\Gamma_{u_2}) = 0$$

which implies, in particular, up to a set of N-dimensional measure zero (Note that Q_T is an $N+1$-dimensional set),

$$H_\varepsilon(\overline{z}) = \overline{H_\varepsilon(\overline{z})}, \quad h_\varepsilon(\overline{z}) = \overline{h_\varepsilon(\overline{z})} = \hat{h}_\varepsilon(\overline{z}).$$

Also notice that since $z \in BV(Q_T)$, we must have

$$\left|\frac{\partial}{\partial t}(\varphi z)\right|(G) = 0, \quad \left|\operatorname{div}(\varphi\vec{\beta}\overline{z})\right|(G) = 0$$

for any set G of N-dimensional measure zero. Thus from (3.16) and the fact that $\lim_{\varepsilon\to 0} sh_\varepsilon(s) = 0$, we obtain

$$\begin{aligned} &J(u_1,u_2,\varphi) \\ \ge\ & \lim_{\varepsilon\to 0}\iint_{Q_T} H_\varepsilon(\overline{z})\frac{\partial}{\partial t}(\varphi z) - \lim_{\varepsilon\to 0}\iint_{Q_T} H_\varepsilon(\overline{z})\operatorname{div}(\varphi\vec{\beta}\overline{z}) \\ =\ & \lim_{\varepsilon\to 0}\iint_{Q_T} \overline{H_\varepsilon(\overline{z})}\frac{\partial}{\partial t}(\varphi z) - \lim_{\varepsilon\to 0}\iint_{Q_T} \overline{H_\varepsilon(\overline{z})}\operatorname{div}(\varphi\vec{\beta}\overline{z}) \\ =\ & -\lim_{\varepsilon\to 0}\iint_{Q_T} \varphi\overline{z}\frac{\partial}{\partial t}H_\varepsilon(\overline{z}) + \lim_{\varepsilon\to 0}\iint_{Q_T} \varphi\vec{\beta}\overline{z}\nabla H_\varepsilon(\overline{z}) \\ =\ & -\lim_{\varepsilon\to 0}\iint_{Q_T} \varphi\overline{z}\hat{h}_\varepsilon(\overline{z})\frac{\partial z}{\partial t} + \lim_{\varepsilon\to 0}\iint_{Q_T} \varphi\vec{\beta}\overline{z}\hat{h}_\varepsilon(\overline{z})\nabla z \\ =\ & 0 \end{aligned}$$

and (3.15) is proved.

Now we use (3.15) to complete the proof of our theorem. Note that

$$\begin{aligned}
&\iint_{Q_T} \operatorname{sgn}\overline{z}\nabla(\alpha\overline{z})\nabla\varphi dxdt \\
= &\lim_{\varepsilon\to 0}\iint_{Q_T} H_\varepsilon(\alpha\overline{z})\nabla(\alpha\overline{z})\nabla\varphi dxdt \\
= &\lim_{\varepsilon\to 0}\iint_{Q_T} \nabla\left[\alpha\overline{z}H_\varepsilon(\alpha\overline{z})\right]\nabla\varphi dxdt \\
&\quad -\lim_{\varepsilon\to 0}\iint_{Q_T} \alpha\overline{z}h_\varepsilon(\alpha\overline{z})\nabla(\alpha\overline{z})\nabla\varphi dxdt \\
= &-\lim_{\varepsilon\to 0}\iint_{Q_T} \alpha\overline{z}H_\varepsilon(\alpha\overline{z})\Delta\varphi dxdt \\
= &-\iint_{Q_T} \alpha\overline{z}\operatorname{sgn}(\alpha\overline{z})\Delta\varphi dxdt \\
= &-\iint_{Q_T} \alpha|\overline{z}|\Delta\varphi dxdt.
\end{aligned}$$

From this and (3.15) we see that for $0 \le \varphi \in C_0^\infty(Q_T)$,

$$\iint_{Q_T} |\overline{z}|\frac{\partial\varphi}{\partial t}dxdt \ge \iint_{Q_T} \vec{\beta}|\overline{z}|\nabla\varphi dxdt - \iint_{Q_T} \alpha|\overline{z}|\Delta\varphi dxdt. \tag{3.17}$$

Given τ, $s \in (0,T)$, $\tau < s$. Let

$$\psi_\varepsilon(t) = \int_{\tau-t}^{s-t} \alpha_\varepsilon(\sigma)d\sigma \quad \varepsilon < \min\{\tau, T-s\},$$

where $\alpha_\varepsilon(t)$ is the kernel of a mollifier with $\alpha_\varepsilon(t) = 0$ for $t \notin (-\varepsilon,\varepsilon)$. In particular, choose $\varphi_R \in C_0^\infty(\mathbb{R}^N)$ such that $\varphi_R(x) = 1$ for $|x| < R$, $\varphi_R(x) = 0$ for $|x| > R+1$ and

$$0 \le \varphi_R(x) \le 1, \quad |\nabla\varphi_R(x)| \le C, \quad |\Delta\varphi_R(x)| \le C, \qquad \textit{for} \quad x \in \mathbb{R}^N.$$

Take $\varphi = \psi_\varepsilon(t)\varphi_R(x)\omega_\lambda(x)$ in (3.17) and notice that

$$|\nabla\omega_\lambda(x)| \le C_\lambda\omega_\lambda(x), \quad |\Delta\omega_\lambda(x)| \le C_\lambda\omega_\lambda(x).$$

Then we obtain

$$\int_{\mathbb{R}^N} \varphi_R(x)\omega_\lambda(x)dx \int_0^T |\overline{z}(x,t)| \left[\alpha_\varepsilon(s-t) - \alpha_\varepsilon(\tau - t)\right] dt$$
$$\leq C_\lambda \int_0^{s+\varepsilon} dt \int_{\mathbb{R}^N} |\overline{z}(x,t)|\omega_\lambda(x)dx$$

which implies, by letting $\varepsilon \to 0$ and then $R \to \infty$, that for almost all τ, $s \in (0,T)$,

$$\int_{\mathbb{R}^N} |\overline{z}(x,s)|\omega_\lambda(x)dx$$
$$\leq C_\lambda \int_0^s dt \int_{\mathbb{R}^N} |\overline{z}(x,t)|\omega_\lambda(x)dx + \int_{\mathbb{R}^N} |\overline{z}(x,\tau)|\omega_\lambda(x)dx.$$

Since BV functions have trace on the superplane $t = 0$ and the trace of u_i must be $u_i^0(x)$, we further obtain

$$\int_{\mathbb{R}^N} |\overline{z}(x,s)|\omega_\lambda(x)dx$$
$$\leq C_\lambda \int_0^s dt \int_{\mathbb{R}^N} |\overline{z}(x,t)|\omega_\lambda(x)dx + \int_{\mathbb{R}^N} |\overline{z}^0(x,0)|\omega_\lambda(x)dx$$

by letting $\tau \to 0$, where $z^0(x) = u_1^0(x) - u_2^0(x)$. Finally we may use Gronwall's inequality to complete the proof of Theorem 3.3.2. □

3.3.3 *Existence of BV solutions*

Theorem 3.3.3 *Assume that $A(s)$, $\vec{B}(s)$ and u_0 are appropriately smooth. Then the problem (3.1),(3.2) admits a BV solution.*

Notice that here in Theorem 3.3.3, we do not restrict the equations to be weakly degenerate.

Proof. Consider the regularized problem

$$\frac{\partial u_\varepsilon}{\partial t} = \Delta A_\varepsilon(u_\varepsilon) + \mathrm{div}\vec{B}(u_\varepsilon), \tag{3.18}$$

$$u_\varepsilon(x,0) = u_{0\varepsilon}(x), \tag{3.19}$$

where $A_\varepsilon(s) = \varepsilon s + A(s)$ and $u_{0\varepsilon}$ is a smooth approximation of u_0. Let u_ε be the solution of this problem, which exists from the classical theory of parabolic equations. We need to establish some estimates for u_ε.

First we note that if $\sup|u_0(x)| \le M$, then we can require $u_{0\varepsilon}$ to satisfy $\sup|u_{0\varepsilon}(x)| \le M$. Hence by the maximum principle first we have

$$\sup_{Q_T}|u_\varepsilon(x,t)| \le M.$$

Next, we establish the L^1 estimate for $\dfrac{\partial u_\varepsilon}{\partial t}$ and ∇u_ε. Let v_ε be one of $\dfrac{\partial u_\varepsilon}{\partial t}$ and $\dfrac{\partial u_\varepsilon}{\partial x_i}$ $(i = 1, \cdots, N)$. Differentiating (3.18) gives

$$\frac{\partial v_\varepsilon}{\partial t} = \Delta\left(a_\varepsilon(u_\varepsilon)v_\varepsilon\right) + div\left(\vec{b}(u_\varepsilon)v_\varepsilon\right)$$

where $a_\varepsilon(s) = \varepsilon + a(s)$. Multiply this equality by $\varphi H_\eta(v_\varepsilon)$ with $0 \le \varphi \in C_0^\infty(\mathbb{R}^N)$ and $H_\eta(s)$ being the function introduced in the proof of Theorem 3.3.2 and integrate over $\mathbb{R}^N$.Then we obtain

$$\begin{aligned}&\frac{\partial}{\partial t}\int_{\mathbb{R}^N}\varphi\Theta_\eta(v_\varepsilon(x,t))dx\\ =&\int_{\mathbb{R}^N}\varphi H_\eta(v_\varepsilon)\left(\Delta\left(a_\varepsilon(u_\varepsilon)v_\varepsilon\right) + div\left(\vec{b}(u_\varepsilon)v_\varepsilon\right)\right)dx,\end{aligned} \tag{3.20}$$

where

$$\Theta_\eta(s) = \int_0^s H_\eta(\sigma)d\sigma.$$

Integrating by parts gives

$$\begin{aligned}&\int_{\mathbb{R}^N}\varphi H_\eta(v_\varepsilon)\Delta\left(a_\varepsilon(u_\varepsilon)v_\varepsilon\right)dx\\ =&-\int_{\mathbb{R}^N}\mathrm{div}\left(\varphi H_\eta(v_\varepsilon)\right)\nabla\left(a_\varepsilon(u_\varepsilon)v_\varepsilon\right)dx\\ =&-\int_{\mathbb{R}^N}\varphi h_\eta(v_\varepsilon)\nabla v_\varepsilon\nabla\left(a_\varepsilon(u_\varepsilon)v_\varepsilon\right)dx\\ &+\int_{\mathbb{R}^N}a_\varepsilon(u_\varepsilon)v_\varepsilon\mathrm{div}\left(\nabla\varphi H_\eta(v_\varepsilon)\right)dx\\ =&-\int_{\mathbb{R}^N}\varphi h_\eta(v_\varepsilon)|\nabla v_\varepsilon|^2 a_\varepsilon(u_\varepsilon)dx\\ &-\int_{\mathbb{R}^N}\varphi v_\varepsilon h_\eta(v_\varepsilon)\nabla v_\varepsilon\nabla u_\varepsilon a_\varepsilon'(u_\varepsilon)dx\\ &+\int_{\mathbb{R}^N}a_\varepsilon(u_\varepsilon)v_\varepsilon h_\eta(v_\varepsilon)\nabla v_\varepsilon\nabla\varphi dx\end{aligned}$$

$$+\int_{\mathbb{R}^N} a_\varepsilon(u_\varepsilon)v_\varepsilon H_\eta(v_\varepsilon)\Delta\varphi dx$$

and

$$\begin{aligned}&\int_{\mathbb{R}^N} \varphi H_\eta(v_\varepsilon)\text{div}\left(\vec{b}(u_\varepsilon)v_\varepsilon\right)dx\\ =&\ -\int_{\mathbb{R}^N} \varphi v_\varepsilon h_\eta(v_\varepsilon)\vec{b}(u_\varepsilon)\nabla v_\varepsilon dx - \int_{\mathbb{R}^N} v_\varepsilon H_\eta(v_\varepsilon)\vec{b}(u_\varepsilon)\nabla\varphi dx.\end{aligned}$$

Substituting into (3.20), giving up a nonpositive term and then letting $\eta \to 0$ and noticing that $\lim_{\eta\to 0} sh_\eta(s) = 0$, we obtain

$$\frac{\partial}{\partial t}\int_{\mathbb{R}^N} \varphi|v_\varepsilon(x,t)|dx \le \int_{\mathbb{R}^N} a_\varepsilon(u_\varepsilon)|v_\varepsilon|\Delta\varphi dx - \int_{\mathbb{R}^N} |v_\varepsilon|\vec{b}(u_\varepsilon)\nabla\varphi dx$$

or

$$\begin{aligned}&\int_{\mathbb{R}^N} \varphi|v_\varepsilon(x,t)|dx - \int_{\mathbb{R}^N} \varphi|v_\varepsilon(x,0)|dx.\\ \le&\ \iint_{Q_t} a_\varepsilon(u_\varepsilon)|v_\varepsilon|\Delta\varphi dxds - \iint_{Q_t} |v_\varepsilon|\vec{b}(u_\varepsilon)\nabla\varphi dxds.\end{aligned}$$

By approximation, we can replace φ in the above inequality by

$$\varphi(x) = \omega_\lambda(x) \equiv \exp\left(-\lambda\sqrt{1+|x|^2}\right).$$

Using the estimate for the maximum norm of u_ε and

$$|\nabla\omega_\lambda(x)| \le C_\lambda\omega_\lambda(x), \quad |\Delta\omega_\lambda(x)| \le C_\lambda\omega_\lambda(x),$$

we further obtain

$$\begin{aligned}&\int_{\mathbb{R}^N} |v_\varepsilon(x,t)|\omega_\lambda(x)dx\\ \le&\ \int_{\mathbb{R}^N} |v_\varepsilon(x,0)|\omega_\lambda(x)dx + C_\lambda\int_0^t\int_{\mathbb{R}^N} |v_\varepsilon(x,s)|\omega_\lambda(x)dxds\end{aligned}$$

and then, using Gronwall's inequality gives

$$\sup_{0<t<T}\int_{\mathbb{R}^N} |v_\varepsilon(x,t)|\,\omega_\lambda(x) \le C_\lambda. \tag{3.21}$$

Here we note that the uniform boundedness of the initial value $v_\varepsilon(x,0)$ follows from the initial value condition (3.19) and the equation (3.18). (3.21) implies that

$$\sup_{0<t<T} \int_{\mathbb{R}^N} \left| \frac{\partial u_\varepsilon}{\partial t} \right| \omega_\lambda(x) dx \le C,$$

$$\sup_{0<t<T} \int_{\mathbb{R}^N} |\nabla u_\varepsilon| \, \omega_\lambda(x) dx \le C. \tag{3.22}$$

Finally, we estimate the L^2 norm of $\nabla A_\varepsilon(u_\varepsilon)$. To this purpose we multiply (3.18) by $A_\varepsilon(u_\varepsilon)\omega_\lambda(x)$, integrate over Q_t and then obtain

$$\begin{aligned} \iint_{Q_t} \frac{\partial u_\varepsilon}{\partial t} A_\varepsilon(u_\varepsilon)\omega_\lambda(x) dx ds &= \iint_{Q_t} \Delta A_\varepsilon(u_\varepsilon) A_\varepsilon(u_\varepsilon)\omega_\lambda(x) dx ds \\ &+ \iint_{Q_t} \operatorname{div}\vec{B}(u_\varepsilon) A(u_\varepsilon)\omega_\lambda(x) dx ds. \end{aligned} \tag{3.23}$$

Let

$$\Psi_\varepsilon(s) = \int_0^s A_\varepsilon(\sigma) d\sigma.$$

Then by integrating by parts, (3.23) can be written as

$$\begin{aligned} &\int_{\mathbb{R}^N} \Psi_\varepsilon(u_\varepsilon(x,t))\omega_\lambda(x) dx - \int_{\mathbb{R}^N} \Psi_\varepsilon(u_{0\varepsilon})\omega_\lambda(x) dx \\ = & -\iint_{Q_t} |\nabla A_\varepsilon(u_\varepsilon)|^2 \omega_\lambda(x) dx ds + \frac{1}{2} \iint_{Q_t} |A_\varepsilon(u_\varepsilon)|^2 \Delta\omega_\lambda(x) dx ds \\ & - \iint_{Q_t} \vec{B}(u_\varepsilon) \nabla A_\varepsilon(u_\varepsilon)\omega_\lambda(x) dx ds - \iint_{Q_t} \vec{B}(u_\varepsilon) A_\varepsilon(u_\varepsilon) \nabla\omega_\lambda(x) dx ds. \end{aligned}$$

Using Young's inequality to the third term on the right hand side and noticing the uniform boundedness of u_ε, we then obtain

$$\begin{aligned} &\int_{\mathbb{R}^N} \Psi_\varepsilon(u_\varepsilon(x,t))\omega_\lambda(x) dx + \iint_{Q_t} |\nabla A_\varepsilon(u_\varepsilon)|^2 \omega_\lambda(x) dx ds \\ \le & \frac{1}{2} \iint_{Q_t} |\nabla A_\varepsilon(u_\varepsilon)|^2 \omega_\lambda(x) dx ds + C \end{aligned}$$

which implies, in particular,

$$\iint_{Q_T} |\nabla A_\varepsilon(u_\varepsilon)|^2 \omega_\lambda(x) dx dt \le C. \tag{3.24}$$

By virtue of (3.22), (3.24) and the uniform boundedness of u_ε, we can find a subsequence of $\{u_\varepsilon\}$, supposed to be $\{u_\varepsilon\}$itself, such that

$$\lim_{\varepsilon\to 0} u_\varepsilon = u, \quad \text{in} \quad L^1_{loc}(Q_T);$$

the limit function $u \in BV(Q_T) \cap L^\infty(Q_T)$ and

$$\iint_{Q_T} |\nabla A(u)|^2 \omega_\lambda(x) dx dt < +\infty.$$

Furthermore, for any $\varphi \in C_0^\infty(Q_T)$, from (3.18), (3.19) we have

$$\iint_{Q_T} u_\varepsilon \frac{\partial \varphi}{\partial t} dx dt + \iint_{Q_T} A_\varepsilon(u_\varepsilon)\Delta\varphi dx dt - \iint_{Q_T} \vec{B}(u_\varepsilon)\nabla\varphi dx dt = 0$$

and hence

$$\iint_{Q_T} u \frac{\partial \varphi}{\partial t} dx dt + \iint_{Q_T} A(u)\Delta\varphi dx dt - \iint_{Q_T} \vec{B}(u)\nabla\varphi dx dt = 0.$$

This means that u is a generalized solution of (3.1). It is easy to see that u satisfies (3.2) in the sense of Definition 3.3.1. Theorem 3.3.3 is thus proved. $\square$

3.3.4 *Some extensions*

The argument developed in Theorem 3.3.2 and Theorem 3.3.3 is applicable to equations more general in form. We can also treat the first boundary value problem for (3.1) in a similar way. For weakly degenerate equations the first boundary value problem can be treated without any essential difficulty. However, things are quite different for strongly degenerate equations. In this case, one has to overcome some difficulties in dealing with the boundary value condition. We will refer to this point in §3.3.6 of this chapter. Readers who are interested, may consult [WZ1],[WZ2]. Uniqueness theorem can also be proved for solutions in a more general class of functions, $BV_x(Q_T)$, by means of the method used in the proof of Theorem 3.2.1 (see[ZH10]). In [OK] the theory of semigroups is applied to the existence of BV solutions.

Moreover the argument on the existence and uniqueness can be extended to equations with double degeneracy,

$$\frac{\partial u}{\partial t} = \mathrm{div}\vec{F}(\nabla A(u)) + \mathrm{div}\vec{B}(u)$$

which are the extension of non-Newtonian filtration equations. Some special cases, especially equations with a single point of degeneracy, have been studied by many authors For general weakly degenerate equations, Yin has studied

$$\frac{\partial u}{\partial t} = \operatorname{div}\left(a(u)|\nabla u|^{p-2}\nabla u\right) + \vec{b}(u)\nabla u$$

in [YI7] and proved the existence of continuous solutions for the first boundary value problem provided $p > N$. In [YI7], uniqueness is established for continuous solutions with some regularity.

3.4 Strongly Degenerate Equations in One Dimension

From this section on we study equations with strong degeneracy. We begin with one dimensional case in this section. For simplicity, we consider equations of the form

$$\frac{\partial u}{\partial t} = \frac{\partial^2 A(u)}{\partial x^2} + \frac{\partial B(u)}{\partial x} \tag{4.1}$$

where $A(s)$ and $B(s)$ are appropriately smooth with $A(0) = B(0) = 0$ and $A'(s) \geq 0$. Strong degeneracy means that $E = \{s; A'(s) = 0\}$ may have interior points.

Denote $Q_T = I \times (0, T)$ with $I = \mathbb{R} = (-\infty, +\infty)$ for the Cauchy problem and $I = (0, 1)$ for the boundary value problem.

3.4.1 *Definitions of solutions with discontinuity*

The remarkable situation in treating strongly degenerate equations is that the solutions of such equations might be discontinuous. This can be exposed in the following consideration. Suppose $E \supset [a, b] (a < b)$. Then for $u \in [a, b]$, (4.1) becomes the first order quasilinear equation

$$\frac{\partial u}{\partial t} = \frac{\partial B(u)}{\partial x} \tag{4.2}$$

whose solutions, as is well-known, might have discontinuity, even if the initial value is smooth enough.

The first problem is how to define solutions with discontinuity for (4.1). Motivated by the theory of shock waves, a meaningful discontinuous solution u of (4.2) should satisfy the so-called entropy condition

$$(\overline{u}-k)\gamma_t \leq (\overline{B(u)}-B(k))\gamma_x, \qquad \forall \quad k \in \mathbb{R} \tag{4.3}$$

at the points of discontinuity of u in addition to the integral equality

$$\iint_{Q_T} \left(u\frac{\partial \varphi}{\partial t} - B(u)\frac{\partial \varphi}{\partial x} \right) dxdt = 0, \qquad \forall \quad \varphi \in C_0^\infty(Q_T). \tag{4.4}$$

Here and below, $\overline{u} = \frac{1}{2}(u^+ + u^-)$ denotes the symmetric mean value and $u^\pm$ the approximate limits of u at the points of discontinuity. It is not difficult to see that (4.3) and (4.4) imply

$$\begin{aligned} &\iint_{Q_T} \operatorname{sgn}(u-k) \left\{ (u-k)\frac{\partial \varphi}{\partial t} - (B(u)-B(k))\frac{\partial \varphi}{\partial x} \right\} dxdt \geq 0 \\ &\qquad \forall \quad 0 \leq \varphi \in C_0^\infty(Q_T), \quad k \in \mathbb{R}. \end{aligned} \tag{4.5}$$

In fact, at least for piecewise continuous functions (4.3),(4.4) are equivalent to (4.5). However the integral in (4.5) makes sense for any $u \in L^1_{loc}(Q_T)$, so we can use (4.5) to define more general solutions. It was Kruzhkov who first defined generalized solutions of (4.2) in this way and proved the solvability of the Cauchy problem in $L^\infty(Q_T)$.

Inspired by Kruzhkov's idea, Vol'pert and Hudjaev [VH1] defined generalized solutions for (4.1) as follows.

Definition 3.4.1 A function $u \in BV(Q_T) \cap L^\infty(Q_T)$ is called a generalized solution of the equation (4.1), if $\frac{\partial A(u)}{\partial x} \in L^1_{loc}(Q_T)$ and

$$\begin{aligned} &\iint_{Q_T} \operatorname{sgn}(u-k) \left\{ (u-k)\frac{\partial \varphi}{\partial t} - \frac{\partial A(u)}{\partial x}\frac{\partial \varphi}{\partial x} \right\} dxdt \\ &\quad - \iint_{Q_T} \operatorname{sgn}(u-k) \left\{ (B(u)-B(k))\frac{\partial \varphi}{\partial x} \right\} dxdt \geq 0, \\ &\quad \forall \quad 0 \leq \varphi \in C_0^\infty(Q_T), k \in \mathbb{R}. \end{aligned} \tag{4.6}$$

The existence and uniqueness of generalized solutions thus defined is discussed in [VH1]. However, as pointed out by Wu in [WZQ1], the proof of uniqueness given there is incorrect due to the adoption of the wrong form of a discontinuity condition for the solutions considered, which plays an essential role in the proof. Wu and Yin [WY1] revised this condition. On the basis of the correct form of the discontinuity condition and a deep study of the properties of BV functions and BV_x functions, they finally completed the proof of uniqueness.

Afterwards, related problems are studied in [BG], [BK1], [BK2], [BAD], [BU], [BW1], [BW2], [BWC], [CB], [EK1], [EK2], [EK3], [EK4], [EK5], [GK], [JA], [KBDE]. Theory and numerical analysis are applied to those equations modelling gravtational solid-liquid separation processes, such as sedimentation-consolidation processes.

3.4.2 *Interior discontinuity condition*

Theorem 3.4.1 *Let u be a BV solution of (4.1). Then H-almost everywhere on Γ_u^*,*

$$(u^+ - u^-)\gamma_t - (B(u) - B(k))\gamma_x - (w^r - w^l)|\gamma_x| = 0, \tag{4.7}$$

$$A'(s) = 0, \quad \forall s \in [u_*, u^*], \tag{4.8}$$

where

$$w = \frac{\partial A(u)}{\partial x}, \quad u_* = \min\{u^+, u^-\}, \quad u^* = \max\{u^+, u^-\}.$$

In this section, by a BV solution we always mean a generalized solution in the sense of Definition 3.4.1.

Proof. Taking $k > |u|_{L^\infty}$ and $k < -|u|_{L^\infty}$ respectively, we are led from the inequality (4.6) to the measure equality

$$\frac{\partial u}{\partial t} = \frac{\partial w}{\partial x} + \frac{\partial B(u)}{\partial x} \tag{4.9}$$

which implies, in particular, that $\dfrac{\partial w}{\partial x}$ is a Randon measure on Q_T. The discontinuity conditions (4.7), (4.8) will be proved based on the measure equality (4.9).

Without loss of generality, we may assume that $I = (0,1)$. Denote

$$E_N = \left\{ t \in (0,T);\ \int_I |w(x,t)|\, dx + \int_I \left| \frac{\partial w(\cdot,t)}{\partial x} \right| \le N \right\}.$$

From the properties of functions of bounded variation of one variable it follows that

$$\sup_{x\in I} |w(x,t)| \le \int_I |w(x,t)|\, dx + \int_I \left| \frac{\partial w(\cdot,t)}{\partial x} \right|$$

and there exists a set $F \subset (0,T)$ with mes$F = 0$, such that for $t \in (0,T)\backslash F$,

$$\lim_{N\to\infty} \chi_{E_N}(t) = 1.$$

By Lemma 3.7.7, we have $H[(I \times F) \cap \Gamma_u^*] = 0$. Set

$$D_N = I \times E_N.$$

We are ready to show that (4.7) holds H-almost everywhere on $D_N \cap \Gamma_u^*$ for any N. Once this is done, since the set $D_N \cap \Gamma_u^*$ is increasing with N and tends to the set $(I \times ((0,T) \setminus F)) \cap \Gamma_u^*$ as $N \to \infty$, using the fact $H((I \times F) \cap \Gamma_u^*) = 0$, we conclude that (4.7) holds H-almost everywhere on Γ_u^*.

Let S be an arbitrary bounded and measurable subset of $D_N \cap \Gamma_u^*$. Integrating (4.9) on S and using Lemma 3.7.4 yield

$$\int_S (u^+ - u^-)\gamma_t dH = \iint_S \frac{\partial w}{\partial x} + \int_S (B(u^+) - B(u^-))\gamma_x dH.$$

Since by Lemma 3.7.8 and Corollary 3.7.3, we have

$$\begin{aligned}\iint_S \frac{\partial w}{\partial x} &= \int_0^T dt \int_{S^t} \frac{\partial w(\cdot,t)}{\partial x}\\ &= \int_0^T dt \sum_{x\in S^t} (w^r(x,t) - w^l(x,t)) = \int_S (w^r(x,t) - w^l(x,t))\, |\gamma_x| dH,\end{aligned}$$

from the arbitrariness of S, we obtain the desired conclusion.

Now we prove (4.8). Since $w = \dfrac{\partial A(u)}{\partial x} \in L^1_{loc}(Q_T)$, for any bounded and measurable subset S of Γ_u^*, we have

$$\iint_S w(x,t)dxdt = \iint_S \frac{\partial A(u)}{\partial x} = 0.$$

On the other hand, using Lemma 3.7.4 gives

$$\iint_S \frac{\partial A(u)}{\partial x} = \int_S \int_{u^-}^{u^+} A'(s)ds\gamma_x dH.$$

Thus, from the arbitrariness of S, there holds H-almost everywhere on Γ_u^*,

$$\gamma_x \int_{u^-}^{u^+} A'(s)ds = 0$$

which proves (4.8), since (4.7) implies, in particular, that $\gamma_x \neq 0$ H-almost everywhere on Γ_u^*. The theorem is proved. □

Theorem 3.4.2 *Let u be a BV solution of (4.1). Then there exists a subset $G \subset \Gamma_u$ with $H(G) = 0$ such that for any $(x,t) \in \Gamma_u \setminus G$ and $k \in \mathbb{R}$,*

$$\left(sgn(u^+ - k) - sgn(u^- - k)\right)\left((\overline{u} - k)\gamma_t - (\overline{B(u)} - B(k))\gamma_x - \widetilde{w}\gamma_x\right) \leq 0. \tag{4.10}$$

Proof. Denote $z = u - k$. Notice that we may replace sgn$z$ by

$$\sigma = \frac{1}{2}\left(\text{sgn}z^+ + \text{sgn}z^-\right).$$

Using the measure equality (4.9) we can derive from the inequality (4.6),

$$J(u,k,\varphi) \equiv \iint_{Q_T} \sigma\left(\frac{\partial \varphi z}{\partial t} - \frac{\partial \varphi \beta z}{\partial x}\right) - \iint_{Q_T} \sigma \frac{\partial}{\partial x}(\varphi w) \geq 0,$$

where

$$\beta = \int_0^1 B'(\lambda u + (1-\lambda)k)d\lambda.$$

By Lemma 3.7.5 we have

$$\begin{aligned} &\iint_{Q_T} \sigma\left(\frac{\partial \varphi z}{\partial t} - \frac{\partial \varphi \beta z}{\partial x}\right) \\ = \ &-\int_{\Gamma_u} \varphi\left(\text{sgn}z^+ - \text{sgn}z^-\right)\left(\overline{z}\gamma_t - (\overline{B(u)} - B(k))\gamma_x\right)dH \\ \equiv \ &-\int_{\Gamma_u} \varphi h(k,x,t)dH. \end{aligned}$$

Thus

$$\int_{\Gamma_u} \varphi h(k,x,t)dH + \iint_{Q_T} \sigma \frac{\partial}{\partial x}(\varphi w) \leq 0. \tag{4.11}$$

For fixed N, choose a sequence $\{g_j(t)\} \subset C_0^\infty(0,T)$ such that

$$0 \leq g_j(t) \leq 1, \quad \lim_{j\to\infty} g_j(t) = \psi_N(t) = \chi_{E_N}(t)$$

for almost all $t \in (0,T)$. Replacing φ by φg_j in (4.11) and letting $j \to \infty$, we obtain, by Lemma 3.7.7 and the dominated convergence theorem,

$$\int_{\Gamma_u} \psi_N \varphi h(k,x,t)dH + \iint_{Q_T} \psi_N \sigma \frac{\partial}{\partial x}(\varphi w) \leq 0.$$

Multiplying this inequality by $f(k)$ with $0 \le f(\lambda) \in C_0^\infty(\mathbb{R})$ and integrating the resulting inequality with respect to k over R yield

$$\begin{aligned} &\int_{\Gamma_u} \psi_N \varphi \left(\int_{\mathbb{R}} f(k) h(k,x,t) dk \right) dH \\ &\quad + \iint_{Q_T} \psi_N \left(\int_{\mathbb{R}} f(k) \sigma dk \right) \frac{\partial}{\partial x} (\varphi w) \le 0. \end{aligned} \tag{4.12}$$

Noticing that

$$\int_{\mathbb{R}} f(k)\sigma dk = 2\overline{F(u)} - F(\infty), \quad F(s) = \int_{-\infty}^{s} f(\lambda) d\lambda)$$
$$\int_I \frac{\partial}{\partial x} (\varphi w(\cdot,t)) = 0$$

and using Lemma 3.7.3 and Lemma 3.7.8, we obtain

$$\begin{aligned} &\iint_{Q_T} \psi_N \left(\int_{\mathbb{R}} f(k)\sigma dk \right) \frac{\partial}{\partial x} (\varphi w) \\ = \;& 2 \iint_{Q_T} \psi_N \overline{F(u)} \frac{\partial}{\partial x} (\varphi w) \\ = \;& 2 \int_0^T \psi_N(t) dt \int_I \overline{F(u(x,t))} \frac{\partial}{\partial x} (\varphi w(\cdot,t)) \\ = \;& 2 \int_0^T \psi_N(t) dt \int_I \widetilde{F(u(x,t))} \frac{\partial}{\partial x} (\varphi w(\cdot,t)). \end{aligned}$$

Integrating by parts and using Lemma 3.7.8 yield

$$\begin{aligned} & 2 \int_0^T \psi_N(t) dt \int_I \widetilde{F(u(x,t))} \frac{\partial}{\partial x} (\varphi w(\cdot,t)) \\ = \;& -2 \int_0^T \psi_N(t) dt \int_I \varphi(x,t) \widetilde{w(x,t)} \frac{\partial F(u(\cdot,t))}{\partial x} \\ = \;& -2 \iint_{Q_T} \psi_N \varphi \widetilde{w} \frac{\partial F(u)}{\partial x}. \end{aligned}$$

Thus (4.12) can be written as

$$\int_{\Gamma_u} \psi_N \varphi \left(\int_{\mathbb{R}} f(k) h(k,x,t) dk \right) dH - 2 \iint_{Q_T} \psi_N \varphi \widetilde{w} \frac{\partial F(u)}{\partial x} \le 0. \tag{4.13}$$

For any measurable subset S of Γ_u, similar to Lemma 3.7.6, we can select a sequence $\{\varphi_j\} \subset C_0^\infty(Q_T)$ such that

$$|\varphi_j(x,t)| \leq 1, \quad \lim_{j\to\infty} \varphi_j = \chi_S \quad \text{in } L^1\left(Q_T, \left|\frac{\partial u}{\partial x}\right| + \left|\frac{\partial u}{\partial t}\right|\right).$$

From this and Corollary 3.7.1 (for both $|\frac{\partial u}{\partial x}|$ and $|\frac{\partial u}{\partial t}|$) it follows that (precisely along a subsequence)

$$\lim_{j\to\infty} \varphi_j(x,t) = \chi_S(x,t)$$

H-almost everywhere on Γ_u. Then replacing φ by φ_j in (4.13), letting $j \to \infty$ and using the dominated convergence theorem and Lemma 3.7.4, we obtain

$$\int_S \psi_N \left(\int_{\mathbb{R}} f(k)h(k,x,t)dk\right) dH - 2\int_S \psi_N \widetilde{w} \left(\int_{u^-}^{u^+} f(k)dk\right) \gamma_x dH \leq 0.$$

By the arbitrariness of N and S, there exists a set $G_f \subset \Gamma_u$ with $H(\Gamma_u \setminus G_f) = 0$, such that for each $(x,t) \in G_f$,

$$-\int_{\mathbb{R}} f(k)h(k,x,t)dk + 2\widetilde{w}\int_{u^-}^{u^+} f(k)dk\gamma_x \geq 0.$$

Choose $f(\lambda) \geq 0$ to be a smooth function with $f(0) = 1$ and $\text{supp} f = [-1,1]$. Set

$$f_{s,\rho}(\lambda) = f\left(\frac{s-\lambda}{\rho}\right),$$

where s, ρ are rational numbers with $s \in \mathbb{R}$, $0 < \rho < 1$. Denote

$$G_0 = \bigcap_{s,\rho} G_{f_{s,\rho}}.$$

Clearly, $H(\Gamma_u \setminus G_0) = 0$, and for any $(x,t) \in G_0$,

$$-\int_{\mathbb{R}} f_{s,\rho}(k)h(k,x,t)dk + 2\widetilde{w}\int_{u^-}^{u^+} f_{s,\rho}(k)dk\gamma_x \geq 0.$$

Noticing that

$$h(k,x,t) = 0, \quad \forall k \overline{\in} [u_*, u^*],$$

we derive at once the following relations: for any $(x,t) \in G_0$,

$$-h(k,x,t) + 2\widetilde{w}(x,t)\gamma_x \geq 0, \qquad \text{if} \quad k \in (u^-, u^+), u^- < u^+,$$

$$-h(k,x,t) - 2\widetilde{w}(x,t)\gamma_x \geq 0, \qquad \text{if} \quad k \in (u^+, u^-), u^+ < u^-.$$

From this it follows that (4.10) holds for any $k \neq u^+, u^-$ with $G = \Gamma_u \setminus G_0$. To prove (4.10) for $k = u^+$ and $k = u^-$, it suffices to let k tend to u^- and u^+ from the interval with u^- and u^+ as endpoints. The proof of Theorem 3.4.2 is complete. □

As an immediate consequence of (4.7) and (4.10), we have

Corollary 3.4.1 *There exists a set $G \subset \Gamma_u^*$ with $H(G) = 0$, such that for any $(x,t) \in \Gamma_u^* \setminus G$ and any $k \in \mathbb{R}$, there holds*

$$\begin{aligned} & sgn(u^+ - k)\{(u^+ - k)\gamma_t - (B(u^+) - B(k))\gamma_x \\ & \qquad - (w^r sgn^+\gamma_x - w^l sgn^-\gamma_x)\gamma_x\} \\ \leq\ & sgn(u^- - k)\{(u^- - k)\gamma_t - (B(u^-) - B(k))\gamma_x \\ & \qquad - (w^l sgn^+\gamma_x - w^r sgn^-\gamma_x)\gamma_x\} \end{aligned} \tag{4.14}$$

Corollary 3.4.2 *H-almost everywhere on Γ_u,*

$$\gamma_x \neq 0. \tag{4.15}$$

Proof. Let

$$\Gamma_u^0 = \{(x,t) \in \Gamma_u; \gamma_x = 0\}.$$

If (4.15) were not true, then from (4.10), we would have $H(\Gamma_u^0) > 0$ and for any $k \in \mathbb{R}$,

$$(\text{sgn}(u^+ - k) - \text{sgn}(u^- - k))(\bar{u} - k)\gamma_t \leq 0$$

holds H-almost everywhere on Γ_u^0, which is impossible due to the arbitrariness of k and the fact that $\gamma_t \neq 0$ H-almost everywhere on Γ_u^0. □

Remark 3.4.1 From Corollary 3.4.2 it follows that $\Gamma_u = \Gamma_u^*$ except for a set of Hausdorff measure zero and hence (4.7), (4.8) and (4.14) hold H-almost everywhere on Γ_u.

3.4.3 Uniqueness of BV solutions of the Cauchy problem

Consider the Cauchy problem for (4.1) with initial value condition

$$u(x,0) = u_0(x). \tag{4.16}$$

Definition 3.4.2 A function u is called a generalized solution of the Cauchy problem (4.1), (4.16), if u is a generalized solution of (4.1) on Q_T in the sense of Definition 3.4.1 satisfying

$$\text{ess} \lim_{t\to 0^+} \overline{u}(x,t) = u_0(x)$$

for almost all $x \in \mathbb{R}$.

We will always simply call a generalized solution thus defined a *BV* solution of the Cauchy problem (4.1), (4.16).

Theorem 3.4.3 *Let u_1 and u_2 be BV solutions of the Cauchy problem for (4.1) with initial value u_1^0 and u_2^0 respectively. Then for almost all $t \in (0,T)$,*

$$\int_{\mathbb{R}} |u_1(x,t) - u_2(x,t)|\, \omega_\lambda(x) dx \le e^{k_\lambda t} \int_{\mathbb{R}} |u_{10}(x) - u_{20}(x)|\, \omega_\lambda(x) dx, \tag{4.17}$$

where $\lambda > 0, K_\lambda$ is a constant depending only on λ and the bound of u_1 and u_2, and

$$\omega_\lambda(x) = exp\left(-\lambda\sqrt{1+x^2}\right).$$

The proof is quite long and difficult. Denote

$$J(u_1,u_2,\varphi) \equiv \iint_{Q_T} \text{sgn} z \left(z\frac{\partial \varphi}{\partial t} - \beta z \frac{\partial \varphi}{\partial x} - w\frac{\partial \varphi}{\partial x}\right) dxdt$$

where $0 \le \varphi \in C_0^\infty(Q_T)$, $z = u_1 - u_2$, $w = w_1 - w_2$,

$$w_1 = \frac{\partial A(u_1)}{\partial x}, \quad w_2 = \frac{\partial A(u_2)}{\partial x},$$

$$\beta = \int_0^1 B'(\tau u_1 + (1-\tau)u_2) d\tau.$$

Clearly, sgn$z$ can be replaced by

$$\sigma = \frac{1}{2}(\text{sgn} z^+ + \text{sgn} z^-).$$

Using the measure equality

$$\frac{\partial u}{\partial t} = \frac{\partial}{\partial x}\left(\frac{\partial A(u)}{\partial x}\right) + \frac{\partial B(u)}{\partial x}$$

derived from the integral inequality in Definition 3.4.1, which both u_1 and u_2 satisfy, $J(u_1, u_2, \varphi)$ can be written as

$$J(u_1, u_2, \varphi) = \iint_{Q_T} \sigma\left(\frac{\partial \varphi z}{\partial t} - \frac{\partial \varphi \beta z}{\partial x} - \frac{\partial \varphi \omega}{\partial x}\right),$$

All efforts we make in the following is to prove

$$J(u_1, u_2, \varphi) \geq 0 \qquad \forall 0 \leq \varphi \in C_0^\infty(Q_T).$$

Once this is done, the proof of Theorem 3.4.3 can be completed in the same way just as we did in Theorem 3.2.1.

For fixed $0 \leq \varphi \in C_0^\infty(Q_T)$, denote by $D_\varphi \equiv I_\varphi \times J_\varphi$ the minimal open rectangle containing suppφ and set

$$\begin{aligned} E_\varphi^N = \Bigg\{ & t \in J_\varphi; \int_{I_\varphi} |w_1(x,t)|\,dx + \int_{I_\varphi} |w_2(x,t)|\,dx \\ & + \int_{I_\varphi} \left|\frac{\partial}{\partial x} w_2(x,t)\right| + \int_{I_\varphi} \left|\frac{\partial}{\partial x} w_1(x,t)\right| < N \Bigg\}. \end{aligned}$$

Then from Lemma 3.7.2, we have

$$\begin{aligned} & \sup_{I_\varphi} |w_i(x,t)| \\ \leq \; & C_\varphi \int_{I_\varphi} |w_i(x,t)|\,dx + C_\varphi \int_{I_\varphi} \left|\frac{\partial}{\partial x} w_i(\cdot,t)\right| \quad (i = 1, 2) \end{aligned} \tag{4.18}$$

and

$$\lim_{N\to\infty} \psi_N(t) \equiv \lim_{N\to\infty} \chi_{E_\varphi^N}(t) = 1$$

for almost all $t \in (0, T)$.

By Lemma 3.7.5,

$$\iint_{Q_T} \sigma\left(\frac{\partial \varphi z}{\partial t} - \frac{\partial \varphi \beta z}{\partial x}\right) = -\int_{\Gamma_z} \varphi(\text{sgn}z^+ - \text{sgn}z^-)(\overline{z}\gamma_t - \overline{\beta z}\gamma_x)dH.$$

Thus if we denote

$$
\begin{aligned}
&J_N(u_1, u_2, \varphi) \\
&= -\int_{\Gamma_z} \psi_N \varphi(\mathrm{sgn} z^+ - \mathrm{sgn} z^-)(\overline{z}\gamma_t - \overline{\beta z}\gamma_x)dH \\
&\quad - \iint_{Q_T} \psi_N \sigma \frac{\partial}{\partial x}(\varphi\omega)
\end{aligned} \tag{4.19}
$$

then, by Lemma 3.7.7, corollary 4.2 and the dominated convergence theorem, we have

$$
\lim_{N\to\infty} J_N(u_1, u_2, \varphi) = J(u_1, u_2, \varphi). \tag{4.20}
$$

Lemma 3.4.1 *For any* $0 \le \varphi \in C_0^\infty(Q_T)$

$$
\begin{aligned}
&J_N(u_1, u_2, \varphi) \\
&\ge -\int_{\Gamma_z} \psi_N \varphi(sgnz^+ - sgnz^-)\left(\overline{z}\gamma_t - \overline{\beta z}\gamma_x - \widetilde{w}\gamma_x\right) dH \\
&\quad + \int_{\Gamma_0} \psi_N \varphi\left(|w^l\, sgnz^r| + |w^r\, sgnz^l|\right)|\gamma_x| dH,
\end{aligned} \tag{4.21}
$$

where

$$
\Gamma_0 = \left\{(x,t) \in \Gamma_z; z^+(x,t)\cdot z^-(x,t) = 0\right\}.
$$

Proof. Since Lemma 3.7.5 can not be applied, the last term in (4.19) is quite difficult to treat. The difficulty comes from the fact that w is only a member of $BV_x(Q_T)$ and not of $BV(Q_T)$. w is the weak derivative of the BV function $\alpha z \equiv A(u_1) - A(u_2)$, its status at the points of discontinuity of u_1, u_2 is rather complicated. We first classify the points of discontinuity and then analyze the integrals over each set.

Denote

$$
\Gamma_- = \left\{(x,t) \in \Gamma_z; z^r(x,t) z^l(x,t) < 0\right\},
$$

$$
\Gamma_-^t = \{x; (x,t) \in \Gamma_-\}, \quad \Gamma_0^t = \{x; (x,t) \in \Gamma_0\},
$$

$$
\Gamma_{r,+}^t = \{x \in \Gamma_0^t; z^r(x,t) > 0\}, \quad \Gamma_{r,-}^t = \{x \in \Gamma_0^t; z^r(x,t) < 0\},
$$

$$
\Gamma_{l,+}^t = \left\{x \in \Gamma_0^t; z^l(x,t) > 0\right\}, \quad \Gamma_{l,-}^t = \left\{x \in \Gamma_0^t; z^l(x,t) < 0\right\}.
$$

Since $H(S) = 0$ implies that

$$
\mathrm{mes}\,\{t; (x,t) \in S \quad \text{for some x}\} = 0,
$$

we have, for almost all $t \in (0,T)$,

$$\Gamma_0^t = \Gamma_{r,+}^t \cup \Gamma_{r,-}^t \cup \Gamma_{l,+}^t \cup \Gamma_{l,-}^t .$$

Let

$$z_*(x,t) = \min(z^+(x,t), z^-(x,t)),$$

$$z^*(x,t) = \max(z^+(x,t), z^-(x,t))$$

and denote

$$E^{t,+} = \{x; z_*(x,t) > 0\}, \quad E^{t,-} = \{x; z^*(x,t) < 0\}.$$

Since the functions $z_*(\cdot,t)$ and $z^*(\cdot,t)$ are lower continuous and upper continuous respectively, we see that the sets $E^{t,+}$ and $E^{t,-}$ can be expressed as the unions of at most countable open intervals. Denote by $E_r^{t,+}$, $E_r^{t,-}$, and $E_l^{t,+}$, $E_l^{t,-}$ the sets of right endpoints and left endpoints of these intervals respectively. We decompose these sets as follows:

$$E_r^{t,+} = E_{r,+}^{t,+} \cup E_{r,0}^{t,+}, \quad E_l^{t,+} = E_{l,+}^{t,+} \cup E_{l,0}^{t,+},$$

$$E_r^{t,-} = E_{r,-}^{t,-} \cup E_{r,0}^{t,-}, \quad E_l^{t,-} = E_{l,-}^{t,-} \cup E_{l,0}^{t,-},$$

where

$$E_{r,+}^{t,+} = \{x \in E_r^{t,+}; z^l(x,t) > 0\}, \quad E_{r,0}^{t,+} = \{x \in E_r^{t,+}; z^l(x,t) = 0\},$$

$$E_{l,+}^{t,+} = \{x \in E_l^{t,+}; z^r(x,t) > 0\}, \quad E_{l,0}^{t,+} = \{x \in E_l^{t,+}; z^r(x,t) = 0\},$$

$$E_{r,-}^{t,-} = \{x \in E_r^{t,-}; z^l(x,t) < 0\}, \quad E_{r,0}^{t,-} = \{x \in E_r^{t,-}; z^l(x,t) = 0\},$$

$$E_{l,-}^{t,-} = \{x \in E_l^{t,-}; z^r(x,t) < 0\}, \quad E_{l,0}^{t,-} = \{x \in E_l^{t,-}; z^r(x,t) = 0\}.$$

From the facts

$$(z^r(x,t))^r = (z^l(x,t))^r = (z(x,t))^r,$$

$$(z^r(x,t))^l = (z^l(x,t))^l = (z(x,t))^l,$$

it is easy to see that for almost all $t \in (0,T)$,

$$\Gamma_-^t = \left(E_{l,+}^{t,+} \cap E_{r,-}^{t,-}\right) \cup \left(E_{r,+}^{t,+} \cap E_{l,-}^{t,-}\right),$$

$$\Gamma_{r,+}^t = E_{l,+}^{t,+} \setminus E_{r,-}^{t,-}, \quad \Gamma_{r,-}^t = E_{l,-}^{t,-} \setminus E_{r,+}^{t,+},$$

$$\Gamma_{l,+}^t = E_{r,+}^{t,+} \setminus E_{l,-}^{t,-}, \quad \Gamma_{l,-}^t = E_{r,-}^{t,-} \setminus E_{l,+}^{t,+}.$$

Now we are ready to investigate the last term in (4.19), namely, the term

$$\Lambda \equiv -\iint_{Q_T} \psi_N \sigma \frac{\partial}{\partial x}(\varphi w).$$

To this purpose, we write it as

$$\Lambda = \Lambda_1 + \Lambda_2$$

with

$$\Lambda_1 = -\iint\limits_{z^+ z^- = 0} \psi_N \sigma \frac{\partial}{\partial x}(\varphi w), \Lambda_2 = -\iint\limits_{z^+ z^- > 0} \psi_N \sigma \frac{\partial}{\partial x}(\varphi w),$$

and calculate Λ_1 and Λ_2 separately.

First, by Lemma 3.7.8, we have

$$\Lambda_1 = \int_0^T \psi_N(t)\lambda_1(t)dt,$$

where

$$\begin{aligned}
\lambda_1(t) = & -\frac{1}{2}\sum_{x\in\Gamma_0^t}(\text{sgn}z^+ + \text{sgn}z^-)\varphi(w^r - w^l) \\
= & -\frac{1}{2}\sum_{x\in\Gamma_{r,+}^t}\varphi(w^r - w^l) + \frac{1}{2}\sum_{x\in\Gamma_{r,-}^t}\varphi(w^r - w^l) \\
& -\frac{1}{2}\sum_{x\in\Gamma_{l,+}^t}\varphi(w^r - w^l) + \frac{1}{2}\sum_{x\in\Gamma_{l,-}^t}\varphi(w^r - w^l)
\end{aligned}$$

for almost all $t \in (0, T)$. Similarly, we have

$$\Lambda_2 = \int_0^T \psi_N(t)\lambda_2(t)dt,$$

where

$$\begin{aligned}
\lambda_2(t) = & -\sum_{x\in E_r^{t,+}}\varphi w^l + \sum_{x\in E_l^{t,+}}\varphi w^r + \sum_{x\in E_r^{t,-}}\varphi w^l - \sum_{x\in E_l^{t,-}}\varphi w^r \\
= & \sum_{x\in E_{l,+}^{t,+}\cap E_{r,-}^{t,-}}\varphi(w^r + w^l) - \sum_{x\in E_{r,+}^{t,+}\cap E_{l,-}^{t,-}}\varphi(w^r + w^l) \\
& - \sum_{x\in(E_{r,+}^{t,+}\setminus E_{l,-}^{t,-})\cup E_{r,0}^{t,+}}\varphi w^l + \sum_{x\in(E_{l,+}^{t,+}\setminus E_{r,-}^{t,-})\cup E_{l,0}^{t,+}}\varphi w^r
\end{aligned}$$

$$+ \sum_{x\in(E^{t,-}_{r,-}\setminus E^{t,+}_{l,+})\cup E^{t,-}_{r,0}} \varphi w^l - \sum_{x\in(E^{t,-}_{l,-}\setminus E^{t,+}_{r,+})\cup E^{t,-}_{l,0}} \varphi w^r$$

$$= \sum_{x\in\Gamma^t_-} (\text{sgn} z^r - \text{sgn} z^l)\varphi\widetilde{w}$$

$$- \sum_{x\in\Gamma^t_{l,+}\cup E^{t,+}_{r,0}} \varphi w^l + \sum_{x\in\Gamma^t_{r,+}\cup E^{t,+}_{l,0}} \varphi w^r$$

$$+ \sum_{x\in\Gamma^t_{l,-}\cup E^{t,-}_{r,0}} \varphi w^l - \sum_{x\in\Gamma^t_{r,-}\cup E^{t,-}_{l,0}} \varphi w^r$$

for almost all $t \in (0,T)$. Hence

$$\Lambda \equiv \iint_0^T \psi_N(t)(\lambda_1(t)+\lambda_2(t))dt$$

$$= \int_0^T \psi_N \sum_{x\in\Gamma^t_z} (\text{sgn} z^r - \text{sgn} z^l)\varphi\widetilde{w}dt + \Lambda_3,$$

where

$$\Lambda_3 = \int_0^T \psi_N(t)\lambda_3(t)dt,$$

$$\lambda_3(t) = -\sum_{x\in E^{t,+}_{r,0}} \varphi w^l + \sum_{x\in E^{t,+}_{l,0}} \varphi w^r + \sum_{x\in E^{t,-}_{r,0}} \varphi w^l - \sum_{x\in E^{t,-}_{l,0}} \varphi w^r,$$

and using Corollary 3.7.3, we further obtain

$$\Lambda = \int_{\Gamma_z} \psi_N\varphi(\text{sgn} z^+ - \text{sgn} z^-)\widetilde{w}\gamma_x dH + \Lambda_3.$$

Substituting this into (4.19), we see that to prove (4.21), it remains to verify that

$$\Lambda_3 \geq \int_{\Gamma_0} \psi_N\varphi\left(|w^l \text{sgn} z^r| + |w^r \text{sgn} z^l|\right)|\gamma_x|dH. \tag{4.22}$$

Let

$$F^t_{r,+} = \{x; w^r(x,t) > 0\}, \quad F^t_{r,-} = \{x; w^r(x,t) < 0\},$$

$$F^t_{l,+} = \left\{x; w^l(x,t) > 0\right\}, \quad F^t_{l,-} = \left\{x; w^l(x,t) < 0\right\}.$$

We may assert

$$\begin{cases} \Gamma^t_{r,+} \cap F^t_{l,+} \subset E^{t,-}_{r,0}, & \Gamma^t_{r,+} \cap F^t_{l,-} \subset E^{t,+}_{r,0}, \\ \Gamma^t_{r,-} \cap F^t_{l,+} \subset E^{t,-}_{r,0}, & \Gamma^t_{r,-} \cap F^t_{l,-} \subset E^{t,+}_{r,0}, \\ \Gamma^t_{l,+} \cap F^t_{r,+} \subset E^{t,+}_{l,0}, & \Gamma^t_{l,+} \cap F^t_{r,-} \subset E^{t,-}_{l,0}, \\ \Gamma^t_{l,-} \cap F^t_{r,+} \subset E^{t,+}_{l,0}, & \Gamma^t_{l,-} \cap F^t_{r,-} \subset E^{t,-}_{l,0}. \end{cases} \tag{4.23}$$

As an example, we prove

$$\Gamma^t_{r,+} \cap F^t_{l,+} \subset E^{t,-}_{r,0}. \tag{4.24}$$

Let $x_0 \in \Gamma^t_{r,+} \cap F^t_{l,+}$, and assume that $z^r(x_0,t)$, $z^l(x_0,t)$ and $w^l(x_0,t)$ exist. Then by definition, we have

$$z^r(x_0,t) > 0, \quad z^l(x_0,t) = 0, \quad w^l(x_0,t) > 0.$$

Hence there exists a left neighborhood V of x_0, such that for any $x \in V$,

$$A(u_1(x,t)) - A(u_2(x,t)) < 0.$$

By virtue of the discontinuity condition (4.8), we have for $x \in V$,

$$A(u_i(x,t)) = A(u_i^r(x,t)) = A(u_i^l(x,t)) \quad (i = 1, 2).$$

Since $A(s)$ is nondecreasing, it follows that for $x \in V$

$$z^r(x,t) = u_1^r(x,t) - u_2^r(x,t) < 0,$$

$$z^l(x,t) = u_1^l(x,t) - u_2^l(x,t) < 0,$$

which imply that $x_0 \in E^{t,-}_{r,0}$, namely, (4.24) holds.

From the definition of $E^{t,+}_{r,0}$, $E^{t,-}_{r,0}$, $E^{t,+}_{l,0}$ and $E^{t,-}_{l,0}$, it is easily seen that for almost all $t \in (0,T)$,

$$w^l(x,t) \le 0, \quad \text{if} \quad x \in E^{t,+}_{r,0},$$

$$w^r(x,t) \ge 0, \quad \text{if} \quad x \in E^{t,+}_{l,0},$$

$$w^l(x,t) \ge 0, \quad \text{if} \quad x \in E^{t,-}_{r,0},$$

$$w^r(x,t) \le 0, \quad \text{if} \quad x \in E^{t,-}_{l,0}.$$

Using these and noticing that for almost all $t \in (0,T)$ and any $x \in \Gamma_0^t \setminus (F_{r,+}^t \cup F_{r,-}^t \cup F_{l,+}^t \cup F_{l,-}^t)$,

$$w^r(x,t) = w^l(x,t) = 0,$$

we derive (4.22) by Corollary 3.7.3 and (4.23), and thus complete the proof of Lemma 3.4.1. □

Lemma 3.4.2 *For any $0 \le \varphi \in C_0^\infty(Q_T)$, $J(u_1,u_2,\varphi) \ge 0$.*

Proof. Since from (4.20), (4.21) we have

$$\begin{aligned} J(u_1,u_2,\varphi) \ge & -\int_{\Gamma_z} \varphi(\text{sgn}z^+ - \text{sgn}z^-)(\overline{z}\gamma_t - \overline{\beta z}\gamma_x - \widetilde{w}\gamma_x)dH \\ & + \int_{\Gamma_0} \varphi(|w^l \text{sgn}z^r| + |w^r \text{sgn}z^l|)|\gamma_x| dH, \end{aligned}$$

it suffices to prove that

$$(\text{sgn}z^+ - \text{sgn}z^-)(\overline{z}\gamma_t - \overline{\beta z}\gamma_x - \widetilde{w}\gamma_x) \le 0 \tag{4.25}$$

holds H-almost everywhere on Γ_- and

$$(\text{sgn}z^+ - \text{sgn}z^-)(\overline{z}\gamma_t - \overline{\beta z}\gamma_x - \widetilde{w}\gamma_x) - (|w^l \text{sgn}z^r| + |w^r \text{sgn}z^l|)|\gamma_x| \le 0 \tag{4.26}$$

holds H-almost everywhere on Γ_0.

First we rewrite (4.7) as

$$\begin{aligned} &(u^+ - u^-)\gamma_t - (B(u^+) - B(u^-))\gamma_x - \left(w^r \text{sgn}^+ \gamma_x - w^l \text{sgn}^- \gamma_x\right)\gamma_x \\ &\quad - \left(w^r \text{sgn}^- \gamma_x - w^l \text{sgn}^+ \gamma_x\right)\gamma_x = 0, \end{aligned} \tag{4.27}$$

which holds H-almost everywhere on Γ_u (Notice Remark 3.4.1). Using (4.27), we see that (4.25) is equivalent to

$$\begin{aligned} & \text{sgn}z^+ \left\{z^+\gamma_t - (\beta z)^+\gamma_x - (w_1^r \text{sgn}^+\gamma_x - w_1^l \text{sgn}^-\gamma_x)\gamma_x \right. \\ & \quad \left. + (w_2^r \text{sgn}^+\gamma_x - w_2^l \text{sgn}^-\gamma_x)\gamma_x \right\} \\ \le\ & \text{sgn}z^- \left\{z^-\gamma_t - (\beta z)^-\gamma_x - (w_1^l \text{sgn}^+\gamma_x - w_1^r \text{sgn}^-\gamma_x)\gamma_x \right. \\ & \quad \left. + (w_2^l \text{sgn}^+\gamma_x - w_2^r \text{sgn}^-\gamma_x)\gamma_x \right\}. \end{aligned} \tag{4.28}$$

In order to prove (4.28), we apply the discontinuity condition (4.14) presented in Corollary 3.4.1 to u_1 with $k = u_2^+$, and then to u_2, with $k = u_1^-$. Thus we obtain

$$\text{sgn}z^+ \left\{z^+\gamma_t - (\beta z)^+\gamma_x - (w_1^r \text{sgn}^+\gamma_x - w_1^l \text{sgn}^-\gamma_x)\gamma_x\right.$$

$$
\begin{aligned}
&\quad + (w_2^r \mathrm{sgn}^+\gamma_x - w_2^l \mathrm{sgn}^-\gamma_x)\gamma_x\} \\
\le\ & \mathrm{sgn}(u_1^- - u_2^+)\left\{(u_1^- - u_2^+)\gamma_t - (B(u_1^-) - B(u_2^+))\gamma_x\right\} \\
&-\mathrm{sgn}(u_1^- - u_2^+)(w_1^l \mathrm{sgn}^+\gamma_x - w_1^r \mathrm{sgn}^-\gamma_x)\gamma_x \\
&+\mathrm{sgn}(u_1^+ - u_2^+)(w_2^r \mathrm{sgn}^+\gamma_x - w_2^l \mathrm{sgn}^-\gamma_x)\gamma_x \\
=\ & \mathrm{sgn}(u_2^+ - u_1^-)\left\{(u_2^+ - u_1^-)\gamma_t - (B(u_2^+) - B(u_1^-))\gamma_x\right. \\
&-\mathrm{sgn}(u_2^+ - u_1^-)(w_2^r \mathrm{sgn}^+\gamma_x - w_2^l \mathrm{sgn}^-\gamma_x)\gamma_x \\
&+\mathrm{sgn}(u_1^+ - u_2^+)(w_2^r \mathrm{sgn}^+\gamma_x - w_2^l \mathrm{sgn}^-\gamma_x)\gamma_x \\
&+\mathrm{sgn}(u_2^+ - u_1^-)(w_2^r \mathrm{sgn}^+\gamma_x - w_2^l \mathrm{sgn}^-\gamma_x)\gamma_x \\
&-\mathrm{sgn}(u_1^- - u_2^+)(w_1^l \mathrm{sgn}^+\gamma_x - w_1^r \mathrm{sgn}^-\gamma_x)\gamma_x \\
\le\ & \mathrm{sgn}(u_2^- - u_1^-)\left\{(u_2^- - u_1^-)\gamma_t - (B(u_2^-) - B(u_1^-))\gamma_x\right\} \\
&-\mathrm{sgn}(u_2^- - u_1^-)(w_2^l \mathrm{sgn}^+\gamma_x - w_2^r \mathrm{sgn}^-\gamma_x)\gamma_x \\
&+\mathrm{sgn}(u_1^+ - u_2^+)(w_2^r \mathrm{sgn}^+\gamma_x - w_2^l \mathrm{sgn}^-\gamma_x)\gamma_x \\
&+\mathrm{sgn}(u_2^+ - u_1^-)(w_2^r \mathrm{sgn}^+\gamma_x - w_2^l \mathrm{sgn}^-\gamma_x)\gamma_x \\
&-\mathrm{sgn}(u_1^- - u_2^+)(w_1^l \mathrm{sgn}^+\gamma_x - w_1^r \mathrm{sgn}^-\gamma_x)\gamma_x \\
=\ & \mathrm{sgn} z^- \left\{z^-\gamma_t - (\beta z)^-\gamma_x - (w_1^l \mathrm{sgn}^+\gamma_x - w_1^r \mathrm{sgn}^-\gamma_x)\gamma_x\right. \\
&\left. + (w_2^l \mathrm{sgn}^+\gamma_x - w_2^r \mathrm{sgn}^-\gamma_x)\gamma_x\right\} + \Lambda,
\end{aligned}
$$

where

$$
\begin{aligned}
\Lambda =\ & \left(\mathrm{sgn}(u_2^+ - u_1^-) + \mathrm{sgn}(u_1^+ - u_2^+)\right)(w_2^r \mathrm{sgn}^+\gamma_x - w_2^l \mathrm{sgn}^-\gamma_x)\gamma_x \\
& + \left(\mathrm{sgn}(u_1^- - u_2^-) + \mathrm{sgn}(u_2^+ - u_1^-)\right)(w_1^l \mathrm{sgn}^+\gamma_x - w_1^r \mathrm{sgn}^-\gamma_x)\gamma_x.
\end{aligned}
$$

We assert that $\Lambda = \Lambda(x,t) \le 0$ holds H-almost everywhere on Γ_-. Once this is done, we get (4.17) and hence (4.25) immediately. For simplicity, in the following discussion, we omit the argument (x,t) and assume that for definiteness, $z^+ > 0$, $z^- < 0$ at the point considered. We conclude that

(1) if $u_1^- \ne u_2^+$, then

$$u_1^- > u_2^+, \quad \gamma_x > 0 \quad \text{imply} \quad w_1^l = 0,$$

$$u_1^- < u_2^+, \quad \gamma_x > 0 \quad \text{imply} \quad w_2^r = 0,$$

$$u_1^- > u_2^+, \quad \gamma_x < 0 \quad \text{imply} \quad w_1^r = 0,$$

$$u_1^- < u_2^+, \quad \gamma_x < 0 \quad \text{imply} \quad w_2^l = 0,$$

(2) if $u_1^- = u_2^+$, then

$$\gamma_x > 0, \quad \text{imply} \quad w_1^l \geq 0,\ w_2^r \leq 0,$$

$$\gamma_x < 0, \quad \text{imply} \quad w_1^r \leq 0,\ w_2^l \geq 0.$$

As an example, we prove the first conclusion of (1), In case $u_1^- > u_2^+, \gamma_x > 0$, it is certainly $u_2^+ < u_1^- < u_2^-$. By the discontinuity condition (4.8), we see that $A(u_1)$ must be a constant in a certain neighborhood of x and hence $w_1^l = 0$. Thus, $\Lambda = -2w_1^l\gamma_x = 0$. We can prove $\Lambda \leq 0$ in all other cases similarly and hence complete the proof of (4.25).

To prove (4.26), we rewrite the discontinuity condition (4.7) as

$$(u^r - u^l)\gamma_t - (B(u^r) - B(u^l))\gamma_x - (w^r - w^l)\gamma_x = 0. \tag{4.29}$$

If $z^r \neq 0$, then $z^l = 0$, and hence

$$z^r\gamma_t - (\beta z)^r\gamma_x = (w^r - w^l)\gamma_x.$$

It follows that

$$\begin{aligned}
&(\text{sgn}z^+ - \text{sgn}z^-)(\overline{z}\gamma_t - \overline{\beta z}\gamma_x - \widetilde{w}\gamma_x) - (|w^l\text{sgn}z^r| + |w^r\text{sgn}z^l|)|\gamma_x| \\
= \ &(\text{sgn}z^+ - \text{sgn}z^-)\left(\frac{1}{2}(w^r - w^l)\gamma_x - \widetilde{w}\gamma_x\right) - |w^l||\gamma_x| \\
= \ &|\gamma_x||w^l|(-1 - \text{sgn}z^r\text{sgn}w^l) \leq 0.
\end{aligned}$$

This means that (4.26) is true in case $z^r \neq 0$. Similarly we can treat the case $z^l \neq 0$. The proof of Lemma 3.4.2 and hence the proof of Theorem 3.4.3 is completed. □

3.4.4 *Formulation of the boundary value problem*

Let $I = (0,1)$. Consider the first boundary value problem for (4.1) on $Q_T = I \times (0,T)$ with boundary value conditions

$$u(0,t) = \psi_1(t), \quad u(1,t) = \psi_2(t).$$

For weakly degenerate equations, the formulation of the boundary value condition is almost the same as that for equations without any degeneracy. However things are quite different for equations with strong degeneracy. This can be directly seen by observing the extreme case (4.2), In this case, it is well-known that one can not require the value of the solutions to be prescribed on the whole boundaries $x = 0$ and $x = 1$. For linear degenerate equations, on which part of the boundary the value of the solution can be prescribed, is determined in advance by the coefficients of the equations (see[OR_1]). For quasilinear equations with weak degeneracy, the situation is similar. However, it is impossible to pose the boundary value problem for quasilinear equations with strong degeneracy in this way. Instead, we use an integral inequality similar to (4.6) but more complicated, which involves part of the boundary value condition in an implicit manner. (see[WZ_1]). This leads to the following definition of generalized solutions of the first boundary value problem for (4.1), in which, for simplicity, only the homogeneous boundary condition

$$u(0,t) = u(1,t) = 0 \tag{4.30}$$

is concerned.

Definition 3.4.3 A function $u \in L^\infty(Q_T) \cap BV(Q_T)$ is called a generalized solution (We will simply call it a BV solution) of the boundary value problem (4.1), (4.16), (4.30), if the following conditions are fulfilled:

(1) $w \equiv \dfrac{\partial A(u)}{\partial x} \in L^\infty(Q_T)$;

(2) $\operatorname{ess}\lim\limits_{t\to 0+} \overline{u}(x,t) = u_0(x)$ for almost all $x \in (0,1)$ and $A(u^r(0,t)) = A(u^l(1,t)) = 0$ for almost all $t \in (0,T)$;

(3) for any $0 \le \varphi_1, \varphi_2 \in C^\infty(\overline{Q_T})$ with $\varphi_1(0,t) = \varphi_2(0,t) = \varphi_1(1,t) = \varphi_2(1,t)$, $\operatorname{supp}\varphi_1, \operatorname{supp}\varphi_2 \subset [0,1] \times (0,T)$, and any $k \in R$, there holds

$$\begin{aligned}&\iint_{Q_T} \operatorname{sgn}(u-k)\left((u-k)\frac{\partial \varphi_1}{\partial t} - (B(u)-B(k)+w)\frac{\partial \varphi_1}{\partial x}\right) dxdt \\ &+ \iint_{Q_T} \operatorname{sgn}k\left((u-k)\frac{\partial \varphi_2}{\partial t} - (B(u)-B(k)+w)\frac{\partial \varphi_2}{\partial x}\right) dxdt \ge 0.\end{aligned} \tag{4.31}$$

3.4.5 *Boundary discontinuity condition*

Denote by $\dot{E}$ the subset of all interior points of $E = \{s; A'(s) = 0\}$. Then, since for $s\overline{\in}\dot{E}$, $A(s)$ is strictly increasing, we see from Definition 3.4.3 that if the boundary value $u^r(0,t)(u^l(1,t))$ of the solution happens to fall outside of $\dot{E}$, then the solution satisfies the boundary value condition in classical sense. When the boundary value falls in $\dot{E}$, the boundary value condition is implicitly involved in the integral inequality (4.31). In what follows, we will reveal the concrete form of this part of the boundary value condition.

First from $A(u^r(0,t)) = A(u^l(1,t)) = 0$ and the monotonicity of $A(s)$, we have

Lemma 3.4.3 *Let u be a BV solution of the problem (4.1), (4.16), (4.30). Then*

$$A'(s) = 0 \qquad \forall s \in I(u^r(0,t), u^l(1,t)), \tag{4.32}$$

where

$$\begin{aligned} & I(u^r(0,t), u^l(1,t)) \\ = \; & \left(\min(u^r(0,t), u^l(1,t), 0), \max(u^r(0,t), u^l(1,t), 0)\right). \end{aligned}$$

Lemma 3.4.4 *Assume that $u \in BV_x(Q_T)$. Then there exists a set K with mes$K = 0$, such that for any fixed $k \in \mathbb{R} \setminus K$, the limits*

$$\lim_{x \to 0^+} sgn(\widetilde{u}(x,t) - k) = sgn(u^r(0,t) - k), \tag{4.33}$$

$$\lim_{x \to 1^-} sgn(\widetilde{u}(x,t) - k) = sgn(u^l(1,t) - k) \tag{4.34}$$

exist for almost all $t \in (0,T)$.

Proof. Using the properties of BV_x function, we have, for almost all $t \in (0,T)$, (4.33) holds provided $k \neq u^r(0,t)$. Choose $f(k) \in C_0^\infty(\mathbb{R})$, $\psi(t) \in C_0^\infty(0,T)$ arbitrarily and consider the integral

$$\int_0^T \int_{\mathbb{R}} \text{sgn}(\widetilde{u}(x,t) - k) f(k) \psi(t) dk dt.$$

By the dominated convergence theorem,

$$\begin{aligned} & \int_0^T \int_{\mathbb{R}} \lim_{x \to 0^+} \text{sgn}(\widetilde{u}(x,t) - k) f(k) \psi(t) dk dt \\ = & \int_0^T \int_{\mathbb{R}} \text{sgn}(u^r(0,t) - k) f(k) \psi(t) dk dt. \end{aligned}$$

It follows from the arbitrariness of $f(k)$ that

$$\int_0^T \lim_{x \to 0^+} \text{sgn}(\widetilde{u}(x,t) - k) \psi(t) dt = \int_0^T \text{sgn}(u^r(0,t) - k) \psi(t) dt$$

for almost all $k \in \mathbb{R}$ and from the arbitrariness of $\psi(t)$ that (4.33) holds. (4.34) can be proved similarly. □

Lemma 3.4.5 *Let u be a BV solution of the problem (4.1), (4.30), (4.16). Then there exists a subset $G \subset (0,T)$ with $mes((0,T)\setminus G)=0$, such that for any $t\in G$, $k\in\mathbb{R}$,*

$$(sgn(u^r(0,t)-k)+sgnk)\,(B(u^r(0,t))-B(k)+w^r(0,t))\geq 0, \quad (4.35)$$

$$\big(sgn(u^l(1,t)-k)+sgnk\big)\,\big(B(u^l(1,t))-B(k)+w^l(1,t)\big)\leq 0. \quad (4.36)$$

Proof. Let $h_\varepsilon(s)$ be the kernel of a mollifier in one dimension and denote

$$\varphi_\varepsilon(x)=1-\int_{-\infty}^{x-2\varepsilon}h_\varepsilon(s)ds.$$

Choose $\varphi_1(x,t)=\varphi_2(x,t)=\varphi_\varepsilon(x)\psi(t)$ in (4.31) with $0\leq\psi(t)\in C_0^\infty(0,T)$. Then we obtain

$$\iint_{Q_T}(\text{sgn}(u-k)+\text{sgn}k)\,((u-k)\varphi_\varepsilon\psi'-(B(u)-B(k)+w)\varphi_\varepsilon'\psi)\,dxdt\geq 0.$$

Since $w\in BV_x(Q_T)$ which follows from the measure equality (4.9) (Notice that obviously (4.31) implies (4.5) and hence (4.9)), we have $w(\cdot,t)\in BV(I)$ by Lemma 3.7.8. Letting $\varepsilon\to 0$ and using the properties of one-dimensional functions of bounded variation, we see that (4.35) holds. (4.36) can be proved similarly. □

Remark 3.4.2 If $A(s)\equiv 0$, then the discontinuity conditions (4.35) and (4.36) become

$$(\text{sgn}(u^r(0,t)-k)+\text{sgn}k)\,(B(u^r(0,t))-B(k))\geq 0,$$

$$\big(\text{sgn}(u^l(1,t)-k)+\text{sgn}k\big)\,\big(B(u^l(1,t))-B(k)\big)\leq 0.$$

Let $k=0$. Then

$$\text{sgn}\,u^r(0,t)B(u^r(0,t))\geq 0, \qquad \text{sgn}\,u^l(1,t)B(u^l(1,t))\leq 0.$$

From these it is seen that if $B'(s)<(>)0$, $u^r(0,t)=0(u^l(1,t)=0)$, then the boundary value condition is really satisfied only on the right (left) boundary.

3.4.6 Uniqueness of BV solutions of the first boundary value problem

Theorem 3.4.4 *Let u_1 and u_2 be BV solutions of the first boundary value problem (4.1), (4.16), (4.30) with initial value $u_0 = u_1^0$ and $u_0 = u_2^0$ respectively. Then for almost all $t \in (0,T)$,*

$$\int_0^1 |u_1(x,t) - u_2(x,t)|\,dx \leq \int_0^1 |u_{10}(x) - u_{20}(x)|\,dx \tag{4.37}$$

Similar to the Cauchy problem, the crucial step of the proof is to verify the following lemma

Lemma 3.4.6 *For any $0 \leq \varphi \in C^\infty(\overline{Q_T})$ with supp$\varphi \subset [0,1] \times (0,T)$,*

$$J(u_1,u_2,\varphi) \equiv \iint_{Q_T} sgnz\left(z\frac{\partial \varphi}{\partial t} - \beta z\frac{\partial \varphi}{\partial x} - w\frac{\partial \varphi}{\partial x}\right) dxdt \geq 0, \tag{4.38}$$

where $z = u_1 - u_2$, $w = w_1 - w_2$,

$$w_1 = \frac{\partial A(u_1)}{\partial x}, \quad w_2 = \frac{\partial A(u_2)}{\partial x},$$

$$\beta = \int_0^1 B'(\lambda u_1 + (1-\lambda)u_2)d\lambda.$$

Proof. For $0 \leq \varphi \in C_0^\infty(Q_T)$, the proof of (4.38) is just the same as the Cauchy problem. Now given $0 \leq \varphi \in C^\infty(\overline{Q_T})$ with supp$\varphi \subset [0,1] \times (0,T)$. Let

$$\varphi_\varepsilon(x) = 1 - \int_{-\infty}^{x-2\varepsilon} h_\varepsilon(s)ds + \int_{-\infty}^{x-1+2\varepsilon} h_\varepsilon(s)ds,$$

where $h_\varepsilon(s)$ is the function in the proof of Lemma 3.4.5. Then

$$0 \leq \varphi_\varepsilon(x) \leq 1, \qquad 1 - \varphi_\varepsilon \in C_0^\infty(0,1)$$

and

$$J(u_1,u_2,\varphi) = J(u_1,u_2,\varphi_\varepsilon\varphi) + J(u_1,u_2,(1-\varphi_\varepsilon)\varphi) \geq J(u_1,u_2,\varphi_\varepsilon\varphi).$$

Letting $\varepsilon \to 0$ yields

$$\begin{aligned} J(u_1,u_2,\varphi) \geq & -\int_0^T \left((\beta|z|)^l(1,t) - (\beta|z|)^r(0,t)\right) dt \\ & - \int_0^T \left((w\text{sgn}z)^l(1,t) - (w\text{sgn}z)^r(0,t)\right) dt. \end{aligned} \tag{4.39}$$

It remains to prove that for almost all $t \in (o, T)$,

$$(\beta|z|)^l(1,t) + (w\mathrm{sgn}z)^l(1,t) \le 0, \tag{4.40}$$

$$(\beta|z|)^r(0,t) + (w\mathrm{sgn}z)^r(0,t) \ge 0. \tag{4.41}$$

To prove (4.40), it suffices to verify

$$u_1^l > u_2^l \ge 0, \quad \text{implies} \quad B(u_1^l) - B(u_2^l) + w_1^l \le 0,\ w_2^l \ge 0,$$

$$u_1^l > 0 > u_2^l, \quad \text{implies} \quad B(u_1^l) + w_1^l \le 0,\ B(u_2^l) + w_2^l \ge 0,$$

$$0 \ge u_1^l > u_2^l, \quad \text{implies} \quad w_1^l \le 0,\ B(u_1^l) - B(u_2^l) - w_2^l \le 0,$$

$$0 \le u_1^l < u_2^l, \quad \text{implies} \quad w_1^l \ge 0,\ B(u_1^l) - B(u_2^l) - w_2^l \ge 0,$$

$$u_1^l < 0 < u_2^l, \quad \text{implies} \quad B(u_1^l) + w_1^l \ge 0,\ B(u_2^l) + w_2^l \le 0,$$

$$u_1^l < u_2^l \le 0, \quad \text{implies} \quad B(u_1^l) - B(u_2^l) + w_1^l \ge 0,\ w_2^l \le 0,$$

$$u_1^l = u_2^l,\ w^l = 0, \quad \text{implies} \quad (w\mathrm{sgn}z)^l = 0,$$

$$u_1^l = u_2^l,\ w^l \ne 0, \quad \text{implies} \quad (w\mathrm{sgn}z)^l < 0.$$

Here we have omitted the argument $(1,t)$. As an example, we verify the first conclusion. To this purpose, it suffices to take $u = u_1^l$, $k = u_2^l$ in (4.37) and using (4.32). Similarly we can prove (4.41). □

Proof of Theorem 3.4.4. Having Lemma 3.4.6 in our hands, the proof of the theorem becomes very simple. In fact, it suffices to choose $\varphi(x,t) = \psi(t)$ with $0 \le \psi(t) \in C_0^\infty(0,T)$ in (4.38) to obtain

$$\iint_{Q_T} |u_1(x,t) - u_2(x,t)|\, \psi'(t)dxdt \ge 0$$

from which the conclusion of Theorem 3.4.4 follows by the arbitrariness of $\psi(t)$. □

3.4.7 *Existence of BV solutions of the first boundary value problem*

Theorem 3.4.5 *Assume that $A(s)$, $B(s)$ and $u_0(x)$ are appropriately smooth with $u_0(0) = u_0(1) = 0$. Then the first boundary value problem (4.1), (4.16), (4.30) admits a BV solution.*

Proof. Similar to the proof of Theorem 3.2.2, consider the approximate problem

$$\frac{\partial u_\varepsilon}{\partial t} = \frac{\partial^2 A_\varepsilon(u_\varepsilon)}{\partial x^2} + \frac{\partial B(u_\varepsilon)}{\partial x}, \tag{4.42}$$

$$u_\varepsilon(0,t) = u_\varepsilon(1,t) = 0, \tag{4.43}$$

$$u_\varepsilon(x,0) = u_{0\varepsilon}(x), \tag{4.44}$$

where $A_\varepsilon(s) = \varepsilon s + A(s)$ and $u_{0\varepsilon}$ is a smooth uniform approximation of u_0 with $u_{0\varepsilon}^{(k)}(0) = u_{0\varepsilon}^{(k)}(1) = 0 \quad (k = 0,1,2)$ and $u'_{0\varepsilon}$, $u''_{0\varepsilon}$ being uniformly bounded.

Let u_ε be a solution of the problem (4.42)–(4.44). From the maximum principle, we first have

$$\sup_{Q_T} |u_\varepsilon(x,t)| \le \sup_{0<x<1} |u_{0\varepsilon}(x)| \le M \tag{4.45}$$

with constant M independent of ε.

Using the method of §3.3.3, we can obtain the following estimates for u_ε

$$\sup_{0<t<T} \int_0^1 \left|\frac{\partial u_\varepsilon}{\partial t}\right| dx \le C, \tag{4.46}$$

$$\sup_{0<t<T} \int_0^1 \left|\frac{\partial u_\varepsilon}{\partial x}\right| dx \le C. \tag{4.47}$$

Integrating (4.42) with respect to x and using Lemma 3.2.1 (which clearly holds in the present case) and (4.45)–(4.47) we further obtain

$$\sup_{Q_T} \left|\frac{\partial}{\partial x} A_\varepsilon(u_\varepsilon(x,t))\right| \le C. \tag{4.48}$$

On the basis of these estimates, we can conclude that there exists a subsequence of $\{u_\varepsilon\}$, denoted still by $\{u_\varepsilon\}$, such that

$$\lim_{\varepsilon\to 0} u_\varepsilon(x,t) = u(x,t), \qquad \text{a.e in} Q_T,$$

$$\lim_{\varepsilon\to 0} \frac{\partial}{\partial x} A_\varepsilon(u_\varepsilon) = \frac{\partial}{\partial x} A(u), \quad \text{weakly in } L^\infty(Q_T).$$

In addition, $u \in L^\infty(Q_T) \bigcap \mathrm{BV}(Q_T)$ with $w = \dfrac{\partial A(u)}{\partial x} \in L^\infty(Q_T)$ and clearly u satisfies the condition $A(u^r(0,t)) = A(u^l(1,t)) = 0$ for almost all $t \in (0,T)$. It is also easy to see that

$$\text{ess}\lim_{t\to 0^+} \overline{u}(x,t) = u_0(x)$$

for almost all $x \in (0,T)$.

It remains to prove that u satisfies the inequality (4.31). Let $0 \le \varphi_1 \in C^\infty(\overline{Q_T})$ with $\text{supp}\varphi_1 \subset [0,1] \times (0,T)$ and $H_\eta(s)$ be the function in the proof of Theorem 3.3.1. Multiply (4.42) by $\varphi_1 H_\eta(u_\varepsilon - k)$ and integrate over Q_T:

$$\begin{aligned}&\iint_{Q_T} \varphi_1 H_\eta(u_\varepsilon - k)\frac{\partial u_\varepsilon}{\partial t} dxdt \\ = &\iint_{Q_T} \varphi_1 H_\eta(u_\varepsilon - k)\frac{\partial^2 A_\varepsilon(u_\varepsilon)}{\partial x^2} dxdt + \iint_{Q_T} \varphi_1 H_\eta(u_\varepsilon - k)\frac{\partial B(u_\varepsilon)}{\partial x} dxdt.\end{aligned}$$

By integrating by parts, it is easily seen that

$$\begin{aligned}&\iint_{Q_T} \varphi_1 H_\eta(u_\varepsilon - k)\frac{\partial u_\varepsilon}{\partial t} dxdt \\ = &-\iint_{Q_T} (u_\varepsilon - k) H_\eta(u_\varepsilon - k)\frac{\partial \varphi_1}{\partial t} dxdt - \iint_{Q_T} \varphi_1 (u_\varepsilon - k) h_\eta(u_\varepsilon - k)\frac{\partial u_\varepsilon}{\partial t} dxdt\end{aligned}$$

and hence

$$\lim_{\eta\to 0}\iint_{Q_T} \varphi_1 H_\eta(u_\varepsilon - k)\frac{\partial u_\varepsilon}{\partial t} dxdt = -\iint_{Q_T} |u_\varepsilon - k| \frac{\partial \varphi_1}{\partial t} dxdt. \tag{4.49}$$

Similarly, we have

$$\begin{aligned}&\lim_{\eta\to 0}\iint_{Q_T} \varphi_1 H_\eta(u_\varepsilon - k)\frac{\partial B(u_\varepsilon)}{\partial x} dxdt \\ = &-\iint_{Q_T} \text{sgn}(u_\varepsilon - k)(B(u_\varepsilon) - B(k))\frac{\partial \varphi_1}{\partial x} dxdt \\ &-\text{sgn}k \int_0^T \varphi_1 \left(B(u_\varepsilon) - B(k)\right)\Big|_{x=0}^{x=1} dt.\end{aligned} \tag{4.50}$$

In addition, since

$$\begin{aligned}&\iint_{Q_T} \varphi_1 H_\eta(u_\varepsilon - k)\frac{\partial^2 A_\varepsilon(u_\varepsilon)}{\partial x^2} dxdt \\ = &-\iint_{Q_T} H_\eta(u_\varepsilon - k)\frac{\partial A_\varepsilon(u_\varepsilon)}{\partial x}\frac{\partial \varphi_1}{\partial x} dxdt\end{aligned}$$

$$\begin{aligned}
&-\iint_{Q_T}\varphi_1 h_\eta(u_\varepsilon-k)A'_\varepsilon(u_\varepsilon)\left(\frac{\partial u_\varepsilon}{\partial x}\right)^2 dxdt\\
&+\int_0^T H_\eta(u_\varepsilon-k)\frac{\partial A_\varepsilon(u_\varepsilon)}{\partial x}\varphi_1\Big|_{x=0}^{x=1}dt\\
\le\;&-\iint_{Q_T} H_\eta(u_\varepsilon-k)\frac{\partial A_\varepsilon(u_\varepsilon)}{\partial x}\frac{\partial\varphi_1}{\partial x}dxdt\\
&+\int_0^T H_\eta(u_\varepsilon-k)\frac{\partial A_\varepsilon(u_\varepsilon)}{\partial x}\varphi_1\Big|_{x=0}^{x=1}dt,
\end{aligned}$$

there holds

$$\begin{aligned}
&\lim_{\eta\to 0}\iint_{Q_T}\varphi_1 H_\eta(u_\varepsilon-k)\frac{\partial^2 A_\varepsilon(u_\varepsilon)}{\partial x^2}dxdt\\
\le\;&-\iint_{Q_T}\operatorname{sgn}(u_\varepsilon-k)\frac{\partial A_\varepsilon(u_\varepsilon)}{\partial x}\frac{\partial\varphi_1}{\partial x}dxdt \qquad (4.51)\\
&+\int_0^T\operatorname{sgn}(u_\varepsilon-k)\frac{\partial A_\varepsilon(u_\varepsilon)}{\partial x}\varphi_1\Big|_{x=0}^{x=1}dt.
\end{aligned}$$

Therefore

$$\begin{aligned}
&\iint_{Q_T}\operatorname{sgn}(u_\varepsilon-k)\Big\{(u_\varepsilon-k)\frac{\partial\varphi_1}{\partial t}-\\
&\quad-(B(u_\varepsilon)-B(k))\frac{\partial\varphi_1}{\partial x}-\frac{\partial}{\partial x}A_\varepsilon(u_\varepsilon)\frac{\partial\varphi_1}{\partial x}\Big\}dxdt \qquad (4.52)\\
\ge\;&\operatorname{sgn}k\int_0^T\varphi_1\left(B(u_\varepsilon)-B(k)+\frac{\partial}{\partial x}A_\varepsilon(u_\varepsilon)\right)\Big|_{x=0}^{x=1}dt.
\end{aligned}$$

Finally, for any $0\le\varphi_2\in C^\infty(\overline{Q_T})$ with $\varphi_2(0,t)=\varphi_1(0,t)$, $\varphi_1(1,t)=\varphi_2(1,t)$ and $\operatorname{supp}\varphi_2\subset[0,1]\times(0,T)$,

$$\begin{aligned}
&\iint_{Q_T}\Big\{(u_\varepsilon-k)\frac{\partial\varphi_2}{\partial t}-\\
&\quad-(B(u_\varepsilon)-B(k))\frac{\partial\varphi_2}{\partial x}-\frac{\partial}{\partial x}A_\varepsilon(u_\varepsilon)\frac{\partial\varphi_2}{\partial x}\Big\}dxdt\\
=\;&-\int_0^T\varphi_2\left(B(u_\varepsilon)-B(k)+\frac{\partial}{\partial x}A_\varepsilon(u_\varepsilon)\right)\Big|_{x=0}^{x=1}dt\\
&-\iint_{Q_T}\varphi_2\left\{\frac{\partial u_\varepsilon}{\partial t}-\frac{\partial B(u_\varepsilon)}{\partial x}-\frac{\partial^2 A_\varepsilon(u_\varepsilon)}{\partial x^2}\right\}dxdt\\
=\;&-\int_0^T\varphi_1\left(B(u_\varepsilon)-B(k)+\frac{\partial}{\partial x}A_\varepsilon(u_\varepsilon)\right)\Big|_{x=0}^{x=1}dt.
\end{aligned}$$

Combining this with (4.52) yields

$$\iint_{Q_T} \operatorname{sgn}(u_\varepsilon - k)\left\{(u_\varepsilon - k)\frac{\partial \varphi_1}{\partial t} - \right.$$
$$\left. - (B(u_\varepsilon) - B(k))\frac{\partial \varphi_1}{\partial x} - \frac{\partial}{\partial x}A_\varepsilon(u_\varepsilon)\frac{\partial \varphi_1}{\partial x}\right\} dxdt$$
$$+ \iint_{Q_T} \operatorname{sgn} k \left\{(u_\varepsilon - k)\frac{\partial \varphi_2}{\partial t} - \right.$$
$$\left. - (B(u_\varepsilon) - B(k))\frac{\partial \varphi_2}{\partial x} - \frac{\partial}{\partial x}A_\varepsilon(u_\varepsilon)\frac{\partial \varphi_2}{\partial x}\right\} dxdt \geq 0.$$

From this is follows by letting $\varepsilon \to 0$ that u satisfies (4.31). The proof of Theorem 3.4.5 is complete. □

3.4.8 *Some extensions*

The argument developed above can be extended to the strongly degenerate equations more general in form

$$\frac{\partial u}{\partial t} = \frac{\partial}{\partial x}\left(a(x,t,u)\frac{\partial u}{\partial x}\right) + \frac{\partial}{\partial x}b(x,t,u) + c(x,t,u)$$

with $a(x,t,u) \geq 0$ and the double strongly degenerate equations

$$\frac{\partial u}{\partial t} = \frac{\partial}{\partial x}\left(F\left(\frac{\partial}{\partial x}A(u)\right)\right) + \frac{\partial B(u)}{\partial x}$$

with $F'(s) \geq 0$, $A'(s) \geq 0$ or equations of this kind more general in form. In doing this, especially in proving the existence of solutions, some new difficulties occur because of the higher nonlinearity. Since the compactness for the approximate solutions which we can obtain is not strong enough, we must treat a weak limit process.

An interesting result is that when

$$\lim_{s \to \pm\infty} F(s) = \pm\infty$$

and $E_A = \{s; A'(s) = 0\}$ has no interior point, the problem for the equations considered admits a continuous solution, no matter whether $E_F = \{s; F'(s) = 0\}$ has interior point or not (see[YI1]).

3.4.9 *Equations with degeneracy at infinity*

As indicated in the introduction of this chapter, there is a class of equations with degeneracy at infinity, which is also said to be strongly degenerate. It was Bertsch and Passo[BP] who first introduced such kind of equations in consideration. They studied equations of the form

$$\frac{\partial u}{\partial t} = \frac{\partial}{\partial x}\left(a(u)F\left(\frac{\partial u}{\partial x}\right)\right), \tag{4.53}$$

where $a(s)$ and $F(s)$ are appropriately smooth, $a(s) > 0$, $F(-s) = -F(s)$, $F'(s) > 0$ and

$$\lim_{s \to +\infty} F(s) \equiv F_\infty < +\infty. \tag{4.54}$$

An typical example of $F(s)$ is

$$F(s) = \frac{s}{\sqrt{1+s^2}}.$$

Although $E_F = \{s; F'(s) = 0\}$ is empty, (4.53) can be regarded as being degenerate at infinity, since (4.54) implies

$$\underline{\lim}_{s \to +\infty} F'(s) = 0.$$

Degeneracy occurs whenever $\dfrac{\partial u}{\partial x} = +\infty$. One conjectures reasonably that some character of (4.53) might be close to that of the equations

$$\frac{\partial u}{\partial t} = F_\infty \frac{\partial a(u)}{\partial x},$$

especially, the solutions might be discontinuous.

Bertsch and Passo[BP], [PA] discussed the existence, uniqueness and regularity of solutions in certain general sense. They called a function $u \in L^\infty(Q_T) \bigcap BV(Q_T)$ a generalized solution of the Cauchy problem for (4.53) with initial value u_0, if the following conditions are fulfilled:

(1) There exists a continuous function $\overline{F}(x,t)$, such that

$$\overline{F}(x,t) = \lim_{h \to 0} F\left(\frac{u(x+h,t) - u^l(x,t)}{h}\right) = \lim_{h \to 0} F\left(\frac{u(x+h,t) - u^r(x,t)}{h}\right);$$

(2) $\text{ess}\lim_{t \to 0^+} \overline{u}(x,t) = u_0(x)$for almost all $x \in (0,T)$;

(3) For any $\varphi \in C^\infty(\overline{Q_T})$ with compact support in x and $\varphi(x,T) = 0$,

$$\iint_{Q_T} \left(u\frac{\partial \varphi}{\partial t} - a(u)\overline{F}\frac{\partial \varphi}{\partial x} \right) dxdt + \int_{\mathbb{R}} u_0(x)\varphi(x,0)dx = 0.$$

If the generalized solution u satisfies the entropy condition:

$$a(s) \le \frac{a(u^+) - a(u^-)}{u^+ - u^-}(s - u^-) + a(u^-), \quad \forall s \in I(u^-, u^+)$$

at its points of discontinuity where $I(u^-, u^+)$ is the interval with endpoints u^-, u^+, then u is called a entropy solution.

In fact, the entropy solutions can be defined by means of an integral inequality involving an arbitrary nonnegative test function and an arbitrary constant.

3.4.10 *Properties of the curves of discontinuity*

As the generalized solutions of the strongly degenerate equations (4.1) might be discontinuous, it is interesting to know how does the discontinuity emerge and develop and how is the regularity of the curves of discontinuity. We will briefly discuss this problem for equations without terms of lower order,

$$\frac{\partial u}{\partial t} = \frac{\partial^2 A(u)}{\partial x^2}. \tag{4.55}$$

The following situation seems to be typical:

$$A'(s) = 0 \quad \text{for } s < 0, \quad A'(s) > 0 \quad \text{for } s > 0,$$
$$u_0(x) < 0 \quad \text{for } x < 0, \quad u_0(x) > 0 \quad \text{for } x > 0.$$

In this case, one could conjecture that the corresponding BV solution u would have a curve of discontinuity $x = \lambda(t)$ with $\lambda'(t) < 0$ starting from (0,0). Since $u_0(x) < 0$ for $x < 0$ and $A'(s) = 0$ for $s < 0$, we first have

$$u(x,t) = u_0(x) \quad \text{for} \quad x < \lambda(t).$$

Secondly, from the discontinuity condition (4.8) it follows that

$$u^+|_{x=\lambda(t)} = 0,$$

and hence, noticing that $w^-|_{x=\lambda(t)} = 0$, from the discontinuity condition (4.7) we get

$$-u_0(\lambda(t))\lambda'(t) + w^r(\lambda(t), t) = 0.$$

Thus the BV solution $u(x,t)$ and the curve of discontinuity $x = \lambda(t)$ should be a solution of the following free boundary value problem

$$\frac{\partial u}{\partial t} = \frac{\partial^2 A(u)}{\partial x^2}, \quad 0 < t < T, x < \lambda(t), \tag{4.56}$$

$$u(\lambda(t), t) = 0, \quad 0 < t < T, \tag{4.57}$$

$$u(x,0) = u_0(x), \quad x > 0. \tag{4.58}$$

It is easily seen that if $(u(x,t), \lambda(t))$ is a solution of the free boundary value problem (4.56)–(4.58), then

$$\hat{u}(x,t) = \begin{cases} u(x,t), & \text{for } x > \lambda(t), \\ u_0(x), & \text{for } x < \lambda(t) \end{cases}$$

is a generalized solution of the Cauchy problem (4.55), (4.16). The free boundary value problem (4.56)–(4.58) is first studied in [WZQ1] for a special case and then in [LH1] for general case.

3.5 Degenerate Equations in Higher Dimension without Terms of Lower Order

This and the next section are devoted to strongly degenerate equations in higher dimension. The study of equations in higher dimension with strong degeneracy is rather difficult. This can be imagined from the discussions in one dimension presented in §3.3.4. Only the equations without any terms of lower order are relatively easy to treat. So far only such kind of equations have been studied thoroughly (see[BC]). We will start our discussion from this special case.

Consider equations of the form

$$\frac{\partial u}{\partial t} = \Delta A(u) \tag{5.1}$$

with $A'(s) \geq 0$ and $E = \{s; A'(s) = 0\}$ being allowed to have interior points. Assume the initial value condition

$$u(x,0) = u_0(x). \tag{5.2}$$

The solutions of the Cauchy problem (5.1), (5.2) is defined by Definition 3.3.1. We will see below that, just as in the case of weakly degenerate equations, the formulation of the Cauchy problem for strongly degenerate equations of the form (5.1) with solutions defined by Definition 3.3.1 is correct, namely, solutions of the problem thus defined are uniquely determined.

3.5.1 *Uniqueness of bounded and integrable solutions*

Theorem 3.5.1 *Assume that $A(s) \in C^1(\mathbb{R})$ and $A'(s) \geq 0$. Then the Cauchy problem (5.1), (5.2) admits at most one bounded and integrable generalized solution.*

Proof. Let u_1 and u_2 be bounded and integrable generalized solutions of (5.1), (5.2). Denote $z = u_1 - u_2$, $v = A(u_1) - A(u_2)$. Then from Definition 3.3.1, we have

$$\iint_{Q_T} \left(z\frac{\partial \varphi}{\partial t} + v\Delta\varphi \right) dxdt = 0 \tag{5.3}$$

for any $\varphi \in C_0^\infty(Q_T)$.

We wish to assert that $z \equiv 0$ from this equality by choosing special test functions φ.

From the L^p theory for elliptic equations (see for example [SW]), for any $f \in L^p(\mathbb{R}^N)$ $(1 \leq p \leq \infty)$, the problem

$$-\Delta u + \lambda u = f, \quad \text{in } \mathcal{D}'(\mathbb{R}^N) \quad (\lambda > 0) \tag{5.4}$$

has a unique solution $u_\lambda \in L^p(\mathbb{R}^N)$. Defining an operator T_λ by $u_\lambda = T_\lambda f$, one has the estimate

$$\lambda \|T_\lambda f\|_p \leq \|f\|_p, \tag{5.5}$$

where $\|\cdot\|_p$ will denote either the norm of $L^p(\mathbb{R}^N)$ or the norm of $L^p(Q_T)$ depending on the context. Because of (5.5), T_λ also defines an operator $T_\lambda : L^p(Q_T) \to L^p(Q_T)$ $(1 \leq p \leq \infty)$ and (5.5) holds equally for $f \in L^p(\mathbb{R}^N)$ and $f \in L^p(Q_T)$. For any $\gamma \in C_0^\infty(Q_T)$, we wish to set $\varphi = T_\lambda\gamma$ in (5.3).

Since T_λ commutes with differentiations, we have $T_\lambda\gamma \in C^\infty(Q_T)$. Clearly $(T_\lambda\gamma)(x,t) = 0$ for t near 0 and T. Moreover, since $z, v \in L^\infty(Q_T)$, (5.3) clearly continues to hold for $\varphi \in C^\infty(Q_T)\bigcap L^1(Q_T)$ with $\varphi(x,t) = 0$ for t near 0 and T provided that $\dfrac{\partial\varphi}{\partial t}$, $\Delta\varphi$, $|\nabla\varphi| \in L^1(Q_T)$. $\varphi = T_\lambda\gamma$ has these properties. Moreover, $\Delta T_\lambda\gamma = \lambda T_\lambda\gamma - \gamma$, $\dfrac{\partial T_\lambda\gamma}{\partial t} = T_\lambda\dfrac{\partial\gamma}{\partial t}$. Thus, from (5.3), it follows that for $\gamma \in C_0^\infty(Q_T)$,

$$\begin{aligned}&\iint_{Q_T}\left(zT_\lambda\frac{\partial\gamma}{\partial t} + v\left(\lambda T_\lambda\gamma - \gamma\right)\right)dxdt\\ = &\iint_{Q_T}\left(T_\lambda z\frac{\partial\gamma}{\partial t} + \left(\lambda T_\lambda v - v\right)\gamma\right)dxdt = 0,\end{aligned}$$

where the first equality is due to the obvious symmetry of T_λ and the absolute convergence of all integrals involved. This means that in the sense of distributions,

$$\frac{\partial T_\lambda z}{\partial t} = \lambda T_\lambda v - v. \tag{5.6}$$

Denote

$$g_\lambda(t) = \int_{\mathbb{R}^N} z(x,t)T_\lambda z(x,t)dx.$$

Since $z, T_\lambda z \in L^\infty(Q_T)\cap L^1(Q_T)$, $g_\lambda(t)$ is well defined for almost all $t \in (0,T)$. If we can prove that

$$\lim_{\lambda\to 0} g_\lambda(t) = 0, \quad \text{a.e. } t \in (0,T), \tag{5.7}$$

then, since

$$g_\lambda(t) = \int_{\mathbb{R}^N}\left(\lambda T_\lambda z - \Delta T_\lambda z\right)T_\lambda z dx = \lambda\|T_\lambda z(\cdot,t)\|_2^2 + \|\nabla T_\lambda z(\cdot,t)\|_2^2,$$

we have, for almost all $t \in (0,T)$,

$$\lim_{\lambda\to 0}\lambda T_\lambda z(\cdot,t) = 0, \quad \lim_{\lambda\to 0}\nabla T_\lambda z(\cdot,t) = 0 \quad \text{in} \quad L^2(\mathbb{R}^N).$$

It follows that, in particular, in the sense of distributions,

$$\lim_{\lambda\to 0}\lambda T_\lambda z(\cdot,t) = 0, \quad \lim_{\lambda\to 0}\Delta T_\lambda z(\cdot,t) = 0$$

and hence

$$z(\cdot,t)=\lim_{\lambda\to 0}(\lambda T_\lambda z(\cdot,t)-\Delta T_\lambda z(\cdot,t))=0.$$

This shows that $z=u_1-u_2=0$ a.e in Q_T.

Now we turn to the proof of (5.7). Let J_ε be the standard mollifier in t. Then from (5.6) and the symmetry of T_λ, we have

$$\begin{aligned}&\frac{d}{dt}\int_{\mathbb{R}^N}T_\lambda(J_\varepsilon z)J_\varepsilon z dx=2\int_{\mathbb{R}^N}\frac{\partial}{\partial t}T_\lambda(J_\varepsilon z)J_\varepsilon z dx\\ =\;& 2\int_{\mathbb{R}^N}J_\varepsilon\left(\frac{\partial}{\partial t}T_\lambda z\right)J_\varepsilon z dx=2\int_{\mathbb{R}^N}J_\varepsilon\left(\lambda T_\lambda v-v\right)J_\varepsilon z dx.\end{aligned}$$

Integrating with respect to t gives

$$\begin{aligned}&\int_{\mathbb{R}^N}T_\lambda(J_\varepsilon z(x,t))J_\varepsilon z(x,t)dx\\ =\;&\int_{\mathbb{R}^N}T_\lambda(J_\varepsilon z(x,0))J_\varepsilon z(x,0)dx+2\iint_{Q_t}J_\varepsilon\left(\lambda T_\lambda v-v\right)J_\varepsilon z dx ds.\end{aligned}$$

Letting $\varepsilon\to 0$ and noticing that the nondecreasingness of $A(s)$ implies $z(x,t)w(x,t)\ge 0$, we derive

$$\begin{aligned}g_\lambda(t)=\;&2\iint_{Q_t}\left(\lambda T_\lambda v-v\right)z dx ds\\ &\le 2\lambda\iint_{Q_t}zT_\lambda v=2\lambda\iint_{Q_t}vT_\lambda z dx ds.\end{aligned}\tag{5.8}$$

Since $z\in L^\infty(Q_T)$ and $A(s)\in C^1(\mathbb{R})$ imply $|v|\le C|z|$, we further obtain

$$g_\lambda(t)\le C\lambda\iint_{Q_t}|zT_\lambda z|dx ds.\tag{5.9}$$

From (5.4) it is easily seen that

$$\int_{\mathbb{R}^N}|\nabla T_\lambda f|^2dx+\lambda\int_{\mathbb{R}^N}|T_\lambda f|^2dx\le C\int_{\mathbb{R}^N}|f|^2dx.$$

Using Sobolev's inequality gives

$$\int_{\mathbb{R}^N}|T_\lambda f|^2dx\le C\int_{\mathbb{R}^N}|f|^2dx.$$

Thus from (5.9) it follows that

$$g_\lambda(t) \le C\lambda \int_{\mathbb{R}^N} |z|^2 dx$$

which and $z \in L^2(Q_T)$ imply (5.7). The proof of Theorem 3.5.1 is complete. □

In order to prove the uniqueness of bounded and integrable solutions of (5.1), (5.2) under the assumption weaker than $A(s) \in C^1(\mathbb{R})$, we will apply the following lemma.

Lemma 3.5.1 *Let T_λ be the operator introduced in the proof of Theorem 3.5.1 and $f \in L^\infty(\mathbb{R}^N) \cap L^1(\mathbb{R}^N)$. Then*

$$\lim_{\lambda\to 0} \lambda \|T_\lambda f\|_\infty = 0.$$

Proof. Let $K(x)$ be a solution of the equation

$$-\Delta u + u = \delta(x)$$

where $\delta(x)$ is the Dirac function. Then (see[SW])

$$C(r) \equiv \sup_{|x|\ge r} K(x) < \infty, \quad K \in L^1(B),$$

where $B = \{x; |x| \le 1\}$. Using $K(x)$, it is not difficult to verify that the solution $T_\lambda f$ of (5.4) can be expressed by

$$\lambda(T_\lambda f)(x) = \lambda^{n/2} \int_{\mathbb{R}^N} K(\sqrt{\lambda}(x-y)) f(y) dy.$$

Thus, for any fixed $r > 0$,

$$\begin{aligned} \lambda|(T_\lambda f)(x)| \le \; & \lambda^{n/2} \int_{\{\sqrt{\lambda}|x-y|\ge r\}} |K(\sqrt{\lambda}(x-y)) f(y)| dy \\ & + \lambda^{n/2} \int_{\{\sqrt{\lambda}|x-y|\le r\}} |K(\sqrt{\lambda}(x-y)) f(y)| dy \\ \le \; & \lambda^{n/2} C(r) \|f\|_1 + \|f\|_\infty \int_{\{|y|\le r\}} K(y) dy. \end{aligned}$$

Letting $\lambda \to 0$ gives

$$\overline{\lim_{\lambda\to 0}} \lambda |(T_\lambda f)(x)| \le \|f\|_\infty \int_{\{|y|\le r\}} K(y) dy,$$

and the conclusion of Lemma 3.5.1 by setting $r \to 0$. □

Theorem 3.5.2 *Assume that $A(s) \in C(\mathbb{R})$ and $A(s)$ is nondecreasing. Then the Cauchy problem (5.1), (5.2) admits at most one bounded and integrable generalized solution.*

Proof. Checking the proof of Theorem 3.5.1, we see that, it suffices to verify (5.7) under the assumption of the present theorem.

From the boundedness of u_1, u_2 and the continuity of $A(s)$, it is easy to see that for any $\xi > 0$, there exists an $\eta > 0$ such that

$$\{(x,t); |w(x,t)| > \xi\} \subset \{(x,t); |z(x,t)| > \eta\}$$

up to a set of measure zero. This and $z \in L^1(Q_T)$ imply

$$C(\xi,t) \equiv \text{mes}\,\{(x,t); |w(x,t)| > \xi\} < \infty$$

for almost all $t \in (0,T)$.

Now we estimate the right hand side of (5.8) which still holds under the assumption $A(s) \in C(\mathbb{R})$. For fixed $\xi > 0$, we have

$$\begin{aligned} &\left|\lambda \int_{\mathbb{R}^N} wT_\lambda z dx\right| \\ \leq\ & \left|\lambda \iint_{\{x;|w(x,t)|>\xi\}} wT_\lambda z\right| \left|\lambda \iint_{\{x;|w(x,t)|\leq\xi\}} wT_\lambda z\right| \\ \leq\ & C(\xi,t)\|w(\cdot,t)\|_\infty \|\lambda T_\lambda z(\cdot,t)\|_\infty + \xi\lambda\|T_\lambda z(\cdot,t)\|_1 \end{aligned}$$

and hence, by Lemma 3.5.1,

$$\overline{\lim_{\lambda\to 0}} \left|\lambda \int_{\mathbb{R}^N} wT_\lambda z dx\right| \leq \xi\lambda\|T_\lambda z(\cdot,t)\|_1.$$

Thus

$$\overline{\lim_{\lambda\to 0}} \left|\lambda \int_{\mathbb{R}^N} wT_\lambda z dx\right| = 0 \tag{5.10}$$

for almost all $t \in (0,T)$.

On the other hand,

$$\begin{aligned} &\left|\lambda \int_{\mathbb{R}^N} w(x,t)T_\lambda z(x,t) dx\right| = \left|\lambda \int_{\mathbb{R}^N} z(x,t)T_\lambda w(x,t) dx\right| \\ \leq\ & \|\lambda T_\lambda w(x,t)\|_\infty \|z(x,t)\|_1 \leq \|w(x,t)\|_\infty \|z(x,t)\|_1. \end{aligned} \tag{5.11}$$

Letting $\lambda \to 0$ in (5.8) and using (5.10), (5.11) and the dominated convergence theorem yield (5.7) and complete the proof of our theorem. □

3.5.2 *A lemma on weak convergence*

First we introduce a lemma whose proof can be found in [TAR].

Lemma 3.5.2 *Let $\Omega \subset \mathbb{R}^N$ be a domain, $\{u_k\}$ be a uniformly bounded sequence in $L^\infty(\Omega)$, namely, $|u_k(x)| \leq M$ a.e. in Ω, with M independent of k. Then there exist a subsequence $\{u_{k_j}\}$ of $\{u_k\}$, and a family of probability measures $\{\mu_x\}_{x\in\Omega}$ with supp$\mu_x \subset [-M, M]$ such that for any $F \in C(\mathbb{R})$,*

$$F(u_{k_j}(x)) \stackrel{*}{\rightharpoonup} \int_{\mathbb{R}} F(\lambda)\mu_x(d\lambda), \qquad in \quad L^\infty(\Omega).$$

Using this lemma, we can prove the following result on the weak convergence.

Lemma 3.5.3 *Let $\Omega \subset \mathbb{R}^N$ be a bounded domain, $|u_k(x)| \leq M$ a.e in Ω and*

$$u_k(x) \stackrel{*}{\rightharpoonup} u(x), \qquad in \quad L^\infty(\Omega).$$

Assume that $A(s)$, $B(s) \in C(\mathbb{R})$ and $A(s)$ is nondecreasing. If for any $\alpha \in A(R)$, $B(A^{-1}(\alpha))$ contains only a single point, and

$$A(u_k(x)) \to w(x), \quad B(u_k(x)) \to v(x), \quad a.e.\ in \quad \Omega,$$

then

$$A(u(x)) = w(x), \quad B(u(x)) = v(x), \quad a.e.\ in \quad \Omega.$$

Proof. Applying Lemma 3.5.2, we assert that there exist a family of measures $\{\mu_x\}_{x\in\Omega}$ with supp$\mu_x \subset [-M, M]$, sup$|u_k(x)| \leq M$, and a subsequence $\{u_{k_j}\}$, such that for any $\varphi(s) \in C(\mathbb{R})$,

$$\varphi(w(x)) = \lim_{j\to\infty} \varphi(A(u_{k_j}(x))) = \int_{\mathbb{R}} \varphi(A(\lambda))\mu_x(d\lambda) \qquad (5.12)$$

holds for $x \in \Omega \setminus E_\varphi$ with mes$E_\varphi = 0$. Denote

$$K = \sup_{|s|\leq M} |A(s)|.$$

Choose a sequence of function $\{\varphi_k\}$, which is dense in $C[-K, K]$, and denote

$$E = \bigcup_{k=1}^{\infty} E_{\varphi_k}.$$

It is easy to see that mes$E = 0$ and for any $\varphi \in C(\mathbb{R})$, (5.12) holds for $x \in \Omega \setminus E$.

Using Lemma 3.5.2 again yields

$$u_{k_j}(x) \stackrel{*}{\rightharpoonup} \int_{\mathbb{R}} \lambda \mu_x(d\lambda) = u(x), \quad \text{in } L^\infty(\Omega). \tag{5.13}$$

Without loss of generality, we may suppose that for any $x \in \Omega \setminus E$,

$$|u_{k_j}(x)| \le M, \quad \int_{\mathbb{R}} \lambda \mu_x(d\lambda) = u(x).$$

For any fixed $x \in \Omega \setminus E$, we assert

$$F_x \equiv \{s \in R; A(s) = w(x)\} \ne \emptyset.$$

In fact, if for some $x \in \Omega \setminus E$, $F_x = \emptyset$, then by the nondecreasingness of $A(s)$, we must have $w(x)\overline{\in}[A(-M), A(M)]$. However

$$A(u_{k_j}(x)) \in [A(-M), A(M)],$$

which implies that

$$w(x) = \lim_{j\to\infty} A(u_{k_j}(x)) \in [A(-M), A(M)],$$

and we get a contradiction. Since $F_x \ne \emptyset$ and $A(s)$ is nondecreasing and continuous, F_x must be a closed interval or a set of single point.

Now we prove that

$$\text{supp}\mu_x \subset F_x, \qquad \text{for} \quad x \in \Omega \setminus E. \tag{5.14}$$

Suppose for some $x_0 \in \Omega \setminus E$,

$$\text{supp}\mu \equiv \text{supp}\mu_{x_0} \not\subset F_{x_0} \equiv F,$$

namely,

$$\mu([-M, M] \setminus F) > 0.$$

Let $\varphi(s) = -|s - w_0|$, $w_0 = w(x_0)$. From (5.12) and the definition of F, we see that

$$\begin{aligned}\varphi(w(x_0)) = 0 &= \int_{\mathbb{R}} \varphi(A(\lambda))\mu(d\lambda)\\ &= -\int_{-M}^{M} |A(\lambda) - w_0|\mu(d\lambda)\\ &= -\int_{[-M,M]\setminus F} |A(\lambda) - w_0|\mu(d\lambda) < 0.\end{aligned}$$

This contradiction shows the validity of (5.14).

From (5.13), (5.14) it is easy to see that

$$\inf F_x \le u(x) = \int_{\mathbb{R}} \lambda\mu_x(d\lambda) = \int_{F_x} \lambda\mu_x(d\lambda) \le \sup F_x,$$

and hence

$$u(x) = \int_{\mathbb{R}} \lambda\mu_x(d\lambda) \in F_x.$$

This proves $A(u(x)) = w(x)$, the first equality in our lemma.

To prove the second equality, let

$$E_x = \{s \in R; B(s) = v(x)\}\,.$$

It suffices to show that $F_x \subset E_x$ a.e. in Ω, since $u(x) \in F_x$ implies $B(u(x)) = v(x)$. For fixed $x_0 \in \Omega \setminus E$, let

$$\varepsilon_k = |A(u_k(x_0)) - w(x_0)|\,.$$

Then

$$[a_k, b_k] \equiv A^{-1}\left([w(x_0) - \varepsilon_k, w(x_0) + \varepsilon_k]\right) \to A^{-1}(w(x_0)).$$

Hence, by the continuity of $B(s)$,

$$[\alpha_k, \beta_k] \equiv B([a_k, b_k]) \to B(A^{-1}(w(x_0))).$$

Since $u_k(x_0) \in [a_k, b_k]$, we have $B(u_k(x_0)) \in [\alpha_k, \beta_k]$ and hence $v(x_0) \in B(A^{-1}(w(x_0)))$. However by the assumption of the lemma, $B(A^{-1}(w(x_0)))$ is a one point set. Thus $\{v(x_0)\} = B(A^{-1}(w(x_0)))$, or

$$A^{-1}(w(x_0)) \subset B^{-1}(v(x_0)),$$

namely, $F_{x_0} \subset E_{x_0}$. The proof is complete. □

3.5.3 Existence of solutions

As indicated in §3.3.3, the Cauchy problem (5.1), (5.2) admits BV solutions provided that $A(s)$ and $u_0(x)$ are appropriately smooth (Theorem 3.3.3). For initial value with weaker regularity, we can also establish the existence of solutions, however what we can obtain are not BV solutions in general. To this purpose, we will apply the theory of compensated compactness.

Theorem 3.5.3 *Assume that $A(s) \in C^1(\mathbb{R})$ and $u_0(x) \in L^\infty(\mathbb{R}^N)$ with $A'(s) \geq 0$ and $A(u_0) \in W^{1,2}_{loc}(\mathbb{R}^N)$. Then the Cauchy problem (5.1), (5.2) admits a bounded generalized solution.*

Proof. As in §3.3.3, we consider the regularized problem

$$\frac{\partial u_\varepsilon}{\partial t} = \Delta A_\varepsilon(u_\varepsilon), \tag{5.15}$$

$$u_\varepsilon(x,0) = u_{0\varepsilon}(x), \tag{5.16}$$

where $A_\varepsilon(s)$ and $u_{0\varepsilon}$ are the locally uniform smooth approximation of $A(s)$ and u_0 respectively, with $A'_\varepsilon(s) > 0$, $A_\varepsilon(0) = 0$ and $\sup |u_{0\varepsilon}(x)| \leq M$.

From the maximum principle, we first have

$$\sup_{Q_T} |u_\varepsilon(x,t)| \leq M. \tag{5.17}$$

Next, multiply (5.15) by $\dfrac{\partial}{\partial t} A_\varepsilon(u_\varepsilon)\omega_\lambda(x)$ and integrate over Q_t, where

$$\omega_\lambda(x) = \exp\left(-\lambda\sqrt{1+|x|^2}\right).$$

Then we obtain

$$\iint_{Q_t} A'_\varepsilon(u_\varepsilon)\left(\frac{\partial u_\varepsilon}{\partial t}\right)^2 \omega_\lambda(x)dxds = \iint_{Q_t} \frac{\partial A_\varepsilon(u_\varepsilon)}{\partial t} \triangle A_\varepsilon(u_\varepsilon)\omega_\lambda(x)dxds.$$

Integrating by parts, noticing that $|\nabla \omega_\lambda(x)| \leq C\omega_\lambda(x)$ and using Young's inequality yield

$$\begin{aligned}
&\iint_{Q_t} A'_\varepsilon(u_\varepsilon)\left(\frac{\partial u_\varepsilon}{\partial t}\right)^2 \omega_\lambda(x)dxds \\
= \ & -\frac{1}{2}\iint_{Q_t} \frac{\partial}{\partial t}\left(|\nabla A_\varepsilon(u_\varepsilon)|^2 \omega_\lambda(x)\right) dxds \\
& -\iint_{Q_t} A'_\varepsilon(u_\varepsilon)\frac{\partial u_\varepsilon}{\partial t} \nabla A_\varepsilon(u_\varepsilon) \nabla \omega_\lambda(x)dxds
\end{aligned}$$

$$\le -\frac{1}{2}\int_{\mathbb{R}^N} |\nabla A_\varepsilon(u_\varepsilon)|^2\omega_\lambda(x)dx + \frac{1}{2}\int_{\mathbb{R}^N} |\nabla A_\varepsilon(u_{0\varepsilon})|^2\omega_\lambda(x)dx$$
$$+\frac{1}{2}\iint_{Q_t} A'_\varepsilon(u_\varepsilon)\left(\frac{\partial u_\varepsilon}{\partial t}\right)^2 \omega_\lambda(x)dxds + C\iint_{Q_t} |\nabla A_\varepsilon(u_\varepsilon)|^2\omega_\lambda(x)dxds,$$

namely,

$$\iint_{Q_t} A'_\varepsilon(u_\varepsilon)\left(\frac{\partial u_\varepsilon}{\partial t}\right)^2 \omega_\lambda(x)dxds + \int_{\mathbb{R}^N} |\nabla A_\varepsilon(u_\varepsilon)|^2 \omega_\lambda(x)dx$$
$$\le \int_{\mathbb{R}^N} |\nabla A_\varepsilon(u_{0\varepsilon})|^2 \omega_\lambda(x)dx + C\iint_{Q_t} |\nabla A_\varepsilon(u_\varepsilon)|^2 \omega_\lambda(x)dxds,$$

from which, using Gronwall's inequality, it follows that

$$\iint_{Q_T} A'_\varepsilon(u_\varepsilon)\left(\frac{\partial u_\varepsilon}{\partial t}\right)^2 \omega_\lambda(x)dxdt \le C, \tag{5.18}$$

$$\iint_{Q_T} |\nabla A_\varepsilon(u_\varepsilon)|^2 \omega_\lambda(x)dxdt \le C. \tag{5.19}$$

Denote $v_\varepsilon = A_\varepsilon(u_\varepsilon)$. From (5.17), (5.18), (5.19) we see that there exists a subsequence of $\{u_\varepsilon\}$, denoted still by $\{u_\varepsilon\}$, such that

$$u_\varepsilon(x,t) \stackrel{*}{\rightharpoonup} u(x,t), \quad \text{in } L^\infty(Q_T), \tag{5.20}$$

$$v_\varepsilon(x,t) \to w(x,t), \quad \text{a.e. in } Q_T, \tag{5.21}$$

where $u \in L^\infty(Q_T)$, $w \in W^{1,2}_{loc}(Q_T)$ with

$$\iint_{Q_T} |\nabla w|^2 \omega_\lambda(x)dxdt < +\infty.$$

Since $A_\varepsilon(s)$ approximates $A(s)$ locally uniformly, from the uniform boundedness of $\{u_\varepsilon\}$ and (5.21), we have

$$w_\varepsilon \equiv A(u_\varepsilon) \to w(x,t), \quad \text{a.e in } Q_T. \tag{5.22}$$

Using Lemma 3.5.3, we then conclude from (5.21) and (5.22) that $w(x,t) = A(u(x,t))$ a.e. in Q_T.

Finally, it is easy to check that u satisfies the integral equality in Definition 3.3.1. and the initial value condition in the sense of Definition 3.3.1. The proof is complete. □

Theorem 3.5.4 *If in addition to the conditions in Theorem 3.5.3, assume that $u_0 \in L^1(\mathbb{R}^N)$. Then the Cauchy problem (5.1), (5.2) admits a generalized solution $u \in L^1(Q_T) \bigcap L^\infty(Q_T)$.*

Proof. Since $u_0 \in L^1(\mathbb{R}^N)$, we may choose the approximate initial value $u_{0\varepsilon}$ to have compact support. Instead of the regularized Cauchy problem (5.15), (5.16), we consider the first boundary value problem for (5.15) with initial value condition (5.16) and boundary value condition

$$u_\varepsilon|_{\partial B_{R_\varepsilon}} = 0 \tag{5.23}$$

where $R_\varepsilon > 0$ is a constant such that $\text{supp} u_{0\varepsilon} \subset B_{R_\varepsilon}$ with $\lim\limits_{\varepsilon\to 0} R_\varepsilon = \infty$.

From the maximum principle, we have

$$\sup_{Q^\varepsilon} |u_\varepsilon(x,t)| \le M \tag{5.24}$$

for a certain constant M independent of ε, where $Q^\varepsilon = B_{R_\varepsilon} \times (0,T)$.

Similar to the proof of Theorem 3.4.3, but instead of multiplying (5.15) by $\dfrac{\partial A_\varepsilon(u_\varepsilon)}{\partial t}\omega_\lambda(x)$, using $\dfrac{\partial A_\varepsilon(u_\varepsilon)}{\partial t}$ to multiply (5.15), we may obtain, instead of (5.18), (5.19),

$$\iint_{Q^\varepsilon} A'_\varepsilon(u_\varepsilon)\left(\frac{\partial u_\varepsilon}{\partial t}\right)^2 dxdt \le C, \tag{5.25}$$

$$\iint_{Q^\varepsilon} |\bigtriangledown A_\varepsilon(u_\varepsilon)|^2 dxdt \le C. \tag{5.26}$$

Here we notice that to calculate the integral $\displaystyle\iint_{Q_t^\varepsilon} \frac{\partial A_\varepsilon(u_\varepsilon)}{\partial t} \triangle A_\varepsilon(u_\varepsilon)dxds$ with $Q_t^\varepsilon = B_{R_\varepsilon} \times (0,t)$, by integrating by parts, the fact that the boundary integral vanishes follows from the boundary value condition (5.23) and $A_\varepsilon(0) = 0$.

The compactness which follows from (5.24), (5.25), (5.26) implies the existence of a subsequence of $\{u_\varepsilon\}$, converging in the sense as in the proof of Theorem 3.5.3. The limit function u is the generalized solution of the Cauchy problem (5.1), (5.2).

To prove that $u \in L^1(Q_T)$ in addition to $u \in L^\infty(Q_T)$, we multiply (5.16) by $H_\eta(u_\varepsilon)$ and integrate over Q_t^ε. Then we obtain

$$\frac{\partial}{\partial t}\int_{B_{R_\varepsilon}} \theta_\eta(u_\varepsilon)dx = \iint_{Q_t^\varepsilon} H_\eta(u_\varepsilon) \triangle A_\varepsilon(u_\varepsilon)dxds.$$

Here $H_\eta(s)$ is the function introduced in the proof of Theorem 3.3.2 and

$$\theta_\eta(s) = \int_0^s H_\eta(\tau)d\tau.$$

Integrating by parts and using (5.24) again yield

$$\frac{\partial}{\partial t}\int_{B_{R_\varepsilon}} \theta_\eta(u_\varepsilon)dx = -\iint_{Q_t^\varepsilon} A'_\varepsilon(u_\varepsilon)h_\eta(u_\varepsilon)|\bigtriangledown u_\varepsilon|^2dxds \le 0.$$

Hence

$$\int_{B_{R_\varepsilon}} \theta_\eta(u_\varepsilon)dx \le \int_{B_{R_\varepsilon}} \theta_\eta(u_{0\varepsilon})dx.$$

Letting $\eta \to \infty$ gives

$$\int_{B_{R_\varepsilon}} |u_\varepsilon(x,t)|dx \le \int_{B_{R_\varepsilon}} |u_{0\varepsilon}(x)|dx.$$

From this and $u_0 \in L^1(\mathbb{R}^N)$ we may conclude that $u \in L^1(Q_T)$. The proof of Theorem 3.5.4 is complete. □

In Theorem 3.5.3 and Theorem 3.5.4, the existence of solutions is established under the assumption $A(s) \in C^1(\mathbb{R})$. In case that $A(s)$ is merely a nondecreasing and continuous function, we can also obtain a solution provided the condition on the initial value u_0 is a little strengthened. To this purpose we will use the following result whose proof can be found in [KRU].

Lemma 3.5.4 *Let $Q = (a,b) \times (0,T)$. Assume that $u \in L^\infty(Q)$ satisfies*

$$\iint_Q |u(x+\Delta x,t) - u(x,t)|\, dxdt \le C|\Delta x|$$

and for any $\varphi \in C_0^\infty(Q)$, there holds

$$\begin{aligned}&\left|\iint_Q \varphi(u(x,t+\Delta t) - u(x,t))dxdt\right| \\ \le\ & C|\Delta t| \max_Q (|\varphi| + |\nabla\varphi|).\end{aligned}$$

Then

$$\iint_Q |u(x,t+\Delta t) - u(x,t)|\, dxdt \le C|\Delta t|^{1/2}.$$

Here the value of u outside Q is regarded as zero.

Theorem 3.5.5 *Assume that $A(s)$ is nondecreasing and continuous and $u_0 \in L^\infty(\mathbb{R}^N) \cap BV(\mathbb{R}^N)$. Then the Cauchy problem (5.1), (5.2) admits a BV_x solution.*

Proof. As in §3.3.3, under the assumption $u_0 \in BV(\mathbb{R}^N) \cap L^\infty(\mathbb{R}^N)$, for the approximate solution u_ε, we have the estimate (5.17) on the maximum norm and the estimate

$$\sup_{0<t<T} \int_{\mathbb{R}^N} |\nabla u_\varepsilon(x,t)|\, \omega_\lambda(x) dx \le C,$$

which implies that for any bounded domain $G \subset \mathbb{R}^N$ and any constant $0 < \delta < T$, there holds

$$\iint_Q |u_\varepsilon(x+\Delta x, t) - u_\varepsilon(x,t)|\, dxdt \le C|\Delta x|, \tag{5.27}$$

where $Q = G \times (\delta, T)$ and the constant C depends only on G.

Furthermore, from the equation (5.15) we see that for any $\varphi \in C_0^\infty(Q)$,

$$\begin{aligned} &\iint_Q \varphi(u_\varepsilon(x,t+\Delta t) - u_\varepsilon(x,t)) dxdt \\ = \;& \Delta t \iint_Q \varphi \int_0^1 \frac{\partial}{\partial t} u_\varepsilon(x, t+\sigma\Delta t) d\sigma dxdt \\ = \;& \Delta t \int_0^1 d\sigma \iint_Q \varphi \Delta A_\varepsilon(u_\varepsilon) dxdt \\ = \;& -\Delta t \int_0^1 d\sigma \iint_Q \nabla\varphi \nabla A_\varepsilon(u_\varepsilon) dxdt. \end{aligned}$$

Hence, using (5.19) gives

$$\begin{aligned} &\left| \iint_Q \varphi(u_\varepsilon(x,t+\Delta t) - u_\varepsilon(x,t)) dxdt \right| \\ \le \;& C|\Delta t| \int_0^1 d\sigma \left| \iint_Q \nabla\varphi \nabla A_\varepsilon(u_\varepsilon) dxdt \right| \le C|\Delta t| \end{aligned} \tag{5.28}$$

with constant C depending only on G.

Combining (5.28) with (5.27) and using Lemma 3.5.4, we derive

$$\iint_Q |u_\varepsilon(x, t + \Delta t) - u_\varepsilon(x, t)|\, dxdt \le C|\Delta t|^{1/2}. \tag{5.29}$$

From (5.27) and (5.29), it follows that there exists a subsequence of $\{u_\varepsilon\}$, denoted still by $\{u_\varepsilon\}$, such that

$$\lim_{\varepsilon\to 0} u_\varepsilon(x, t) = u(x, t), \quad \text{a.e. in } Q_T.$$

It is easy to verify that the limit function u is a generalized solution of the Cauchy problem (5.1), (5.2) and $u \in L^\infty(Q_T) \cap BV_x(Q_T)$. The proof is complete. □

3.5.4 *Finite propagation of disturbances*

There are a lot of papers discussing the properties of solutions of weakly degenerate equations. However the study in this aspect for strongly degenerate equations is very few. In what follows we present a result on the propagation of disturbances for such equations. To prove the result, we need the following embedding theorem, for the proof, we refer to [LU2].

Lemma 3.5.5 *Let Ω be a bounded domain of $\mathbb{R}^N$ with smooth boundary and $N > 2$. Then for any $u \in H_0^1(\Omega)$, there holds*

$$\left(\int_\Omega |u|^m dx\right)^{1/m} \le C|\Omega|^{1/m-1/2^*} \left(\int_\Omega |\nabla u|^2 dx\right)^{1/2},$$

where $|\Omega|$ denotes the measure of Ω, C is a constant independent of Ω and u, and

$$2^* = \frac{2N}{N-2}, \quad m \in [1, 2^*].$$

We need also the following iteration lemma.

Lemma 3.5.6 *Assume that $F(s)$ is a nonnegative and bounded function on $[0, +\infty)$ satisfying*

$$F(R) \le \frac{C}{R^\alpha} F(2R)^\beta, \qquad \forall 0 < R \le R_0,$$

where $\alpha > 0$, $\beta > 1$ and C is a constant independent of R. Then for large R_0,

$$\lim_{m\to\infty} F\left(\frac{R_0}{2^m}\right) = 0.$$

Proof. From the assumption, we have

$$F\left(\frac{R_0}{2^m}\right) \le C\frac{2^{m\alpha}}{R_0^\alpha} F\left(\frac{R_0}{2^{m-1}}\right)^\beta.$$

By iteration we obtain

$$F\left(\frac{R_0}{2^m}\right) \le \left(\frac{C}{R_0^\alpha}\right)^{\alpha_m} (2^\alpha)^{\beta_m} F(R_0)^{\beta^m},$$

where

$$\alpha_m = \sum_{i=0}^{m-1} \beta^i = \frac{\beta^m - 1}{\beta - 1} \ge \beta^{m-1},$$

$$\beta_m = \sum_{i=0}^{m-1} (m-i)\beta^i = \frac{\beta(\beta^m - 1) - m(\beta - 1)}{(\beta-1)^2} \le \beta^m \frac{\beta}{(\beta-1)^2}.$$

If we choose R_0 such that

$$\lambda \equiv \left(\frac{C}{R_0^\alpha}\right)^{1/\beta} (2^\alpha)^{\beta/(\beta-1)^2} F(R_0) < 1,$$

then we have

$$F\left(\frac{R_0}{2^m}\right) \le \lambda^{\beta^m} \to 0, \qquad \text{as} \quad m \to \infty$$

and thus the lemma is proved. □

Theorem 3.5.6 *Assume that $A(s)$ and $u_0(x)$ are appropriately smooth with $A'(s) \ge 0$ and*

$$\Psi(s) \equiv \int_0^s A(\sigma)d\sigma \ge \gamma |A(s)|^p \tag{5.30}$$

for some constants $\gamma > 0$ and $1 \le p < 2$. Let u be a BV solution of the Cauchy problem (5.1), (5.2). If

$$supp u_0 \subset B_L$$

for some constant L, then there exists a constant L_1, such that

$$\mathrm{supp}u(0,t) \subset B_{L_1}$$

for almost all $t \in (0,T)$.

Proof. From the definition of generalized solutions, we have

$$\iint_{Q_T} \varphi \frac{\partial u}{\partial t} + \iint_{Q_T} \nabla w \nabla \varphi dxdt = 0 \tag{5.31}$$

for any $\varphi \in C_0^\infty(Q_T)$, where $w = A(u)$. Denote by $B_R = B_R(x_0)$ the ball of radius R centered at x_0. Let $0 \le \psi(t) \in C_0^\infty(0,T)$ and $0 \le h(x) \in C_0^\infty(B_{2R})$, $h(x) = 1$ in B_R and

$$\left|\frac{\partial h}{\partial x_i}\right| \le \frac{C}{R}, \quad \left|\frac{\partial^2 h}{\partial x_i \partial x_j}\right| \le \frac{C}{R^2}, \quad (i,j = 1,\cdots,N).$$

Choosing $\varphi = h^2\psi\hat{w}$ in (5.31), we derive

$$\iint_{Q_T} h^2\psi \frac{\partial \Psi(u)}{\partial t} + \iint_{Q_T} \psi \left|\nabla w\right|^2 h^2 dxdt - \iint_{Q_T} \psi|w|^2 \mathrm{div}(h\nabla h)dxdt = 0,$$

from which it follows by the arbitrariness of $\psi(x)$ that for any $t \in (0,T)$,

$$\iint_{Q_t} h^2 \frac{\partial \Psi(u)}{\partial t} + \iint_{Q_t} \left|\nabla w\right|^2 h^2 dxds - \iint_{Q_t} |w|^2 \mathrm{div}(h\nabla h)dxdt = 0,$$

or

$$\int_{B_{2R}} h^2\Psi(u)dx - \int_{B_{2R}} h^2\Psi(u_0)dx + \int_0^t\int_{B_{2R}} |\nabla w|^2 h^2 dxds$$
$$- \int_0^t\int_{B_{2R}} |w|^2 \mathrm{div}(h \nabla h)dxds = 0.$$

Thus, if $B_{2R} \cap \mathrm{supp}u_0 = \emptyset$, then $\displaystyle\int_{B_{2R}} h^2\Psi(u_0)dx = 0$ and using (5.30) yields

$$\sup_{0<t<T}\int_{B_{2R}} |w|^p h^2 dx + \int_0^T\int_{B_{2R}} \left|\nabla w\right|^2 h^2 dxdt$$
$$\le C\int_0^T\int_{B_{2R}} |w|^2 \left|\mathrm{div}(h\nabla h)\right| dxdt \le \frac{C}{R^2}\int_0^T\int_{B_{2R}} |w|^2 dxdt,$$

which implies, in particular, that

$$\sup_{0<t<T}\int_{B_R}|w|^p dx \le \frac{C}{R^2}\int_0^T\int_{B_{2R}}|w|^2 dxdt, \tag{5.32}$$

$$\int_0^T\int_{B_{2R}}|\nabla w|^2 h^2 dxdt \le \frac{C}{R^2}\int_0^T\int_{B_{2R}}|w|^2 dxdt. \tag{5.33}$$

By Lemma 3.5.5, we have

$$\begin{aligned}\left(\int_{B_R}|w|^{2^*}dx\right)^{1/2^*} &\le \left(\int_{B_{2R}}|wh|^{2^*}dx\right)^{1/2^*}\\ \le\ & C\left(\int_{B_{2R}}|\nabla(wh)|^2 dx\right)^{1/2}\\ \le\ & C\left(\int_{B_{2R}}|\nabla w|^2 h^2 dx\right)^{1/2}+\left(\frac{C}{R^2}\int_{B_{2R}}|w|^2 dx\right)^{1/2}.\end{aligned}$$

From this, using Hölder's inequality and (5.32), (5.33) we derive

$$\begin{aligned}&\int_0^T\int_{B_R}|w|^{2+2p/N}dxdt\\ \le\ & \int_0^T\left(\int_{B_R}|w|^p dx\right)^{2/N}\left(\int_{B_R}|w|^{2^*}dx\right)^{2/2^*}dt\\ \le\ & \left(\sup_{0<t<T}\int_{B_R}|w|^p dx\right)^{2/N}\int_0^T\left(\int_{B_R}|w|^{2^*}dx\right)^{2/2^*}dt\\ \le\ & \left(\frac{C}{R^2}\int_0^T\int_{B_{2R}}|w|^2 dxdt\right)^{1+2/N}.\end{aligned}$$

Using Hölder's inequality again gives

$$\begin{aligned}&\int_0^T\int_{B_R}|w|^2 dxdt\\ \le\ & \left(\int_0^T\int_{B_R}|w|^{2+2p/N}dxdt\right)^{N/(N+p)}\left(\int_0^T\int_{B_R}dxdt\right)^{p/(N+p)}\\ \le\ & CR^{(Np-2N-4)/(N+p)}\left(\int_0^T\int_{B_{2R}}|w|^2 dxdt\right)^{(N+2)/(N+p)},\end{aligned}$$

or

$$F(R) \leq CR^{-4/(N+p)} F(2R)^{(N+2)/(N+p)},$$

where

$$F(R) = \frac{1}{R^N} \int_0^T \int_{B_R} |w|^2 dxdt.$$

Since $1 \leq p < 2$ implies that $\frac{N+2}{N+p} > 1$, from Lemma 3.5.6 we see that for large R_0,

$$\lim_{m\to\infty} F\left(\frac{R_0}{2^m}\right) = 0. \tag{5.34}$$

From (5.34) and the definition of $F(R)$, using Lebesgue's Theorem, we conclude that for large R_0, $w(x,t) = 0$ almost everywhere in the set $Q_T \backslash (B_{L+2R_0}(0) \times (0,T))$. This and the equality

$$\iint_{Q_T} u \frac{\partial \varphi}{\partial t} dxdt + \iint_{Q_T} w \triangle \varphi dxdt = 0$$

holding for any $\varphi \in C_0^\infty(B_{L+2R_0}(0) \times (0,T))$ imply that for large R_0, $u(x,t) = 0$ almost everywhere in the set $Q_T \backslash (B_{L+2R_0}(0) \times (0,T))$. The proof of Theorem 3.5.6 is thus completed. □

Remark 3.5.1 The result stated in Theorem 3.5.6 can also be proved by other methods (see[YI2], [YU3]).

Moreover, one can extend the result to equations with double degeneracy of the form

$$\frac{\partial u}{\partial t} = \text{div} \vec{F} \left(\nabla A(u) \right).$$

Under certain conditions, the existence and uniqueness of solutions with compact support has been established in [YI6].

3.6 General Strongly Degenerate Equations in Higher Dimension

Now we turn to equations of the form

$$\frac{\partial u}{\partial t} = \Delta A(u) + \text{div} \vec{B}(u), \tag{6.1}$$

where

$$A(u) = \int_0^u a(s)ds, \quad \vec{B}(u) = \int_0^u \vec{b}(s)ds$$

with $a(s) \geq 0$ and $a(s)$ and $\vec{b}(s)$ being continuous functions, but $E = \{s; a(s) = 0\}$ being arbitrary. Contrary to §3.3.5, in this section, we are concerned only the first boundary value problem. Let Ω be a bounded domain of $\mathbb{R}^N$ with smooth boundary $\partial\Omega$. Assume the boundary value condition and initial value condition as follows:

$$u(x,t) = 0 \qquad (x,t) \in \partial\Omega \times (0,T), \tag{6.2}$$

$$u(x,0) = u_0(x) \quad x \in \Omega. \tag{6.3}$$

Denote $Q_T = \Omega \times (0,T)$. The following definition of generalized solutions is similar to that given in §3.4.4.

Definition 3.6.1 A function $u \in L^\infty(Q_T) \cap BV(Q_T)$ is called a generalized solution (simply called a BV solution) of the boundary value problem (6.1)–(6.3), if the following conditions are fulfilled:

(1) $\nabla A(u) \in L^2(Q_T)$;

(2) ess $\lim\limits_{t\to 0+} \overline{u}(x,t) = u_0(x)$ for almost all $x \in \Omega$ and $A(\gamma u(x,t)) = 0$ for almost all $(x,t) \in \partial\Omega \times (0,T)$, where $\gamma u(x,t)$ is the trace of u at (x,t);

(3) For any $0 \leq \varphi_1, \varphi_2 \in C^\infty(\overline{Q_T})$ with supp$\varphi_1, \varphi_2 \subset \overline{\Omega} \times (0,T)$, $\varphi_1\Big|_{\partial\Omega} = \varphi_2\Big|_{\partial\Omega}$ and $k \in \mathbb{R}^1$, there holds

$$\iint_{Q_T} \operatorname{sgn}(u-k)\left[(u-k)\frac{\partial\varphi_1}{\partial t} - (\vec{B}(u) - \vec{B}(k) + \nabla A(u))\nabla\varphi_1\right] dxdt$$
$$+ \iint_{Q_T} \operatorname{sgn}k\left[(u-k)\frac{\partial\varphi_2}{\partial t} - (\vec{B}(u) - \vec{B}(k) + \nabla A(u))\nabla\varphi_2\right] dxdt \geq 0.$$

3.6.1 *Existence of BV solutions*

In one dimensional case, the proof of the existence of BV solutions for the boundary value problem is close to that for the Cauchy problem, no notable difference is there. However things are quite different in higher dimensional case. Roughly speaking, for the Cauchy problem , the L^1 norm of every derivative of first order of the approximate solutions u_ε can be estimated separately, however, for the boundary value problem, we can only estimate the L^1 norm of $|\nabla u_\varepsilon|$ as a whole. This situation will cause us more difficulty and make things more complicated.

We first prove the following lemmas.

Lemma 3.6.1 *Assume that $A(s)$ and $f(x)$ are appropriately smooth with $A'(s) > 0$. Let u be a solution of the problem*

$$\Delta A(u) = f, \qquad x \in \Omega,$$

$$u\Big|_{\partial\Omega} = 0.$$

Then

$$A'(0) \int_{\partial\Omega} \left|\frac{\partial u}{\partial n}\right| d\sigma \le \int_\Omega |f(x)|\, dx,$$

where $\dfrac{\partial u}{\partial n}$ is the derivative of u along the outward normal vector.

Proof. Consider the problem

$$\operatorname{div}(A'(u(x))\nabla u_1) = f^+(x), \qquad x \in \Omega,$$

$$u_1\Big|_{\partial\Omega} = 0.$$

Here and below, as before, $f^+ = \max\{f, 0\}$, $f^- = \max\{-f, 0\}$. The maximum principle shows that $u_1(x) \le 0$ in Ω and hence $\dfrac{\partial u_1}{\partial n} \ge 0$ on $\partial\Omega$. Integrating the equation that u_1 satisfies, we obtain

$$\int_\Omega f^+(x)dx = A'(0)\int_{\partial\Omega} \frac{\partial u_1}{\partial n} d\sigma = A'(0)\int_{\partial\Omega}\left|\frac{\partial u_1}{\partial n}\right| d\sigma.$$

Similarly, if we consider the problem

$$\operatorname{div}(A'(u(x))\nabla u_2) = f^-(x), \qquad x \in \Omega,$$

$$u_2\Big|_{\partial\Omega} = 0,$$

then we obtain

$$\int_\Omega f^-(x)dx = A'(0)\int_{\partial\Omega} \frac{\partial u_2}{\partial n} d\sigma = A'(0)\int_{\partial\Omega}\left|\frac{\partial u_2}{\partial n}\right| d\sigma.$$

By the uniqueness of solutions, we have $u = u_1 - u_2$. Thus

$$\begin{aligned}
& A'(0) \int_{\partial\Omega} \left| \frac{\partial u}{\partial n} \right| d\sigma \\
\leq \; & A'(0) \int_{\partial\Omega} \left| \frac{\partial u_1}{\partial n} \right| d\sigma + A'(0) \int_{\partial\Omega} \left| \frac{\partial u_2}{\partial n} \right| d\sigma \\
= \; & \int_\Omega f^+(x)dx + \int_\Omega f^-(x)dx = \int_\Omega |f(x)|dx.
\end{aligned}$$

The proof of Lemma 3.6.1 is complete. □

Lemma 3.6.2 *Denote* $\xi = (\xi_1, \cdots, \xi_N)$, $|\xi| = \left(\xi_1^2 + \cdots + \xi_N^2\right)^{1/2}$ *and*

$$I_\eta(\xi) = \sqrt{\eta + |\xi|^2}, \qquad (\eta > 0).$$

Then $I_\eta(\xi)$ *is a strictly concave function, namely, the matrix* $\left(\dfrac{\partial^2 I_\eta}{\partial \xi_i \partial \xi_j} \right)$ *is positively definite, and for any* $j = 1, \cdots, N$, $\xi \in \mathbb{R}^N$,

$$\lim_{\eta \to 0} \xi_i \frac{\partial^2 I_\eta}{\partial \xi_i \partial \xi_j} = 0. \tag{6.4}$$

Proof. By an immediate calculation, we first obtain

$$\begin{aligned}
\frac{\partial I_\eta}{\partial \xi_i} &= \frac{\xi_i}{\sqrt{\eta + |\xi|^2}}, \\
\frac{\partial^2 I_\eta}{\partial \xi_i \partial \xi_j} &= \frac{\delta_{ij}}{\sqrt{\eta + |\xi|^2}} - \frac{\xi_i \xi_j}{(\eta + |\xi|^2)^{3/2}}.
\end{aligned}$$

Hence, for any $\alpha = (\alpha_1, \cdots, \alpha_N) \neq 0$,

$$\begin{aligned}
& \sum_{i,j=1}^N \frac{\partial^2 I_\eta}{\partial \xi_i \partial \xi_j} \alpha_i \alpha_j = \frac{1}{\sqrt{\eta + |\xi|^2}} \left(|\alpha|^2 - \frac{|\alpha \cdot \xi|^2}{\eta + |\xi|^2} \right) \\
\geq \; & \frac{1}{\sqrt{\eta + |\xi|^2}} \left(|\alpha|^2 - \frac{|\alpha|^2 |\xi|^2}{\eta + |\xi|^2} \right) \\
= \; & \frac{|\alpha|^2}{\sqrt{\eta + |\xi|^2}} \left(1 - \frac{|\xi|^2}{\eta + |\xi|^2} \right) > 0,
\end{aligned}$$

which shows the strict convexity of $I_\eta(\xi)$.

Next we have

$$\begin{aligned}
& \lim_{\eta\to 0} \xi_i \frac{\partial^2 I_\eta}{\partial \xi_i \partial \xi_j} \\
= & \lim_{\eta\to 0} \frac{\xi_j}{\sqrt{\eta+|\xi|^2}} - \lim_{\eta\to 0} \frac{|\xi|^2 \xi_j}{(\eta+|\xi|^2)^{3/2}} \\
= & \lim_{\eta\to 0} \frac{\xi_j}{\sqrt{\eta+|\xi|^2}} \left(1 - \frac{|\xi|^2}{\eta+|\xi|^2}\right) \\
= & \ 0,
\end{aligned}$$

which shows (6.4). □

Theorem 3.6.1 *Assume that $A(s)$, $\vec{B}(s)$ and u_0 are appropriately smooth and $u_0(x)$ satisfies suitable compatibility conditions on $\partial\Omega \times \{t = 0\}$. Then the boundary value problem (6.1)–(6.3) admits a BV solution.*

Proof. Consider the regularized problem

$$\frac{\partial u_\varepsilon}{\partial t} = \Delta A_\varepsilon(u_\varepsilon) + \mathrm{div}\vec{B}(u_\varepsilon), \tag{6.5}$$

$$u_\varepsilon(x,t) = 0, \qquad (x,t) \in \partial\Omega \times (0,T), \tag{6.6}$$

$$u_\varepsilon(x,0) = u_0(x), \quad x \in \Omega \tag{6.7}$$

where $A_\varepsilon(s) = \varepsilon s + A(s) \quad (\varepsilon > 0)$. The existence of a appropriately smooth solution u_ε follows from the standard theory.

First, by the maximum principle, we have

$$\sup_{Q_T} |u_\varepsilon(x,t)| \leq M \tag{6.8}$$

provided $\sup\limits_{\mathbb{R}^N} |u_0(x)| \leq M$.

Next, we estimate the L^2 norm of $\nabla A_\varepsilon(u_\varepsilon)$. To this purpose, multiply (6.5) by $A_\varepsilon(u_\varepsilon)$ and integrate over Q_t,

$$\begin{aligned}
& \iint_{Q_t} A_\varepsilon(u_\varepsilon) \frac{\partial u_\varepsilon}{\partial t} dxds \\
= & \iint_{Q_t} \Delta A_\varepsilon(u_\varepsilon) A_\varepsilon(u_\varepsilon) dxds + \iint_{Q_t} \mathrm{div}\vec{B}(u_\varepsilon) A_\varepsilon(u_\varepsilon) dxds.
\end{aligned}$$

Integrating by parts yields

$$\int_\Omega \Psi_\varepsilon(u_\varepsilon) dx + \iint_{Q_t} |\nabla A_\varepsilon(u_\varepsilon)|^2 dxds$$

$$
\begin{aligned}
&= \int_\Omega \Psi_\varepsilon(u_0)dx - \iint_{Q_t} \vec{B}(u_\varepsilon)\nabla A_\varepsilon(u_\varepsilon)dxds \\
&\le \frac{1}{2}\iint_{Q_t} |\nabla A_\varepsilon(u_\varepsilon)|^2 dxds + \frac{1}{2}\iint_{Q_t} \left|\vec{B}(u_\varepsilon)\right|^2 dxds + C \\
&\le \frac{1}{2}\iint_{Q_t} |\nabla A_\varepsilon(u_\varepsilon)|^2 dxds + C,
\end{aligned}
$$

where

$$\Psi_\varepsilon(s) = \int_0^s A_\varepsilon(s)ds.$$

Hence

$$\iint_{Q_T} |\nabla A_\varepsilon(u_\varepsilon)|^2 dxdt \le C. \tag{6.9}$$

Finally, we estimate the L^1 norm of the derivatives of u_ε. Doing this for $\frac{\partial u_\varepsilon}{\partial t}$ is easy. To this purpose, differentiate (6.5) with respect to t and denote $v_\varepsilon = \frac{\partial u_\varepsilon}{\partial t}$. Then

$$\frac{\partial v_\varepsilon}{\partial t} = \Delta\left(a_\varepsilon(u_\varepsilon)v_\varepsilon\right) + \operatorname{div}\left(\vec{b}(u_\varepsilon)v_\varepsilon\right).$$

Let $H_\eta(s)$ be the function defined and used in the proof of Theorem 3.3.1. Multiplying the above equality by $H_\eta(v_\varepsilon)$ and then integrating over Ω with respect to x, yield

$$
\begin{aligned}
&\frac{\partial}{\partial t}\int_\Omega \theta_\eta(v_\varepsilon(x,t))dx \\
&= \int_\Omega H_\eta(v_\varepsilon)\left(\Delta\left(a_\varepsilon(u_\varepsilon)v_\varepsilon\right) + \operatorname{div}\left(\vec{b}(u_\varepsilon)v_\varepsilon\right)\right)dx \\
&= -\int_\Omega a_\varepsilon(u_\varepsilon)H'_\eta(v_\varepsilon)\left(\nabla v_\varepsilon\right)^2 dx \qquad (6.10)\\
&\quad -\int_\Omega a'_\varepsilon(u_\varepsilon)v_\varepsilon H'_\eta(v_\varepsilon)\nabla v_\varepsilon \nabla u_\varepsilon dx \\
&\quad -\int_\Omega \vec{b}(u_\varepsilon)v_\varepsilon H'_\eta(v_\varepsilon)\nabla v_\varepsilon dx,
\end{aligned}
$$

where

$$\theta_\eta(s) = \int_0^s H_\eta(\sigma)d\sigma.$$

Remove the first term on the right hand side of (6.10), which is nonpositive and then integrate with respect to t and let $\eta \to 0$. We obtain

$$\sup_{0<t<T} \int_\Omega \left| \frac{\partial u_\varepsilon(x,t)}{\partial t} \right| dx \le C. \tag{6.11}$$

The L^1 norm of $|\nabla u_\varepsilon|$ is a little difficult to estimate. Let $I_\eta(\xi)$ be the function defined in Lemma 3.6.2. Differentiate (6.5) with respect to x_i multiply the resulting equality by $\frac{\partial I_\eta(\nabla u_\varepsilon)}{\partial \xi_i}$ and sum up for i from 1 up to N, and then integrate with respect to x on Ω,

$$\begin{aligned}
&\frac{\partial}{\partial t}\int_\Omega I_\eta(\nabla u_\varepsilon)dx = \int_\Omega \frac{\partial I_\eta(\nabla u_\varepsilon)}{\partial \xi_i}\frac{\partial u_{\varepsilon x_i}}{\partial t}dx \\
= &\int_\Omega \frac{\partial I_\eta(\nabla u_\varepsilon)}{\partial \xi_i}\frac{\partial}{\partial x_i}\triangle A_\varepsilon(u_\varepsilon)dx + \int_\Omega \frac{\partial I_\eta(\nabla u_\varepsilon)}{\partial \xi_i}\frac{\partial}{\partial x_i}\mathrm{div}\vec{B}(u_\varepsilon)dx.
\end{aligned}$$

Integrating by parts gives

$$\begin{aligned}
\frac{\partial}{\partial t}\int_\Omega I_\eta(\nabla u_\varepsilon)dx = & -\int_\Omega \frac{\partial^2 I_\eta(\nabla u_\varepsilon)}{\partial \xi_i \partial \xi_j}\frac{\partial^2 u_\varepsilon}{\partial x_j \partial x_p}\frac{\partial^2 A_\varepsilon(u_\varepsilon)}{\partial x_i \partial x_p}dx \\
& -\int_\Omega \frac{\partial^2 I_\eta(\nabla u_\varepsilon)}{\partial \xi_i \partial \xi_j}\frac{\partial^2 u_\varepsilon}{\partial x_j \partial x_p}\frac{\partial B_p(u_\varepsilon)}{\partial x_i}dx \\
& +\int_{\partial\Omega} \frac{\partial I_\eta(\nabla u_\varepsilon)}{\partial \xi_i}\vec{n}\cdot\nabla\frac{\partial}{\partial x_i}A_\varepsilon(u_\varepsilon)d\sigma \\
& +\int_{\partial\Omega} \frac{\partial I_\eta(\nabla u_\varepsilon)}{\partial \xi_i}\vec{n}\cdot\frac{\partial}{\partial x_i}\vec{B}(u_\varepsilon)d\sigma \\
= & \; J_{1\eta} + J_{2\eta} + J_{3\eta} + J_{4\eta}.
\end{aligned} \tag{6.12}$$

By Lemma 3.6.2, we have

$$\begin{aligned}
&\lim_{\eta\to 0} J_{1\eta} \\
= & -\lim_{\eta\to 0}\int_\Omega a_\varepsilon(u_\varepsilon)\frac{\partial^2 I_\eta(\nabla u_\varepsilon)}{\partial \xi_i \partial \xi_j}\frac{\partial^2 u_\varepsilon}{\partial x_j \partial x_p}\frac{\partial^2 u_\varepsilon}{\partial x_i \partial x_p}dx \\
& -\lim_{\eta\to 0}\int_\Omega a'_\varepsilon(u_\varepsilon)\frac{\partial^2 I_\eta(\nabla u_\varepsilon)}{\partial \xi_i \partial \xi_j}\frac{\partial^2 u_\varepsilon}{\partial x_j \partial x_p}\frac{\partial u_\varepsilon}{\partial x_i}\frac{\partial u_\varepsilon}{\partial x_p}dx \\
\le & -\lim_{\eta\to 0}\int_\Omega a'_\varepsilon(u_\varepsilon)\frac{\partial^2 I_\eta(\nabla u_\varepsilon)}{\partial \xi_i \partial \xi_j}\frac{\partial^2 u_\varepsilon}{\partial x_j \partial x_p}\frac{\partial u_\varepsilon}{\partial x_i}\frac{\partial u_\varepsilon}{\partial x_p}dx = 0.
\end{aligned} \tag{6.13}$$

Similarly

$$\lim_{\eta\to 0} J_{2\eta} = -\lim_{\eta\to 0}\int_\Omega b_p(u_\varepsilon)\frac{\partial^2 I_\eta(\nabla u_\varepsilon)}{\partial\xi_i\partial\xi_j}\frac{\partial^2 u_\varepsilon}{\partial x_j\partial x_p}\frac{\partial u_\varepsilon}{\partial x_i}dx = 0. \tag{6.14}$$

To estimate $J_{3\eta}$ and $J_{4\eta}$, notice that $u_\varepsilon|_{\partial\Omega} = 0$ implies $\dfrac{\partial u_\varepsilon}{\partial x_i} = n_i\dfrac{\partial u_\varepsilon}{\partial n}$ on $\partial\Omega$, where n_i is the i-component of n. Hence

$$\begin{aligned}
\vec{n}\cdot\frac{\partial}{\partial x_i}\vec{B}(u_\varepsilon) &= n_p\frac{\partial}{\partial x_i}B_p(u_\varepsilon)\\
&= n_p b_p(u_\varepsilon)\frac{\partial u_\varepsilon}{\partial x_i} = b_p(u_\varepsilon)\frac{\partial u_\varepsilon}{\partial n}n_i n_p,\\
\vec{n}\cdot\nabla\frac{\partial}{\partial x_i}A_\varepsilon(u_\varepsilon) &= n_p\frac{\partial^2 A_\varepsilon(u_\varepsilon)}{\partial x_p\partial x_i}\\
&= n_p\frac{\partial}{\partial x_p}\left(n_i\frac{\partial A_\varepsilon(u_\varepsilon)}{\partial n}\right)\\
&= \frac{\partial}{\partial x_p}\left(\frac{\partial A_\varepsilon(u_\varepsilon)}{\partial n}\right)n_i n_p + \frac{\partial A_\varepsilon(u_\varepsilon)}{\partial n}n_p\frac{\partial n_i}{\partial x_p}.
\end{aligned}$$

On the other hand, since

$$\Delta A_\varepsilon(u_\varepsilon) + \nabla\vec{B}(u_\varepsilon) = \frac{\partial u_\varepsilon}{\partial t} = 0, \qquad \text{on} \quad \partial\Omega,$$

we have

$$\begin{aligned}
b_p(u_\varepsilon)\frac{\partial u_\varepsilon}{\partial n}n_p &= -\Delta A_\varepsilon(u_\varepsilon) = -\frac{\partial}{\partial x_p}\left(n_p\frac{\partial A_\varepsilon(u_\varepsilon)}{\partial n}\right)\\
&= -\frac{\partial}{\partial x_p}\left(\frac{\partial A_\varepsilon(u_\varepsilon)}{\partial n}\right)n_p - \frac{\partial A_\varepsilon(u_\varepsilon)}{\partial n}\frac{\partial n_p}{\partial x_p}.
\end{aligned}$$

Hence for $(x,t)\in\Omega$,

$$\vec{n}\cdot\frac{\partial}{\partial x_i}\vec{B}(u_\varepsilon) + \vec{n}\cdot\nabla\frac{\partial}{\partial x_i}A_\varepsilon(u_\varepsilon) = \frac{\partial A_\varepsilon(u_\varepsilon)}{\partial n}\left(n_p\frac{\partial n_i}{\partial x_p} - n_i\frac{\partial n_p}{\partial x_p}\right).$$

Therefore

$$\begin{aligned}
&|J_{3\eta} + J_{4\eta}|\\
\le\ &\int_{\partial\Omega}\left|\frac{\partial I_\eta(\nabla u_\varepsilon)}{\partial\xi_i}\right|\left|\vec{n}\cdot\nabla\frac{\partial}{\partial x_i}A_\varepsilon(u_\varepsilon) + \vec{n}\cdot\frac{\partial}{\partial x_i}\vec{B}_\varepsilon(u_\varepsilon)\right|d\sigma\\
\le\ &\int_{\partial\Omega}\left|\frac{\partial A_\varepsilon(u_\varepsilon)}{\partial n}\right|\left|n_p\frac{\partial n_i}{\partial x_p} - n_i\frac{\partial n_p}{\partial x_p}\right|d\sigma
\end{aligned}$$

$$\leq \; CA'_\varepsilon(0)\int_{\partial\Omega}\left|\frac{\partial u_\varepsilon}{\partial n}\right|d\sigma.$$

Using Lemma 3.6.1, (6.5) and (6.8), (6.11), we further obtain

$$\begin{aligned}|J_{3\eta}+J_{4\eta}| &\leq \int_\Omega |\Delta A_\varepsilon(u_\varepsilon)|dx \\ &\leq \int_\Omega\left|\frac{\partial u_\varepsilon}{\partial t}\right|dx+\int_\Omega\left|\nabla\vec{B}(u_\varepsilon)\right|dx \\ &\leq C+C\int_\Omega|\nabla u_\varepsilon|\,dx.\end{aligned} \tag{6.15}$$

Combining (6.13), (6.14), (6.15) with (6.12) we are led to

$$\int_\Omega|\nabla u_\varepsilon(x,t)|\,dx\leq C+C\iint_{Q_t}|\nabla u_\varepsilon(x,s)|dxds,$$

and then using Gronwall's inequality, finally obtain

$$\sup_{0<t<T}\int_\Omega|\nabla u_\varepsilon(x,t)|\,dx\leq C. \tag{6.16}$$

Just as in one dimensional case, we may apply the estimates (6.8), (6.10), (6.11), (6.16) to conclude the existence of a subsequence of $\{u_\varepsilon\}$, whose limit function u is a BV solution of the problem (6.1)–(6.3), and thus complete the proof of our theorem. □

Remark 3.6.1 Compared with one dimensional case, the uniqueness problem is much more difficult in higher dimension, which is remained unsolved up to now. The reason seems to be that one can not prove that for the BV solution u, each $\frac{\partial^2}{\partial x_i^2}A(u)$ $(i=1,\cdots,N)$ is a Radon measure, although one can do that for $\triangle A(u)$.

For equations with strong degeneracy, the solutions are impossible to have better regularity in general. The following result is reasonable: For any point of approximate continuity (x_0,t_0) of the BV solution u, such that $a(u(x_0,t_0))>0$, there exists a neighborhood of (x_0,t_0) in which u is equivalent to a classical solution (see [ZH8]).

3.6.2 *Some extensions*

The argument presented in this section can be extended to more general equations of the form

$$\frac{\partial u}{\partial t} = \frac{\partial}{\partial x_i}\left(a^{ij}(u,x,t)\frac{\partial u}{\partial x_j}\right) + \frac{\partial}{\partial x_i}b_i(u,x,t) + c(u,x,t), \tag{6.17}$$

where $a^{ij} = a^{ji} \quad (i,j = 1,\cdots,N)$ satisfy the condition

$$a^{ij}(u,x,t)\xi_i\xi_j \geq 0, \quad \forall \xi = (\xi_1,\ldots,\xi_N) \in \mathbb{R}^N.$$

It should be pointed out that the explicit dependent on (x,t) in the nonlinear coefficient causes not only more complicated calculation, but also some essential difficulty in the proof of existence. In estimating the L^1 norm of the derivatives of the approximate solution u_ε, one need to treat

$$\left|\frac{\partial u_\varepsilon}{\partial t}\right| + |\nabla u_\varepsilon|$$

as a whole. It seems difficult to treat $\dfrac{\partial u_\varepsilon}{\partial t}$ and $\dfrac{\partial u_\varepsilon}{\partial x_i} \quad (i = 1,\cdots,N)$ separately.

It has been proved in [WZ1] that the first boundary value problem for (6.17) admits a BV solution, provided that for any $M > 0$, there exists a $\delta > 0$, such that

$$\begin{aligned} a^{ij}(u,x,t)\xi_i\xi_j - \delta\sum_{s,j=1}^{N}(a^{ij}_{x_s}(u,x,t)\xi_i)^2 & \\ -\delta\sum_{j=1}^{N}(a^{ij}_t(u,x,t)\xi_i)^2 \geq 0. \quad \forall |u| \leq M, (x,t) \in \overline{Q_T}, \xi \in \mathbb{R}^N. & \end{aligned} \tag{6.18}$$

If we replace the condition (6.18) by

$$a^{ij}(u,x,t)\xi_i\xi_j - \delta\sum_{s,j=1}^{N}(a^{ij}_{x_s}(u,x,t)\xi_i)^2 \geq 0. \quad \forall |u| \leq M, (x,t) \in \overline{Q_T}, \xi \in \mathbb{R}^N,$$

then we can merely obtain a $BV_{1,1/2}$ solution, a little weaker solution (see[WZ2]). By $BV_{1,1/2}(Q_T)$, we mean the class of all integrable function

u satisfying

$$\iint_{Q_T} |u(x+\Delta x, t) - u(x,t)|\, dxdt \le C|\Delta x|,$$

$$\iint_{Q_T} |u(x, t+\Delta t) - u(x,t)|\, dxdt \le C|\Delta t|^{1/2}.$$

For more related papers, we refer to [Dd], [Dp] and [CAR].

3.7 Appendix Classes BV and BV_x

In this section we list a series of properties of BV functions and BV_x functions without proofs, which are needed in our book.

Let Ω be a domain of $\mathbb{R}^N$, $Q_T = \Omega \times (0,T)$. Denote by $BV(Q_T)$ the set of all functions of locally bounded variation, namely, a subset of $L^1_{loc}(Q_T)$, in which the weak derivatives of each function are Radon measures on Q_T. A little general class, denoted by $BV_x(Q_T)$, is another subset of $L^1_{loc}(Q_T)$, in which only the derivatives in x of each function are Radon measures on Q_T. Clearly

$$BV(Q_T) \subset BV_x(Q_T).$$

The following lemma is a basic result in measure theory.

Lemma 3.7.1 *Let μ be a Radon measure in the measure space X. For any Borel set G, define*

$$\nu(G) = \int_G f d\mu,$$

where $f \in L^1(X, |\mu|)$. Then ν is also a finite Radon measure on X and for any function g which is essentially bounded and measurable with respect to ν, there holds

$$\int_X g d\nu = \int_X g f d\mu.$$

Lemma 3.7.2 *[KRI] Assume that $u \in BV_x(Q_T)$. Then for almost all $t \in (0,T)$, $u(\cdot,t) \in BV(\Omega)$. If we define*

$$\mu_i(t) \equiv \frac{\partial u(\cdot,t)}{\partial x_i}$$

and use $|\mu_i(t)|(G)$ to denote the total variation of the measure $|\mu_i(t)|$ on a rectangle G in Ω, then

$$|\mu_i(t)|(G) \equiv \int_G \left| \frac{\partial u(\cdot, t)}{\partial x_i} \right|, \quad (i = 1, \cdots, n)$$

is a locally integrable function on $(0, T)$.

Now we introduce some related notations. For $u \in BV(Q_T)$, denote by Γ_u the set of all point of discontinuity of u, $\gamma = (\gamma_1, \cdots, \gamma_n, \gamma_t)$ the unit normal vector to Γ_u, $u^+(x_0, t_0)$ and $u^-(x_0, t_0)$ the approximate limits of u at $(x_0, t_0) \in \Gamma_u$ taking from the half space

$$(x_1 - x_1^0)\gamma_1 + \cdots + (x_n - x_n^0)\gamma_n + (t - t_0)\gamma_t > 0$$

and

$$(x_1 - x_1^0)\gamma_1 + \cdots + (x_n - x_n^0)\gamma_n + (t - t_0)\gamma_t < 0$$

respectively, $\overline{u}(x,t) \equiv \frac{1}{2}(u^+(x,t) + u^-(x,t))$ the symmetric mean value of u at the regular point (x,t). By a regular point, we mean either a point of approximate continuity or a point of discontinuity.

For BV functions we have the following formula for derivatives of composites: If $f \in C^1(\mathbb{R})$ and $u \in L^\infty(Q_T) \cap BV(Q_T)$, then $f(u) \in BV(Q_T)$ and

$$\frac{\partial f(u(x,t))}{\partial \xi} = \hat{f}'(u(x,t)) \frac{\partial u(x,t)}{\partial \xi}$$

where ξ denotes t or $x_i \quad (i = 1, \cdots, N)$, $\hat{g}(u(x,t))$ denote the functional superposition of $g(u)$ and $u(x,t)$ which is defined by

$$\hat{g}(u(x,t)) = \int_0^1 g(\tau u^+(x,t) + (1-\tau)u^-(x,t))d\tau.$$

Also we have the product rule for BV functions: If $u, v \in L^\infty(Q_T) \cap BV(Q_T)$, then $u, v \in BV(Q_T)$ and

$$\frac{\partial uv}{\partial \xi} = \overline{u}\frac{\partial v}{\partial \xi} + \overline{v}\frac{\partial u}{\partial \xi}$$

where ξ denotes t or $x_i \quad (i = 1, \cdots, N)$.

The Hausdorff measure of a subset S of Γ_u is denoted by $H(S)$.

For fixed t, we use Γ_u^t, H^t, $(\gamma_1^t, \cdots, \gamma_n^t)$ and $u_\pm^t$ to denote the set of points of discontinuity of $u(\cdot, t)$, the Hausdorff measure on Γ_u^t, the unit normal vector to Γ_u^t and the approximate limits of $u(\cdot, t)$.

For simplicity, all results are stated merely in one space variable in the sequel and Ω is assumed to be an open interval which may be infinite. For fixed $t \in (0, T)$, we use $\tilde{u}(\cdot, t)$, $u^r(\cdot, t)$ and $u^l(\cdot, t)$ to denote the symmetric mean value, the right approximate limit and left approximate limit of $u(\cdot, t)$. From Lemma 3.7.2, we see that if $u \in BV_x(Q_T)$, then for almost all $t \in (0, T)$ and any $x \in I$, $\tilde{u}(x, t)$, $u^r(x, t)$ and $u^l(x, t)$ exist. The following result explains the relation between $\overline{u}(x, t)$ and $\tilde{u}(x, t)$ and the relation among $u^+(x, t)$, $u^r(x, t)$, $u^-(x, t)$, and $u^l(x, t)$.

Lemma 3.7.3 *[VO] Assume $u \in BV(Q_T)$. Then there exists a subset $E_t \subset (0, T)$ with $mesE_t = 0$, such that for any $(x, t) \in I \times [(0, T) \setminus E_t]$,*

$$\overline{u}(x, t) = \widetilde{u}(x, t), \quad \left|u^+(x, t) - u^-(x, t)\right| = \left|u^r(x, t) - u^l(x, t)\right|.$$

Lemma 3.7.4 *[VH2] Assume that $u \in BV(Q_T)$ and S is a bounded and H-measurable subset of Γ_u with $\overline{S} \subset \Gamma_u$. Then S is measurable with respect to the measure $\dfrac{\partial u}{\partial x}$ and*

$$\iint_S \frac{\partial u}{\partial x} = \int_S (u^+ - u^-)\gamma_x dH.$$

Corollary 3.7.1 *Assume that $u \in BV(Q_T)$ and*

$$S \subset \Gamma_u^* \equiv \{(x, t) \in \Gamma_u; \gamma_x \neq 0\}.$$

Then

$$\left|\frac{\partial u}{\partial x}\right|(S) = 0 \quad \textit{iff} \quad H(S) = 0.$$

Corollary 3.7.2 *Assume that $u \in BV(Q_T)$. If $f(x, t)$ is measurable with respect to $\dfrac{\partial u}{\partial x}$ on Q_T. Then $f(x, t)$ is H-measurable on Γ_u^*.*

Lemma 3.7.5 *[VH1] Assume that $u, v \in BV(Q_T)$ and $|v(x, t)| \leq K|u(x, t)|$ for almost all $(x, t) \in Q_T$, where K is a constant. Then for any $\varphi \in$*

$C_0^\infty(Q_T)$,

$$\iint_{Q_T} \sigma \frac{\partial}{\partial t}(\varphi v) = -\int_{\Gamma_u} \varphi(sgnu^+ - sgnu^-)\overline{v}\gamma_t dH,$$
$$\iint_{Q_T} \sigma \frac{\partial}{\partial x}(\varphi v) = -\int_{\Gamma_u} \varphi(sgnu^+ - sgnu^-)\overline{v}\gamma_x dH,$$

where

$$\sigma = \frac{1}{2}(sgnu^+ + sgnu^-).$$

Lemma 3.7.6 *[WY1] Assume that $u \in BV_x(Q_T)$, G is a bounded rectangle with $\overline{G} \subset Q_T$ and $f(x,t)$ is bounded and measurable with respect to $|\frac{\partial u}{\partial x}|$ on G. Then there exist a sequence $\{f_n\} \subset C_0^\infty(G)$ and a constant M, such that*

$$|f_n(x,t)| \le M, \quad \lim_{n\to\infty} \|f_n - f\|_X = 0,$$

where $X = L^1\left(G, \left|\frac{\partial u}{\partial x}\right|\right)$.

Lemma 3.7.7 *[WY1] Assume that $u \in BV_x(Q_T)$, $E_t \subset (0,T)$ is a set of Lebesgue measure zero and $G = I \times E_t$. Then $\left|\frac{\partial u}{\partial x}\right|(G) = 0$. If $u \in BV(Q_T)$, then $H(G \cap \Gamma_u^*) = 0$.*

Lemma 3.7.8 *[WY1] Assume that $u \in BV_x(Q_T)$ and $f(x,t)$ is measurable with respect to the measure $\frac{\partial u}{\partial x}$, with compact support. Then for almost all $t \in (0,T)$, $f(x,t)$ is measurable with respect to the measure $\frac{\partial u(\cdot,t)}{\partial x}$. Moreover,*

$$\int_I f(x,t)\frac{\partial u(\cdot,t)}{\partial x} \quad and \quad \int_I f(x,t)\left|\frac{\partial u(\cdot,t)}{\partial x}\right|$$

are Lebesgue integrable on $(0,T)$ and there hold

$$\begin{aligned} \iint_{Q_T} f(x,t)\frac{\partial u}{\partial x} &= \int_0^T dt \int_I f(x,t)\frac{\partial u(\cdot,t)}{\partial x}, \\ \iint_{Q_T} f(x,t)\left|\frac{\partial u}{\partial x}\right| &= \int_0^T dt \int_I f(x,t)\left|\frac{\partial u(\cdot,t)}{\partial x}\right|. \end{aligned} \tag{7.1}$$

As formulas transforming the double integral into an iterated integral, (7.1) are very useful in the study of discontinuity conditions. For convenience of applications we need to discuss the integrability of the integrand $f(x,t)$. First let us recall the definition of the Caratheodory functions. $f(x,t)$ is called a Caratheodory function, if

(1) for almost all $t \in (0,T)$, $f(x,t)$ is continuous in x,

(2) for any $x \in I$, $f(x,t)$ is Lebesgue measurable in t.

Now we introduce a new class of functions, denote by $\overline{C}_a(Q_T)$: $f \in \overline{C}_a(Q_T)$ if and only if there exist a sequence $\{f_n(x,t)\}$ of Caratheodory functions and a constant $M > 0$, such that

$$|f_n(x,t)| \le M, \quad \lim_{n\to\infty} f_n(x,t) = f(x,t),$$

for almost all $t \in (0,T)$ and any $x \in I$.

Remark 3.7.1 If $u \in L^\infty(Q_T) \cap BV_x(Q_T)$, then

$$\widetilde{u}(x,t), u^r(x,t), u^l(x,t) \in \overline{C}_a(Q_T).$$

If, in addition, $w \in BV_x(Q_T)$, then $\text{sgn}\widetilde{u}$, $\text{sgn}u^r$ and $\text{sgn}u^l$ are measurable with respect to the measure $\dfrac{\partial w}{\partial x}$.

Lemma 3.7.9 *[WY1] Assume that $u \in BV_x(Q_T)$ and $f \in \overline{C}_a(Q_T)$ with compact support. Then $f(x,t)$ is integrable on Q_T with respect to the measure $\dfrac{\partial u}{\partial x}$.*

In applications, the following formula which transforms the double integral into a curve integral, is also very useful.

Lemma 3.7.10 *[WY1] Assume that $u \in L^\infty(Q_T) \cap BV(Q_T)$ and S is a subset of Γ_u^*, bounded and H-measurable. Then for any bounded and measurable (with respect to $\dfrac{\partial u}{\partial x}$) function $f(x,t)$ on Q_T with compact support, there holds*

$$\iint_S f(x,t)\frac{\partial u}{\partial x} = \int_0^T dt \sum_{x\in S^t} \left[u^r(x,t) - u^l(x,t)\right] f(x,t)$$
$$= \int_S f(x,t)\left[u^r(x,t) - u^l(x,t)\right] |\gamma_x| dH.$$

Corollary 3.7.3 *Under the assumptions of Lemma 3.7.10, there holds*

$$\int_0^T dt \sum_{x\in S^t} f(x,t) = \int_S f(x,t)|\gamma_x| dH.$$

Corollary 3.7.4 *If* $u \in L^\infty(Q_T) \cap BV(Q_T)$, *then*

$$u^+(x,t) - u^-(x,t) = \left[u^r(x,t) - u^l(x,t)\right] sgn\gamma_x$$

holds H-almost everywhere on Γ_u^*

Corollary 3.7.5 *If* $u \in L^\infty(Q_T) \cap BV(Q_T)$, *then*

$$\begin{cases} u^+(x,t) = u^r(x,t) sgn^+\gamma_x - u^l(x,t) sgn^-\gamma_x, \\ u^-(x,t) = u^l(x,t) sgn^+\gamma_x - u^r(x,t) sgn^-\gamma_x \end{cases}$$

holds H-almost everywhere on Γ_u^*, *where*

$$sgn^+ s = \begin{cases} 1, & for \quad s > 0, \\ 0, & for \quad s \le 0, \end{cases} \qquad sgn^- s = \begin{cases} 0, & for \quad s \ge 0, \\ -1, & for \quad s < 0. \end{cases}$$

Chapter 4

Nonlinear Diffusion Equations of Higher Order

4.1 Introduction

In this chapter, we are concerned with nonlinear diffusion equations of higher order. Particular attention will be paid to those equations with degeneracy. However, some related topics for equations without degeneracy are also considered. The investigation of higher order parabolic equations with degeneracy began almost at the same time as the second order equations. Most works are devoted to linear equations, among them are those by means of the method based on the theory of pseudo differential operators, for an overview we refer to the summary paper by Glushko and Savchenko [GS]. However, the study for higher order equations, especially for quasilinear equations is far from completion, compared with the second order case. The main reason is that many effective methods used in treating second order equations such as those based on maximum principle, are no longer effective for higher order equations. In addition, the occurrence of degeneracy makes things more difficult. Early works for higher order equations such as [DU1], [DU2], [VI], are mainly based on energy estimates. In the 1980's, Soltanov [SO1], [SO2] introduced the so called "Nonlinear Sobolev Spaces", and discussed a class of higher order equations with degeneracy depending only on the derivatives of unknown functions. Mkrtychyan [MK] successfully extended the related results of Ivanov [IV1] for $(A, \vec{b})$ type degenerate parabolic equations to the higher order case. The relatively systematic work appeared in recent ten years, which is devoted mainly to two kinds of typical equations. The one is the fourth order equations with the structure similar to the Newtonian and non-Newtonian filtration equations, namely,

$$\frac{\partial u}{\partial t} + \frac{\partial^4}{\partial x^4}\left(|u|^{m-1}u\right) = 0, \quad m > 0 \tag{1.1}$$

and

$$\frac{\partial u}{\partial t} + \frac{\partial^2}{\partial x^2}\left(\left|\frac{\partial^2 \Phi(u)}{\partial x^2}\right|^{p-2} \frac{\partial^2 \Phi(u)}{\partial x^2}\right) = 0, \tag{1.2}$$

where $p > 1$, $\Phi(u) = |u|^{q-2}u$, $q > 1$. The other is the Cahn-Hilliard equation

$$\frac{\partial u}{\partial t} + D\left[m(u)(kD^3u - DA(u))\right] = 0. \tag{1.3}$$

Bernis and McLeod [BM] have ever investigated systematically the similarity solutions of the equation (1.1). In particular, they pointed out that the similarity solutions do not preserve their sign in general and may have infinite number of zeros, which reveal the main difference between the higher order equations and second order equations. These properties have also been verified for the general equation (1.2). It was Bernis [BE2] who first proved the nonexistence of nonnegative solutions for (1.1). He also discussed the propagation of disturbances by means of the weighted Nirenberg inequality, the Hardy inequality and the Kjellberg inequality.

As another important kind of fourth order equation, the Cahn-Hilliard equation, was originally proposed by Cahn and Hilliard as a model of spinodal decomposition for a binary mixtures [CH]. It was also derived from the diffusive processes of populations, see for example [CM], [HKL]. The equation can also be used to describe the diffusive process of an oil film spreading over a solid surface, see [TAY]. Elliott and Zheng [EZ] first investigated the Cahn-Hilliard equation with constant mobility, and discussed the global existence of classical solutions and the Blow-up phenomenon. Since then, many authors have discussed the properties of solutions in this case, see Zheng [ZE], Alikakos, Bates and Fusco [ABF], Alikakos and McKinney [AM], Carr, Gurtin and Slemrod [CGS], Elliott and French [EF], Novick-Cohen [NC], etc. The Cahn-Hilliard equation with concentration dependent mobility was studied in recent years, see for example, Novick-Cohen and Segel [NS], Tayler [TAY], Bernis and Friedman [BF], Yin [YI8], [YI9], Elliott and Garcke [EG], Elliott and Mikelic [EM], among them are

[TAY], [BF], [YI9] and [EG] which are devoted to the equation with degenerate mobility.

In this chapter, we will introduce some basic results and methods for typical equations of higher order, such as (1.1), (1.2), (1.3) and others, from which, in particular, we will find the common points and differences between the fourth order parabolic equations and the second order equations. This chapter is arranged as follows. Section 2 is devoted to a discussion of similarity solutions of the equation (1.1), while Section 3 is devoted to the first boundary value problem for the equation (1.2), presenting some results on the basic existence, uniqueness and some properties of solutions such as the propagation of disturbances, asymptotic behavior and the non existence of nonnegative solutions. From Section 4 on, we turn to the study of the Cahn-Hilliard equation with mobility in different cases. We first study the global existence of classical solutions, and the Blow-up phenomenon of the local classical solutions in Section 4 and Section 5. Subsequently, we consider a special case of the Cahn-Hilliard equation, namely, the thin film equation in Section 6, and study the existence of nonnegative solutions and the properties of such solutions. Finally, in Section 7, we consider the Cahn-Hilliard equation with degenerate mobility, and discuss the existence of physical solutions, i.e., solutions u with the property that $0 \leq u \leq 1$.

4.2 Similarity Solutions of a Fourth Order Equation

As a typical example of quasilinear diffusion equations of second order, the Newtonian filtration equation

$$\frac{\partial u}{\partial t} = \frac{\partial^2}{\partial x^2}\left(|u|^{m-1}u\right), \quad m > 0$$

has been well-studied. In order to find some information for higher order diffusion equations, it seems to be natural to investigate the following fourth order equation

$$\frac{\partial u}{\partial t} + \frac{\partial^4}{\partial x^4}\left(|u|^{m-1}u\right) = 0, \quad m > 0. \tag{2.1}$$

This section is devoted to a discussion of the properties of similarity solutions of the equation (2.1).

4.2.1 *Definition of similarity solutions*

For convenience, we change the argument (t, x) to (t, y), and rewrite (2.1) as an equation for the unknown function $w(t, y)$

$$w_t + D_y^4(|w|^{m-1}w) = 0, \quad m > 0. \tag{2.2}$$

We try to seek its similarity solutions of the following form

$$w(y,t) = t^{-k\beta} v\left(\frac{\beta^{1/4} y}{t^\beta}\right),$$

where $\beta > 0$ is a given constant, k is a constant depending on β to be specified later. A direct calculation shows that

$$\begin{aligned} w_t &= -k\beta t^{-k\beta-1} v(x) - \beta t^{-k\beta-\beta-1} \beta^{1/4} y v'(x) \\ &= -k\beta t^{-k\beta-1} v(x) - \beta t^{-k\beta-1} x v'(x), \\ D_y^4\left(|w|^{m-1}w\right) &= D_x^4\left(|v|^{m-1}v\right)\left(\frac{\beta^{1/4}}{t^\beta}\right)^4 t^{-k\beta m}, \end{aligned}$$

where $x = \dfrac{\beta^{1/4} y}{t^\beta}$. Thus

$$D_x^4\left(|v|^{m-1}v\right)\beta t^{-4\beta-k\beta m} = k\beta t^{-k\beta-1} v(x) + \beta t^{-k\beta-1} x v'(x).$$

Choose k such that $4\beta + k\beta m = k\beta + 1$, i.e., $k \equiv \dfrac{1-4\beta}{\beta(m-1)}$ (if $m = 1$, then k can be chosen arbitrarily, but β must be specified as $\dfrac{1}{4}$). Setting $u(x) = |v(x)|^{m-1} v(x)$, we see that $u(x)$ satisfies the following ordinary differential equation

$$u^{(iv)} = x(g(u))' + kg(u), \tag{2.3}$$

where $g(u) = |u|^{1/m}\mathrm{sgn}u$.

Definition 4.2.1 A function u is said to be a solution of the equation (2.3) on the interval J, if $u \in C^3(J)$, and for any $a, x \in J$,

$$u'''(x) - u'''(a) = xg(u(x)) - ag(u(a)) + (k-1)\int_a^x g(u(\tau))d\tau.$$

If $u(x)$ is a solution of the equation (2.3) on the interval J, then

$$w(y,t) = t^{-k\beta} v\left(\frac{\beta^{1/4} y}{t^{\beta}}\right)$$

is called a similarity solution of the equation (2.2) on $D_J = \{(y,t); t > 0, \frac{\beta^{1/4} y}{t^{\beta}} \in J\}$.

Remark 4.2.1 The solution u of (2.3) defined in Definition 4.2.1 is equivalent to a function $u \in C^3(J)$ satisfying (2.3) on J in the sense of distributions. If $0 < m \leq 1$, then $u \in C^4(J)$ and u is a classical solution of (2.3). While if $m > 1$, then u may not belong to $C^4(J)$, and $(g(u))'$ has no meaning at the point where $u = 0$. The similarity solution $w(y,t)$ of (2.2) is equivalent to a function $w \in C^3(D_J)$ satisfying the equation (2.2) in D_J in the sense of distributions. If $0 < m \leq 1$, then $w \in C^4(D_J)$ and w is a classical solution of the equation (2.2).

4.2.2 *Existence and uniqueness of global solutions of the Cauchy problem*

Consider the Cauchy problem for the equation (2.3) with the following initial value conditions

$$u^{(j)}(a) = \alpha_j, \qquad j = 0,1,2,3. \tag{2.4}$$

It is clear that the initial value problem (2.3)–(2.4) in the sense of Definition 4.2.1 is equivalent to the following problem

$$\begin{cases} u'''(x) = \alpha_3 - a g(\alpha_0) + x g(u(x)) + (k-1) w(x), \\ w'(x) = g(u(x)), \\ u^{(j)}(a) = \alpha_j, \quad j = 0,1,2, \\ w(a) = 0. \end{cases} \tag{2.5}$$

By Taylor's formula, this problem is equivalent to the integral equation

$$u(x) = P_a(x) + \int_a^x H(x,t) g(u(t)) dt, \tag{2.6}$$

where

$$H(x,t) = \frac{1}{2}t(x-t)^2 + \frac{1}{6}(k-1)(x-t)^3,$$

$$P_a(x) = \alpha_0 + \alpha_1(x-a) + \frac{\alpha_2}{2}(x-a)^2 + \frac{\alpha_3 - ag(\alpha_0)}{6}(x-a)^3.$$

From the standard theory of ordinary differential equations, we see that for $m > 0$, the problem (2.5) is solvable in some neighborhood of $x = a$. Let (u, w) be a solution. If $0 < m \le 1$, then the solution is unique, since the Lipschitz condition is satisfied. For $m > 1$, the right hand side of (2.5) is merely continuous in (x, u, w). However, if we can prove that the solution of the initial value problem is unique, then the solution is continuously dependent on the initial data. The following theorem gives an incomplete answer to the uniqueness.

Theorem 4.2.1 *Let $k \in \mathbb{R}$, $m > 1$. If $\sum_{j=0}^{3} |\alpha_j| > 0$, then the problem (2.3), (2.4) has at most one solution in a neighborhood of the point $x = a$.*

Proof. As an example, we prove the uniqueness in a right neighborhood of the point $x = a$. We need only to prove the uniqueness of solutions of the problem (2.6). Let u_1, u_2 be two solutions of the problem (2.6). Then from (2.6), we see that in a right neighborhood of $x = a$,

$$|u_1(x) - u_2(x)| \le C(x-a)^2 \int_a^x |g(u_1(t)) - g(u_2(t))|dt. \tag{2.7}$$

Let α_j be the first non-zero number in $\{\alpha_i\}$ and define the function

$$f_n(x) = \begin{cases} \dfrac{u_n(x)}{(x-a)^j}, & \text{if } x > a, \\ \dfrac{\alpha_j}{j!} \ne 0, & \text{if } x = a. \end{cases} \tag{2.8}$$

From (2.7) we have

$$|f_1(x) - f_2(x)| \le C(x-a)^{2-j+j/m} \int_a^x |g(f_1(t)) - g(f_2(t))|dt.$$

Because the function $f_n(x)$ is non-zero in the small right neighborhood of the point $x = a$, and $g(s)$ is Lipschitz continuous away from the point $s = 0$, we get

$$|f_1(x) - f_2(x)| \leq C(x-a)^{2-j+j/m} \int_a^x |f_1(t) - f_2(t)|dt.$$

Here and below, we always use the same notation C to denote different constants. Set

$$F(x) = \max_{a \leq t \leq x} |f_1(t) - f_2(t)|.$$

Then

$$F(x) \leq C(x-a)^{3-j+j/m} F(x).$$

Notice that for $j \leq 3$, $3 - j + j/m > 0$. So, if $x > a$ and $x - a$ is sufficiently small, then $F(x) = 0$, which implies $u_1(x) = u_2(x)$. The proof is complete. □

Theorem 4.2.2 *Let $k \in \mathbb{R}$, $m > 1$. Then any solution $u(x)$ of the equation (2.3) exists globally on $\mathbb{R}$ and*

$$u(x) = O(|x|^{4m/(m-1)}), \qquad \textit{as } |x| \to \infty.$$

Proof. Let u be a solution of the equation (2.3) in a neighborhood of the point $x = a$. As an example, we prove that u can be extended to $x > a$. It suffices to prove that u is bounded on any bounded interval, since from (2.5) the boundedness of u implies the boundeness of u', u'', u''' and w. Notice that in the interval where u exists, u satisfies the integral equation (2.6). Set

$$U(x) = \max_{a \leq t \leq x} |u(t)|.$$

Since for $x > a$,

$$|P_a(x)| \leq C(1 + (x-a)^3), \qquad \int_a^x |H(x,t)|dt \leq C(1 + (x-a)^4),$$

we have

$$U(x) \leq C(1 + (x-a)^3) + CU(x)^{1/m}(1 + (x-a)^4).$$

Noticing that $m > 1$ and using Young's inequality yield

$$U(x) \leq \frac{1}{2}U(x) + C(1 + (x-a)^4)^{m/(m-1)},$$

which implies the boundedness of the solution $u(x)$ on any bounded interval, and the estimates for the growth order of $u(x)$. The proof is complete. □

4.2.3 *Regularity of solutions*

Theorem 4.2.3 *Let $J \subset \mathbb{R}$ be an open interval, u be a solution of the equation (2.3) on the interval J, $m > 1$, $k \in \mathbb{R}$. Assume that*

$$\sum_{j=0}^{3} |u^{(j)}(t)| > 0, \qquad t \in J. \tag{2.9}$$

Then u''' and $g(u)$ are absolutely continuous on any bounded closed interval S of J, and except for a finite number of points in S,

$$u^{(iv)}(t) = tg'(u(t))u'(t) + kg(u(t)), \quad g'(u) = \frac{1}{m}|u|^{1/m-1}. \tag{2.10}$$

Proof. If u is identically equal to some constant, the conclusion of the theorem is trivial. In what follows, we assume that u is not a constant.

We need only to prove that the zero points of u' are isolated. In fact, if we have proved this fact, then u is piecewise monotone on S. Since $g(s)$ is absolutely continuous, the composition $g(u(x))$ is absolutely continuous on S. Furthermore, from the definition of solutions, we see that u''' is absolutely continuous, and except for a finite number of points in S, (2.10) is valid.

Let $u'(c) = 0$. We want to prove that c is an isolated zero point of u'. If $u''(c) \neq 0$ or $u'''(c) \neq 0$, then the conclusion is obvious. Now, we assume that $u''(c) = 0$, $u'''(c) = 0$. From (2.9), we see that $u(c) \neq 0$. So, in a neighborhood of the point $t = c$, $u \neq 0$, and (2.10) holds. We conclude that $k \neq 0$. If this were not true, then we would have, in a neighborhood of the point $t = c$,

$$u^{(iv)}(t) = tg'(u(t))u'(t).$$

On the other hand, at the point $t = c$, $u' = u'' = u''' = 0$. From this we see that in the neighborhood of $t = c$, $u' \equiv 0$, and further $u' = 0$ everywhere, which contradicts the assumption that u is not a constant. Therefore, $k \neq 0$, and consequently $u^{(iv)}(c) = kg(u(c)) \neq 0$, which together with $u''(c) = 0$, $u'''(c) = 0$ implies that c is an isolated zero point of u'. The proof of the theorem is complete. □

Corollary 4.2.1 *Under the assumption of Theorem 4.2.3 ($m > 1$ is not required, and in the case $0 < m \le 1$, (2.9) is not required), for $a, x \in \overline{J}$, we have*

$$\left[u'(t)u'''(t) - \frac{1}{2}u''(t)^2 - \frac{km}{m+1}|u(t)|^{1+1/m}\right]_{t=a}^{t=x} = \frac{1}{m}\int_a^x t|u(t)|^{1/m-1}u'(t)^2 dt. \tag{2.11}$$

Proof. If $0 < m \le 1$, $u \in C^4$, then (2.10) holds in classical sense. While if $m > 1$, then from Theorem 4.2.3, (2.10) holds almost everywhere. So, for any $m > 0$, we may multiply both sides of (2.10) by $u'(t)$ and integrate over (a, x) to see that (2.11) holds for $a, x \in J$. By continuity, we see that (2.11) holds for any $a, x \in \overline{J}$, and complete the proof. □

4.2.4 *Properties of solutions at zero points*

Theorem 4.2.4 *Let u be a solution of the equation (2.3) on $\mathbb{R}^+ = [0, +\infty)$. Then*

$$N = \{t \ge 0; u^{(j)}(t) = 0,\ j = 0, 1, 2, 3\}$$

is a closed and connected set. In particular, if $0 < m \le 1$, then N is an empty set or a half line $\mathbb{R}^+$.

Proof. If $0 < m \le 1$, then $g(s)$ is Lipschitz continuous, and the uniqueness of solutions ensures that null initial data imply null solution, and the conclusion follows.

Let $m > 1$. First, $u \in C^3$ implies that the set N is closed. We prove the connectness by contradiction. Assume that for some $c_1, c_3 \in N$ and $c_2 \notin N$, $c_1 < c_2 < c_3$. Then we may choose the maximal interval (a, b) such that $c_2 \in (a, b)$,

$$\sum_{j=0}^{3} |u^{(j)}(t)| > 0, \qquad t \in (a, b)$$

and $a, b \in N$. Applying the corollary (i.e. (2.11)) of Theorem 4.2.3 for the interval (a, b), and setting $x = b$, we obtain

$$\int_a^b t|u(t)|^{1/m-1}u'(t)^2 dt = 0,$$

which implies that on (a,b), $u'(t) \equiv 0$, $u(t) \equiv 0$. A contradiction. The proof is complete. □

Theorem 4.2.5 *Let $k \geq 0$. Then for any bounded solution u of the equation (2.3) on $\mathbb{R}^+$, the set*

$$N = \{t \geq 0; u^{(j)}(t) = 0,\ j = 0,1,2,3\}$$

is empty or a half line $[a,\infty)$ with $a \geq 0$.

Proof. From Theorem 4.2.4, we see that N is a closed and connected set. If the conclusion were false, then we would have

$$N = [b,a], \quad a < +\infty.$$

Applying the corollary (i.e. (2.11)) of Theorem 4.2.3 on the interval $(a,+\infty)$, and noticing that $k \geq 0$, we see that

$$u'(x)u'''(x) \geq \frac{1}{m}\int_a^{a+1} t|u(t)|^{1/m-1}u'(t)^2 dt > 0 \tag{2.12}$$

holds for $x \geq a+1$. In particular, this inequality implies that $u'(x)$ and $u'''(x)$ have the same sign on $x \geq a+1$. For definiteness, we assume that

$$u'(x) > 0, \quad u'''(x) > 0, \quad \text{if } x \geq a+1.$$

Taking this into account and using the boundedness of $u(x)$ (and hence $u'(x)$), it is easily seen that $\lim_{x\to\infty} u'(x)$ exists, and the value is nonnegative. Moreover, by (2.12), it follows that for x sufficiently large, $u'''(x) > \lambda$ for some constant λ, which implies $\lim_{x\to\infty} u''(x) = +\infty$. This contradicts the non existence of the limit $\lim_{x\to\infty} u'(x)$. The proof is complete. □

4.2.5 *Properties of unbounded solutions*

Theorem 4.2.6 *Let u be a solution of the equation (2.3) on the interval $[a,b)$, $a \geq 0$, $k,m > 0$ and let*

$$u^{(j)}(a) \geq 0, \quad j = 0,1,2,3, \quad \sum_{j=0}^{3} u^{(j)}(a) > 0.$$

Then for any $t \in (a,b)$, we have

$$u^{(j)}(t) > 0, \quad j = 0,1,2,3. \tag{2.13}$$

Proof. We first show that (2.13) is valid in some neighborhood (a, δ) of the point a. We need only to consider the case $u'(a) > 0$, since the other cases can be proved similarly. In this case, in some neighborhood of a, $u'(t) > 0$, $u(t) > 0$. However, from the equation (2.3), we also have $u^{(iv)}(t) > 0$, and hence for all $j = 0, 1, 2, 3$, $u^{(j)}(t) > 0$. Moreover, we may prove that for all $t \in (a, b)$, (2.13) holds. The proof is complete. □

Theorem 4.2.7 *Let u be a unbounded solution of the equation (2.3) on the interval $[a, b)$, $k > 0$. Then one of the following conclusions is valid*

(1) $\lim\limits_{x \to b^-} u^{(j)}(x) = +\infty$, $j = 0, 1, 2, 3$,

(2) $\lim\limits_{x \to b^-} u^{(j)}(x) = -\infty$, $j = 0, 1, 2, 3$.

Proof. We first check the availability of (2.11). If $0 < m \le 1$, then u is a classical solution, and (2.11) is obvious. Now, we assume that $m > 1$. By Theorem 4.2.2 and the unboundedness of u, we may conclude that $b = +\infty$. On the other hand, from Theorem 4.2.4,

$$N = \{t \ge 0; u^{(j)}(t) = 0,\ j = 0, 1, 2, 3\}$$

is a closed and connected set. Since u is unbounded, the set N can not be a half line. Without loss of generality, we may assume that $[a, b)$ does not intersect with N. Then the assumption of Theorem 4.2.3 is satisfied with $J = (a, b)$. Therefore, for any $m > 0$, we may always use Theorem 4.2.3 and its corollary. It follows from (2.11) that

$$u'(x)u'''(x) + C \ge \frac{km}{m+1}|u(x)|^{1+1/m}, \qquad x \in (a, b), \tag{2.14}$$

from which and the unboundedness of u it is not difficult to prove that u' can not have infinite number of zero points near the point b, i.e., $u'(x) > 0$ or $u'(x) < 0$ near b. As an example, we assume that $u'(x) > 0$. Therefore $\lim\limits_{x \to b^-} u(x) = +\infty$, and from the equation (2.3), we also have $\lim\limits_{x \to b^-} u^{(iv)}(x) = +\infty$, and $\lim\limits_{x \to b^-} u^{(j)}(x) = +\infty (j = 1, 2, 3)$. The proof is complete. □

4.2.6 *Bounded solutions on the half line*

Theorem 4.2.8 *Let $m > 1$, $k \ge 0$. Then the bounded solution of the equation (2.3) on $[0, +\infty)$ with initial value $u^{(j)}(0) = \alpha_j, j = 0, 1, 2, 3$ is unique.*

Proof. Let u_1, u_2 be two solutions and set

$$N_i = \{t \geq 0; u_i^{(j)}(t) = 0, \ j = 0, 1, 2, 3\}.$$

If N_1 is an empty set, then the conditions in Theorem 4.2.1 are satisfied, and the conclusion follows at once. Now, we assume that N_1 is not an empty set. Then by Theorem 4.2.5 (and $k \geq 0$), N_1 is the half line $[a, +\infty)$. However, we may apply Theorem 4.2.1 on the interval $[0, a)$ to obtain $u_1(x) = u_2(x)$ on $[0, a)$. On the other hand, by continuity, $a \in N_2$, and hence N_2 is not an empty set too. Using Theorem 4.2.5 again, we see that $N_1 = N_2$. The proof is complete. □

Theorem 4.2.9 *Let $k, m > 0$. Then for any $\alpha_0, \alpha_1, \beta \in \mathbb{R}$, the equation (2.3) admits a bounded solution u on $[0, +\infty)$ satisfying*

$$u(0) = \alpha_0, \quad u'(0) = \alpha_1, \quad u'''(0) = \beta (\text{or } u''(0) = \beta). \tag{2.15}$$

Proof. In this theorem, the solution is required to satisfy only three initial value conditions. The fourth one is in fact the boundedness condition. We try to find a bounded solution by selecting suitable value $u_\lambda''(0) = \lambda$. Let u_λ be the solution of the equation (2.3) satisfying (2.15) and $u_\lambda''(0) = \lambda$, and $[0, b_\lambda)$ be the maximal right interval of existence.

If $0 < m \leq 1$, then u_λ is uniquely determined, and hence u_λ depends continuously on λ. If $m > 1$, then we consider the following two cases separately.

Case 1. There exists $\lambda \in \mathbb{R}$, $t_0 \geq 0$ and a solution u_λ such that

$$u_\lambda^{(j)}(t_0) = 0, \quad j = 0, 1, 2, 3.$$

The bounded solution $\hat{u}(t)$ can be obtained by zero extension

$$\hat{u}(t) = u_\lambda(t) \quad \text{if } 0 \leq t \leq t_0, \quad \hat{u}(t) = 0 \quad \text{if } t > t_0.$$

The conclusion then follows.

Case 2. For all $\lambda \in \mathbb{R}$ and $t \geq 0$,

$$\sum_{j=0}^{3} \left| u_\lambda^{(j)}(t) \right| > 0.$$

Notice that, by Theorem 4.2.2, if $m > 1$, then $b_\lambda = +\infty$. In this case, the conclusion of Theorem 4.2.1 implies that u_λ is uniquely determined, and hence depends continuously on λ.

Set

$$S_1 = \{\lambda; \text{for some } t \in [0, b_\lambda] \text{such that } u_\lambda^{(j)}(t) > 0, j = 0, 1, 2, 3\},$$

$$S_1 = \{\lambda; \text{for some } t \in [0, b_\lambda] \text{such that } u_\lambda^{(j)}(t) < 0, j = 0, 1, 2, 3\}.$$

By the continuous dependence of the solution u_λ on λ, we see that S_1 and S_2 are all open sets. From Theorem 4.2.6, the set S_1 does not intersect with S_2. If we can prove that S_1 and S_2 are both non empty, then there exists $\lambda \in \mathbb{R} \setminus (S_1 \cup S_2)$, such that the corresponding solution u_λ is bounded (by Theorem 4.2.7). In case $m > 1$, we have seen that $b_\lambda = +\infty$. If $0 < m \leq 1$, $b_\lambda < +\infty$, then by (2.5), u_λ''', and hence u_λ'' and u_λ' are bounded. This shows that $[0, b_\lambda)$ can not be the maximal existence interval. Therefore, we have $b_\lambda = +\infty$.

Now, we turn to the proof of the fact that S_1 and S_2 are non empty. As an example, we consider the set S_1. For $\lambda > 0$, set

$$v_\lambda(x) = \lambda^{-2m/(m+1)} u_\lambda(\lambda^{(m-1)/2(m+1)} x).$$

Then v_λ is still a solution of the equation (2.3), satisfying the following initial value conditions

$$v_\lambda(0) = \alpha_0 \lambda^{-2m/(m+1)}, \quad v_\lambda'(0) = \alpha_1 \lambda^{-(3m+1)/2(m+1)}$$

$$v_\lambda''(0) = 1, \quad v_\lambda'''(0) = \beta \lambda^{-(m+3)/2(m+1)}.$$

Setting $\lambda \to +\infty$, taking the limit in some small neighborhood J of the point $x = 0$, and using the continuous dependence, we see that (in $C^3(J)$) the limit function v is a solution of the equation (2.3) too, which satisfies

$$v(0) = v'(0) = v'''(0) = 0, \qquad v''(0) = 1.$$

By Theorem 4.2.1, even in case that $m > 1$, such solution v is unique too. So, we also have the continuous dependence near the initial data. Applying Theorem 4.2.6, we have

$$v^{(j)}(t) > 0, \qquad j = 0, 1, 2, 3.$$

Since $u_\lambda^{(j)}$ converges as $\lambda \to +\infty$ in $C^3(J)$, we see that for λ sufficiently large, $\lambda \in S_1$, namely, the set S_1 is empty.

To prove the theorem in the case with $u'''(0) = \beta$ replaced by $u''(0) = \beta$, we need only to select the value of $u_\lambda'''(0) = \lambda$. The proof is complete. □

Lemma 4.2.1 *(Interpolation inequality, [RW]) Let* $J = (a, b)$, $h = b - a$, $u \in C^n([a, b])$. *Then*

$$U_i \le C(n, i) U_0^{1-i/n} (U_n^*)^{i/n}, \quad 0 < i < n,$$

where

$$U_j = \sup |u^{(j)}(x)|, \quad U_n^* = \max\{U_0 h^{-n}, U_n\}.$$

Lemma 4.2.2 *Let* $k > 0, m > 1, b \in \mathbb{R}$, $u \in C([b, \infty))$, $u(x) \ge C_0 > 0$, *and*

$$u(x) \ge \frac{k}{6} \int_b^x (x-t)^3 u(t)^{1/m} dt, \quad x \ge b. \tag{2.16}$$

Then

$$u(x) \ge \left(\frac{k^{1/4}}{\sigma}\right)^{\sigma} (x-b)^{\sigma}, \quad x \ge b,$$

where $\sigma = \dfrac{4m}{m-1}$.

Proof. Substituting $u(x) \ge C_0$ into (2.16), we see that there exists $C_1 > 0$, such that

$$u(x) \ge C_1 (x-b)^4, \quad x \ge b,$$

which combining with (2.16) yields

$$u(x) \ge C_2 (x-b)^{4(1+1/m)}, \quad x \ge b.$$

By induction, if

$$u(x) \ge C_n (x-b)^{\alpha_n}, \quad x \ge b,$$

then

$$\begin{aligned}
u(x) \ge & \ \frac{k}{6} C_n^{1/m} (x-b)^{4+\alpha_n/m} \int_0^1 (1-s)^3 s^{\alpha_n/m} ds \\
\ge & \ \frac{k}{6} C_n^{1/m} B(1 + \alpha_n/m, 4)(x-b)^{4+\alpha_n/m} \\
\ge & \ \frac{k}{(\alpha_n/m + 4)^4} C_n^{1/m} (x-b)^{4+\alpha_n/m} \\
\ge & \ \frac{k}{\alpha_{n+1}^4} C_n^{1/m} (x-b)^{4+\alpha_n/m},
\end{aligned}$$

where $B(\cdot,\cdot)$ is the Beta function,

$$\alpha_n = 4\left(1 + \frac{1}{m} + \cdots + \frac{1}{m^{n-1}}\right),$$

and C_n is determined by

$$C_{n+1} = \frac{kC_n^{1/m}}{\alpha_{n+1}^4}.$$

Obviously, α_n is increasing and the limit $\lim_{n\to\infty} \alpha_n$ is $\frac{4m}{m-1}$. Thus

$$C_{n+1} \geq MC_n^{1/m}, \qquad M = k\left(\frac{m-1}{4m}\right)^4,$$

and hence

$$\log C_n \geq \left(1 + \frac{1}{m} + \cdots + \frac{1}{m^{n-1}}\right)\log M + \frac{1}{m^n}\log C_0.$$

Letting $n \to \infty$, we get the desired conclusion and hence complete the proof. □

Theorem 4.2.10 *Let u be a solution of the equation (2.3) on the interval $[0, +\infty)$, $k > 0$. Then*

(1) if $0 < m < 1$, then $u \in L^\infty$,

(2) if $m > 1$, then $u \in L^\infty$, or for some $a > 0$, such that

$$|u(x)| \geq Cx^{4m/(m-1)}, \qquad \text{if } x \geq a > 0. \tag{2.17}$$

Proof. (1) We argue by contradiction. If the conclusion were false, then from Theorem 4.2.7, we might assume that $\lim_{x\to+\infty} u^{(j)}(x) = +\infty (j = 0, 1, 2, 3)$ (otherwise, we consider $-u$). It follows from the equation (2.3) that for some b_1,

$$u^{(j)}(x) \geq 1 (0 \leq j \leq 4) \quad \text{and } u^{(j)}(x) \text{ is strictly increasing on } (b_1, +\infty).$$

Applying Lemma 4.2.1 on the interval (b_1, x) gives

$$u'''(x) \leq Cu(x)^{1/4}(\max\{u(x)(x - b_1)^{-4}, u^{(iv)}(x)\})^{3/4}, \quad x > b_1.$$

From the equation (2.3), we have $u^{(iv)} \geq kg(u)$, i.e., $u(x) \leq C(u^{(iv)})^m$, $x > b_1$. Using the above inequality on (b_1+1, x), and noticing that $u^{(iv)} \geq 1$

and $m < 1$, we have

$$u'''(x) \leq C(u^{(iv)})^{(m+3)/4}, \quad x \geq b,$$

where $b = b_1 + 1$. This is a differential inequality for u'''. Integrating it over (b, x) yields

$$x - b \leq C\left(\frac{1}{u'''(b)^{\mu-1}} - \frac{1}{u'''(x)^{\mu-1}}\right), \quad x \geq b,$$

where $\mu = \dfrac{4}{m+3} > 1$. In particular,

$$x - b \leq \frac{C}{u'''(b)^{\mu-1}}, \quad x \geq b,$$

which is impossible.

(2) We need only to prove that if u is unbounded, then (2.17) is valid. From Theorem 4.2.7, we see that for some b,

$$u^{(j)}(x) \geq 1 (0 \leq j \leq 3), \qquad x \geq b.$$

Taylor's formula implies

$$u(x) = \sum_{j=0}^{3} \frac{u^{(j)}(b)}{j!}(x-b)^j + \frac{1}{6}\int_b^x (x-t)^3 u^{(iv)}(t)dt.$$

Noticing that from the equation (2.3), we have $u^{(iv)} \geq kg(u) = ku^{1/m}$ and hence

$$u(x) \geq \frac{k}{6}\int_b^x (x-t)^3 u(t)^{1/m} dt, \quad x \geq b,$$

from which (2.17) follows by using Lemma 4.2.2. The proof is complete. □

Lemma 4.2.3 *Let u_1 and u_2 be two solutions of the equation (2.3) on $[0, +\infty)$, $k \geq 1$, $U(x) = u_1(x) - u_2(x)$. If*

$$U^{(j)}(0) \geq 0 (j = 0, 1, 2, 3) \quad \textit{and} \sum_{j=0}^{3} U^{(j)}(0) > 0,$$

then

$$U^{(j)}(x) > 0 (j = 0, 1, 2, 3).$$

Proof. From (2.6),

$$U(x) = \sum_{j=0}^{3} \frac{1}{j!} U^{(j)}(0) x^j + \int_0^x H(x,t)[g(u_1(t)) - g(u_2(t))]dt. \tag{2.18}$$

If $x > t > 0$, then, since $k \geq 1$, we have

$$H(x,t) = \frac{1}{2} t(x-t)^2 + \frac{1}{6}(k-1)(x-t)^3 > 0.$$

By the assumption of the theorem and $U \in C^3[0, +\infty)$, we see that for some $\delta > 0$, $U(x) > 0$ holds in $(0, \delta)$. We conclude that

$$U(x) > 0, \qquad \forall x > 0.$$

If not, then there would exist some $c > 0$, such that $U(x) > 0$ in $(0, c)$ but $U(c) = 0$. Letting $x = c$ in (2.18), and using the assumption of the theorem and the continuity of g, we have $U(c) > 0$, a contradiction. Differentiating (2.18) three times, we see that $U''' > 0$ for $x \in (0, +\infty)$, and hence $U'' > 0$, $U' > 0$ for $x \in (0, +\infty)$. The proof is complete. □

Lemma 4.2.4 *Let u_1 and u_2 be bounded solutions of the equation (2.3) on $[0, +\infty)$, $k \geq 1$, $U(x) = u_1(x) - u_2(x)$. If two of $U^{(j)}(0)(j = 0, 1, 2, 3)$ are zero, for example, $U(0) = 0$, $U'(0) = 0$, then one of the following conclusions is valid:*

(1) $U''(0) = 0$, $U'''(0) = 0$,
(2) $U''(0)U'''(0) < 0$.

Proof. If (1) and (2) were all false, then $U''(0)$ and $U'''(0)$ would not equal to zero at the same time and $U''(0)U'''(0) \geq 0$. If one of $U''(0)$ and $U'''(0)$ is equal to zero, then the other is non zero. Applying Lemma 4.2.3 to U or $-U$, we see that $U^{(j)}(x) > 0 (j = 0, 1, 2, 3)$, which implies the unboundedness of U. If both $U''(0)$ and $U'''(0)$ are non zero, then they have the same sign, and an application of Lemma 4.2.3 to U or $-U$ leads to the unboundedness of U. This contradiction shows that one of the conclusions of the lemma is valid. The proof is complete. □

Theorem 4.2.11 *Let $k \geq 1$, $m > 0$. Then for any $\alpha_0, \alpha_1, \beta \in \mathbb{R}$, the equation (2.3) admits at most one bounded solution u on $[0, +\infty)$ satisfying*

$$u(0) = \alpha_0, \quad u'(0) = \alpha_1, \quad u'''(0) = \beta (\text{or } u''(0) = \beta).$$

Proof. We consider only the first case. Let $u_1(x)$ and $u_2(x)$ be two solutions, and set $U(x) = u_1(x) - u_2(x)$. Then $U(0) = U'(0) = U'''(0) = 0$. Lemma 4.2.4 implies $U''(0) = 0$. Thus using the uniqueness for bounded solutions and Theorem 4.2.8, we have $U(x) \equiv 0$ and hence complete the proof of the theorem. □

4.2.7 *Bounded solutions on the whole line*

Theorem 4.2.12 *Let $k \geq 1$.*

1) If $0 < m \leq 1$, then the equation (2.3) admits a unique solution on $\mathbb{R}$ satisfying the initial value conditions

$$u(0) = \alpha_0, \qquad u'(0) = \alpha_1. \tag{2.19}$$

2) If $m > 1$, then the equation (2.3) admits a unique solution on $\mathbb{R}$ satisfying (2.19) and

$$u(x) = o(|x|^{4m/(m-1)}), \quad |x| \to \infty. \tag{2.20}$$

In both cases, we have $u \in L^\infty(\mathbb{R})$.

Proof. We first prove the uniqueness. From the assumptions and Theorem 4.2.10, we see that $u_1, u_2 \in L^\infty(\mathbb{R})$. An application of Lemma 4.2.4 to $U_+(x) = u_1(x) - u_2(x)$ yields

$$U_+''(0) = U_+'''(0) = 0 \qquad \text{or} \qquad U_+''(0)U_+'''(0) < 0.$$

Using Lemma 4.2.4 again to $U_-(x) = u_1(-x) - u_2(-x)$, we have

$$U_-''(0) = U_-'''(0) = 0 \qquad \text{or} \qquad U_-''(0)U_-'''(0) > 0.$$

Notice that $U_+''(0) = U_-''(0)$, $U_+'''(0) = U_-'''(0)$. Thus $U_+''(0) = U_+'''(0) = 0$, and hence $u_1^{(j)}(0) = u_2^{(j)}(0) (j = 0, 1, 2, 3)$. It follows from Theorem 4.2.8 that $u_1(x) \equiv u_2(x)$.

We notice that in this theorem, to determine the solution, only two initial value conditions are required, which seems to contradict the fact that (2.3) is a fourth order equation. However, since the solution considered belongs to L^∞, some restrictions at infinity are implicitly imposed.

Now, we prove the existence. The proof is based on the existence of L^∞ solutions of the initial value problem on the half line. We try to choose the value of the third order derivative $u'''(0) = \beta$ to achieve the connection of solutions on two half lines. Denote by

$$u = u(x; \alpha_0, \alpha_1, \beta)$$

the bounded solution of the equation (2.3) on the half line $[0, +\infty)$ with the following conditions

$$u(0) = \alpha_0, \quad u'(0) = \alpha_1, \quad u'''(0) = \beta,$$

whose existence and uniqueness are guaranteed by Theorem 4.2.9 and Theorem 4.2.11. The main step for the connection is to coincide the second order derivatives at the point $x = 0$. Consider the function

$$f(\alpha_0, \alpha_1, \beta) = u''(0; \alpha_0, \alpha_1, \beta).$$

If we can prove that for some β^*,

$$f(\alpha_0, \alpha_1, \beta^*) = f(\alpha_0, -\alpha_1, -\beta^*), \tag{2.21}$$

then the function defined below is the desired bounded solution of the equation (2.3) satisfying the condition (2.19):

$$w(x) = \begin{cases} u(x; \alpha_0, \alpha_1, \beta^*), & x \geq 0, \\ u(-x; \alpha_0, -\alpha_1, -\beta^*), & x \leq 0. \end{cases}$$

In fact, such function $w(x)$ satisfies

$$w(0) = \alpha_0, \qquad w'(0^+) = w'(0^-),$$
$$w'''(0^+) = w'''(0^-) = \beta^*, \qquad w''(0^+) = w''(0^-).$$

It remains to show that the constant β^* satisfying (2.21) exists. According to Lemma 4.2.4, when $\beta \neq \hat{\beta}$, for $U(x) = u(x; \alpha_0, \alpha_1, \beta) - u(x; \alpha_0, \alpha_1, \hat{\beta})$, we have

$$U''(0)U'''(0) < 0.$$

In particular,

$$(\beta - \hat{\beta})(f(\alpha_0, \alpha_1, \beta) - f(\alpha_0, \alpha_1, \hat{\beta})) < 0,$$

which shows that $f(\alpha_0, \alpha_1, \beta)$ is strictly increasing in β. Let $\gamma \in \mathbb{R}$. By Theorem 4.2.9, the equation (2.3) admits an L^∞ solution on the half line $[0, \infty)$ satisfying

$$v(0) = \alpha_0, \quad v'(0) = \alpha_1, \quad v''(0) = \gamma.$$

Let $\beta = v'''(0)$. From the uniqueness, we have $\gamma = f(\alpha_0, \alpha_1, \beta)$, which shows that the range of the map $\beta \to f(\alpha_0, \alpha_1, \beta)$ is $\mathbb{R}$.

The fact that $f(\alpha_0, \alpha_1, \beta)$ is strictly increasing in β and the range is the whole line implies that $f(\alpha_0, \alpha_1, \beta)$ is continuous and

$$\lim_{\beta \to +\infty} f(\alpha_0, \alpha_1, \beta) = -\infty, \qquad \lim_{\beta \to -\infty} f(\alpha_0, \alpha_1, \beta) = +\infty.$$

Similarly,

$$\lim_{\beta \to +\infty} f(\alpha_0, -\alpha_1, -\beta) = +\infty, \qquad \lim_{\beta \to -\infty} f(\alpha_0, -\alpha_1, -\beta) = -\infty.$$

Therefore, there exists β^* such that (2.21) holds. The proof is complete.□

4.2.8 *Properties of solutions in typical cases $k = 1, 2, 3, 4$*

As a special case of Theorem 4.2.12, we have

Theorem 4.2.13 *Let $k = 1, 2, 3, 4$.*

1) If $0 < m \leq 1$, then the equation (2.3) admits a unique solution on $\mathbb{R}$ satisfying the following conditions

$$\begin{aligned} &u(0) = 1, \quad u'(0) = 0, \quad \text{if } k = 1 \text{ or } k = 3, \\ &u(0) = 0, \quad u'(0) = 1, \quad \text{if } k = 2 \text{ or } k = 4. \end{aligned} \tag{2.22}$$

2) If $m > 1$, then the equation (2.3) admits a unique solution on $\mathbb{R}$ satisfying (2.22)and

$$u(x) = o(|x|^{4m/(m-1)}), \quad \text{if } |x| \to \infty. \tag{2.23}$$

In both cases, $u \in L^\infty(\mathbb{R})$.

In what follows, we discuss the properties of solutions. We first have

Lemma 4.2.5 *Let $m > 0$, and u be a solution of the equation (2.3) on $[0,+\infty)$. Then*

$$x^{k-1}u^{(iv)} = (x^k g(u))',$$

$$\left(\frac{u^{(4-k)}(x)}{x}\right)^{(k-1)} = g(u(x)) + (-1)^{k-1}(k-1)!u^{(4-k)}(0)x^{-k}.$$

Proof. The first equality can be obtained by multiplying the equation (2.3) by x^{k-1}. The second equality can be derived by a simple calculation. In fact, from the first equality, we have

$$\sum_{j=0}^{k-1}(-1)^j \frac{(k-1)!}{(k-1-j)!}\left(x^{k-1-j}u^{(3-j)}\right)' = \left(x^k g(u)\right)'.$$

Integrating over $(0, x)$ gives

$$\begin{aligned}&\sum_{j=0}^{k-1}(-1)^j \frac{(k-1)!}{(k-1-j)!}x^{k-1-j}u^{(3-j)}(x)\\ =\ & x^k g(u) + (-1)^{k-1}(k-1)!u^{4-k}(0),\end{aligned}$$

i.e.,

$$\begin{aligned}&\sum_{j=0}^{k-1}(-1)^j \frac{(k-1)!}{(k-1-j)!j!}(j!)x^{-1-j}u^{(3-j)}(x)\\ =\ & g(u) + (-1)^{k-1}(k-1)!u^{4-k}(0)x^{-k}.\end{aligned}$$

Since $(x^{-1})^{(j)} = (-1)^j j! x^{-j-1}$, the left hand side of the above equality becomes

$$\sum_{j=0}^{k-1}\frac{(k-1)!}{(k-1-j)!j!}(x^{-1})^{(j)}(u^{(4-k)})^{(k-1-j)} = \left(x^{-1}u^{(4-k)}\right)^{(k-1)}.$$

The proof is complete. □

Theorem 4.2.14 *Assume that u is the solution obtained in Theorem 4.2.9 with $u^{(4-k)}(0) = 0$ and $u \not\equiv 0$. Let A be the supremum of the support of u. Then all the zero points of u on $(0, A)$ form an increasing sequence $\{a_n\}$, such that $\lim_{n\to\infty} a_n = A$ and $u'(a_n) \neq 0$ for any n.*

To prove the theorem, we need the following lemma.

Lemma 4.2.6 *Under the assumptions of Theorem 4.2.14, for any $b \in [0, A)$, u has at least one zero point on the interval (b, A).*

Proof. By the assumptions $u^{(4-k)}(0) = 0$ and Lemma 4.2.5, we have

$$\left(\frac{u^{(4-k)}(x)}{x}\right)^{(k-1)} = g(u(x)). \tag{2.24}$$

Since the solution we consider is bounded, it is easily seen that both the limits

$$\lim_{x\to+\infty} u^{(j)}(x), \quad j = 0, 1, 2, 3,$$

and

$$\lim_{x\to+\infty} \left(\frac{u^{(4-k)}(x)}{x}\right)^{(i)}, \quad i = 0, 1, \cdots, k-1$$

are equal to zero, if exist.

Suppose the contrary. Without loss of generality, we assume that $u(x) > 0$ in (b, A). Then from the definition of A, we see that $u(x) \geq 0$ in $S = (b, \infty)$. When $k = 1$, from (2.24), $u'''(x) \geq 0$ on S, and hence $u''(x)$ is increasing. However, $\lim\limits_{x\to\infty} u''(x) = 0$, so, $u''(x) \leq 0$ on S. Repeating the above argument, we see that $u'(x) \geq 0$ and $u(x) \leq 0$ on S, which contradicts the fact that $u > 0$ on (b, A). As for the cases $k = 2, 3, 4$, the proofs are similar. □

Lemma 4.2.7 *Under the assumptions of Theorem 4.2.14, let $y > x \geq 0$, $s \geq 2$,*

$$\Phi(x, y) = s(y - x)^{s-1} u'(y)^2 - (y - x)^s u'(y) u''(y). \tag{2.25}$$

Then

$$\Phi(x, y) \geq 0, \qquad \textit{if } u(y) = 0.$$

Proof. If $u'(y) = 0$, then the conclusion is obvious. We may assume that $u'(y) > 0$ (otherwise, consider $-u$). Since $u^{(4-k)}(0) = 0$, $u(y) = 0$, integrating the second equality in Lemma 4.2.5 over $(0, y)$ yields

$$\sum_{j=0}^{k-1} (-1)^j \frac{(k-1)!}{(k-1-j)!} y^{k-1-j} u^{(3-j)}(y) = 0$$

i.e.,

$$u'''(y) = 0, \quad \text{if } k = 1 \tag{2.26}$$

$$u''(y) = yu'''(y), \quad \text{if } k = 2 \tag{2.27}$$

$$u''(y) = \frac{y}{2}u'''(y) + \frac{1}{y}u'(y), \qquad \text{if } k = 3 \tag{2.28}$$

$$u''(y) = \frac{y}{3}u'''(y) + \frac{2}{y}u'(y), \quad \text{if } k = 4. \tag{2.29}$$

We split into two cases.

(1) The case $k = 1$ or $k = 2$. In this case, it suffices to prove that $u''(y) < 0$. If not, then $u''(y) \geq 0$, and from (2.26) or (2.27), $u'''(y) \geq 0$. By Theorem 4.2.6, for any $t > y$, $u^{(j)}(t) > 0 (j = 0, 1, 2, 3)$, which contradicts the boundedness of u.

(2) The case $k = 3$ or $k = 4$. Substituting (2.28) or (2.29) into (2.25), we have

$$\Phi(x,y) = \left(s - 1 + \frac{x}{y}\right)(y-x)^{s-1}u'(y)^2 - \frac{y}{2}(y-x)^s u'(y)u'''(y),\ k = 3$$

$$\Phi(x,y) = \left(s - 2 + \frac{2x}{y}\right)(y-x)^{s-1}u'(y)^2 - \frac{y}{3}(y-x)^s u'(y)u'''(y),\ k = 4.$$

It remains to prove $u'''(y) < 0$, which is similar to that for the first case, and we omit the details. The proof is complete. □

Proof of Theorem 4.2.14. By virtue of Lemma 4.2.6 and the fact that the zero points of u on $(0, A)$ are isolated, we need only to show that if $0 < y < A$, $u(y) = 0$, then $u'(y) \neq 0$. Suppose the contrary, namely, $u(y) = u'(y) = 0$. From (2.26)–(2.29), we see that $u''(y)$ and $u'''(y)$ have the same sign or equal to zero. Since $-u$ is also a solution of the equation (2.3), we may assume that $u''(y) \geq 0$, $u'''(y) \geq 0$. However, from the boundedness of u, and Theorem 4.2.7, we conclude that $u''(y) = u'''(y) = 0$. For the case $0 < m \leq 1$, this implies $u \equiv 0$, which contradicts the assumption $u \not\equiv 0$. While for the case $m > 1$, from Theorem 4.2.5, we see that $u(x) \equiv 0$ for any $x > y$, which contradicts the definition of A. The proof is complete. □

The following two theorems show that if $0 < m < 1$, then the solution decays at infinity with negative power order. While if $m > 1$, then the solution has compact support.

Theorem 4.2.15 *Let $k = 1, 2, 3, 4$, $0 < m < 1$ and u be the solution of the equation (2.3) on $\mathbb{R}$ satisfying (2.22). Then for any $j = 0, 1, 2, 3, 4$, we have*

$$u^{(j)}(x) = O(|x|^{\sigma - j}), \quad |x| \to \infty, \tag{2.30}$$

where

$$\sigma = -\frac{4m}{1-m}.$$

Theorem 4.2.16 *Let $k = 1, 2, 3, 4$, $m > 1$ and u be the solution of the equation (2.3) on $\mathbb{R}$ satisfying (2.22) and (2.23). Then u has compact support.*

To prove the above theorems, we need some lemmas.

Lemma 4.2.8 *Let $s > 3$, $0 \le x < y < \infty$, $q = 1 + \dfrac{1}{m}$. Then*

$$\begin{aligned}
&\int_x^y (t-x)^s u''(t)^2 dt + \left(\frac{s+1}{m+1} - k\right) \int_x^y (t-x)^s |u(t)|^q dt \\
&\quad + \frac{sx}{m+1} \int_x^y (t-x)^{s-1} |u(t)|^q dt + \Phi(x,y) \\
= \quad & \Psi(x,y) - 2s(s-1) \int_x^y (t-x)^{s-2} u(t) u''(t) dt \\
&\quad + \frac{1}{2} s(s-1)(s-2)(s-3) \int_x^y (t-x)^{s-4} u(t)^2 dt,
\end{aligned}$$

where $\Phi(x,y)$ is given by (2.25), and

$$\begin{aligned}
\Psi(x,y) = \; & \frac{y}{m+1}(y-x)^s |u(y)|^q - (y-x)^s u(y) u'''(y) + \\
& \qquad + s(y-x)^{s-1} u(y) u''(y) \\
& + s(s-1)(y-x)^{s-2} u(y) u'(y) - \\
& - \frac{s}{2}(s-1)(s-2)(y-x)^{s-3} u(y)^2.
\end{aligned}$$

Proof. It suffices to multiply (2.10) by $(t-x)^s u(t)$, and integrate the resulting relation over (x, y) with respect to t. □

Now, we introduce the following notations

$$E_s(x,y) = \int_x^y (t-x)^s u''(t)^2 dt, \quad F_s(x,y) = \int_x^y (t-x)^s u(t)^2 dt,$$

$$G_s(x,y) = \int_x^y (t-x)^s u(t)^q dt, \quad q = 1 + \frac{1}{m},$$

$$E_s(x) = E_s(x,+\infty), \quad G_s(x) = G_s(x,+\infty),$$

$$I_s(x) = E_s(x) + G_s(x).$$

Lemma 4.2.9 *Let* $q = 1 + \frac{1}{m}$, $m > 0$, $s = 5m+3$. *Then for any* $\varepsilon > 0$, *there exists a constant* C *depending only on* ε *and* m, *such that*

$$\int_x^\infty (t-x)^{s-4} u(t)^2 dt \le \varepsilon \int_x^\infty (t-x)^s |u(t)|^q dt + C \sup_{t>x} |u(t)|^{s/4}.$$

Proof. Without loss of generality, we assume that $x = 0$. First, we have

$$\int_0^1 t^{s-4} u(t)^2 dt \le \frac{1}{s-3} \sup u(t)^2.$$

If $q \le 2$, then

$$\begin{aligned}\int_1^\infty t^{s-4} u(t)^2 dt &\le \sup |u(t)|^{2-q} \int_1^\infty t^s |u(t)|^q dt \\ &\le C\left(\sup u(t)^2 + \left(\int_1^\infty t^s |u(t)|^q dt\right)^{2/q}\right).\end{aligned}$$

While if $q > 2$, then Hölder's inequality implies

$$\begin{aligned}\int_1^\infty t^{s-4} u(t)^2 dt &= \int_1^\infty t^{2s/q} u(t)^2 t^{s-4-2s/q} dt \\ &\le \left(\int_1^\infty t^s |u(t)|^q dt\right)^{2/q} \left(\int_1^\infty t^b dt\right)^{(q-2)/q} \\ &\le \left(\int_1^\infty t^s |u(t)|^q dt\right)^{2/q},\end{aligned}$$

where $b = \dfrac{m - 5m^2 - 1}{1-m}$. Here we have noticed the fact that $b < -1$ if $0 < m < 1$.

Summing up, for any q, we have

$$\int_0^\infty t^{s-4}u(t)^2dt \le C\left(\sup u(t)^2 + \left(\int_0^\infty t^s|u(t)|^q dt\right)^{2/q}\right).$$

Replacing $u(t)$ by $u(t/\lambda)$, and optimizing λ, we get the inequality

$$\int_0^\infty t^{s-4}u(t)^2dt \le C\left(\int_0^\infty t^s|u(t)|^q dt\right)^{2a/q} \sup|u(t)|^{2(1-a)},$$

where $a = \dfrac{m+1}{2(5m+4)}$. Young's inequality then implies the desired conclusion. The proof is complete. □

Lemma 4.2.10 *(Gagliardo-Nirenberg's inequality, [NI])*

$$\sup_{t>0}|u(t)| \le C\left(\int_0^\infty |u''(t)|^2dt\right)^{b/2}\left(\int_0^\infty |u(t)|^q dt\right)^{(1-b)/q},$$

where $b = \dfrac{2}{3q+2}$.

Lemma 4.2.11 *Let* $s = 5m+3$, $0 \le x < y < \infty$, $u(y) = 0$. *Then*

$$E_s(x,y) + G_s(x,y) \le C \sup_{x<t<y} |u(t)|^{s/4}.$$

Proof. Since $u(y) = 0$ implies $\Psi(x,y) = 0$, from Lemma 4.2.7, we see that $\Phi(x,y) \ge 0$. Notice that

$$\frac{s+1}{m+1} - k \ge \frac{m}{m+1}, \qquad s = 5m+3, k \le 4.$$

It follows from Lemma 4.2.8 that

$$E_s(x,y) + G_s(x,y) \le C\int_x^y (t-x)^{s-2}|u(t)u''(t)|dt + CF_{s-4}(x,y).$$

Cauchy-Schwarz's inequality gives

$$\begin{aligned}\int_x^y (t-x)^{s-2}|u(t)u''(t)|dt \le{}& (E_s(x,y))^{1/2}(F_{s-4}(x,y))^{1/2}\\ \le{}& \varepsilon E_s(x,y) + C_\varepsilon F_{s-4}(x,y).\end{aligned}$$

An application of Lemma 4.2.9 then implies the desired conclusion and the proof if complete. □

Lemma 4.2.12

$$I_1(x) \le CI_0(x)^{1+(m-1)/(5m+3)}, \qquad x \ge 0.$$

Proof. Let $q = 1 + \dfrac{1}{m}$. By Gagliardo-Nirenberg's inequality

$$\sup_{t>x} |u(t)| \le CE_0(x)^{b/2} G_0(x)^{(1-b)/q}, \quad b = \frac{2}{3q+2} = \frac{2m}{5m+3}$$

and the inequality

$$A^\alpha B^\alpha \le C(\alpha, \beta)(A+B)^{\alpha+\beta},$$

we have

$$\sup_{t>x} |u(t)| \le CI_0(x)^{4m/(5m+3)}, \quad x \ge 0. \tag{2.31}$$

It follows from Theorem 4.2.16 that all zero points of $u(x)$ form an increasing sequence $\{a_n\}$ with $\lim_{n\to\infty} a_n = +\infty$. Letting $y = a_n \to +\infty$ in Lemma 4.2.11, we obtain

$$E_s(x) + G_s(x) \le C \sup_{t>x} |u(t)|^{s/4}.$$

Therefore

$$I_s(x) \le CI_0(x)^m, \quad s = 5m+3, \quad x \ge 0. \tag{2.32}$$

Notice that

$$I_1(x) = \int_x^\infty (t-x) f(t) dt, \quad f(t) \ge 0.$$

By virtue of Hölder's inequality, we see that for any $s > 1$,

$$I_1(x) \le I_s(x)^{1/s} I_0(x)^{1-1/s}, \quad x \ge 0. \tag{2.33}$$

The desired inequality follows immediately from (2.32) and (2.33). The proof is complete. □

Proof of Theorem 4.2.15. Let $0 < m < 1$. We only consider the case for the right half line. Since $I_1'(x) = -I_0(x)$, Lemma 4.2.12 presents a first order differential inequality. The integration implies

$$I_1(x) = O(x^{-(6m+2)/(1-m)}), \quad x \to +\infty. \tag{2.34}$$

Since $I_0(x)$ is decreasing, we have

$$I_1\left(\frac{x}{2}\right) = \int_{x/2}^{\infty} I_0(t)dt \geq \int_{x/2}^{x} I_0(t)dt \geq \frac{x}{2} I_0(x). \tag{2.35}$$

It follows from (2.31), (2.34) and (2.35),

$$u(x) = O(x^\sigma), \quad x \to +\infty, \sigma = -\frac{4m}{1-m},$$

which implies (2.30) for $j = 0$. Therefore

$$g(u(x)) = O(x^{\sigma-4}), \quad x \to +\infty. \tag{2.36}$$

Thus from Definition 4.2.1, we see that $\lim_{a\to\infty} u'''(a)$ exists, and must be zero (noticing the boundedness of u). Therefore

$$u'''(x) = xg(u(x)) - (k-1)\int_x^{\infty} g(u(t))dt, \quad x \geq 0,$$

which together with (2.36) implies (2.30) for $j = 3$. By integrating and using the equation (2.3), it is easily seen that (2.30) holds for $j = 1, 2, 4$. The proof is complete. □

Proof of Theorem 4.2.16. Let $m > 1$. We only consider the case for the right half line. Suppose the contrary, namely, u does not have compact support. Then from Theorem 4.2.5, the set

$$N = \{t \geq 0; u^{(j)}(t) = 0, \ j = 0, 1, 2, 3\}$$

is empty, and hence $I_1(x) > 0$ for all $x \geq 0$. It follows from Lemma 4.2.12 that

$$\begin{aligned} x \leq & \ C(I_1(0)^{(m-1)/(6m+2)} - I_1(x)^{(m-1)/(6m+2)}) \\ \leq & \ CI_1(0)^{(m-1)/(6m+2)} \end{aligned}$$

holds for any x. This contradiction shows that N is a half line $[a, +\infty)$, i.e., u has compact support. The proof is complete. □

4.2.9 *Behavior of similarity solutions as* $t \to 0^+$

According to Definition 4.2.1, if $u(x)$ is a solution of the equation (2.3), then

$$w(y,t) = t^{-k\beta} g(u(\frac{\beta^{1/4} y}{t^\beta})) \tag{2.37}$$

is a similarity solution of the equation (2.2). Now, we discuss the behavior of $w(y,t)$ as $t \to 0^+$.

Theorem 4.2.17 *Let $k = 1,2,3,4$, and u be the solution given in Theorem 4.2.13, which satisfies $u(0) = 1$, $u'(0) = 0$ when $k = 1,3$ and $u(0) = 0$, $u'(0) = 1$ when $k = 2,4$. Then the function $w(y,t)$ defined by (2.37), satisfies*

$$w(\cdot,t) \to B_k \delta^{(k-1)}, \qquad \text{as } t \to 0^+ \tag{2.38}$$

in the sense of distributions, where δ is the Dirac function and

$$B_k = \frac{(-1)^k}{(k-1)!\beta^{k/4}} \int_{-\infty}^{\infty} x^{k-1} g(u(x))dx. \tag{2.39}$$

To prove the theorem, we first verify

Lemma 4.2.13 *Under the assumptions of Theorem 4.2.17, let*

$$M_j(k) \equiv \int_{-\infty}^{\infty} x^j g(u(x))dx. \tag{2.40}$$

If $0 \le j \le 3$ and $j \neq k-1$, then $M_j(k) = 0$.

Proof. Multiply the equation (2.3) by $x^j (0 \le j \le 3)$, and then integrate over $(0, +\infty)$,

$$\int_0^\infty x^j u^{(iv)}(x)dx = \int_0^\infty x^{j+1}(g(u(x)))'dx + \int_0^\infty kx^j g(u(x))dx.$$

Integrating by parts and using Theorem 4.2.15 or 2.16, we may easily check that

$$-u'''(0) = (k-1)\int_0^\infty g(u(x))dx,$$

$$u''(0) = (k-2)\int_0^\infty xg(u(x))dx,$$

$$-2u'(0) = (k-3)\int_0^\infty x^2 g(u(x))dx,$$
$$6u(0) = (k-4)\int_0^\infty x^3 g(u(x))dx.$$

As an example, we consider the case $k = 1$, when $u(x)$ is an even function. So, $xg(u(x))$ and $x^3g(u(x))$ are all odd functions, and hence $M_1(1) = 0, M_3(1) = 0$. Noticing that in the case $k = 1$, the boundary value conditions become $u(0) = 1$, $u'(0) = 0$ and we have $M_2(1) = 0$, which shows that apart from $j = 0$, $M_j(1) = 0$. The proof is complete. □

Proof of Theorem 4.2.17. It suffices to show that for any $\varphi \in C_0^\infty(\mathbb{R})$,

$$\lim_{t\to 0^+}\int_{-\infty}^\infty w(y,t)\varphi(y)dy = (-1)^{k-1}B_k\varphi^{(k-1)}(0). \tag{2.42}$$

Taylor's formula yields

$$\varphi(y) = \sum_{j=0}^{k-2}\frac{1}{j!}\varphi^{(j)}(0)y^j + \frac{1}{(k-1)!}\psi(y)y^{k-1}.$$

Since $\varphi(y)$ has compact support, we have $\lim\limits_{|y|\to\infty}\psi(y) = 0$, and $\psi \in L^\infty(\mathbb{R})$. Substituting the formula for φ into (2.41), and using Lemma 4.2.13, we obtain

$$\begin{aligned}
&\int_{-\infty}^\infty w(y,t)\varphi(y)dy \\
= &\sum_{j=0}^{k-2}\frac{1}{j!}\varphi^{(j)}(0)\int_{-\infty}^\infty y^j w(y,t)dy \\
&+\frac{1}{(k-1)!}\int_{-\infty}^\infty y^{k-1}w(y,t)\psi(y)dy \\
= &\frac{1}{(k-1)!}\int_{-\infty}^\infty y^{k-1}w(y,t)\psi(y)dy.
\end{aligned}$$

Set $x = \dfrac{\beta^{1/4}y}{t^\beta}$. Then the above equality turns out to be

$$\begin{aligned}
&\int_{-\infty}^\infty w(y,t)\varphi(y)dy \\
= &\frac{1}{(k-1)!}\frac{1}{\beta^{k/4}}\int_{-\infty}^\infty x^{k-1}g(u(x))\psi(\frac{xt^\beta}{\beta^{1/4}})dx.
\end{aligned}$$

Noticing that $\psi(0) = \varphi^{(k-1)}(0)$, letting $t \to 0^+$ in the above formula, and applying the Lebesgue dominated convergence theorem, we immediately obtain (2.41) and the proof is complete. □

4.3 Equations with Double-Degeneracy

In this section, we consider an equation with double degeneracy, namely,

$$\frac{\partial u}{\partial t} + \frac{\partial^2}{\partial x^2}\left(\left|\frac{\partial^2 \Phi(u)}{\partial x^2}\right|^{p-2} \frac{\partial^2 \Phi(u)}{\partial x^2}\right) = 0, \tag{3.1}$$

where $p > 1$, $\Phi(u) = |u|^{q-2}u$, $q > 1$, and discuss the existence, uniqueness and the properties of solutions, such as the propagation of disturbances and the nonexistence of nonnegative solutions. The equation (2.1) is a special case of the equation (3.1) for $p = 2$. If we regard the equation (2.1) as an analogous version of the Newtonian porous medium equation, then the equation (3.1) can be thought of as an analogous version of the non-Newtonian porous medium equation.

4.3.1 *Existence of solutions*

For convenience of treatment, we rewrite (3.1) as the form

$$\frac{\partial B(u)}{\partial t} + \frac{\partial^2}{\partial x^2} A\left(\frac{\partial^2 u}{\partial x^2}\right) = 0, \qquad (x.t) \in Q_T, \tag{3.2}$$

where $A(s) = |s|^{p-2}s$, $B(s) = |s|^{q-2}s (p, q > 1)$, $Q_T = (0,1) \times (0,T)$. We will consider the first boundary value problem as an example. The corresponding initial and boundary value conditions are

$$u(x,0) = u_0(x), \tag{3.3}$$

$$u(0,t) = u(1,t) = u_x(0,t) = u_x(1,t) = 0. \tag{3.4}$$

Definition 4.3.1 A function u is said to be a generalized solution of the problem (3.2)–(3.4), if the following conditions are fulfilled:

(1) $u \in L^\infty(0,T; W_0^{2,p}(I))$, $B(u) \in C(0,T; L^{q'}(I))$,

$$\frac{\partial B(u)}{\partial t} \in L^\infty(0,T; W^{-2,p'}(I)), \qquad B(u(x,0)) = B(u_0),$$

where $I = (0,1)$ and p' and q' are the conjugate exponents of p, q respectively;

(2) For any $\varphi \in C_0^\infty(Q_T)$, the following integral equality holds

$$-\iint_{Q_T} B(u)\frac{\partial \varphi}{\partial t}dxdt + \iint_{Q_T} A\left(\frac{\partial^2 u}{\partial x^2}\right)\frac{\partial^2 \varphi}{\partial x^2}dxdt = 0.$$

Theorem 4.3.1 *Let $u_0 \in W_0^{2,p}(I)$, $q \le p$. Then the boundary value problem (3.2)–(3.4) admits at least one generalized solution.*

We adopt the time discrete method to construct an approximate solution. Divide the interval $(0, T)$ into N equal segments and denote by $h = \dfrac{T}{N}$. Consider the problem

$$\frac{1}{h}(B(u_{k+1}) - B(u_k)) + \frac{d^2}{dx^2}A\left(\frac{d^2 u_{k+1}}{dx^2}\right) = 0, \tag{3.5}$$

$$u_{k+1}(0) = u_{k+1}(1) = u'_{k+1}(0) = u'_{k+1}(1) = 0, \tag{3.6}$$

$$k = 0, 1, \cdots, N-1,$$

where u_0 is the initial datum.

Lemma 4.3.1 *For fixed k, if $B(u_k) \in L^q(I)$, then the problem (3.5), (3.6) admits a generalized solution u_{k+1} in the space $W_0^{2,p}(I)$, namely, there exists a function $u_{k+1} \in W_0^{2,p}(I)$, such that for any $\varphi \in C_0^\infty(I)$,*

$$\frac{1}{h}\int_0^1 (B(u_{k+1}) - B(u_k))\varphi dx + \int_0^1 A\left(\frac{d^2 u_{k+1}}{dx^2}\right)\frac{d^2 \varphi}{dx^2} = 0. \tag{3.7}$$

Proof. Consider the functionals

$$\Phi_A[u] = \frac{1}{p}\int_0^1 \left|\frac{\partial^2 u}{\partial x^2}\right|^p dx,$$

$$\Phi_B[u] = \frac{1}{q}\int_0^1 |u|^q dx,$$

$$\Psi[u] = \Phi_A[u] + \frac{1}{h}\Phi_B[u] - \int_0^1 fudx,$$

where $f \in L^{q'}(I)$ is a given function. By virtue of Young's inequality, we see that for some constants $C_1, C_2 > 0$,

$$\Psi[u] \ge C_1 \int_0^1 \left|\frac{\partial^2 u}{\partial x^2}\right|^p dx - C_2 \int_0^1 |f|^{q'} dx.$$

Therefore

$$\Psi[u] \to +\infty, \qquad \text{if } \|u\|_{2,p} \to +\infty.$$

Here we use the notation $\|u\|_{2,p}$ to denote the norm of u in $W_0^{2,p}(I)$. Noticing that $\Psi[u]$ is obviously a weakly semi-lower continuous functional, we conclude that there exists $u_* \in W_0^{2,p}(I)$, such that

$$\Psi(u_*) = \inf \Psi[u],$$

and u_* is the solution of the Euler equation corresponding to $\Psi[u]$

$$\frac{1}{h}B(u) + \frac{d^2}{dx^2}A\left(\frac{d^2u}{dx^2}\right) = f.$$

Choosing $f = \dfrac{1}{h}B(u_k)$, we then obtain a generalized solution u_{k+1} of the problem (3.5), (3.6). The proof is complete. □

Now, we construct an approximate solution u^h of the problem (3.2)–(3.4) by defining

$$u^h(x,t) = u_k(x), \quad kh < t \le (k+1)h, k = 0, 1, \cdots, N-1,$$
$$u^h(x,0) = u_0(x).$$

The desired solution of the problem (3.2)–(3.4) will be obtained as the limit of some subsequence of $\{u^h\}$. For this purpose, we need some uniform estimates on u^h.

Lemma 4.3.2 *For the generalized solution u_k of the problem (3.5),(3.6), the following estimates hold*

$$h\sum_{k=1}^{N}\int_0^1 \left|\frac{d^2u_k(x)}{dx^2}\right|^p dx \le C, \tag{3.8}$$

$$\sup_{0<t<T}\int_0^1 \left|\frac{d^2u^h(x,t)}{dx^2}\right|^p dx \le C, \tag{3.9}$$

where C is a constant independent of h, k.

Proof. We first prove (3.8). Notice that, we may choose $\varphi \in W_0^{2,p}(I)$ as the test function in (3.7). In particular, by the choice $\varphi = u_{k+1}$, we have

$$\frac{1}{h}\int_0^1 |u_{k+1}|^q dx - \frac{1}{h}\int_0^1 |u_k|^{q-2}u_k u_{k+1} dx + \int_0^1 \left|\frac{d^2 u_{k+1}}{dx^2}\right|^p dx = 0.$$

Then by Hölder's inequality

$$\begin{aligned} &\frac{1}{h}\int_0^1 |u_{k+1}|^q dx + \int_0^1 \left|\frac{d^2 u_{k+1}}{dx^2}\right|^p dx \\ \le\ & \frac{1}{hq'}\int_0^1 |u_k|^q dx + \frac{1}{hq}\int_0^1 |u_{k+1}|^q dx, \end{aligned}$$

i.e.,

$$\frac{1}{q'}\int_0^1 |u_{k+1}|^q dx + h\int_0^1 \left|\frac{d^2 u_{k+1}}{dx^2}\right|^p dx \le \frac{1}{q'}\int_0^1 |u_k|^q dx. \tag{3.10}$$

Summing up these inequalities for k from 0 up to $N-1$, we have

$$h\sum_{k=0}^{N-1}\int_0^1 \left|\frac{d^2 u_{k+1}}{dx^2}\right|^p dx \le \frac{1}{q'}\int_0^1 |u_0|^q dx.$$

So, (3.8) holds.

To prove (3.9), we choose $\varphi = u_{k+1} - u_k$ in (3.7) and obtain

$$\begin{aligned} &\frac{1}{h}\int_0^1 (B(u_{k+1}) - B(u_k))(u_{k+1} - u_k)\,dx + \\ &\qquad + \int_0^1 A\left(\frac{d^2 u_{k+1}}{dx^2}\right)\left(\frac{d^2 u_{k+1}}{dx^2} - \frac{d^2 u_k}{dx^2}\right) dx = 0. \end{aligned}$$

Since the first term of the left hand side of the above equality is nonnegative, it follows that

$$\int_0^1 A\left(\frac{d^2 u_{k+1}}{dx^2}\right)\left(\frac{d^2 u_{k+1}}{dx^2} - \frac{d^2 u_k}{dx^2}\right) dx \le 0.$$

Hölder's inequality then yields

$$\int_0^1 \left|\frac{d^2u_{k+1}}{dx^2}\right|^p dx \le \int_0^1 \left|\frac{d^2u_{k+1}}{dx^2}\right|^{p-2} \frac{d^2u_{k+1}}{dx^2}\frac{d^2u_k}{dx^2}dx$$
$$\le \frac{1}{p'}\int_0^1 \left|\frac{d^2u_{k+1}}{dx^2}\right|^p dx + \frac{1}{p}\int_0^1 \left|\frac{d^2u_k}{dx^2}\right|^p dx,$$

i.e.,

$$\int_0^1 \left|\frac{d^2u_{k+1}}{dx^2}\right|^p dx \le \int_0^1 \left|\frac{d^2u_k}{dx^2}\right|^p dx.$$

For any m with $1 \le m \le N-1$, summing up the above inequality for k from 0 up to $m-1$, we have

$$\int_0^1 \left|\frac{d^2u_m}{dx^2}\right|^p dx \le \int_0^1 \left|\frac{d^2u_0}{dx^2}\right|^p dx.$$

Therefore (3.9) holds and the proof is complete. □

Lemma 4.3.3 *For the generalized solution u_{k+1} of the problem (3.5), (3.6), the following estimate holds*

$$-Ch \le \int_0^1 |u_{k+1}|^q dx - \int_0^1 |u_k|^q dx \le 0, \tag{3.11}$$

where C is a constant independent of h.

Proof. The second inequality of (3.11) is an immediate consequence of (3.10). To prove the first inequality, we choose $\varphi = u_k$ in (3.7) and obtain

$$\frac{1}{h}\int_0^1 (B(u_{k+1}) - B(u_k))\, u_k dx +$$
$$+\int_0^1 \left|\frac{d^2u_{k+1}}{dx^2}\right|^{p-2} \frac{d^2u_{k+1}}{dx^2}\frac{d^2u_k}{dx^2}dx = 0.$$

Applying Hölder's inequality and the estimate (3.9) yields

$$\int_0^1 B(u_k)u_k dx - \int_0^1 B(u_{k+1})u_k dx \le Ch.$$

Therefore

$$\int_0^1 |u_k|^q dx \le Ch + \int_0^1 |u_{k+1}|^{q-2} u_{k+1} u_k dx$$
$$\le Ch + \frac{1}{q'} \int_0^1 |u_{k+1}|^q dx + \frac{1}{q} \int_0^1 |u_k|^q dx.$$

Here we have used Hölder's inequality again. Thus

$$\frac{1}{q'} \int_0^1 |u_k|^q dx \le Ch + \frac{1}{q'} \int_0^1 |u_{k+1}|^q dx,$$

which shows that the first inequality of (3.11) is valid. The proof is complete. □

Corollary 4.3.1

$$\sup_{0<t<T} \int_0^1 |u^h(x,t)|^q dx \le \int_0^1 |u_0|^q dx. \tag{3.12}$$

Proof. We need only, for any fixed m with $1 \le m \le N-1$, to sum the second inequality in (3.11) for k from 0 up to $m-1$. □

Now, we define operators B^t and A^t by

$$B^t(u^h) = B(u_k),$$
$$A^t\left(\frac{d^2u^h}{dx^2}\right) = A\left(\frac{d^2u_k}{dx^2}\right),$$
$$\Delta^h B^t(u^h) = B(u_{k+1}) - B(u_k),$$

where $kh < t \le (k+1)h$, $h = 0, 1, \cdots, N-1$. From the discrete equations (3.5) and (3.8) in Lemma 4.3.2, we see that

$$\frac{1}{h}\Delta^h B^t(u^h) \qquad \text{is bounded in } L^\infty\left(0, T; W^{-2,p'}(I)\right). \tag{3.13}$$

Using the results obtained above, we can now complete the proof of Theorem 4.3.1.

Proof of Theorem 4.3.1. By (3.9),(3.12),(3.13) and (3.7), we may extract a subsequence from $\{u^h\}$, denoted still by $\{u^h\}$, such that

$$u^h \overset{*}{\rightharpoonup} u, \qquad \text{in } L^\infty\left(0,T;W^{2,p}(I)\right),$$

$$B^t(u^h) \overset{*}{\rightharpoonup} v, \qquad \text{in } L^\infty\left(0,T;L^{q'}(I)\right),$$

$$\frac{1}{h}\Delta^h B^t(u^h) \overset{*}{\rightharpoonup} \frac{\partial v}{\partial t}, \qquad \text{in } L^\infty\left(0,T;W^{-2,p'}(I)\right),$$

$$A^t\left(\frac{\partial^2 u^h}{\partial x^2}\right) \overset{*}{\rightharpoonup} w, \qquad \text{in } L^\infty\left(0,T;L^{p'}(I)\right)$$

hold for some u, v, w. Then in the sense of distributions

$$\frac{\partial v}{\partial t} + \frac{\partial^2 w}{\partial x^2} = 0. \tag{3.14}$$

In fact, from the relation (3.7), we see that for any $\varphi \in C_0^\infty(Q_T)$,

$$\iint_{Q_T} \left(\frac{1}{h}\Delta^h B^t(u^h)\varphi + A^t\left(\frac{\partial^2 u^h}{\partial x^2}\right)\frac{\partial^2 \varphi}{\partial x^2}\right) dxdt = 0,$$

and (3.14) follows by letting $h \to 0$.

Now, we turn to the proof of $B(u) = v$ a.e. in Q_T. Define

$$f_h(t) = \frac{t-kh}{q'h}\left(\int_0^1 |u_{k+1}|^q dx - \int_0^1 |u_k|^q dx\right) + \frac{1}{q'}\int_0^1 |u_k|^q dx,$$
$$kh \le t \le (k+1)h, \quad k = 0, 1, \cdots, N-1.$$

From (3.11),

$$\frac{1}{q'}\int_0^1 |u_k|^q dx - Ch \le f_h(t) \le \frac{1}{q'}\int_0^1 |u_k|^q dx,$$
$$-C \le f_h'(t) \le 0.$$

According to the Ascoli-Arzela theorem, there exists a function $f(t) \in C([0,T])$, such that

$$\lim_{h\to 0} f_h(t) = f(t) \qquad \text{uniformly for } t \in [0,T].$$

Using (3.11) again, we have

$$\lim_{h\to 0} \frac{1}{q'}\int_0^1 |u^h|^q dx = f(t), \qquad \text{uniformly for } t \in [0,T]. \tag{3.15}$$

For fixed $t \in (0,T)$, from $\dfrac{\delta\Phi_B[u]}{\delta u} = B(u)$ and the convexity of the functional $\Phi_B[u]$, we see that for any $g \in L^q(I)$,

$$\frac{1}{q}\int_0^1 |g|^q dx - \frac{1}{q}\int_0^1 |u^h|^q dx \geq \int_0^1 B^t(u^h)(g - u^h)dx. \tag{3.16}$$

Here, we use $\dfrac{\delta\Phi_B[u]}{\delta u}$ to denote the variational derivative of the functional $\Phi_B[u]$. By virtue of the estimate (3.9), we may conclude that for $t \in (0,T)$, as $h \to 0$ (precisely, along a subsequence $\{h_n\}$ with $h_n \to 0$),

$$u^h(\cdot,t) \to u(\cdot,t), \qquad \text{in } L^q(I),$$

$$B^t(u^h(\cdot,t)) \rightharpoonup v(\cdot,t), \qquad \text{in } L^{q'}(I).$$

Letting $h \to 0$ in (3.16) yields

$$\frac{1}{q}\int_0^1 |g|^q dx - \frac{1}{q}\int_0^1 |u|^q dx \geq \int_0^1 v(g-u)dx.$$

Replacing g by $\varepsilon g + u$ leads to

$$\frac{1}{\varepsilon}\left(\Phi_B(u+\varepsilon g) - \Phi_B[u]\right) \geq \int_0^1 vgdx.$$

It follows by letting $\varepsilon \to 0$,

$$\int_0^1 \frac{\delta\Phi_B[u]}{\delta u} gdx \geq \int_0^1 vgdx.$$

Due to the arbitrariness of g, we see that $v(\cdot,t) = \dfrac{\delta\Phi_B[u]}{\delta u} = B(u(\cdot,t))$.

Finally, we prove that $w = A\left(\dfrac{\partial^2 u}{\partial x^2}\right)$, a.e. in Q_T. From (3.10),

$$\frac{1}{q'}\int_0^1 |u_N|^q dx + \iint_{Q_T} \left|\frac{\partial^2 u^h}{\partial^2} x\right|^p dxdt \leq \frac{1}{q'}\int_0^1 |u_0|^q dx.$$

Letting $h \to 0$ and using (3.15), we have

$$\begin{aligned}
&\varliminf_{h\to 0} \iint_{Q_T} \left|\frac{\partial^2 u^h}{\partial x^2}\right|^p dxdt \le f(0) - f(T) \\
= \;& \lim_{\varepsilon\to 0} \frac{1}{\varepsilon}\int_0^{T-\varepsilon} (f(t) - f(t+\varepsilon))\, dt \\
= \;& \lim_{\varepsilon\to 0} \lim_{h\to 0} \frac{1}{\varepsilon q'}\int_0^{T-\varepsilon}\int_0^1 \left(|u^h(x,t)|^q - |u^h(x,t+\varepsilon)|^q\right) dxdt.
\end{aligned}$$

Consider the functional

$$\Psi_B[u] = \frac{1}{q'}\int_0^1 |u|^{q'} dx.$$

Obviously, $\dfrac{\delta \Psi_B[u]}{\delta u} = |u|^{q'-2}u$. Noticing that $|B(u)|^{q'-2}B(u) = u$, and using the convexity of $\Psi_B[u]$, we see that

$$\begin{aligned}
&\frac{1}{q'}\int_0^1 |u^h(x,t)|^q dx - \frac{1}{q'}\int_0^1 |u^h(x,t+\varepsilon)|^q dx \\
= \;& \frac{1}{q'}\int_0^1 |B^t(u^h(x,t))|^{q'} dx - \frac{1}{q'}\int_0^1 |B^t(u^h(x,t+\varepsilon))|^{q'} dx \\
\le \;& \int_0^1 \left(B^t(u^h(x,t)) - B^t(u^h(x,t+\varepsilon))\right) u^h(x,t) dx.
\end{aligned}$$

Thus

$$\begin{aligned}
&\lim_{h\to 0} \frac{1}{\varepsilon q'}\int_0^{T-\varepsilon}\int_0^1 \left(|u^h(x,t)|^q - |u^h(x,t+\varepsilon)|^q\right) dxdt \\
&\quad \le \frac{1}{\varepsilon}\int_0^{T-\varepsilon}\int_0^1 (v(x,t) - v(x,t+\varepsilon))u(x,t)dxdt
\end{aligned}$$

and hence

$$\varliminf_{h\to 0} \iint_{Q_T} \left|\frac{\partial^2 u^h}{\partial x^2}\right|^p dxdt \le -\int_0^T \langle \frac{\partial v}{\partial t}, u\rangle dt,$$

where $\langle\cdot,\cdot\rangle$ denotes the dual product of the functions in $W^{-2,p'}(I)$ and $W^{2,p}(I)$. It follows from (3.14) that

$$\varliminf_{h\to 0} \iint_{Q_T} \left|\frac{\partial^2 u^h}{\partial x^2}\right|^p dxdt \le \iint_{Q_T} w\frac{\partial^2 u}{\partial x^2} dxdt. \tag{3.17}$$

On the other hand, for any $g \in L^\infty(0,T;W_0^{2,p}(I))$, from $\frac{\delta\Phi_A[u]}{\delta u} = \frac{\partial^2}{\partial x^2}A\left(\frac{\partial^2 u}{\partial x^2}\right)$, and the convexity of $\Phi_A[u]$, we have

$$\frac{1}{p}\iint_{Q_T}\left|\frac{\partial^2 g}{\partial x^2}\right|^p dxdt - \frac{1}{p}\iint_{Q_T}\left|\frac{\partial^2 u^h}{\partial x^2}\right|^p dxdt \geq \iint_{Q_T} A\left(\frac{\partial^2 u^h}{\partial x^2}\right)\frac{\partial^2}{\partial x^2}(g-u^h)dxdt.$$

Using (3.17), the weak lower semi-continuity of $\Phi_A[u]$, and letting $h \to 0$ yield

$$\frac{1}{p}\iint_{Q_T}\left|\frac{\partial^2 g}{\partial x^2}\right|^p dxdt - \frac{1}{p}\iint_{Q_T}\left|\frac{\partial^2 u}{\partial x^2}\right|^p dxdt \geq \iint_{Q_T} w\frac{\partial^2}{\partial x^2}(g-u)dxdt.$$

Replacing g by $\varepsilon g + u$ leads to

$$\frac{1}{\varepsilon}\left(\Phi_A(u+\varepsilon g) - \Phi_A(u)\right) \geq \iint_{Q_T} w\frac{\partial^2 g}{\partial x^2}dxdt.$$

Therefore

$$\iint_{Q_T}\frac{\delta\Phi_A[u]}{\delta u}gdxdt = \iint_{Q_T} A\left(\frac{\partial^2 u}{\partial x^2}\right)dxdt \geq \iint_{Q_T} ø\frac{\partial^2 g}{\partial x^2}dxdt.$$

Due to the arbitrariness of g, we see that $w = A\left(\frac{\partial^2 u}{\partial x^2}\right)$ and the proof is complete. □

4.3.2 *Uniqueness of solutions*

So far, there is no complete uniqueness result for the general equation (3.1). In what follows, we discuss the uniqueness for two special cases $p = 2$ and $q = 2$.

We first discuss the case $p = 2$, namely, the equation

$$\frac{\partial u}{\partial t} + \frac{\partial^4\Phi(u)}{\partial x^4} = 0, \tag{3.18}$$

where $\Phi(u) = |u|^{q-2}u$, $q > 1$.

Definition 4.3.2 A function u is said to be an L^2 solution of the problem (3.18), (3.3), (3.4), if $u \in L^2(Q_T)$, $\Phi(u) \in L^2(Q_T)$, and the integral equality

$$-\int_0^1 u_0\varphi(x,0)dx - \iint_{Q_T} u\frac{\partial\varphi}{\partial t}dxdt + \iint_{Q_T} \Phi(u)\frac{\partial^4\varphi}{\partial x^4}dxdt = 0$$

holds for any $\varphi \in C^\infty(Q_T)$ with $\varphi(0,t) = \varphi(1,t) = \varphi_x(0,t) = \varphi_x(1,t) = \varphi(x,T) = 0$(hence for any $\varphi \in H_0^2(Q_T) \cap H^4(Q_T)$).

Remark 4.3.1 The L^2 solution of the problem (3.18),(3.3),(3.4) is a generalized solution in the sense of Definition 4.3.1.

Theorem 4.3.2 *The problem (3.18), (3.3), (3.4) admits at most one L^2 solution.*

Proof. The basic idea is to transform the equation into an ordinary differential inequality (see (3.21)) by means of the inverse operator of a fourth order elliptic operator, which is defined as follows

$$T_\lambda : g \mapsto y, \quad L^2(I) \to H_0^2(I) \cap H^4(I),$$

where y is the solution of the two point boundary value problem

$$\frac{\partial^4 y}{\partial x^4} + \lambda y = g, \quad (\lambda > 0)$$
$$y(0) = y(1) = y'(0) = y'(1) = 0.$$

Obviously, $T_\lambda g$ is uniquely determined by g. It is easily checked that the following estimate holds

$$\lambda^2\int_0^1 (T_\lambda g)^2 dx + 2\lambda\int_0^1 \left(\frac{\partial^2 T_\lambda g}{\partial x^2}\right)^2 dx + \int_0^1 \left(\frac{\partial^4 T_\lambda g}{\partial x^4}\right)^2 dx \le \int_0^1 g^2 dx. \tag{3.19}$$

Moreover, the operator T_λ is symmetric:

$$\int_0^1 (T_\lambda f)g dx = \int_0^1 f(T_\lambda g)dx, \quad \forall\, f, g \in L^2(I).$$

Let u_1 and u_2 be two L^2 solutions of the problem (3.18), (3.3), (3.4). Set $w = u_1 - u_2$ and $v = \Phi(u_1) - \Phi(u_2)$. To prove $w = 0$, a.e. in Q_T, it suffices to verify

$$\iint_{Q_T}\left(\left(\frac{\partial^2 T_\lambda w}{\partial x^2}\right)^2+\lambda\,(T_\lambda w)^2\right)dxdt\to 0\quad(\lambda\to 0).\tag{3.20}$$

In fact, if (3.20) is valid, then, since for any $\varphi\in H_0^2(Q_T)$, integrating the equality

$$\left(\frac{\partial^4 T_\lambda w}{\partial x^4}+\lambda T_\lambda w\right)\varphi=w\varphi$$

over Q_T and integrating by parts yield

$$\iint_{Q_T}\left(\frac{\partial^2 T_\lambda w}{\partial^2}\frac{\partial^2\varphi}{\partial x^2}+\lambda T_\lambda w\varphi\right)dxdt=\iint_{Q_T}w\varphi dxdt,$$

we have

$$\begin{aligned}&\left(\iint_{Q_T}w\varphi dxdt\right)^2\\ \le\ & C_\varphi\iint_{Q_T}\left(\frac{\partial^2 T_\lambda w}{\partial x^2}\right)^2dxdt+C_\varphi\lambda\iint_{Q_T}(T_\lambda w)^2dxdt\\ &\to 0\quad(\lambda\to 0).\end{aligned}$$

To verify (3.20), first we need to check the existence of $\dfrac{\partial}{\partial t}(T_\lambda w)$ in the space $L^2(Q_T)$, and prove

$$\frac{\partial}{\partial t}T_\lambda w(\cdot,t)=\lambda T_\lambda v(\cdot,t)-v(\cdot,t),\quad\text{a.e. in } t\in(0,T).\tag{3.21}$$

From the definition of generalized solutions, we have

$$-\iint_{Q_T}w\frac{\partial\varphi}{\partial t}dxdt+\iint_{Q_T}v\frac{\partial^4\varphi}{\partial x^4}dxdt=0.\tag{3.22}$$

For any $k(x)\in C_0^\infty(I)$ and $\psi(t)\in C_0^\infty(0,T)$, choosing the test function $\varphi(x,t)=(T_\lambda k)\psi(t)$ in (3.22), and applying the symmetric property of T_λ, we obtain

$$\begin{aligned}
0 = & -\iint_{Q_T} (T_\lambda k)\psi'(t)w dxdt + \iint_{Q_T} \frac{\partial^4 T_\lambda k}{\partial x^4}\psi(t)v dxdt \\
= & -\iint_{Q_T} (T_\lambda w)\psi'(t)k(x)dxdt \\
& + \iint_{Q_T} (k - \lambda T_\lambda k)\psi(t)v dxdt \\
= & -\iint_{Q_T} (T_\lambda w)\psi'(t)k(x)dxdt \\
& + \iint_{Q_T} (v - \lambda T_\lambda v)k(x)\psi(t)dxdt,
\end{aligned}$$

from which, it follows that for any $\varphi \in C_0^\infty(Q_T)$,

$$\iint_{Q_T} (T_\lambda w)\frac{\partial \varphi}{\partial t} dxdt = \iint_{Q_T} (v - \lambda T_\lambda v)\varphi dxdt,$$

which shows that $\frac{\partial}{\partial t}(T_\lambda w) \in L^2(Q_T)$ and (3.21) holds.

Set

$$g_\lambda(t) = \int_0^1 (T_\lambda w(x,t))w(x,t)dx.$$

Obviously

$$\begin{aligned}
g_\lambda(t) = & \int_0^1 T_\lambda w \left(\frac{\partial^4 T_\lambda w}{\partial x^4} + \lambda T_\lambda w\right) dx \\
= & \int_0^1 \left(\left(\frac{\partial^2 T_\lambda w}{\partial x^2}\right)^2 + \lambda (T_\lambda w)^2\right) dx.
\end{aligned}$$

To prove (3.20), it remains to show that

$$g_\lambda(t) \to 0 \quad (\lambda \to 0) \tag{3.23}$$

uniformly.

For any $\psi(t) \in C_0^\infty(0,T)$, taking $\varphi = (T_\lambda w)\psi$ in (3.22) and using (3.21), we have

$$\begin{aligned}
0 = & -\iint_{Q_T} (T_\lambda w)\psi'(t)w dxdt - \iint_{Q_T} \psi(t)w\frac{\partial}{\partial t}T_\lambda w dxdt \\
& + \iint_{Q_T} v\frac{\partial^4 (T_\lambda w)}{\partial x^4}\psi(t)dxdt
\end{aligned}$$

$$\begin{aligned}
&= -\iint_{Q_T}(T_\lambda w)\psi'(t)wdxdt-\iint_{Q_T}\psi(t)w\frac{\partial}{\partial t}T_\lambda wdxdt\\
&\quad +\iint_{Q_T}v(w-\lambda T_\lambda w)\psi(t)dxdt\\
&= -\int_0^T\psi'(t)dt\int_0^1 w(T_\lambda w)dx\\
&\quad +2\int_0^T\psi(t)dt\int_0^1 w(v-\lambda T_\lambda v)dx.
\end{aligned}$$

Therefore

$$g_\lambda'(t)=2\int_0^1(\lambda T_\lambda v(x,t)-v(x,t))w(x,t)dx, \qquad \text{a.e. } t\in(0,T), \tag{3.24}$$

which implies $g_\lambda'(t)\in L^1(0,T)$, and so $g_\lambda(t)$ is absolutely continuous on $[0,T]$.

Let $\alpha_\varepsilon(s)$ be the kernel of an one-dimensional mollifier, and denote

$$\psi_\varepsilon(t)=\int_t^\infty\alpha_\varepsilon(s-\varepsilon)ds.$$

Letting $\varphi(x,t)=(T_\lambda w)\psi_\varepsilon(t)$ in (3.22) yields

$$\iint_{Q_T}\alpha_\varepsilon(t-s)wT_\lambda wdxdt+\iint_{Q_T}\psi_\varepsilon(t)v\frac{\partial^4 T_\lambda w}{\partial x^4}dxdt=0.$$

Using the dominated convergence theorem gives

$$\begin{aligned}
g_\lambda(0) &= \lim_{\varepsilon\to0}\int_0^{2\varepsilon}\alpha_\varepsilon(t-\varepsilon)g_\lambda(t)dt\\
&= \lim_{\varepsilon\to0}\iint_{Q_T}\alpha_\varepsilon(t-s)wT_\lambda wdxdt\\
&= -\lim_{\varepsilon\to0}\iint_{Q_T}\psi_\varepsilon(t)v\frac{\partial^4 T_\lambda w}{\partial x^4}dxdt\\
&= 0.
\end{aligned} \tag{3.25}$$

Combining (3.25),(3.24) with (3.14), and noticing that the monotonicity of

$\Phi(s)$ implies that w and v have the same sign, we have

$$\begin{aligned}
0 \leq \ & g_\lambda(t) = g_\lambda(t) - g_\lambda(0) = \int_0^t g_\lambda'(s)ds \\
= \ & 2\int_0^t ds \int_0^1 (\lambda T_\lambda v - v)wdx \\
\leq \ & 2\int_0^t ds \int_0^1 (\lambda T_\lambda v)wdx \\
\leq \ & \sqrt{\lambda} \iint_{Q_T} \lambda (T_\lambda v)^2 dxdt + \sqrt{\lambda} \iint_{Q_T} w^2 dxdt \\
\leq \ & \sqrt{\lambda} \iint_{Q_T} (v^2 + w^2) dxdt \to 0, \quad (\lambda \to 0).
\end{aligned}$$

The proof is complete. □

Now, we consider the case $q = 2$,

$$\frac{\partial u}{\partial t} + \frac{\partial^2}{\partial x^2} A\left(\frac{\partial^2 u}{\partial x^2}\right) = 0, \tag{3.26}$$

where $A(s) = |s|^{p-2}s$, $p > 1$.

Theorem 4.3.3 *The generalized solution u of the problem (3.26), (3.3), (3.4) with $\frac{\partial^2 u}{\partial x^2} \in L^\infty(Q_T)$ (in the sense of Definition 4.3.1) is unique.*

Proof. Assume that the problem (3.26),(3.3),(3.4) admits two generalized solutions u_1, u_2, with $\frac{\partial^2 u_1}{\partial x^2} \in L^\infty(Q_T)$, $\frac{\partial^2 u_2}{\partial x^2} \in L^\infty(Q_T)$. Then for any test function φ,

$$\iint_{Q_T} (u_1 - u_2)\frac{\partial \varphi}{\partial t} dxdt = \iint_{Q_T} \widehat{a}\left(\frac{\partial^2 u_1}{\partial x^2} - \frac{\partial^2 u_2}{\partial x^2}\right)\frac{\partial^2 \varphi}{\partial x^2} dxdt, \tag{3.27}$$

where

$$\widehat{a} = \int_0^1 a\left(\lambda \frac{\partial^2 u_1}{\partial x^2} + (1-\lambda)\frac{\partial^2 u_2}{\partial x^2}\right) d\lambda,$$
$$a(s) = (p-1)|s|^{p-2}.$$

Choose an approximate sequence $\{\widehat{a}_\varepsilon\} \subset C_0^\infty(Q_T)$, such that

$$0 \leq \widehat{a}_\varepsilon(x,t) \leq M, \qquad \iint_{Q_T} |\widehat{a}_\varepsilon - \widehat{a}|^2 \, dxdt < \varepsilon^2, \quad \varepsilon > 0. \tag{3.28}$$

Consider the boundary value problem for the conjugate equation

$$\frac{\partial \varphi_\varepsilon}{\partial t} = \frac{\partial^2}{\partial x^2}\left[(\widehat{a}_\varepsilon + \varepsilon)\frac{\partial^2 \varphi_\varepsilon}{\partial x^2}\right] + f,$$
$$\varphi_\varepsilon(0,t) = \varphi_\varepsilon(1,t) = \varphi_{\varepsilon x}(0,t) = \varphi_{\varepsilon x}(1,t),$$
$$\varphi_\varepsilon(x,T) = 0,$$

where $f \in C_0^\infty(Q_T)$ is an arbitrarily given function. By the classical theory of linear equations, the problem admits a classical solution φ_ε. Moreover, φ_ε satisfies the following estimate

$$\iint_{Q_T} (\widehat{a}_\varepsilon + \varepsilon)\left(\frac{\partial^2 \varphi_\varepsilon}{\partial x^2}\right)^2 dxdt \le \frac{1}{2}e^T \iint_{Q_T} f^2 dxdt. \tag{3.29}$$

In fact, multiplying the equation by φ_ε, and integrating over $(0,1)\times(t,T)$, we have

$$\frac{1}{2}\int_t^T \int_0^1 \frac{\partial \varphi_\varepsilon^2}{\partial t} dxds - \int_t^T \int_0^1 f\varphi_\varepsilon dxds.$$
$$= \int_t^T \int_0^1 \frac{\partial^2}{\partial x^2}\left[(\widehat{a}_\varepsilon + \varepsilon)\frac{\partial^2 \varphi_\varepsilon}{\partial x^2}\right]\varphi_\varepsilon dxds.$$

Integrating by parts yields

$$-\frac{1}{2}\int_0^1 \varphi_\varepsilon^2(x,t)dx - \int_t^T \int_0^1 f\varphi_\varepsilon dxds$$
$$= \int_t^T \int_0^1 (\widehat{a}_\varepsilon + \varepsilon)\left(\frac{\partial^2 \varphi_\varepsilon}{\partial x^2}\right)^2 dxds.$$

Therefore

$$\frac{1}{2}\int_0^1 \varphi_\varepsilon^2(x,t)dx + \int_t^T \int_0^1 (\widehat{a}_\varepsilon + \varepsilon)\left(\frac{\partial^2 \varphi_\varepsilon}{\partial x^2}\right)^2 dxds$$
$$\le \frac{1}{2}\iint_{Q_T} f^2 dxds + \frac{1}{2}\int_t^T \int_0^1 \varphi_\varepsilon^2 dxds,$$

from which and Gronwall's inequality, we easily see that (3.29) holds.

Choose $\varphi = \varphi_\varepsilon$ in the integral equality (3.27) (notice that the smoothness of φ_ε is enough for φ_ε to be a test function, although $\varphi_\varepsilon(x,0)$ may not be equal to zero, which does not prevent us using $\varphi = \varphi_\varepsilon$ in (3.27), since

u_1, u_2 have the same initial value,

$$\begin{aligned}
&\iint_{Q_T}(u_1-u_2)\frac{\partial^2}{\partial x^2}\left[(\widehat{a}_\varepsilon+\varepsilon)\frac{\partial^2\varphi_\varepsilon}{\partial x^2}\right]dxdt\\
&\qquad+\iint_{Q_T}(u_1-u_2)fdxdt\\
&=\iint_{Q_T}\widehat{a}\left(\frac{\partial^2 u_1}{\partial x^2}-\frac{\partial^2 u_2}{\partial x^2}\right)\frac{\partial^2\varphi_\varepsilon}{\partial x^2}dxdt.
\end{aligned}$$

Then by (3.28) and (3.29), we have

$$\begin{aligned}
&\left|\iint_{Q_T}(u_1-u_2)fdxdt\right|\\
\le&\left|\iint_{Q_T}\left(\frac{\partial^2 u_1}{\partial x^2}-\frac{\partial^2 u_2}{\partial x^2}\right)(\widehat{a}_\varepsilon-\widehat{a})\frac{\partial^2\varphi_\varepsilon}{\partial x^2}dxdt\right|\\
&+\left|\iint_{Q_T}\varepsilon\left(\frac{\partial^2 u_1}{\partial x^2}-\frac{\partial^2 u_2}{\partial x^2}\right)\frac{\partial^2\varphi_\varepsilon}{\partial x^2}dxdt\right|\\
\le&\ C\left\{\iint_{Q_T}(\widehat{a}_\varepsilon-\widehat{a})^2dxdt\right\}^{1/2}\left\{\iint_{Q_T}\left(\frac{\partial^2\varphi_\varepsilon}{\partial x^2}\right)^2dxdt\right\}^{1/2}\\
&+C\varepsilon\left\{\iint_{Q_T}\left(\frac{\partial^2\varphi_\varepsilon}{\partial x^2}\right)^2dxdt\right\}^{1/2}\\
\le&\ C\varepsilon\left\{\iint_{Q_T}\left(\frac{\partial^2\varphi_\varepsilon}{\partial x^2}\right)^2dxdt\right\}^{1/2}\\
\le&\ C\sqrt{\varepsilon}\left\{\iint_{Q_T}\varepsilon\left(\frac{\partial^2\varphi_\varepsilon}{\partial x^2}\right)^2dxdt\right\}^{1/2}\to 0,\quad(\varepsilon\to 0).
\end{aligned}$$

Due to the arbitrariness of f, it follows that $u_1(x,t)=u_2(x,t)$. The proof is complete. □

Remark 4.3.2 If u_0 is sufficiently smooth and satisfies some compatibility conditions on $x=0,1$, then the problem (3.26),(3.3),(3.4) admits a generalized solution with $\frac{\partial^2 u}{\partial x^2}\in L^\infty(Q_T)$. For the proof, see [YI11].

4.3.3 Weighted energy equality of solutions

Theorem 4.3.4 *Let u be the solution of the problem (3.2)–(3.4) obtained in Theorem 4.3.1. Then for any $0 \le \rho \in C^2(\bar{I})$,*

$$\begin{aligned}&\frac{1}{q'}\int_0^1 \rho(x)|u(x,t)|^q dx - \frac{1}{q'}\int_0^1 \rho(x)|u_0(x)|^q dx \\ = & -\iint_{Q_t} A\left(\frac{\partial^2 u}{\partial x^2}\right)\frac{\partial^2}{\partial x^2}(\rho(x)u(x,\tau))\,dxd\tau,\end{aligned} \tag{3.30}$$

where $Q_t = (0,1)\times(0,t)$.

Proof. In the proof of Theorem 4.3.1, we have shown

$$f(t) = \frac{1}{q'}\int_0^1 |u(x,t)|^q dx \in C([0,T]).$$

Similarly we can prove that for any $0 \le \rho \in C^2(\bar{I})$,

$$f_\rho(t) = \frac{1}{q'}\int_0^1 \rho(x)|u(x,t)|^q dx \in C([0,T]).$$

Consider the functional

$$\Phi_\rho[v] = \frac{1}{q'}\int_0^1 \rho(x)|v|^q dx \equiv \frac{1}{q'}\int_0^1 \rho(x)|B(v)|^{q'} dx.$$

It is easy to see that $\Phi_\rho[v]$ is a convex functional on $L^q(I)$. So, for any $\tau \in (0,T)$ and $h > 0$,

$$\Phi_\rho[u(\tau+h)] - \Phi_\rho[u(\tau)] \ge \langle B(u(\tau+h)) - B(u(\tau)), \rho u(\tau)\rangle.$$

Here we use $u(\tau)$ to denote $u(x,\tau)$ temporarily.

For any given $t_1, t_2 \in (0,T), t_1 < t_2$, integrating the above inequality over (t_1,t_2) with respect to τ, we have

$$\begin{aligned}&\int_{t_2}^{t_2+h}\Phi_\rho[u(\tau)]d\tau - \int_{t_1}^{t_1+h}\Phi_\rho[u(\tau)]d\tau \\ \ge & \int_{t_1}^{t_2}\langle B(u(\tau+h)) - B(u(\tau)), \rho u(\tau)\rangle d\tau.\end{aligned}$$

Dividing both side of the above inequality by h, noticing that $\frac{\partial}{\partial t}B(u) \in L^\infty\left(0,T;W^{-2,p'}(I)\right)$,and letting $h \to 0$, we obtain

$$\Phi_\rho[u(t_2)] - \Phi_\rho[u(t_1)] \geq \int_{t_1}^{t_2} \langle \frac{\partial}{\partial t}B(u), \rho u\rangle d\tau.$$

Similarly,

$$\Phi_\rho[u(\tau)] - \Phi_\rho[u(\tau - h)] \leq \langle B(u(\tau)) - B(u(\tau-h)), \rho u(\tau)\rangle,$$

which implies that

$$\Phi_\rho[u(t_2)] - \Phi_\rho[u(t_1)] \leq \int_{t_1}^{t_2} \langle \frac{\partial}{\partial t}B(u), \rho u\rangle d\tau.$$

Therefore

$$\Phi_\rho[u(t_2)] - \Phi_\rho[u(t_1)] = \int_{t_1}^{t_2} \langle \frac{\partial}{\partial t}B(u), \rho u\rangle d\rho.$$

Taking $t_1 = 0$, $t_2 = t$, and noticing that u satisfies the equation (3.2) in the sense of distributions, we see that (3.30) holds and the proof is complete□

4.3.4 *Some auxiliary inequalities*

Lemma 4.3.4 *(Nirenberg's inequalily [ADA]) Let $\Omega \subset \mathbb{R}^N$ be a bounded domain with smooth boundary, $u \in W^{m,r}(\Omega)$, Then*

$$\|D^j u\|_{L^p} \leq C_1 \|D^m u\|_{L^r}^a \|u\|_{L^q}^{1-a} + C_2\|u\|_{L^q},$$

where

$$\frac{j}{m} \leq a < 1, \quad \frac{1}{p} = \frac{j}{n} + a\left(\frac{1}{r} - \frac{m}{n}\right) + (1-a)\frac{1}{q}.$$

Lemma 4.3.5 *(Weighted Nirenberg's inequality, [BE1])*

$$\left(\int_{\mathbb{R}} (x)_+^k |Du|^p dx\right)^{1/p}$$
$$\leq C\left(\int_{\mathbb{R}} (x)_+^k |D^2 u|^p dx\right)^{a/p} \left(\int_{\mathbb{R}} (x)_+^k |u|^q dx\right)^{(1-a)/q},$$

where k is a nonnegative integer, $(x)_+ = \max\{x, 0\}$, provided that the integral on the right hand side exists and

$$\frac{1}{2} \le a < 1, \frac{1}{p} = \frac{1}{1+k} + a\left(\frac{1}{p} - \frac{2}{1+k}\right) + (1-a)\frac{1}{q}.$$

Lemma 4.3.6 *(Hardy's inequality [HA])*

$$\int_{\mathbb{R}} x_+^k |u|^p dx \le C \int_{\mathbb{R}} x_+^{k+p} |Du|^p dx,$$

where $k \ge 0, p > 1$, provided that the integrals on both sides exist.

Lemma 4.3.7 *(Kjellberg's inequality [KJ])*

$$\left(\int_{\mathbb{R}} |u|^p dx\right)^{1/p} \le C \left(\int_{\mathbb{R}} x^2 |u|^p dx\right)^{1/p(2p+3)} \left(\int_{\mathbb{R}} |u|^{p+1} dx\right)^{2/(2p+3)},$$

where $p > 0$, provided that the integral on the right hand side exists.

Remark 4.3.3 In Lemma 4.3.5–4.3.7, $\mathbb{R}$ can be replaced by any finite interval, provided that u can be extended via some manner to be defined on $\mathbb{R}$, for example, the functions in $H_0^2(I)$ can be zero extended.

4.3.5 *Finite propagation of disturbances*

Theorem 4.3.5 *Let u be the solution obtained in Theorem 4.3.1 with initial data u_0. If $1 < q < p$ and $\operatorname{supp} u_0 \subset [x_1, x_2], 0 < x_1 < x_2 < 1$, then*

$$\operatorname{supp} u(x, \cdot) \subset [x_1(t), x_2(t)] \cap [0, 1], \qquad a.e.\ t \in (0, T),$$

where $x_1(t), x_2(t)$ can be expressed by

$$x_1(t) = x_1 - C_1 t^\mu, \qquad x_2(t) = x_2 + C_2 t^\mu$$

with positive constants C_1, C_2, μ depending only on p, q and u_0.

Proof. Setting $\rho(x) = (x - y)_+^k$ in Theorem 4.3.4, where $y \in [x_2, 1)$ is any fixed constant, and using the assumption $\operatorname{supp} u_0 \subset [x_1, x_2]$, we have

$$\begin{aligned} &\frac{1}{q'} \int_0^1 (x-y)_+^k |u(x,t)|^q dx \\ = &-\iint_{Q_t} A\left(\frac{\partial^2 u}{\partial x^2}\right) \frac{\partial^2}{\partial x^2}\left((x-y)_+^k u(x,\tau)\right) dx d\tau. \end{aligned}$$

We use Young's inequality to the right hand side to obtain

$$\begin{aligned}
& -\iint_{Q_t} A\left(\frac{\partial^2 u}{\partial x^2}\right)\frac{\partial^2}{\partial x^2}\left((x-y)_+^k u(x,\tau)\right)dxd\tau \\
= & -\iint_{Q_t}(x-y)_+^k\left|\frac{\partial^2 u}{\partial x^2}\right|^p dxd\tau + \\
& -2k\iint_{Q_t}(x-y)_+^{k-1}A\left(\frac{\partial^2 u}{\partial x^2}\right)\frac{\partial u}{\partial x}dxd\tau \\
& -k(k-1)\iint_{Q_t}(x-y)_+^{k-2}A\left(\frac{\partial^2 u}{\partial x^2}\right)udxd\tau \\
\le & -\frac{1}{2}\iint_{Q_t}(x-y)_+^k\left|\frac{\partial^2 u}{\partial x^2}\right|^p dxd\tau - \\
& +C_1\iint_{Q_t}(x-y)_+^{k-p}\left|\frac{\partial u}{\partial x}\right|^p dxd\tau \\
& +C_2\iint_{Q_t}(x-y)_+^{k-2p}|u|^p dxd\tau.
\end{aligned}$$

From this and

$$\iint_{Q_t}(x-y)_+^{k-2p}|u|^p dxd\tau \le C\iint_{Q_t}(x-y)_+^{k-p}\left|\frac{\partial u}{\partial x}\right|^p dxd\tau,$$

which is a consequence of Lemma 4.3.6, we obtain

$$\sup_{0<\tau\le t}\int_0^1 (x-y)_+^k|u(x,\tau)|^q dx \le C\iint_{Q_t}(x-y)_+^k\left|\frac{\partial^2 u}{\partial x^2}\right|^p dxd\tau, \tag{3.31}$$

$$\iint_{Q_t}(x-y)_+^k\left|\frac{\partial^2 u}{\partial x^2}\right|^p dxd\tau \le C\iint_{Q_t}(x-y)_+^{k-p}\left|\frac{\partial u}{\partial x}\right|^p dxd\tau. \tag{3.32}$$

Set

$$\begin{aligned}
f_m(y) &= \iint_{Q_t}(x-y)_+^m\left|\frac{\partial^2 u}{\partial x^2}\right|^p dxd\tau, \qquad m=1,2,\cdots \\
f_0(y) &= \int_0^t\int_y^1\left|\frac{\partial^2 u}{\partial x^2}\right|^p dxd\tau.
\end{aligned}$$

From (3.31),(3.32), Lemma 4.3.5 and Hölder's inequality, we have

$$f_{2p+1}(y) \quad \le C\iint_{Q_t}(x-y)_+^{p+1}\left|\frac{\partial u}{\partial x}\right|^p dxd\tau$$

$$
\begin{aligned}
&\le C\int_0^t\left(\int_0^1 (x-y)_+^{p+1}\left|\frac{\partial^2 u}{\partial x^2}\right|^p dx\right)^a \cdot \\
&\quad \cdot\left(\int_0^1 (x-y)_+^{p+1}|u|^q dx\right)^{(1-a)p/q} d\tau \\
&\le C\,[f_{p+1}(y)]^{(1-a)p/q}\int_0^t\left(\int_0^1 (x-y)_+^{p+1}\left|\frac{\partial^2 u}{\partial x^2}\right|^p dx\right)^a d\tau \\
&\le Ct^{1-a}\left(f_{p+1}(y)\right)^{a+(1-a)p/q},
\end{aligned}
$$

where

$$
a=\frac{\dfrac{1}{q}+\dfrac{1}{p+2}-\dfrac{1}{p}}{\dfrac{1}{q}+\dfrac{2}{p+2}-\dfrac{1}{p}}.
$$

Set $\nu = a + (1-a)\dfrac{p}{q}$. Applying Hölder's inequality to the right hand side of the above inequality, we further obtain

$$
\begin{aligned}
f_{2p+1}(y) &\le Ct^{1-a}\left(\iint_{Q_t}(x-y)_+^{p+1}\left|\frac{\partial^2 u}{\partial x^2}\right|^p dx d\tau\right)^{\nu} \\
&\le Ct^{1-a}\left(\iint_{Q_t}(x-y)_+^{2p+1}\left|\frac{\partial^2 u}{\partial x^2}\right|^p dx d\tau\right)^{(p+1)\nu/(2p+1)} \cdot \\
&\quad \cdot\left(\int_0^t\int_y^1\left|\frac{\partial^2 u}{\partial x^2}\right|^p dx d\tau\right)^{p\nu/(2p+1)} \\
&\le Ct^{1-a}\left(f_{2p+1}(y)\right)^{(p+1)\nu/(2p+1)} f_0(y)^{p\nu/(2p+1)}.
\end{aligned}
$$

Therefore

$$
f_{2p+1}(y) \le Ct^{(1-a)/\sigma}[f_0(y)]^{p\nu/(2p+1)\sigma},
$$

where

$$
\sigma = 1-\frac{p+1}{2p+1}\nu > 0.
$$

Using Hölder's inequality again gives

$$
f_1(y) \le \left(f_{2p+1}(y)\right)^{1/(2p+1)}[f_0(y)]^{(2p)/(2p+1)} \le Ct^{\lambda}[f_0(y)]^{\theta+1},
$$

where

$$\lambda = \frac{1-a}{\sigma(2p+1)}, \qquad \theta = \frac{p\nu}{\sigma(2p+1)^2} - \frac{1}{2p+1} > 0.$$

Since $f_1'(y) = -f_0(y)$, we have

$$f_1'(y) \le -Ct^{-\lambda/(\theta+1)}[f_1(y)]^{1(\theta+1)}.$$

If $f_1(x_2) = 0$, then $\dfrac{\partial^2 u(x,t)}{\partial x^2} = 0$ for $x \in [x_2, 1]$, and hence from the boundary value condition, we see that $u(x,t) = 0$ for $x \in [x_2, 1]$, i.e., $\operatorname{supp} u(\cdot, t) \subset [0, x_2]$. If $f_1(x_2) \ne 0$, then there exists an interval (x_2, x_2^*), such that $f_1(y) > 0$ in (x_2, x_2^*), but $f_1(x_2^*) = 0$. So, for $y \in (x_2, x_2^*)$,

$$\left(f_1(y)^{\theta/(\theta+1)}\right)' = \frac{\theta}{\theta+1}\frac{f_1'(y)}{f_1(y)^{1/\theta}} \le -Ct^{-\lambda/(\theta+1)}.$$

Integrating the above inequality over (x_2, x_2^*), we obtain

$$f_1(x_2^*)^{\theta/(\theta+1)} - f_1(x_2)^{\theta/(\theta+1)} \le -Ct^{-\lambda/(\theta+1)}(x_2^* - x_2).$$

Therefore

$$x_2^* \le x_2 + Ct^{\lambda/(\theta+1)}(f_1(x_2))^{\theta/(\theta+1)} \equiv x_2 + C_2 t^{\mu},$$

which implies

$$\operatorname{supp} u(\cdot, t) \subset [0, x_2 + C_2 t^{\mu}].$$

Similarly

$$\operatorname{supp} u(\cdot, t) \subset [x_1 - C_1 t^{\mu}, 1].$$

The proof is complete. □

4.3.6 *Asymptotic behavior of solutions*

Theorem 4.3.6 *Let u be a generalized solution of the problem (3.2)–(3.4) satisfying the weighted energy equality (3.30). Then*

$$\int_0^1 |u(x,t)|^q dx \le \begin{cases} Ce^{-\mu_1 t}, & \text{if } p = q, \\ \dfrac{1}{(C_1 t + C_2)^{\mu_2}}, & \text{if } 1 < q < p, \end{cases}$$

where $\mu_1, \mu_2, C_1, C_2 > 0$ are constants depending only on p, q and u_0.

Proof. Choosing $\rho(x) \equiv 1$ in (3.30), we have

$$\frac{1}{q'}\int_0^1 |u(x,t)|^q dx - \frac{1}{q'}\int_0^1 |u_0(x)|^q dx = -\iint_{Q_t} \left|\frac{\partial^2 u}{\partial x^2}\right|^p dx d\tau.$$

Denote

$$f(t) = \frac{1}{q'}\int_0^1 |u(x,t)|^q dx.$$

Then

$$f'(t) = -\int_0^1 \left|\frac{\partial^2 u}{\partial x^2}\right|^p dx \leq 0$$

and

$$f(t) \leq C|f'(t)|^{q/p}. \tag{3.33}$$

Integrating (3.33) then gives the desired conclusion and the proof is complete. □

4.3.7 *Extinction of solutions at finite time*

Theorem 4.3.7 *Let $q > p$, and u be a solution of the problem (3.2)–(3.4) satisfying the weighted energy equality (3.30). Then there exists $T^* > 0$, such that*

$$u(x,t) = 0, \qquad \forall t \geq T^*.$$

Proof. Under the assumptions of the theorem, (3.33) is still valid. Let $T^* > 0$, such that $f(t) > 0$ in $(0, T^*)$. We need to prove that $T^* < +\infty$. For this purpose, we notice that $f'(t) \leq 0$, and use (3.33) to obtain

$$f'(t) \leq -Cf(t)^{p/q}.$$

Since $q > p$, for any $t \in (0, T^*)$,

$$f(t)^{1-p/q} - f(0)^{1-p/q} \leq -Ct.$$

Therefore $T^* < +\infty$ and the proof is complete. □

4.3.8 *Nonexistence of nonnegative solutions*

In this subsection, we discuss the nonexistence of nonnegative solutions of the Cauchy problem for the equation (3.2). This will reveal one of the main differences between the fourth order equations and the second order equations. For $0 \le u_0 \in W^{2,p}(\mathbb{R})$ with compact support, the first boundary value problem in the domain $(-X, X) \times (0, T)$ with zero boundary value has solutions with compact support, provided that $X > 0$ is large enough. Therefore, for any fixed T, we may choose X sufficiently large, such that the solution of the corresponding first boundary value problem is just the solution of the Cauchy problem. The following theorem shows that the Cauchy problem does not admit nontrivial global nonnegative solutions.

Theorem 4.3.8 *Let $p = 2, 1 < q < 2, 0 \le u_0 \in W^{2,p}(\mathbb{R}), u_0 \not\equiv 0$, $\operatorname{supp} u_0$ be compact. If u is a nonnegative solution of the Cauchy problem for the equation (3.2) in $G_{T^*} \equiv \mathbb{R} \times (0, T^*)$, then T^* must be a finite number.*

Proof. From the definition of generalized solutions, it is easy to see that

$$\begin{aligned}&\iint_{G_{T^*}} B(u)\frac{\partial \varphi}{\partial t}dxdt + \int_{\mathbb{R}} B(u_0)\varphi(x, 0)dx\\ = &\iint_{G_{T^*}} \frac{\partial^2 u}{\partial x^2}\frac{\partial^2 \varphi}{\partial x^2}dxdt\end{aligned}$$

for any $\varphi \in C^\infty(\overline{G}_{T^*})$, with $\varphi(x, t) = 0$ for sufficiently large $|x|$ and $t = T^*$. By Theorem 4.3.5, for any $t \in (0, T^*)$, $\operatorname{supp} u(\cdot, t)$ is compact, so for any $\psi(t) \in C_0^\infty(0, T^*)$ and any polynomial $\rho(x)$,

$$\iint_{G_{T^*}} B(u)\rho(x)\psi'(t)dxdt = \iint_{G_{T^*}} \frac{\partial^2 u}{\partial x^2}\rho''(x)\psi(t)dxdt.$$

Due to the arbitrariness of $\psi(t)$,

$$\begin{aligned}&\int_{\mathbb{R}} \rho(x)B(u(x, t))dx\\ = &\int_{\mathbb{R}} \rho(x)B(u_0(x))dx + \iint_{G_t} \rho''(x)\frac{\partial^2 u}{\partial x^2}dxd\tau.\end{aligned} \tag{3.34}$$

In particular, by choosing $\rho(x) = 1$ and $\rho(x) = x^2$ respectively, we have

$$\int_{\mathbb{R}} (u(x, t))^{q-1}dx = \int_{\mathbb{R}} B(u_0(x))dx \equiv C_1,$$

$$\int_{\mathbb{R}} x^2(u(x,t))^{q-1}dx = \int_{\mathbb{R}} x^2 B(u_0(x))dx + 2\iint_{G_t} \frac{\partial^2 u}{\partial x^2} dx d\tau$$
$$= \int_{\mathbb{R}} x^2 B(u_0(x))dx \equiv C_2.$$

Applying Lemma 4.3.6 and Theorem 4.3.6 yields

$$\begin{aligned} C_1 &= \int_{\mathbb{R}} (u(x,t))^{q-1}dx \\ &\le C\left(\int_{\mathbb{R}} x^2 u^{q-1}dx\right)^{1/(2q+1)}\left(\int_{\mathbb{R}} u^q dx\right)^{(2q-2)/(2q+1)} \\ &\le C_2^{1/(2q+1)}\left(\int_{\mathbb{R}} u^q dx\right)^{(2q-2)/(2q+1)} \\ &\le \frac{C_2^{1/(2q+1)}}{(C_3 t + C_4)^{(2q-2)\mu/(2q+1)}}. \end{aligned}$$

Letting $t \to \infty$, we derive

$$C_1 = \int_{\mathbb{R}} u^{q-1}dx = \int_{\mathbb{R}} u_0^{q-1}dx = 0.$$

This is a contradiction. The proof is complete. □

4.3.9 *Infinite propagation case*

Theorem 4.3.9 *Let $q > p$, $\int_0^1 B(u_0)dx \neq 0$ and u be a generalized solution of the Cauchy problem for the equation (3.2). Then u must have the property of infinite propagation, namely, there are no functions $x_1(t)$, $x_2(t)$, $-\infty < x_1(t) < x_2(t) < +\infty$, such that*

$$supp u(\cdot,t) \subset [x_1(t), x_2(t)], \quad \forall t \ge 0.$$

Proof. If not, then similar to (3.34), we can prove

$$\int_0^1 B(u(x,t))dx = \int_0^1 B(u_0)dx, \quad \forall t \ge 0.$$

On the other hand, from Theorem 4.3.7, there exists $T^* > 0$, such that $u(x,t) = 0$ for $t \ge T^*$, which together with the above equality contradicts the assumption $\int_0^1 B(u_0)dx \neq 0$. The proof is complete. □

4.4 Cahn-Hilliard Equation with Constant Mobility

As indicated in the introduction of this chapter, a lot of diffusive processes, such as phase separation in binary alloys, growth and dispersal in population, can be described by the Cahn-Hilliard equation, see for example [CH]. A special case of such equation in one space variable is of the form

$$\frac{\partial u}{\partial t} + \gamma \frac{\partial^4 u}{\partial x^4} = \frac{\partial^2 \varphi(u)}{\partial x^2}, \tag{4.1}$$

where u denotes the concentration of one of the phases, $\gamma > 0$ the mobility and $\varphi(u) \equiv H'(u)$ with an important typical case (double well potential)

$$H(u) = -\frac{1}{2}u^2 + \frac{1}{3}\gamma_1 u^3 + \frac{1}{4}\gamma_2 u^4.$$

Consider the boundary value problem for the equation (4.1) in the domain $Q_T \equiv (0,1) \times (0,T)$. Based on physical consideration, the equation is supplemented with the zero flux boundary value condition

$$J\Big|_{x=0,1} = -\frac{\partial \varphi(u)}{\partial x} + \gamma \frac{\partial^3 u}{\partial x^3}\Big|_{x=0,1} = 0,$$

the natural boundary value condition

$$\frac{\partial u}{\partial x}\Big|_{x=0,1} = 0,$$

and the initial value condition

$$u(x,0) = u_0(x). \tag{4.2}$$

Since γ is a positive constant, the boundary value conditions can be replaced by

$$\frac{\partial u}{\partial x}\Big|_{x=0,1} = \frac{\partial^3 u}{\partial x^3}\Big|_{x=0,1} = 0. \tag{4.3}$$

4.4.1 *Existence of classical solutions*

Now, we take $H(u)$ as the double well potential, and discuss the existence of classical solutions of the problem (4.1)–(4.3). In this case,

$$\varphi(u) = H'(u) = -u + \gamma_1 u^2 + \gamma_2 u^3.$$

Set

$$H_E^2(0,1) = \left\{ v \in H^2(0,1); \left.\frac{\partial v}{\partial x}\right|_{x=0,1} = 0 \right\},$$

$$H^{4,1}(Q_T) = \left\{ v; \frac{\partial v}{\partial t} \in L^2(Q_T), \frac{\partial^i v}{\partial x^i} \in L^2(Q_T), 0 \le i \le 4 \right\}.$$

Theorem 4.4.1 *If $\gamma_2 > 0$, then for any initial value $u_0 \in H_E^2(0,1)$, the problem (4.1)–(4.3) admits a unique solution $u \in H^{4,1}(Q_T)$. Moreover, if $u_0 \in H^6(0,1) \cap H_E^2(0,1)$, $D^2 u_0 \in H_E^2(0,1)$, then the solution is classical.*

Proof. We need some a priori estimates. Multiplying both sides of the equation (4.1) by u, and integrating over $(0,1)$ with respect to x, we have

$$\frac{1}{2}\frac{d}{dt}\|u\|^2 + \gamma\|D^2u\|^2 + \int_0^1 \varphi'(u)(Du)^2 dx = 0,$$

where $\|\cdot\|$ denote the norm in $L^2(0,1)$. Since $\gamma_2 > 0$, a simple calculation shows that

$$\varphi'(u) = 3\gamma_2 u^2 + 2\gamma_1 u - 1 \ge -C_0 = -\frac{\gamma_1^2}{3\gamma_2} - 1.$$

Thus

$$\frac{1}{2}\frac{d}{dt}\|u\|^2 + \gamma\|D^2u\|^2 \le C_0\|Du\|^2$$

$$\le \quad C_0\|D^2u\| \cdot \|u\| \le \frac{\gamma}{2}\|D^2u\|^2 + \frac{C_0^2}{\gamma}\|u\|^2,$$

and hence

$$\frac{d}{dt}\|u\|^2 + \|D^2u\|^2 \le C\|u\|^2.$$

Using Gronwall's inequality we obtain

$$\|u(t)\|^2 \le C\|u_0\|^2, \qquad 0 \le t \le T,$$

$$\int_0^t \|D^2u(s)\|^2 ds \le C\|u_0\|^2, \qquad 0 \le t \le T. \tag{4.4}$$

Here and below, we always use $u(t)$ to denote $u(x,t)$, and C to denote a universal constant.

Define

$$F(t) = \int_0^1 \left(H(u(x,t)) + \frac{\gamma}{2}(Du(x,t))^2 \right) dx.$$

After differentiation, we have

$$\frac{dF}{dt} = \int_0^1 \left(\varphi(u)\frac{\partial u}{\partial t} + \gamma DuD\frac{\partial u}{\partial t} \right) dx.$$

Integrating by parts and using the equation (4.1) itself and the boundary value condition (4.3), we see that

$$\begin{aligned}\frac{dF}{dt} &= \int_0^1 [\varphi(u)(-\gamma D^4u + D^2\varphi) - \gamma D^2u(-\gamma D^4u + D^2\varphi)]\, dx \\ &= -\int_0^1 [\gamma^2(D^3u)^2 - 2\gamma D^3uD\varphi + (D\varphi)^2]\, dx \\ &= -\int_0^1 [\gamma D^3u - D\varphi]^2\, dx \le 0.\end{aligned}$$

Therefore

$$F(t) \le F(0) = \int_0^1 \left(H(u_0) + \frac{\gamma}{2}(Du_0)^2 \right) dx. \tag{4.5}$$

Notice that Young's inequality implies

$$u^2 \le \varepsilon u^4 + C_{1\varepsilon}, \qquad |u^3| \le \varepsilon u^4 + C_{2\varepsilon},$$

which will be used to estimate $F(t)$. Clearly

$$\begin{aligned}&\frac{\gamma}{2}\|Du(t)\|^2 + \frac{\gamma_2}{2}\left\{\int_0^1 u^4(t)dx + \int_0^1 u^2(t)dx\right\} \\ \le\ & C_3 + F(0) = C.\end{aligned} \tag{4.6}$$

The Sobolev embedding theorem yields

$$\|u(t)\|_\infty \le C, \qquad 0 \le t \le T. \tag{4.7}$$

Now, we multiply both sides of the equation (4.1) by D^4u and integrate the resulting relation over $(0,1)$ with respect to x. Then we obtain

$$\frac{1}{2}\frac{d}{dt}\|D^2u\|^2+\gamma\|D^4u\|^2=\int_0^1 D^2\varphi(u)D^4udx$$

$$=\int_0^1 \varphi'(u)D^2uD^4udx+\int_0^1 \varphi''(u)(Du)^2D^4udx,$$

where

$$\varphi'(u)=3\gamma_2u^2+2\gamma_1u-1$$

$$\varphi''(u)=6\gamma_2u+2\gamma_1.$$

By Nirenberg's inequality (Lemma 4.3.3), we have

$$\|Du\|_\infty\le C\left(\|D^4u\|^{3/8}\|u\|^{5/8}+\|u\|\right),$$

which combining with the estimates (4.6) and (4.7) gives

$$\begin{aligned}&\left|\int_0^1 \varphi'(u)D^2uD^4udx\right|\\ \le\ & C(\|u(t)\|_\infty^2+\|u(t)\|_\infty+1)\|D^2u(t)\|\cdot\|D^4u(t)\|\\ \le\ & \frac{\gamma}{4}\|D^4u(t)\|^2+C\|D^2u(t)\|^2,\end{aligned}$$

$$\begin{aligned}&\left|\int_0^1 \varphi''(u)(Du)^2D^4udx\right|\\ \le\ & C(\|u(t)\|_\infty+1)\|Du(t)\|_\infty\|Du(t)\|\cdot\|D^4u(t)\|\\ \le\ & C(\|D^4u(t)\|^{3/8}+1)\|D^4u(t)\|\\ \le\ & \frac{\gamma}{4}\|D^4u(t)\|^2+C.\end{aligned}$$

Thus

$$\begin{aligned}&\frac{1}{2}\frac{d}{dt}\|D^2u\|^2+\gamma\|D^4u\|^2\\ \le\ & \frac{\gamma}{2}\|D^4u\|^2+C\left(\|D^2u\|^2+1\right),\end{aligned}$$

namely,

$$\frac{d}{dt}\|D^2u\|^2+\|D^4u\|^2\le C\left(\|D^2u\|^2+1\right),$$

and then Gronwall's inequality gives

$$\|D^2u(t)\|^2 \le C, \qquad 0 \le t \le T, \tag{4.8}$$

$$\int_0^t \|D^4u(s)\|^2 ds \le C, \qquad 0 \le t \le T. \tag{4.9}$$

Using the estimates (4.7), (4.8) and (4.9), it is not difficult to prove that the problem (4.1)–(4.3) admits at least one solution $u \in H^{4,1}(Q_T)$. The uniqueness is easy to prove and we omit the details.

Now, we investigate the regularity of the solution. First, $u \in H^{4,1}(Q_T)$ implies

$$Du \in L^\infty(Q_T), \quad D^2u \in L^2(0,T;L^\infty(0,1)).$$

Let

$$f(x,t) \equiv D^2\varphi(u(x,t)).$$

Then

$$Df \in L^2(Q_T), \qquad D^2f \in L^2(Q_T).$$

Consider the problem

$$\frac{\partial v}{\partial t} + \gamma D^4 v = h,$$

$$Dv\Big|_{x=0,1} - D^3v\Big|_{x=0,1} = 0,$$

$$v\Big|_{t=0} = v_0,$$

where $v_0 \in H^2_E(0,1)$. It is well known (see [LM]) that if $h \in L^2(Q_T)$, then the above problem admits a unique solution $v \in H^{4,1}(Q_T)$. Let $h = D^4\varphi(u)$, $v_0 = D^2u_0$. Then $v = D^2u \in H^{4,1}(Q_T)$. Therefore, $\frac{\partial}{\partial t}D^2\varphi(u) \in L^2(Q_T)$. If we assume that $D^5u_0\Big|_{x=0,1} = 0$, then

$$v_0 = -\gamma D^4u_0 + D^2\varphi(u_0) \in H^2_E(0,1).$$

Taking $h = \frac{\partial}{\partial t}D^2\varphi(u)$ we can prove

$$v = \frac{\partial u}{\partial t} \in H^{4,1}(Q_T).$$

From the embedding theorem, we see that $\frac{\partial u}{\partial t} \in C(\overline{Q}_T)$. Therefore, $D^4 u \in C(\overline{Q}_T)$, which shows that u is a classical solution of the problem (4.1)–(4.3). The proof is complete. □

4.4.2 *Blowing-up of solutions*

In Theorem 4.4.1, the coefficient γ_2 is assumed to be a positive constant. In order to obtain a global solution, this condition can not be removed in general. In other words, if $\gamma_2 < 0$, then the solution of the problem (4.1)–(4.3) may blow up at a finite time.

Theorem 4.4.2 *Let $u_0 \not\equiv 0$. Then there exists a constant $\Gamma > 0$ depending only on u_0, such that for $\gamma_2 < -\Gamma$, the solution u of the problem (4.1)–(4.3) blows up at a finite time, namely for some $T^* > 0$,*

$$\lim_{t \to T^*} \|u(t)\| = +\infty.$$

Proof. Without loss of generality, we assume that

$$\int_0^1 u_0(x)dx = 0.$$

Otherwise, we may consider the equation satisfied by $v = u - M$ with $M = \int_0^1 u_0(x)dx$. From the proof of Theorem 4.4.1, we see that (4.5) is still valid. Therefore

$$2\int_0^1 H(u)dx - 2F(0) \leq -\gamma \|Du\|^2, \tag{4.10}$$

where

$$F(0) = \int_0^1 \left(H(u_0) + \frac{\gamma}{2}|Du_0|^2 \right) dx.$$

Choose a function w, such that

$$D^2 w = u, \quad Dw\Big|_{x=0,1} = 0, \quad \int_0^1 w dx = 0.$$

From the equation (4.1), it is easy to see that

$$\int_0^1 u(x,t)dx = \int_0^1 u_0(x)dx = 0,$$

which shows that such a function w exists and satisfies

$$\|Dw\|^2 \le \|u\|^2. \tag{4.11}$$

Here, we have used the fact that $\int_0^1 w dx = 0$. Multiplying both sides of the equation (4.1) by w, integrating the resulting relation over $(0,1)$ with respect to x, and then using the equation satisfied by w and (4.10), we have

$$\begin{aligned}
\frac{d}{dt}\|Dw\|^2 &= -2\int_0^1 \varphi(u)udx - 2\gamma\|Du\|^2 \\
&\ge 4\int_0^1 H(u)dx - 4F(0) - 2\int_0^1 \varphi(u)udx \\
&= \int_0^1 \left(-\gamma_2 u^4 - \frac{2}{3}\gamma_1 u^3\right) dx - 4F(0) \\
&\ge -\frac{\gamma_2}{4}\int_0^1 u^4 dx - 4F(0) - C_0 \\
&\ge -\frac{\gamma_2}{4}\left(\int_0^1 u^2 dx\right)^2 - 4F(0) - C_0,
\end{aligned}$$

where $C_0 = \dfrac{4\gamma_1^4}{81\gamma_2^3}$, which combining with (4.11) leads to the differential inequality

$$\frac{d}{dt}\|Dw\|^2 \ge -\frac{\gamma_2}{4}\|Dw\|^4 - 4F(0) - C_0.$$

Since

$$-4F(0) = \int_0^1 \left(-\gamma_2 u_0^4 + 2u_0^2 - \frac{4}{3}\gamma_1 u_0^3 - 2\gamma|Du_0|^2\right) dx,$$

if we choose γ_2, such that

$$-\gamma_2 \ge 1 + \frac{\gamma_1^4 + \int_0^1 (2\gamma|Du_0| + 2|\gamma_1||u_0|^3)dx}{\int_0^1 u_0^4 dx},$$

then

$$-4F(0) - C_0 \ge 0.$$

Thus

$$\frac{d}{dt}\|Dw\|^2 \geq -\frac{\gamma_2}{4}\|Dw\|^4$$

and hence

$$\|Dw(t)\|^2 \geq \frac{\|Dw(0)\|^2}{1-\frac{\gamma_2}{4}t\|Dw(0)\|^2},$$

whenever $1-\frac{\gamma_2}{4}t\|Dw(0)\|^2 > 0$, from which, noticing that (4.11) and $u_0 \not\equiv 0$, imply $Dw(0) \not\equiv 0$, we assert that u must blow up at a finite time T^*. The proof is complete. □

4.4.3 *Global existence of solutions for small initial value*

From Theorem 4.4.1 and Theorem 4.4.2, we see that the coefficient γ_2 plays an important role for the existence of global solutions. In particular, if $\gamma_2 < 0$, then the solution may blow up at a finite time. In case $\gamma_2 > 0$, in order that the problem (4.1)–(4.3) has a global solution, one need to restrict the initial value to be "small".

As mentioned above, if we set

$$\int_0^1 u(x,t)dx = \int_0^1 u_0(x)dx \equiv M$$

and

$$v(x,t) = u(x,t) - M,$$

then

$$\int_0^1 v(x,t)dx = 0.$$

So, we may change the problem (4.1)–(4.3) into an equivalent one:

$$\frac{\partial v}{\partial t} + \gamma D^4 v = D^2\widetilde{\varphi}(v), \tag{4.12}$$

$$Dv\Big|_{x=0,1} = D^3 v\Big|_{x=0,1} = 0, \tag{4.13}$$

$$v(x,0) = u_0(x) - M, \tag{4.14}$$

where

$$\widetilde{\varphi}(v) = \gamma_2 v^3 + (3\gamma_2 M + \gamma_1)v^2 + (3\gamma_2 M^2 + 2\gamma_1 M - 1)v.$$

Theorem 4.4.3 *If $\gamma > \dfrac{1}{\pi^2}$, $u_0 \in H^2_E(0,1)$, and $\|u_0\|_2$ is sufficiently small, then the problem (4.1)–(4.3) admits a unique global solution $u \in H^{4,1}(Q_T)$. Moreover, there holds*

$$\lim_{t\to\infty} \|u(t) - M\|_\infty = \lim_{t\to\infty} \|Du(t)\|_\infty = \lim_{t\to\infty} \|D^2 u(t)\|_\infty = 0.$$

Here $\|\cdot\|_2$ denotes the norm in $H^2(0,1)$.

Proof. Under the conditions of the theorem, the uniqueness and existence of local solutions are obvious. To prove the global existence, we need some a priori estimates. Let

$$\gamma_0 = 3\gamma_2 M^2 + 2\gamma_1 M - 1, \quad \widetilde{\gamma}_1 = 3\gamma_2 M + \gamma_1, \quad f \equiv D^2(\gamma_2 v^3 + \widetilde{\gamma}_1 v^2).$$

Rewrite the equation (4.12) as

$$\frac{\partial v}{\partial t} + \gamma D^4 v - \gamma_0 D^2 v = f. \tag{4.15}$$

By the assumption, $\gamma > \dfrac{1}{\pi^2}$ and $\|u_0\|$ is sufficiently small, so we have

$$|\gamma_0| < \gamma\pi^2. \tag{4.16}$$

Our purpose is to estimate the following function

$$N(t) = \sup_{0<s<t} \|v(s)\|_2^2 + \int_0^t \|v(s)\|_2^2 ds.$$

First, multiplying both sides of the equation (4.15) by v and then integrating with respect to x, we have

$$\frac{1}{2}\frac{d}{dt}\|v\|^2 + \gamma\|D^2 v\|^2 + \gamma_0\|Dv\|^2 = \int_0^1 f v dx.$$

Since $Dv \in H^1_0(0,1)$, from Friedrich's inequality

$$\|Dv\|^2 \le \frac{1}{\pi^2}\|D^2 v\|^2 \tag{4.17}$$

we obtain

$$\frac{1}{2}\frac{d}{dt}\|v\|^2 + C_1\|D^2v\|^2 \le \int_0^1 fvdx,$$

where (from (4.16))

$$C_1 = \gamma - \frac{|\gamma_0|}{\pi^2} > 0. \tag{4.18}$$

Since $\int_0^1 v(x,t)dx = 0$, we have $\|v\|^2 \le \|Dv\|^2$. Then using (4.17) implies

$$\|v\|^2 \le \frac{1}{\pi^2}\|D^2v\|^2.$$

Thus

$$\frac{d}{dt}\|v\|^2 + C_2\|D^2v\|^2 \le C_3\|f\|^2$$

and hence

$$\|v\|^2 + C_2\int_0^t |D^2v|^2 ds \le \|v_0\|^2 + C_3\int_0^t \|f\|^2 ds. \tag{4.19}$$

Multiplying both sides of the equation (4.15) by $\frac{\partial v}{\partial t}$, and then integrating with respect to x, we have

$$\left\|\frac{\partial v}{\partial t}\right\|^2 + \gamma\frac{d}{dt}\|D^2v\|^2 + \gamma_0\frac{d}{dt}\|Dv\|^2 \le \|f\|^2.$$

It follows from (4.17) that

$$\int_0^t \left\|\frac{\partial v}{\partial t}\right\|^2 ds + C_1\|D^2v\|^2 \le \gamma\|D^2v_0\|^2 + |\gamma_0|\cdot\|Dv_0\|^2 + \int_0^t \|f\|^2 ds. \tag{4.20}$$

Combining (4.19) with (4.20), we immediately obtain

$$N(t) \le C_4\left(\|v_0\|_2^2 + \int_0^t \|f\|^2 ds\right). \tag{4.21}$$

Since

$$f \equiv D^2(\gamma_2 v^3 + \widetilde{\gamma}_1 v^2) = (3\gamma_2 v^2 + 2\widetilde{\gamma}_1 v)D^2v + (6\gamma_2 v + 2\widetilde{\gamma}_1)(Dv)^2,$$

we have

$$\|f\|^2 \le C_5(\|v\|_\infty^4 + \|v\|_\infty^2)\|D^2v\|^2 + C_5(\|v\|_\infty^2\|Dv\|_\infty^2 + \|Dv\|_\infty^2)\|Dv\|^2.$$

Applying

$$\|v\|_\infty \le C_6\|Dv\|, \qquad \|Dv\|_\infty \le C_7\|D^2v\|$$

we obtain

$$\|f\|^2 \le C_8(\|D^2v\|^4 + \|D^2v\|^6).$$

Therefore

$$\int_0^t \|f\|^2 ds \le C_8 \sup_{0<s<t} \|v\|_2^2 \left[1 + \sup_{0<s<t} \|v\|_2^2\right] \int_0^t \|v\|_2^2 ds.$$

Substituting this into the right hand side of (4.21), we see that

$$N(t) \le C_9\left(\|v_0\|_2^2 + N(t)^2 + N(t)^3\right), \qquad t > 0. \tag{4.22}$$

We conclude that, if $\|v_0\|_2$ is sufficiently small, then there exists some constant C_{10}, such that

$$N(t) \le C_{10}\|v_0\|_2^2, \qquad \forall t > 0. \tag{4.23}$$

In fact, for (4.23) to be valid, we need only to require

$$4(C_9+1)^2\|v_0\|_2^2 + 8(C_9+1)^3\|v_0\|_2^4 < 1. \tag{4.24}$$

To see this, we set

$$M(t) = C_9N(t) + C_9N(t)^2$$

and prove

$$M(t) < \frac{1}{2}, \qquad \forall t \ge 0. \tag{4.25}$$

Obviously, $M(0) < \frac{1}{2}$. By the continuity of $M(t)$, if (4.25) is not true, then there exists $t_0 > 0$, such that $M(t_0) = \frac{1}{2}$, while for $t \in (0, t_0)$, $M(t) < \frac{1}{2}$.

Thus from (4.22), we obtain

$$\begin{aligned} N(t_0) &\leq \frac{C_9\|v_0\|_2^2}{1 - C_9 N(t_0) - C_9 N(t_0)^2} \\ &\leq \frac{C_9\|v_0\|_2^2}{1 - M(t_0)} \leq 2C_9\|v_0\|_2^2. \end{aligned} \tag{4.26}$$

This and (4.24) imply

$$\begin{aligned} M(t_0) &= C_9 N(t_0) + C_9 N(t_0)^2 \\ &\leq 2C_9^2\|v_0\|_2^2 + 4C_9^3\|v_0\|_2^4 < \frac{1}{2}. \end{aligned}$$

This contradiction shows that (4.25) is valid. Similar to the proof of (4.26), we may easily derive (4.23) from (4.25).

The estimate (4.23) implies that the problem (4.12)–(4.14) admits a solution in $H^{2,1}(Q_T)$. To complete the proof, we multiply both sides of the equation (4.15) by $-D^2v$ and D^4v, and integrate to obtain

$$\|Dv\|^2 + \int_0^t \|D^3v\|^2 ds \leq C_{11}\left(\|v_0\|_1^2 + \int_0^t \|f\|^2 ds\right), \tag{4.27}$$

$$\|D^2v\|^2 + \int_0^t \|D^4v\|^2 ds \leq C_{12}\left(\|v_0\|_2^2 + \int_0^t \|f\|^2 ds\right), \tag{4.28}$$

which imply $v \in H^{4,1}(Q_T)$.

Finally, we discuss the asymptotic behavior of the solution. According to (4.22), for any $\varepsilon > 0$, we have for all $t \geq 0$, $N(t) \leq \varepsilon$, so long as $\|v_0\|_2$ is sufficiently small. Therefore for all $t \geq 0$

$$\|f(t)\|^2 \leq \varepsilon \|D^2v(t)\|^2.$$

It follows from (4.18) that

$$\frac{1}{2}\frac{d}{dt}\|v\|^2 + (C_2 - \varepsilon C_3)\|v\|_2^2 \leq 0$$

and hence from (4.19)

$$\|v\|^2 \leq \|v_0\|^2 - (C_2 - \varepsilon C_3)\int_0^t \|v\|_2^2 ds.$$

In particular, if ε is sufficiently small such that $C_2 - \varepsilon C_3 > 0$, then we have

$$\|v\|^2 \le \|v_0\|^2 - (C_2 - \varepsilon C_3) \int_0^t \|v\|^2 ds,$$

which shows that $\|v(t)\|$ decays to zero with an exponential rate provided that $\|v_0\|$ is sufficiently small. Similarly, from the differential inequality (4.27) and (4.28), we conclude that $\|v(t)\|_2$ decays to zero too with an exponentially rate. The proof is complete. □

Remark 4.4.1 For the problem in two and three dimensional case, similar conclusions are valid. For the details, the readers may refer to [EZ]. Moreover, there are many recent papers devoted to the properties of solutions. For example, we refer to [GW], [LZ], [WAW] for the investigation of attractors, [WEW1], [WEW2], [WEW3] for stationary solutions, and [WN], [CX] for singular limits and asymptotic limits. In addition, some variant of such kind of equations has been studied, see for example, [GN1], [MSW], [DAD].

4.5 Cahn-Hilliard Equations with Positive Concentration Dependent Mobility

In the previous section, we have discussed the Cahn-Hilliard equation with positive constant mobility. Such equation is, in fact, derived from the original physical model by linearizing the principal part. In many physical diffusive processes, the mobility is concentration dependent, namely, $\gamma = m(u)$. In this case, the equation becomes

$$\frac{\partial u}{\partial t} + D\left[m(u)(kD^3u - DA(u))\right] = 0, \tag{5.1}$$

where $k > 0$ is a physical constant.

This section is devoted to the case of positive concentration dependent mobility, namely, $m(u) > 0$. We still consider the boundary value problem in the domain $Q_T = (0,1) \times (0,T)$ with initial value condition

$$u(x,0) = u_0(x), \tag{5.2}$$

and boundary value conditions

$$Dv\Big|_{x=0,1} = D^3v\Big|_{x=0,1} = 0, J\Big|_{x=0,1} = Du\Big|_{x=0,1} = 0, \tag{5.3}$$

where

$$J = m(u)\left(kD^3u - DA(u)\right)$$

is the net flux.

Since the principal part of the equation (5.1) is nonlinear, the investigation of the problem is more difficult. First of all, the energy method used in the previous section is not enough for establishing the required a priori estimates. We will use the frame based on the Campanato spaces together with the energy method to establish the Schauder type a priori estimates, and discuss the existence of solutions of the problem.

4.5.1 *A modified Campanato space*

Let $y_i = (x_i, t_i) \in \overline{Q}_T (i = 0, 1, 2)$, $R > 0$ and denote

$$B_R = B_R(x_0) = (x_0 - R, x_0 + R),$$

$$I_R = I_R(t_0) = (t_0 - R^4, t_0 + R^4),$$

$$Q_R = Q_R(y_0) = B_R(x_0) \times I_R(t_0),$$

$$SR = Q_R \cap Q_T,$$

$$ER = E_R(x_0) = B_R(x_0) \cap (0, 1),$$

$$J_R = J_R(t_0) = I_R(t_0) \cap (0, +\infty),$$

$$d(y_1, y_2) = |x_1 - x_2| + |t_1 - t_2|^{1/4}.$$

For a function u defined on Q_T, define the average by

$$u_R = u_{y_0,R} = \frac{1}{|S_R|} \iint_{S_R} u dx dt.$$

Modify u_R as follows

$$\widehat{u}_R = \widehat{u}_{y_0,R} = \begin{cases} u_R, & \text{if } Q_R \cap \partial_p Q_T = \emptyset, \\ 0, & \text{if } Q_R \cap \partial_p Q_T \neq \emptyset, \end{cases}$$

where $\partial_p Q_T$ is the parabolic boundary of Q_T and $|S_R| = \text{mes} S_R$.

Definition 4.5.1 Let $\lambda > 0$,

$$C_*(\overline{Q}_T) = \{u \in C(\overline{Q}_T);\ u = 0 \ \text{ in } \ \partial_p Q_T\}.$$

For any $u \in C_*(\overline{Q}_T)$, set

$$[u]_{2,\lambda} = \left(\sup_{\substack{y_0 \in \overline{Q}_T \\ 0 < R \le R_0}} \frac{1}{R^\lambda} \iint_{S_R(y_0)} |u(x,t) - \widehat{u}_{y_0,R}|^2 dxdt \right)^{1/2},$$

where $R_0 = \operatorname{diam} Q_T$.

Define the set $\mathcal{L}_0^{2,\lambda}(Q_T)$ as a subspace of $C_*(\overline{Q}_T)$, whose element u satisfies $[u]_{2,\lambda} < +\infty$. For $u \in \mathcal{L}_0^{2,\lambda}(Q_T)$, define the norm as

$$\|u\|_{2,\lambda} = \sup_{Q_T} |u(x,t)| + [u]_{2,\lambda}.$$

Then $\mathcal{L}_0^{2,\lambda}(Q_T)$ becomes a Banach space.

Lemma 4.5.1 *Let $\lambda > 5$. Then there is an embedding*

$$\mathcal{L}_0^{2,\lambda}(Q_T) \subset C^{\alpha,\alpha/4}(\overline{Q}_T),$$

where $\alpha = \dfrac{\lambda - 5}{2}$. Moreover, the embedding operator is continuous, and

$$[u]_\alpha \le C(\lambda)[u]_{2,\lambda},$$

where $[u]_\alpha$ denotes the seminorm of u in $C^{\alpha,\alpha/4}(\overline{Q}_T)$, and $C(\lambda)$ depends only on λ.

The proof of the lemma is similar to that in [CA], and we omit the details.

Remark 4.5.1 For n-dimensional domain Ω, Campanato [CA] has ever defined a similar space $\mathcal{L}^{2,\lambda}(\Omega)$, which is the set of all functions u in $L^2(\Omega)$ satisfying

$$[u]_{\mathcal{L}^{2,\lambda}(\Omega)} \equiv \left(\sup_{\substack{\rho > 0 \\ z_0 \in \Omega}} \frac{1}{\rho^\lambda} \int_{\Omega(z_0,\rho)} |u(z) - u_{z_0,\rho}|^2 dz \right)^{1/2} < +\infty,$$

where $\Omega(z_0,\rho) = \{z \in \Omega;\ |z - z_0| < \rho\}$ and

$$u_{z_0,\rho} = \frac{1}{|\Omega(z_0,\rho)|} \int_{\Omega(z_0,\rho)} u(z) dz.$$

In particular, if $n = 2$, $\Omega = Q_T$, then the space $\mathcal{L}^{2,\lambda}(Q_T)$ is quite similar to the space $\mathcal{L}_0^{2,\lambda}(Q_T)$. The only difference lies in the description of functions near the parabolic boundary. Using the space $\mathcal{L}_0^{2,\lambda}(Q_T)$, one can estimate the Hölder norm directly and does not need the estimate near the boundary. This is quite important in treating the fourth order parabolic equations for which the maximum principle is no longer valid.

4.5.2 *Hölder norm estimates for a linear problem*

To investigate the solvability of the problem (5.1)–(5.3), we first consider an auxiliary linear problem

$$\frac{\partial u}{\partial t} + D^2\Big(a(x,t)D^2u\Big) = D^2 f, \quad \text{in} \quad Q_T, \tag{5.4}$$

$$u(x,0) = 0, \tag{5.5}$$

$$u(0,t) = u(1,t) = D^2u(0,t) = D^2u(1,t) = 0. \tag{5.6}$$

We do not want to minimize the smoothness of the coefficient $a(x,t)$ and the function $f(x,t)$. Our purpose is to find the relation between the Hölder's norm estimate of u and Hölder's norm of $a(x,t)$ and $f(x,t)$. We always assume that

$$0 < a_0 \le a(x,t) \le A_0. \tag{5.7}$$

Let u be a solution of the problem (5.4)–(5.6). Decompose u in $S_R = S_R(y_0)$ as $u = u_1 + u_2$, where u_1, u_2 satisfy

$$\frac{\partial u_1}{\partial t} + a(x_0,t_0)D^4u_1 = 0, \qquad \text{in} \quad S_R, \tag{5.8}$$

$$u_1\Big|_{\partial_p S_R} = u\Big|_{\partial_p S_R}, \quad B(x,D)u_1\Big|_{\partial E_R} = B(x,D)u\Big|_{\partial E_R}, \tag{5.9}$$

$$\frac{\partial u_2}{\partial t} + a(x_0,t_0)D^4u_2 = D^2\widetilde{f}, \quad \text{in} \quad S_R, \tag{5.10}$$

$$u_2\Big|_{\partial_p S_R} = 0, \quad B(x,D)u_2\Big|_{\partial E_R} = 0, \tag{5.11}$$

where $\widetilde{f} = (a(x_0,t_0) - a(x,t))D^2u + f$ and

$$B(x,D) = \begin{cases} D^2, & x = 0,1, \\ D, & x \neq 0,1. \end{cases}$$

Based on the classical theory, the above decomposition is uniquely determined by u, and u_1, u_2 are sufficiently smooth on

$$S_R^* = \left\{(x,t) \in \overline{S}_R;\ t > \inf J_R\right\}$$

and satisfy $u_i, Du_i \in C(\overline{S}_R)$, $D^4u_i \in L^2(S_R)(i = 1,2)$.

Lemma 4.5.2 *For the solution u_2 of the problem (5.10)–(5.11), we have*

$$\begin{aligned} &\sup_{J_R}\int_{E_R} u_2^2(x,t)dx + \iint_{S_R}(D^2u_2)^2dxdt \\ \le\ & CR^{2\sigma}\iint_{S_R}(D^2u)^2dxdt + C\sup|f|^2R^5, \end{aligned} \tag{5.12}$$

where C depends only on a_0, A_0 and $\|a\|_\sigma$, and $\|a\|_\sigma$ denotes the norm of a in the space $C^{\sigma/4,\sigma}(\overline{Q}_T)$.

Proof. Set

$$Q_t = (0,1)\times(0,t), \quad S_R^t = S_R \cap Q_t, \quad J_R^t = J_R \cap (0,t).$$

Multiplying both sides of the equation (5.10) by u_2, integrating over S_R^t, and then integrating by parts, we have

$$\begin{aligned} &\frac{1}{2}\int_{E_R} u_2^2(x,t)dx + a(x_0,t_0)\iint_{S_R^t}(D^2u_2)^2dxds \\ =\ &\iint_{S_R^t}\Big[a(x_0,t_0) - a(x,t)\Big]D^2uD^2u_2dxds \\ &+ \iint_{S_R^t} fD^2u_2dxds - \int_{J_R^t} f(\beta_R,s)Du_2(\beta_R,s)ds \\ &+ \int_{J_R^t} f(\alpha_R,s)Du_2(\alpha_R,s)ds, \end{aligned}$$

where α_R and β_R denote the left endpoint and right endpoint of the interval E_R. Noticing that

$$\sup_{x\in E_R}|Du_2(x,s)|^2 \le R\int_{E_R}(D^2u_2(x,s))^2dx + \frac{C}{R^3}\int_{E_R}u_2^2(x,s)dx,$$

we have

$$\begin{aligned}
&\left|\int_{J_R^t} f(\beta_R,s)Du_2(\beta_R,s)ds\right| + \left|\int_{J_R^t} f(\alpha_R,s)Du_2(\alpha_R,s)ds\right| \\
\le\ & \varepsilon\iint_{S_R^t}(D^2u_2)^2dxds + \frac{\varepsilon}{R^4}\iint_{S_R^t}u_2^2dxds+ \\
&+C_\varepsilon R\int_{J_R^t}\Big(|f(s,\beta_R)|^2+|f(s,\alpha_R)|^2\Big)ds \\
\le\ & \varepsilon\iint_{S_R^t}(D^2u_2)^2dxds + \varepsilon\sup_{J_R}\int_{E_R}u_2^2(x,s)dx + C_\varepsilon R^5\sup|f|^2.
\end{aligned}$$

By virtue of the fact that

$$\begin{aligned}
&\left|\iint_{S_R^t}\Big[a(x_0,t_0)-a(x,t)\Big]D^2uD^2u_2dxds\right| \\
\le\varepsilon\ & \iint_{S_R^t}(D^2u_2)^2dxds + C_\varepsilon\|a\|_\sigma^2R^{2\sigma}\iint_{S_R^t}(D^2u)^2dxds, \\
&\left|\iint_{S_R^t}fD^2u_2dxds\right| \le \varepsilon\iint_{S_R^t}(D^2u_2)^2dxds + C_\varepsilon R^5\sup|f|^2,
\end{aligned}$$

it is easy to see that the estimate (5.12) holds. The proof is complete. □

Lemma 4.5.3 *For any* $(x_1,t),(x_2,t),(x,t_1),(x,t_1)\in S_\rho$,

$$\Big|u_1(x_1,t)-u_1(x_2,t)\Big|^2 \le CM(u_1,\rho)|x_1-x_2|, \tag{5.13}$$

$$\Big|u_1(x,t_1)-u_1(x,t_1)\Big|^2 \le CM(u_1,\rho)|t_1-t_2|^{1/4}, \tag{5.14}$$

where

$$M(u_1,\rho)=\sup_{J_\rho}\int_{E_\rho}(Du_1(x,t))^2dx+\iint_{S_\rho}(D^3u_1)^2dxdt,$$

and the constant C *depends only on* a_0 *and* A_0.

Proof. The estimate (5.13) is obvious, and we need only to show (5.14). Without loss of generality, we assume that $\Delta t = t_2 - t_1 > 0$, $x, x + 2(\Delta t)^{1/4} \in E_\rho$. Integrating the equation (5.8) over $(y, y+(\Delta t)^{1/4})\times(t_1,t_2)$,

we have

$$
\begin{aligned}
0 = & \int_y^{y+(\Delta t)^{1/4}} \Big[u_1(z,t_2) - u_1(z,t_1)\Big] dx + \\
& + a(x_0,t_0) \int_{t_1}^{t_2} \Big[D^3 u_1(y+(\Delta t)^{1/4}, s) - D^3 u_1(y,s)\Big] ds,
\end{aligned}
$$

i.e.

$$
\begin{aligned}
0 & = (\Delta t)^{1/4} \int_0^1 \Big[u_1(y+\theta(\Delta t)^{1/4}, t_2) - u_1(y+\theta(\Delta t)^{1/4}, t_1)\Big] d\theta + \\
& + a(x_0,t_0) \int_{t_1}^{t_2} \Big[D^3 u_1(y+(\Delta t)^{1/4}, s) - D^3 u_1(y,s)\Big] ds.
\end{aligned}
$$

Integrating the above equality with respect to y over $(x, x+(\Delta t)^{1/4})$, and using the mean value theorem, we see that

$$
\begin{aligned}
& (\Delta t)^{1/2} \Big[u_1(x^*,t_2) - u_1(x^*,t_1)\Big] \\
= & \int_{t_1}^{t_2} \int_x^{x+(\Delta t)^{1/4}} \Big[D^3 u_1(y+(\Delta t)^{1/4}, s) - D^3 u_1(y,s)\Big] dyds.
\end{aligned}
$$

Therefore

$$
\Big|u_1(x^*,t_2) - u_1(x^*,t_1)\Big|^2 \le C|t_1 - t_2|^{1/4} \iint_{S_\rho} (D^3 u_1)^2 dxdt,
$$

where $x^* = y^* + \theta^*(\Delta t)^{1/4})$, $y^* \in (x, x+(\Delta t)^{1/4})$, $\theta^* \in (0,1)$, which together with (5.13) implies (5.14), and the proof is complete. □

Lemma 4.5.4 *(Caccioppoli type inequality)*

$$
\begin{aligned}
& \sup_{J_{R/4}} \int_{E_{R/4}} (u_1(x,t) - \lambda)^2 dx + \iint_{S_{R/4}} (D^2 u_1)^2 dxdt \\
\le & \frac{C}{R^4} \iint_{S_{R/2}} (u_1 - \lambda)^2 dxdt, \qquad (5.15)
\end{aligned}
$$

$$
\begin{aligned}
& \sup_{J_{R/4}} \int_{E_{R/4}} (Du_1(x,t))^2 dx + \iint_{S_{R/4}} (D^3 u_1)^2 dxdt \\
\le & \frac{C}{R^4} \iint_{S_{R/2}} (Du_1)^2 dxdt \le \frac{C}{R^6} \iint_{S_R} (u_1 - \lambda)^2 dxdt, \qquad (5.16)
\end{aligned}
$$

where C is a constant depending only on a_0, A_0,

$$\lambda = \begin{cases} \text{arbitrary constant}, & \text{if} \quad Q_R \cap \partial_p Q_T = \emptyset, \\ 0, & \text{if} \quad Q_R \cap \partial_p Q_T \neq \emptyset. \end{cases}$$

Proof. We discuss the following cases separately.

1. The case $t_0 - R^4 < 0$. In this case, $\lambda = 0$. Choose a C^∞ function $\chi(x)$, such that $0 \le \chi(x) \le 1$, and

$$\left|\chi^{(i)}(x)\right| \le \frac{C}{R^i}, \quad i = 1, 2, 3, 4.$$

If $0 \in E_R$, then we modify $\chi(x) \equiv 1$ for $x \le x_0$, while if $1 \in E_R$, then modify $\chi(x) \equiv 1$ for $x \ge x_0$. Multiplying both sides of the equation (5.8) by $\chi^4 u_1$, and then integrating over S_R^t, we have

$$\iint_{S_R^t} \frac{\partial u_1}{\partial t} \chi^4 u_1 dx ds + a(x_0, t_0) \iint_{S_R^t} D^4 u_1 \chi^4 u_1 dx ds = 0.$$

Using the boundary value conditions (5.5) and (5.9) yields

$$\chi^4 u_1\Big|_{\partial E_R} = 0, \qquad D^2 u_1 D(\chi^4 u_1)\Big|_{\partial E_R} = 0.$$

Therefore

$$\begin{aligned} 0 = & \frac{1}{2}\int_{E_R} \chi^4 u_1^2(x,t) dx + \iint_{S_R^t} a(x_0, t_0) D^2 u_1 D^2(\chi^4 u_1) dx ds \\ = & \frac{1}{2}\int_{E_R} \chi^4 u_1^2(x,t) dx + \iint_{S_R^t} a(x_0, t_0) \chi^4 (D^2 u_1)^2 dx ds + \\ & +8 \iint_{S_R^t} a(x_0, t_0) \chi^3 \chi' D u_1 D^2 u_1 dx ds \\ & + \iint_{S_R^t} a(x_0, t_0)(24\chi^2 \chi'^2 + 8\chi^3 \chi'') u_1 D^2 u_1 dx ds. \end{aligned}$$

The Cauchy inequality is used to deduce

$$\begin{aligned} & \left| \iint_{S_R^t} a(x_0, t_0)(24\chi^2 \chi'^2 + 8\chi^3 \chi'') u_1 D^2 u_1 dx ds \right| \\ \le & \frac{1}{4} a(x_0, t_0) \iint_{S_R^t} \chi^4 (D^2 u_1)^2 dx ds + \frac{C}{R^4} \iint_{S_{R/2}} u_1^2 dx ds, \\ & \left| 8 \iint_{S_R^t} a(x_0, t_0) \chi^3 \chi' D u_1 D^2 u_1 dx ds \right| \end{aligned}$$

$$\leq \frac{1}{4}a(x_0,t_0)\iint_{S_R^t}\chi^4(D^2u_1)^2dxds + C\iint_{S_R^t}\chi^2\chi'^2(Du_1)^2dxds. \quad (5.17)$$

Noticing that

$$\begin{aligned}
&\iint_{S_R^t}\chi^2\chi'^2(Du_1)^2dxds\\
=\ & -\iint_{S_R^t}u_1D(\chi^2\chi'^2Du_1)dxds\\
=\ & -\iint_{S_R^t}\chi^2\chi'^2u_1D^2u_1dxds + \iint_{S_R^t}u_1^2D^2(\chi^2\chi'^2)dxds\\
\leq\ & \frac{1}{4}a(x_0,t_0)\iint_{S_R^t}\chi^4(D^2u_1)^2dxds + \frac{C}{R^4}\iint_{S_{R/2}}u_1^2dxds,
\end{aligned}$$

we have

$$\sup_{J_R}\int_{E_R}\chi^4u_1^2(x,t)dx + \iint_{S_R}\chi^4(D^2u_1)^2dxdt \leq \frac{C}{R^4}\iint_{S_{R/2}}u_1^2dxds,$$

from which we see that (5.15) holds.

Since $w = Du_1$ satisfies the equation

$$\frac{\partial w}{\partial t} + a(x_0,t_0)D^4w = 0, \qquad \text{in } S_R,$$

and

$$Dw(0,t) = D^3w(0,t) = 0, \text{if } 0 \in \partial E_R,$$

$$Dw(1,t) = D^3w(1,t) = 0, \text{if } 1 \in \partial E_R,$$

we may use a similar argument to prove (5.16).

2. The case $t_0 - R^4 \geq 0$. Choose a function $\eta(t) \in C^\infty$ such that $\eta(t) = 1$ in $(t_0 - \left(\frac{R}{4}\right)^4, +\infty)$, $\eta(t) = 0$ in $(-\infty, t_0 - \left(\frac{R}{2}\right)^4)$, $0 \leq \eta(t) \leq 1$, and $|\eta'(t)| \leq \frac{C}{R^4}$ for all $t \in \mathbb{R}$.

Multiplying both sides of the equation (5.8) by $\chi^4\eta(u_1 - \lambda)$, integrating the resulting relation over S_R^t, we may apply the method used in the first case to complete the proof. □

Let $y_0 = (x_0, t_0) \in \overline{Q}_T$ be fixed and define

$$\varphi(u,\rho) = \iint_{S_\rho}\left(|u - \widehat{u}_\rho|^2 + \rho^4|D^2u|^2\right)dxdt, \qquad (\rho > 0).$$

Lemma 4.5.5 *For any* $0 < \rho < R$,

$$\varphi(u_1, \rho) \le C\left(\frac{\rho}{R}\right)^6 \varphi(u_1, R), \tag{5.18}$$

where C *is a constant depending only on* a_0, A_0 *and* $\|a\|_\sigma$.

Proof. We need only to show (5.18) for $\rho \le \dfrac{R}{4}$. From Lemma 4.5.2 and Lemma 4.5.3,

$$\iint_{S_\rho} |u_1 - \widehat{u}_{1\rho}|^2 dxdt \le CM\left(u_1, \frac{R}{4}\right)\rho^6 \le C\left(\frac{\rho}{R}\right)^6 \iint_{S_R} (u_1 - \lambda)^2 dxdt.$$

It follows by setting $\lambda = \widehat{u}_{1R}$ that

$$\iint_{S_\rho} |u_1 - \widehat{u}_{1\rho}|^2 dxdt \le C\left(\frac{\rho}{R}\right)^6 \iint_{S_R} (u_1 - \widehat{u}_{1R})^2 dxdt. \tag{5.19}$$

On the other hand, from (5.16),

$$\begin{aligned}
&\iint_{S_\rho} \rho^4 (D^2 u_1)^2 dxdt \\
\le\ & C_1 \iint_{S_\rho} \rho^6 (D^3 u_1)^2 dxdt + C_2 \iint_{S_\rho} \rho^2 (Du_1)^2 dxdt \\
\le\ & C_1 \rho^6 \iint_{S_{R/4}} (D^3 u_1)^2 dxdt + C_2 \rho^6 \sup_{J_{R/4}} \int_{E_{R/4}} (Du_1(x,t))^2 dx \\
\le\ & C\left(\frac{\rho}{R}\right)^6 \iint_{S_{R/2}} R^2 (Du_1)^2 dxdt \\
\le\ & C\left(\frac{\rho}{R}\right)^6 \left(\iint_{S_R} R^4 (D^2 u_1)^2 dxdt\right) + C\left(\frac{\rho}{R}\right)^6 \left(\iint_{S_R} (u_1 - \widehat{u}_{1R})^2 dxdt\right) \\
=\ & C\left(\frac{\rho}{R}\right)^6 \varphi(u_1, R),
\end{aligned}$$

which together with (5.19) implies (5.18) and the proof is complete. □

To estimate the Hölder norm of u, we need a technical lemma, whose proof can be found in [GA].

Lemma 4.5.6 *Let* $\varphi(\rho)$ *be a nonnegative and nondecreasing function satisfying*

$$\varphi(\rho) \le A\left(\left(\frac{\rho}{R}\right)^\alpha + \varepsilon\right)\varphi(R) + BR^\beta,$$

for all $0 < \rho \le R \le R_0$, where A, B, α, β are positive constants, $\beta < \alpha$. Then there exists a constant $\varepsilon_0 = \varepsilon_0(A, \alpha, \beta)$, such that for all $0 < \rho \le R \le R_0$ and $0 < \varepsilon < \varepsilon_0$, there holds

$$\varphi(\rho) \le C\left(\left(\frac{\rho}{R}\right)^\beta \varphi(R) + BR^\beta\right),$$

where C is a constant depending only on α, β and A.

Now, we present the key lemma in this section.

Lemma 4.5.7 *Let $a(x,t)$ and $f(x,t)$ be appropriately smooth functions, and u be the smooth solution of the problem (5.4)–(5.6). Then for any $\alpha \in (0, \frac{1}{2})$, there exists a constant C depending only on a_0, A_0, α, T, $\|a\|_\sigma$, $\iint_{Q_T} u^2 dxdt$ and $\iint_{Q_T} (D^2u)^2 dxdt$, such that*

$$\begin{aligned}&|u(x_1,t_1) - u(x_2,t_2)| \\ \le\ & C(1 + \sup|f|)\left(|x_1 - x_2|^\alpha + |t_1 - t_2|^{\alpha/4}\right).\end{aligned} \tag{5.20}$$

Proof. For any fixed $(x_0, t_0) \in \overline{Q}_T$, consider the function $\varphi(u, \rho)$ in Lemma 4.5.5, which is nondecreasing with respect to ρ. From Lemma 4.5.5, we see that

$$\begin{aligned}\varphi(u,\rho) &\le \varphi(u_1,\rho) + \varphi(u_2,\rho) \\ &\le C\left(\frac{\rho}{R}\right)^6 \varphi(u_1, R) + \varphi(u_2, R) \\ &\le C\left(\frac{\rho}{R}\right)^6 \varphi(u, R) + C\varphi(u_2, R)\end{aligned}$$

for all $0 < \rho < R$. It follows from Lemma 4.5.2,

$$\begin{aligned}\varphi(u_2, R) &= \iint_{S_R} \left((u_2 - \widehat{u}_{2R})^2 + R^4(D^2u_2)^2\right) dxdt \\ &\le 4\iint_{S_R} u_2^2 dxdt + R^4 \iint_{S_R} (D^2u_2)^2 dxdt \\ &\le 4R^4 \sup_{J_R} \int_{E_R} u_2^2(x,t)dx + R^4 \iint_{S_R} (D^2u_2)^2 dxdt \\ &\le CR^{4+2\sigma} \iint_{S_R} (D^2u)^2 dxdt + C\sup|f|^2 R^9\end{aligned}$$

$$\le CR^{2\sigma}\varphi(u,R) + C\sup|f|^2 R^9.$$

Therefore

$$\varphi(u,\rho) \le C\left(\left(\frac{\rho}{R}\right)^6 + R^{2\sigma}\right)\varphi(u,R) + C\sup|f|^2 R^9.$$

For the constant ε_0 in Lemma 4.5.6, we choose $R_0 > 0$, such that $R^{2\sigma} < \varepsilon_0$ holds for $R \le R_0$. Then from Lemma 4.5.6,

$$\varphi(u,\rho) \le C\left(\left(\frac{\rho}{R_0}\right)^\lambda \varphi(u,R_0) + \sup|f|^2\rho^\lambda\right)$$

holds for some $5 < \lambda < 6$. Therefore

$$[u]_{2,\lambda}^2 \le C\left(\frac{1}{R_0^\lambda}\varphi(u,R_0) + \sup|f|^2\right).$$

By virtue of this, the desired estimate (5.20) follows immediately from Lemma 4.5.1. The proof is complete. □

4.5.3 *Zero potential case*

With the above preparation, we now turn to the discussion of the existence of solutions. We first consider the zero potential case, namely, $A(u) \equiv 0$, and the problem becomes

$$\frac{\partial u}{\partial t} + D\left[m(u)D^3u\right] = 0, \quad \text{in} \quad Q_T, \tag{5.21}$$

$$Du(0,t) = Du(1,t) = D^3u(0,t) = D^3u(1,t) = 0, \tag{5.22}$$

$$u(x,0) = u_0(x). \tag{5.23}$$

Theorem 4.5.1 *Assume that $m(s) \in C^{1+\alpha}(\mathbb{R})$, $u_0 \in C^{4+\alpha}(\overline{I})$, $D^iu_0(0) = D^iu_0(1) = 0 (i = 1,3)$, $m(s) > 0$. Then the problem (5.21)–(5.23) admits a classical solution $u \in C^{4+\alpha,1+\alpha/4}(\overline{Q}_T)$.*

Proof. For simplicity, we assume that $m(s) \in C^\infty(\mathbb{R})$, $u_0 \in C^\infty(\overline{I})$. We will apply the Schauder fixed point theorem to complete the proof of the theorem. To do this, we need a series of a priori estimates.

Multiplying both sides of the equation (5.21) by D^2u, and then integrating the resulting relation over $Q_t = (0,1) \times (0,t)$, we have

$$\iint_{Q_t} \frac{\partial u}{\partial t} D^2 u dx ds + \iint_{Q_t} D(m(u)D^3u)D^2 u dx ds = 0.$$

Integrating by parts and using the boundary value condition (5.22), we obtain

$$\frac{1}{2}\int_0^1 (Du(x,t))^2 dx - \frac{1}{2}\int_0^1 (Du_0)^2 dx + \iint_{Q_t} m(u)(D^3u)^2 dx ds = 0.$$

Therefore

$$\sup_{0\le t\le T} \int_0^1 (Du(x,t))^2 dx \le C, \tag{5.24}$$

$$\iint_{Q_T} m(u)(D^3u)^2 dx dt \le C. \tag{5.25}$$

Integrating both sides of the equation (5.21) with respect to x over the interval $(0,1)$, we obtain

$$\int_0^1 u(x,t)dx = \int_0^1 u_0(x)dx.$$

The mean value theorem implies that for some $x_t^* \in (0,1)$,

$$u(t, x_t^*) = \int_0^1 u_0(x)dx,$$

from which, we see that for any $(x,t) \in Q_T$,

$$|u(x,t)| \le |u(x,t) - u(t,x_t^*)| + |u(t,x_t^*)| \le \left| \int_{x_t^*}^x Du(t,y)dy \right| + \int_0^1 |u_0(x)|dx.$$

It follows from (5.24) that

$$\sup_{Q_T} |u(x,t)| \le C. \tag{5.26}$$

Furthermore, we may obtain the Hölder's norm estimate of solutions, namely,

$$|u(x_1,t) - u(x_2,t)| \le C|x_1 - x_2|^{1/2}, \tag{5.27}$$

$$|u(x,t_1) - u(x,t_1)| \le C|t_1 - t_2|^{1/8}. \tag{5.28}$$

In fact, (5.27) is a direct consequence of (5.24). To prove (5.28), we need only to consider the case that $0 \le x \le \frac{1}{2}$, $\Delta t = t_2 - t_2 > 0$, $\Delta t \le \frac{1}{4}$. Integrating (5.21) over $(y, y+\Delta t) \times (t_1, t_2)$ gives

$$\begin{aligned}
&\int_y^{y+\Delta t} [u(z,t_2) - u(z,t_1)]dz \\
&= -\int_{t_1}^{t_2} \Big[m(u(y+\Delta t, s))D^3u(y+\Delta t, s) - m(u(y,s))D^3u(y,s)\Big]ds,
\end{aligned}$$

i.e.

$$\begin{aligned}
&\Delta t \int_0^1 (u(y+\theta\Delta t, t_2) - u(y+\theta\Delta t, t_1))d\theta \\
&= -\int_{t_1}^{t_2} \Big[m(u(y+\Delta t, s))D^3u(y+\Delta t, s) - m(u(y,s))D^3u(y,s)\Big]ds.
\end{aligned}$$

Integrating the above equality with respect to y over $(x, x+\Delta t)$, we have

$$|u(t_2, x^*) - u(t_1, x^*)| \le C(\Delta t)^{1/8},$$

where $x^* = y^* + \theta^*\Delta t$, $y^* \in (x, x+\Delta t)$, $\theta^* \in (0,1)$, from which and (5.27) we see that (5.28) holds.

The key step is to estimate the Hölder norm of Du, namely, to prove

$$|Du(x_1,t_1) - Du(x_2,t_2)| \le C(|x_1-x_2|^{1/4} + |t_1-t_2|^{1/16}), \tag{5.29}$$

where C is a constant depending only on the known quantities. In fact, by setting $w = Du - Du_0$, we see that w satisfies

$$\begin{aligned}
&\frac{\partial w}{\partial t} + D^2(m(u)D^2w) = -D^2(m(u)D^3u_0) \equiv D^2 f, \qquad \text{in } Q_T, \\
&w(0,t) = w(1,t) = D^2w(0,t) = D^2w(1,t) = 0, \\
&w(x,0) = 0.
\end{aligned}$$

By the estimates (5.25)–(5.28) and Lemma 4.5.7, we see that (5.29) holds.

Now, rewrite the equation (5.21) in the following form

$$\frac{\partial u}{\partial t} + a(x,t)D^4u + b(x,t)D^3u = 0,$$

where

$$a(x,t) = m(u(x,t)) \geq \inf m(u(x,t)) \equiv m_0 > 0,$$
$$b(x,t) = m'(u(x,t))Du(x,t).$$

It follows from the estimates (5.27), (5.28) and (5.29) that

$$|a(x_1,t_1) - a(x_2,t_2)| \leq C(|x_1 - x_2|^{1/2} + |t_1 - t_2|^{1/8}),$$
$$|b(x_1,t_1) - b(x_2,t_2)| \leq C(|x_1 - x_2|^{1/4} + |t_1 - t_2|^{1/16}).$$

Applying the Schauder theory for linear equations (see that [WA]), we have

$$\left|\frac{\partial u}{\partial t}(x_1,t_1) - \frac{\partial u}{\partial t}(x_2,t_2)\right| \leq C(|x_1 - x_2|^{\beta} + |t_1 - t_2|^{\beta/4}), \tag{5.30}$$

$$|D^4u(x_1,t_1) - D^4u(x_2,t_2)| \leq C(|x_1 - x_2|^{\beta} + |t_1 - t_2|^{\beta/4}), \tag{5.31}$$

where $\beta = \min(1/4, \alpha)$, C is a constant depending only on the known quantities.

By virtue of (5.30) and (5.31), we may further improve the estimates (5.27)–(5.29) and obtain

$$|u(x_1,t_1) - u(x_2,t_2)| \leq C(|x_1 - x_2| + |t_1 - t_2|^{1/4}),$$
$$|Du(x_1,t_1) - Du(x_2,t_2)| \leq C(|x_1 - x_2| + |t_1 - t_2|^{1/4}),$$

from which, the exponents in the estimates (5.30) and (5.31) can be replaced by α.

Consider the linear space

$$\mathrm{X} = \{u \in C^{1+\alpha,(1+\alpha)/4}(\overline{Q}_T); Du(0,t) = Du(1,t) = 0,\ u(x,0) = u_0(x)\}$$

and define an operator T on X

$$\mathrm{T} : \mathrm{X} \longrightarrow \mathrm{X}, \qquad u \longmapsto w,$$

where w is the solution of the following linear problem

$$\frac{\partial w}{\partial t} + m(u(x,t))D^4w + m'(u(x,t))Du(x,t)D^3w = 0,$$
$$Dw(0,t) = Dw(1,t) = D^3w(0,t) = D^3w(1,t) = 0,$$
$$w(x,0) = u_0(x).$$

From the Schauder theory for linear equations (see [WA]), we see that the above problem admits a unique smooth solution in $C^{4+\beta,(4+\beta)/4}(\overline{Q}_T)$. So, the operator T is well defined and compact. Moreover, if $u = \sigma \mathrm{T}u$, $\sigma \in (0,1]$, then u satisfies (5.21), (5.22) and the initial value condition $u(x,0) = \sigma u_0(x)$. From the above discussion, we see that the norm of u in the space $C^{4+\alpha,(4+\alpha)/4}(\overline{Q}_T)$ can be estimated by some constant C depending only on the known quantities. Hence from the Leray-Schauder fixed point theorem, the operator T has a fixed point u, which is just the desired solution of the problem (5.21)–(5.23). The proof is complete. □

Theorem 4.5.2 *The problem (5.21)–(5.23) admits at most one solution in the space* $C^{4+\alpha,(4+\alpha)/4}(\overline{Q}_T)$.

Proof. Let u_1 and u_2 be two solutions of the problem (5.21)–(5.23). Then for any $\varphi(x,t) \in C^\infty(\overline{Q}_T)$ with $D\varphi(0,t) = D\varphi(1,t) = D^3\varphi(0,t) = D^3\varphi(1,t) = \varphi(x,T) = 0$, we have

$$\iint_{Q_T} (u_1 - u_2)\frac{\partial \varphi}{\partial t} dxdt + \iint_{Q_T} [m(u_1)D^3 u_1 - m(u_2)D^3 u_2] D\varphi dxdt = 0,$$

i.e.

$$\iint_{Q_T} (u_1 - u_2)\frac{\partial \varphi}{\partial t} dxdt - \iint_{Q_T} (u_1 - u_2) D^3(A(x,t)D\varphi) dxdt + \\ + \iint_{Q_T} (u_1 - u_2) B(x,t) D\varphi dxdt = 0,$$

where

$$A(x,t) = m(u_1(x,t)),$$

$$B(x,t) = \int_0^1 m'(\lambda u_1 + (1-\lambda)u_2) d\lambda \cdot D^3 u_2.$$

For any $f \in C_0^\infty(Q_T)$, consider the linear problem

$$\frac{\partial \varphi}{\partial t} - D^3(A(x,t)D\varphi) + B(x,t)D\varphi = f(x,t),$$

$$D\varphi(0,t) = D\varphi(1,t) = D^3\varphi(0,t) = D^3\varphi(1,t) = 0,$$

$$\varphi(x,T) = 0.$$

Since

$$A(x,t) \in C^{3+\alpha,(3+\alpha)/4}(\overline{Q}_T),$$
$$B(x,t) \in C^{\alpha,\alpha/4}(\overline{Q}_T),$$

from the Schauder theory for linear equations, we see that the above problem admits a unique classical solution $\varphi \in C^{4+\alpha,(4+\alpha)/4}(\overline{Q}_T)$. Therefore

$$\iint_{Q_T} (u_1 - u_2) f dx dt = 0.$$

Due to the arbitrariness of f, we conclude that $u_1(x,t) \equiv u_2(x,t)$. The proof is complete. □

4.5.4 *General case*

Now, we turn to the general equation (5.1).

Theorem 4.5.3 *Let $m(s) \in C^{1+\alpha}(\mathbb{R})$, $A(s) \in C^{2+\alpha}(\mathbb{R})$, $u_0 \in C^{4+\alpha}(\overline{I})$, $D^i u_0(0) = D^i u_0(1) = 0 (i = 1, 3)$, $m(s) > 0$, and for some constant $\mu > 0$*

$$H(s) \equiv \int_0^s A(\sigma) d\sigma \geq -\mu. \tag{5.32}$$

Then the problem (5.1)–(5.3) admits a unique classical solution u in the space $C^{4+\alpha,1+\alpha/4}(\overline{Q}_T)$.

Proof. Most part of the proof is similar to that of Theorem 4.5.1, we will omit the details. The main difference lies in the proofs of the estimates (5.24) and (5.29). We first consider (5.24). Let

$$F(t) = \int_0^1 \left(\frac{k}{2} (Du)^2 + H(u) + \mu \right) dx.$$

By the assumption (5.32) and the equation (5.1),

$$\begin{aligned} F'(t) &= \int_0^1 \left(kDu \frac{\partial Du}{\partial t} + A(u) \frac{\partial u}{\partial t} \right) dx \\ &= -\int_0^1 \left(kD^2 u - A(u) \right) \frac{\partial u}{\partial t} dx \\ &= -\int_0^1 m(u) \left(kD^3 u - DA(u) \right)^2 dx \leq 0. \end{aligned}$$

Thus $F(t) \le F(0)$, and (5.24) holds. Set $w = Du - Du_0$. Then w satisfies

$$\frac{\partial w}{\partial t} + D^2\left(km(u)D^2 w\right) = D^2 \widehat{f},$$

where

$$\widehat{f} = -km(u)D^3 u_0 + m(u)A'(u)Du.$$

By Lemma 4.5.7, we have

$$\begin{aligned} & |Du(x_1,t_1) - Du(x_2,t_2)| \\ \le\ & C(1+\sup|\widehat{f}|)(|x_1-x_2|^{1/4} + |t_1-t_2|^{1/16}) \\ \le\ & C(1+\sup|Du|)(|x_1-x_2|^{1/4} + |t_1-t_2|^{1/16}), \end{aligned}$$

from which and the interpolation inequality, we see that the estimate (5.29) holds.

Using the above estimates, we may complete the proof of the theorem in a similar way as in Theorem 4.5.1. The proof is complete. □

4.6 Thin Film Equation

Consider the equation

$$\frac{\partial u}{\partial t} + \frac{\partial}{\partial x}\left(|u|^n \frac{\partial^3 u}{\partial x^3}\right) = 0, \tag{6.1}$$

with $n > 0$, which can be regarded as a mathematical model describing the spreading of an oil film over a solid surface, where u denotes the height from the surface of the oil to the surface of the solid. The equation (6.1) is a special case of the Cahn-Hilliard equation with $m(u) = |u|^n$. It degenerates whenever $u = 0$. Similar to the previous two sections, we consider the following initial boundary value problem

$$u(x,0) = u_0(x), \tag{6.2}$$

$$Du\Big|_{x=\pm 1} = D^3 u\Big|_{x=\pm 1} = 0. \tag{6.3}$$

4.6.1 *Definition of generalized solutions*

Definition 4.6.1 A function u is said to be a generalized solution of the problem (6.1)–(6.3), if the following conditions are fulfilled:

(1) $u \in C(\overline{Q}_T)$, $|u|^{n/2}D^3u \in L^2(P)$, $\frac{\partial u}{\partial t}$, Du, D^2u, D^3u, $D^4u \in C(P)$, where $P = \overline{Q}_T \setminus (\{u(x,t)=0\} \cup \{t=0\})$;

(2) u satisfies the equation (6.1) in the following sense

$$\iint_{Q_T} u\frac{\partial \varphi}{\partial t}dxdt + \iint_P |u|^n D^3uD\varphi dxdt = 0,$$

where $\varphi \in \mathrm{Lip}(\overline{Q}_T)$ is any function satisfying $\varphi(x,0)=\varphi(x,T)=0$;

(3) u satisfies the initial value condition (6.2) in the usual sense;

(4) For any point such that $u(\pm 1,t) \neq 0$, u satisfies the boundary value conditions (6.3);

(5) $\lim_{t\to 0} Du(\cdot,t) = u_{0x}$ in $L^2(-1,1)$.

4.6.2 *Approximate solutions*

To discuss the solvability, we first consider the following approximate problem

$$\frac{\partial u}{\partial t} + ((|u|^n+\varepsilon)D^3u)_x = 0, \quad (x,t) \in Q_T, \tag{6.4}$$

$$Du\Big|_{x=\pm 1} = D^3u\Big|_{x=\pm 1} = 0, \tag{6.5}$$

$$u(x,0) = u_{0\varepsilon}(x), \tag{6.6}$$

where $Q_T = (-1,1)\times(0,T)$ and $u_{0\varepsilon}$ is a smooth approximation of the initial data $u_0(x)$. According to the discussion on the Cahn-Hilliard equation with positive concentration dependent mobility, the above problem admits a global classical solution u_ε. We need some uniform estimates on u_ε.

First, multiplying both sides of the equation (6.4) by D^2u_ε, and then integrating over $Q_t = (-1,1)\times(0,t)$, we have

$$\frac{1}{2}\int_{-1}^1 (Du_\varepsilon(x,t))^2dx - \frac{1}{2}\int_{-1}^1 (Du_{0\varepsilon})^2dx + \int_0^t\int_{-1}^1 (|u_\varepsilon|^n+\varepsilon)(D^3u)^2dxdt = 0. \tag{6.7}$$

Therefore

$$\int_{-1}^{1}(Du_\varepsilon(x,t))^2dx \le \int_{-1}^{1}(Du_{0\varepsilon})^2dx. \tag{6.8}$$

On the other hand, integrating the equation (6.4) over Q_t yields

$$\int_{-1}^{1}u_\varepsilon(x,t)dx = \int_{-1}^{1}u_{0\varepsilon}dx. \tag{6.9}$$

We may require the approximate sequence $u_{0\varepsilon}$ to satisfy

$$\int_{-1}^{1}(Du_{0\varepsilon})^2dx \le C\int_{-1}^{1}(Du_0)^2dx. \tag{6.10}$$

Then from (6.8), (6.9) and Poincarè's inequality, we may conclude that

$$|u_\varepsilon(x,t)| \le C, \tag{6.11}$$

where C is a constant independent of ε.

By (6.8), (6.10) and Sobolev's inequality, we may obtain the Hölder norm estimate on u_ε with respect to x

$$|u_\varepsilon(x_2,t) - u_\varepsilon(x_1,t)| \le K|x_2 - x_1|^{1/2}. \tag{6.12}$$

Denote

$$h_\varepsilon = (|u_\varepsilon|^n + \varepsilon)D^3u_\varepsilon.$$

Then from (6.7), (6.10) and (6.11), we get the following estimates

$$\iint_{Q_T} h_\varepsilon^2(x,t)dxdt \le C. \tag{6.13}$$

Now, we turn to the Hölder norm estimate on u_ε with respect to t. This can be done by adopting the approach in the previous section combining with the estimate (6.12).

Lemma 4.6.1 *There exists a constant C independent of ε, such that*

$$|u_\varepsilon(x,t_2) - u_\varepsilon(x,t_1)| \le C|t_2 - t_1|^{1/8}. \tag{6.14}$$

Proof. We argue by contradiction. Suppose that for a sufficiently large constant M, there exist x_0, t_2, t_1, such that

$$|u_\varepsilon(x_0,t_2) - u_\varepsilon(x_0,t_1)| > M|t_2 - t_1|^{1/8}.$$

For definiteness, we may assume that $u_\varepsilon(x_0,t_2) > u_\varepsilon(x_0,t_1)$, $T > t_2 > t_1 > 0$. Then the above inequality becomes

$$u_\varepsilon(x_0,t_2) - u_\varepsilon(x_0,t_1) > M(t_2-t_1)^{1/8}. \tag{6.15}$$

From the equation (6.4) and the boundary value condition (6.5), we have, for any $\varphi \in \mathrm{Lip}(\overline{Q}_T)$ with $\varphi(x,T) = \varphi(x,0) = 0$,

$$\iint_{Q_T} u_\varepsilon \frac{\partial \varphi}{\partial t} dxdt = -\iint_{Q_T} h_\varepsilon D\varphi dxdt. \tag{6.16}$$

We choose φ of the form

$$\varphi(x,t) = \xi(x)\theta_\delta(t),$$

where

$$\xi(x) = \xi_0\left(\frac{16K^2(x-x_0)}{M^2(t_2-t_1)^{1/4}}\right),$$

with constants K and M determined in (6.12) and (6.15) respectively, and $\xi_0 \in C_0^\infty(\mathbb{R})$ satisfying $\xi_0(x) = \xi_0(-x)$, and $\xi_0(x) = 1$ for $0 \le x < \frac{1}{2}$, $\xi_0(x) = 0$ for $x \ge 1$, $\xi_0'(x) \le 0$ for $x \ge 0$, so that

$$\xi(x) = \begin{cases} 0, & \text{if } |x-x_0| \ge \dfrac{M^2}{16K^2}(t_2-t_1)^{1/4}, \\ 1, & \text{if } |x-x_0| \le \dfrac{1}{2}\dfrac{M^2}{16K^2}(t_2-t_1)^{1/4}. \end{cases} \tag{6.17}$$

$\theta_\delta(t)$ is chosen as

$$\theta_\delta(t) = \int_{-\infty}^t \theta_\delta'(s)ds,$$

where

$$\theta_\delta'(t) = \begin{cases} \dfrac{1}{\delta}, & |t-t_2| < \delta, \\ -\dfrac{1}{\delta}, & |t-t_1| < \delta, \\ 0, & \text{otherwise}, \end{cases}$$

and $\delta < \frac{1}{2}(t_2-t_1)$. The function θ_δ thus defined is Lipschitz continuous, satisfies $|\theta_\delta(t)| \le 1$, and for δ sufficiently small, $\theta_\delta(0) = \theta_\delta(T) = 0$.

Substituting $\varphi(x,t)=\xi(x)\theta_\delta(t)$ into (6.16), we obtain

$$\iint_{Q_T} u_\varepsilon \xi \theta'_\delta dxdt = -\iint_{Q_T} h_\varepsilon \xi' \theta_\delta dxdt. \tag{6.18}$$

It is easy to see that

$$\lim_{\delta\to 0}\iint_{Q_T} u_\varepsilon \xi \theta'_\delta dxdt = \int_{-1}^{1} \xi(x)(u_\varepsilon(x,t_2)-u_\varepsilon(x,t_1))dx.$$

To estimate the right hand side of (6.18), we need only to consider the point x satisfying

$$|x-x_0| \le \frac{M^2}{16K^2}(t_2-t_1)^{1/4}. \tag{6.19}$$

For such x, we get from (6.12) and (6.15),

$$\begin{aligned} u_\varepsilon(x,t_2)-u_\varepsilon(x,t_1) &= [u_\varepsilon(x,t_2)-u_\varepsilon(x_0,t_2)] \\ &+[u_\varepsilon(x_0,t_2)-u_\varepsilon(x_0,t_1)]+[u_\varepsilon(x_0,t_1)-u_\varepsilon(x,t_1)] \\ &\ge -2K|x-x_0|^{1/2}+M(t_2-t_1)^{1/8} \ge \frac{M}{2}(t_2-t_1)^{1/8}. \end{aligned}$$

Since we may further assume that $\{x\in(-1,1);\xi(x)=1\}\subset(-1,1)$, it follows that

$$\begin{aligned} &\int_{-1}^{1} \xi(x)(u_\varepsilon(x,t_2)-u_\varepsilon(x,t_1))dx \\ \ge&\frac{M}{2}(t_2-t_1)^{1/8}\frac{M^2}{32K^2}(t_2-t_1)^{1/4}=\frac{M^3}{64K^2}(t_2-t_1)^{3/8}. \end{aligned} \tag{6.20}$$

Now, we estimate the right hand side of (6.18). Set

$$E=\{(x,t);\xi(x)\theta_\delta(t)\neq 0\}.$$

Then, from the definition of $\xi(x)$ and $\theta_\delta(t)$,

$$\text{mes}(E) \le \frac{M^2}{32K^2}(t_2-t_1)^{1/4}(t_2-t_1+2\delta).$$

Hence, using (6.13) we obtain

$$\begin{aligned}
&\left|\iint_{Q_T} h_\varepsilon \xi'(x)\theta_\delta dxdt\right| \\
\le\ & \sup|\xi'(x)|\left(\iint_{Q_T} h_\varepsilon^2 dxdt\right)^{1/2}\left(\iint_E \theta_\delta^2 dxdt\right)^{1/2} \\
\le\ & \frac{4C_1K}{\sqrt{2}M(t_2-t_1)^{1/8}}(t_2-t_1+2\delta)^{1/2}.
\end{aligned}$$

Letting $\delta \to 0$, we derive

$$\overline{\lim_{\delta\to 0}}\left|\iint_{Q_T} h_\varepsilon \xi'(x)\theta_\delta dxdt\right| \le 2\sqrt{2}C_1K(t_2-t_1)^{3/8}. \tag{6.21}$$

Combining (6.20), (6.21) with (6.18) leads to

$$M^3 \le 128\sqrt{2}C_1K^3 = C_2.$$

Since the constant C_2 is independent of ε, M, T, such an inequality can not be held for sufficiently large M. The proof is complete. □

4.6.3 *Existence of solutions*

Theorem 4.6.1 *The problem (6.1)–(6.3) admits a least one generalized solution.*

Proof. The estimates (6.11), (6.12) and (6.14) imply the existence of a subsequence of $\{u_\varepsilon\}$, denoted still by $\{u_\varepsilon\}$, such that

$$u_\varepsilon(x,t) \to u(x,t) \qquad \text{uniformly on } Q_T.$$

Now we prove that the limit function u is a generalized solution of the problem (6.1)–(6.3).

For any admissible test function φ in the definition of generalized solutions, from the equation (6.4),

$$\iint_{Q_T} u_\varepsilon \frac{\partial\varphi}{\partial t}dxdt + \iint_{Q_T}(|u_\varepsilon|^n+\varepsilon)D^3u_\varepsilon D\varphi dxdt = 0. \tag{6.22}$$

Since (6.7) implies

$$\varepsilon\iint_{Q_T}(D^3u_\varepsilon)^2dxdt \le C,$$

we have

$$\lim_{\varepsilon\to 0}\varepsilon\iint_{Q_T} D^3u_\varepsilon D\varphi dxdt = 0. \tag{6.23}$$

From (6.13), there exists a function $h \in L^2(Q_T)$, such that

$$h_\varepsilon \to h \text{ weakly in } L^2(Q_T). \tag{6.24}$$

By the regularity theory for linear equations, we see that $\dfrac{\partial u_\varepsilon}{\partial t}$, Du_ε, D^2u_ε, D^3u_ε, D^4u_ε converge uniformly on any compact set of P. Therefore

$$|u|^n D^3 u = h. \qquad \text{in } P. \tag{6.25}$$

Thus the conditions (1),(3) and (4) in the definition of generalized solutions are fulfilled.

Now, we prove that u satisfies the condition (2) in the definition of solutions. For any fixed $\delta > 0$, from (6.25), we have

$$\iint_{|u|>\delta}(|u_\varepsilon|^n+\varepsilon)D^3u_\varepsilon D\varphi dxdt \to \iint_{|u|>\delta}|u|^n D^3uD\varphi dxdt. \tag{6.26}$$

On the other hand, if $\varepsilon = \varepsilon(\delta)$ is sufficiently small, then from (6.13),

$$\left|\iint_{|u|\le\delta}(|u_\varepsilon|^n+\varepsilon)D^3u_\varepsilon D\varphi dxdt\right|$$

$$\le C\delta^{n/2}\left(\iint_{Q_T}(|u_\varepsilon|^n+\varepsilon)(D^3u)^2dxdt\right)^{1/2} \le C\delta^{n/2}. \tag{6.27}$$

Letting $\varepsilon \to 0$ in (6.22), and using (6.23), (6.26) and (6.27), we immediately obtain the integral equality in the definition of solutions.

Finally, we prove that u satisfies (5) in the definition of solutions. From (6.8),

$$\overline{\lim_{t\to 0}}\int_{-1}^{1}(Du(x,t))^2dx \le \int_{-1}^{1}(Du_0)^2dx.$$

On the other hand, as $t \to 0$,

$$Du(\cdot,t) \rightharpoonup u_{0x} \text{ in } L^2(-1,1). \tag{6.28}$$

By the weakly lower semicontinuity of norm, we have

$$\int_{-1}^{1}(Du_0)^2dx \le \varliminf_{t\to 0}\int_{-1}^{1}(Du(x,t))^2dx.$$

Therefore

$$\lim_{t\to 0}\int_{-1}^{1}(Du(x,t))^2dx = \int_{-1}^{1}(Du_0)^2dx,$$

which and (6.28) imply

$$\lim_{t\to 0}(Du(\cdot,t))^2 = (Du_0)^2, \quad \text{in } L^2(-1,1).$$

The proof is complete. □

4.6.4 *Nonnegativity of solutions*

One of the main differences between the fourth order parabolic equations and the second order equations is that for the fourth order equations the nonnegativity of the initial data does not imply the same property of solutions at any time. In order to preserve the nonnegativity of solutions, some additional conditions, which are a little harsh, should be assumed on the initial data.

Theorem 4.6.2 *Assume that* $0 \le u_0(x) \in H^1(-1,1)$, *and*

$$\begin{aligned} &\int_{-1}^{1}|\log u_0(x)|dx < +\infty, && \text{if } n = 2,\\ &\int_{-1}^{1}u_0^{2-n}(x)dx < +\infty, && \text{if } 2 < n < 4,\\ &u_0(x) > 0, \quad \forall x \in [-1,1], && \text{if } n \ge 4. \end{aligned} \tag{6.29}$$

Then the solution obtained in Theorem 4.6.1 satisfies $u(x,t) \ge 0$.

Proof. Let u_ε be the approximate solution constructed in the proof of Theorem 4.6.1. Choose a positive constant A, such that for all $\varepsilon > 0$, $(x,t) \in Q_T$, we have $A > |u_\varepsilon(x,t)|$. Denote

$$g_\varepsilon(s) = -\int_s^A \frac{dr}{|r|^n+\varepsilon}, \qquad G_\varepsilon(s) = -\int_s^A g_\varepsilon(r)dr. \tag{6.30}$$

Suppose that the approximate function $u_{0\varepsilon}(x)$ of the initial data $u_0(x)$ is chosen such that $u_{0\varepsilon}(x) \geq u_0(x)$. A direct calculation shows that

$$G_0(s) = \begin{cases} \dfrac{A^{2-n}}{(2-n)(n-1)} + \dfrac{sA^{1-n}}{n-1} - \dfrac{s^{2-n}}{(2-n)(n-1)}, & \text{if } n \neq 2, \\ \log \dfrac{A}{s} + \dfrac{s}{A} - 1, & \text{if } n = 2. \end{cases}$$

Obviously, $G_\varepsilon(s)$ is decreasing with respect to ε and s. It follows from (6.29) that

$$\int_{-1}^{1} G_\varepsilon(u_{0\varepsilon}(x))dx \leq \int_{-1}^{1} G_0(u_{0\varepsilon}(x))dx \leq \int_{-1}^{1} G_0(u_0(x))dx \leq C.$$

Multiplying both sides of the equation (6.4) by $g_\varepsilon(u_\varepsilon)$, and then integrating over Q_t, we obtain

$$\int_{-1}^{1} G_\varepsilon(u_\varepsilon(x,t))dx + \iint_{Q_t} (D^2 u_\varepsilon)^2 dxds = \int_{-1}^{1} G_\varepsilon(u_{0\varepsilon}(x))dx.$$

Thus, we have the following estimate

$$\int_{-1}^{1} G_\varepsilon(u_\varepsilon(x,t))dx \leq C, \tag{6.31}$$

$$\iint_{Q_T} (D^2 u_\varepsilon)^2 dxdt \leq C. \tag{6.32}$$

If the conclusion of the theorem were false, then there would exist a point $(x_0, t_0) \in Q_T$, such that $u(x_0, t_0) < 0$. By the continuity and the uniform convergence of u_ε, there exist $\delta > 0$ and $\varepsilon_0 > 0$, such that

$$u_\varepsilon(x, t_0) < -\delta, \quad \text{if } |x - x_0| < \delta, x \in (-1,1), \varepsilon < \varepsilon_0,$$

and hence, for such a x,

$$\begin{aligned} G_\varepsilon(u_\varepsilon(x,t_0)) &= -\int_{u_\varepsilon(x,t_0)}^{A} g_\varepsilon(s)ds \\ &\geq -\int_{-\delta}^{0} g_\varepsilon(s)ds = -\int_{-\delta}^{0} \frac{1}{|s|^n + \varepsilon} ds \to +\infty, \qquad \text{as } \varepsilon \to 0, \end{aligned}$$

which contradicts (6.31) and the proof is complete. □

Theorem 4.6.3 *If in addition to the assumption of Theorem 4.6.2, suppose that $n \geq 4$, then $u(x,t) > 0$ for any $(x,t) \in \overline{Q}_T$. Moreover, such solution is unique.*

Proof. We argue by contradiction to prove that $u(x,t) > 0$. Since we have proved that $u(x,t) \geq 0$, if the conclusion were false, then there would exist a point $(x_0, t_0) \in Q_T$, such that $u(x_0, t_0) = 0$. From the Hölder continuity of u, we see that

$$u(x, t_0) \leq C|x - x_0|^{1/2}.$$

Since $n \geq 4$, we have

$$\int_{-1}^{1} u(x,t_0)^{2-n} dx \geq C \int_{-1}^{1} |x - x_0|^{(2-n)/2} dx = \infty.$$

On the other hand, by (6.31)

$$\int_{-1}^{1} G_0(u(x,t_0))dx = \lim_{\varepsilon \to 0} \int_{-1}^{1} G_\varepsilon(u_\varepsilon(x,t_0))dx \leq C.$$

Therefore

$$\int_{-1}^{1} u(x,t_0)^{2-n} dx \leq C,$$

which is a contradiction.

Now, we prove the uniqueness. It worth pointing out that the uniqueness can be proved similar to the discussion in the previous section. In what follows, we give a new approach to prove the uniqueness, which is based on the special structure of the equation (6.1). Assume that v is another solution. From the first conclusion of the theorem, we have

$$0 < C_1 \leq u(x,t), v(x,t) \leq C_2, \qquad (x,t) \in Q_T,$$

which implies, in particular, that both u and v are the classical solutions of (6.1). Set $w = u - v$. We multiply the difference of the two equations which u and v satisfy by w_{xx}, and integrate over $(-1,1) \times (t_0, t)$. Then we obtain

$$\frac{1}{2}\int_{-1}^{1} (Dw(x,t))^2 dx - \frac{1}{2}\int_{-1}^{1} (Dw(x,t_0))^2 dx$$
$$+ \int_{t_0}^{t}\int_{-1}^{1} (|u|^n D^3 u - |v|^n D^3 v) D^3 w dx ds = 0.$$

Since from the definition of solutions, there hold

$$\lim_{t\to 0} Du(\cdot,t) = u_{0x}, \quad \lim_{t\to 0} v_x(\cdot,t) = v_{0x}\text{in } L^2(-1,1),$$

letting $t_0 \to 0$, we obtain

$$\frac{1}{2}\int_{-1}^{1}(Dw(x,t))^2dx + \iint_{Q_t}(|u|^nD^3u - |v|^nD^3v)D^3w dxdt = 0.$$

Rewriting the integrand of the second integral as

$$\left(|u|^nD^3u - |v|^nD^3v\right)D^3w = |u|^n(D^3w)^2 + (|u|^n - |v|^n)D^3vD^3w,$$

and noticing that

$$||u|^n - |v|^n| \le C|w|,$$

we derive

$$\begin{aligned}
&\sup_{0<s<t}\int_{-1}^{1}(Dw(x,s))^2dx + \iint_{Q_t}(D^3w)^2dxds \\
\le\ & C\iint_{Q_t}|wD^3vD^3w|dxds \\
\le\ & \frac{1}{2}\iint_{Q_t}(D^3w)^2dxds + \iint_{Q_t}|w^2(D^3v)^2|dxds.
\end{aligned}$$

Therefore

$$\sup_{0<s<t}\int_{-1}^{1}(Dw(x,s))^2dx + \iint_{Q_t}(D^3w)^2dxds \le \iint_{Q_t}|w^2(D^3v)^2|dxds. \tag{6.33}$$

On the other hand, since

$$\int_{-1}^{1}u(x,t)dx = \int_{-1}^{1}v(x,t)dx = \int_{-1}^{1}u_0(x)dx,$$

we have

$$\int_{-1}^{1}w(x,t)dx = 0,$$

which implies by Poincarè's inequality

$$\sup_{(-1,1)\times(0,t)}w^2(x,s) \le C\sup_{0<s<t}\int_{-1}^{1}(Dw(x,s))^2dx.$$

By virtue of this inequality, we get from (6.33),

$$\sup_{0<s<t}\int_{-1}^{1}(Dw(x,s))^2dx \le C\left(\iint_{Q_t}(D^3v)^2dxds\right)\sup_{0<s<t}\int_{-1}^{1}(Dw(x,s))^2dx.$$

This shows that for small t, we must have $Dw(x,t)\equiv 0$, and hence $u(x,t)\equiv v(x,t)$. The proof is complete. □

4.6.5 *Zeros of nonnegative solutions*

Theorem 4.6.4 *If in addition to the assumptions of Theorem 4.6.2, suppose that* $2\le n<4$, *then the set* $\{(x,t)\in Q_T; u(x,t)=0\}$ *has zero measure, and*

$$\int_{-1}^{1}|\log u(x,t)|dx<\infty, \quad \forall t\in[0,T], \quad n=2, \tag{6.34}$$

$$\int_{-1}^{1}u(x,t)^{2-n}dx<\infty, \quad \forall t\in[0,T], \quad 2<n<4. \tag{6.35}$$

Proof. To prove $\text{mes}\{(x,t)\in Q_T; u(x,t)=0\}=0$, it suffices to verify that for any $t\in(0,T)$, the measure of the one dimensional set $\{x\in(-1,1); u(\cdot,t)=0\}$ is zero. If the conclusion were false, then there would exist a point $t_0\in(0,T)$, such that the set $E=\{x\in(-1,1); u(\cdot,t_0)=0\}$ has positive measure. Since u_ε converges uniformly to u, for any given $\delta>0$, there exists $\varepsilon_0>0$, such that for any $x\in E$ and $\varepsilon<\varepsilon_0$, $u_\varepsilon(x,t_0)<\delta$. Hence for any $x\in E$, as $\varepsilon\to 0$,

$$G_\varepsilon(u_\varepsilon(x,t_0))\ge -\int_\delta^A g_\varepsilon(s)ds\to -\int_\delta^A g_0(s)ds=\lambda(\delta). \tag{6.36}$$

It is easy to check that

$$\lim_{\delta\to 0}\lambda(\delta)=+\infty.$$

Letting $\varepsilon\to 0$ in (6.36), we obtain

$$\overline{\lim_{\varepsilon\to 0}}\int_{-1}^{1}G_\varepsilon(u_\varepsilon(x,t_0))dx\ge\lambda(\delta)\text{meas}E$$

and hence by letting $\delta\to 0$,

$$\overline{\lim_{\varepsilon\to 0}}\int_{-1}^{1}G_\varepsilon(u_\varepsilon(x,t_0))dx=+\infty,$$

which contradicts the estimate (6.31).

Now, we prove the second part of the theorem. If $u(x,t)>0$, then

$$\lim_{\varepsilon\to 0} G_\varepsilon(u_\varepsilon(x,t)) = G_0(u(x,t)). \tag{6.37}$$

Since we have already seen that the set $\{(x,t)\in Q_T; u(x,t)=0\}$ has zero measure, (6.37) holds almost everywhere, and from (6.31) and Fatou's lemma, we obtain

$$\int_{-1}^{1} G_0(u(x,t))dx \le C.$$

The desired conclusions (6.34), (6.35) follow from the definition of $G_0(s)$ and the proof is complete. □

4.6.6 *Regularity of solutions*

Theorem 4.6.5 *Let $u_0(x)$ be as in Theorem 4.6.2, $n>1$, and u be the corresponding solution. Then*

$$Du \in L^2(0,T;H_0^1(-1,1)) \tag{6.38}$$

and u satisfies the equation (6.1) in the following sense

$$\iint_{Q_T} u\frac{\partial\varphi}{\partial t}dxdt = \iint_{Q_T} |u|^n D^2uD^2\varphi dxdt$$

$$+n\iint_{Q_T}(|u|^{n-1}sgnu)DuD^2uD\varphi dxdt, \tag{6.39}$$

where $\varphi\in C^2(\overline{Q}_T)$ with $\varphi(x,0)=\varphi(x,T)=D\varphi(\pm 1,t)=0$.

Moreover, $Du(\pm 1,t)=0$ for almost all $t\in(0,T)$.

To prove the theorem, we need the following lemmas.

Lemma 4.6.2 *([LI]) Let E_0, E and E_1 be reflexive Banach spaces, $E_0\subset E\subset E_1$, the embedding map $E_0\to E$ is compact and $E\to E_1$ is continuous. Let $1<p_0,p_1<\infty$. If $\{v_k\}$ is a bounded sequence in $L^{p_0}(0,T;E_0)$ and $\{\frac{dv_k}{dt}\}$ is a bounded sequence in $L^{p_1}(0,T;E_1)$, then there exists a subsequence of $\{v_k\}$, which converges strongly both in $L^{p_0}(0,T;E)$ and $C([0,T];E_1)$.*

Lemma 4.6.3 *Du_ε converges to Du strongly in $L^2(Q_T)$ as $\varepsilon\to 0$.*

Proof. From (6.32), we see that Du_ε converges weakly in the space $L^2(0,T;H_0^1(-1,1))$. From (6.4) and the estimate (6.13), $\{\frac{\partial Du_\varepsilon}{\partial t}\}$ is a bounded sequence in $L^2(0,T;H^{-2}(-1,1))$. Choose $E_0 = H_0^1(-1,1)$, $E = L^2(-1,1)$, $E_1 = H^{-2}(-1,1)$, $p_0 = p_1 = 2$, $v_k = Du_\varepsilon$. Then by Lemma 4.6.2, Du_ε converges to Du strongly in $L^2(Q_T)$. The proof is complete. □

Proof of Theorem 4.6.5. (6.38) is a direct consequence of (6.32). To prove (6.39), we first notice that for any test function φ,

$$\begin{aligned}&\iint_{Q_T} u_\varepsilon \frac{\partial \varphi}{\partial t} dxdt + \varepsilon \iint_{Q_T} D^3 u_\varepsilon D\varphi dxdt\\ = &\iint_{Q_T} (|u_\varepsilon|^n + \varepsilon) D^2 u_\varepsilon D^2 \varphi dxdt \qquad (6.40)\\ &+n \iint_{Q_T} (|u_\varepsilon|^{n-1} \mathrm{sgn} u_\varepsilon) Du_\varepsilon D^2 u_\varepsilon D\varphi dxdt.\end{aligned}$$

Letting $\varepsilon \to 0$, and using Lemma 4.6.3 and (6.32), we immediate obtain (6.39). The proof is complete. □

Theorem 4.6.6 *Let u_0 be as in Theorem 4.6.2, $n \geq \frac{8}{3}$ and u be the corresponding solution. Then the set*

$$\{t \in (0,T); \exists x \in [-1,1], s.t. u(x,t) = 0\}$$

has zero measure, and the following boundary value condition holds for almost every $t \in (0,T)$

$$D^3 u(\pm 1, t) = 0.$$

Proof. From (6.38), we see that for almost every $t \in (0,T)$,

$$D^2 u(\cdot,t) \in L^2(-1,1), \qquad Du(\pm 1, t) = 0. \qquad (6.41)$$

We want to show that, for such t, $u(x,t) > 0$ for any $x \in [-1,1]$. Suppose the contrary. Then for some t_0 satisfying (6.41), there exists $x_0 \in [-1,1]$, such that $u(x_0,t_0) = 0$. (6.41) implies $u(\cdot,t_0) \in C^{3/2}([-1,1])$. If $x_0 \in (-1,1)$, then, since $u(x,t_0) \geq 0$, we have $Du(x_0,t_0) = 0$. Thus

$$u(x,t_0) \leq C|x - x_0|^{3/2},$$

and hence

$$\int_{-1}^{1} u(x,t_0)^{2-n} dx \geq C \int_{-1}^{1} |x - x_0|^{3(2-n)/2} dx = \infty,$$

which contradicts (6.35) in Theorem 4.6.4. The proof is complete. □

4.6.7 *Monotonicity of the support of solutions*

Let us observe again the physical phenomenon described by the equation (6.1). Suppose that at the initial time, the oil film occupies the domain Ω_0. Then as the time evolves, due to the effect of gravity, a touching domain Ω_t will expand. Returning to the problem (6.1)–(6.3), from the mathematical point of view, it is natural to ask, if $\mathrm{supp}u_0 \subset (-1,1)$, then whether the set $\mathrm{supp}u(\cdot,t)$ is compact for any $t>0$ and whether the set $\mathrm{supp}u(\cdot,t)$ increases with t. The following theorem gives a positive answer to the second question.

Theorem 4.6.7 *Let $0 \le u_0 \in H_0^1(-1,1)$, $n \ge 4$. Then the support of $u(\cdot,t)$ is increasing with respect to t.*

Proof. We first notice that, the conclusion of Theorem 4.6.3 can not be applied, since here the initial data $u_0(x)$ do not satisfy (6.29). To obtain a nonnegative solution, let $\delta>0$ and u_δ be the solution of the equation (6.1) satisfying the boundary value condition (6.2) and the initial value condition $u_\delta(x,0)=u_0(x)+\delta$. Then using Theorem 4.6.3, we have $u_\delta(x,t)>0$. It is easy to see that, $u(x,t)=\lim\limits_{\delta\to0}u_\delta(x,t)$ is a nonnegative solution of the problem (6.1)–(6.3).

To prove the theorem, it suffices to verify that for any $x_0\in(-1,1)$ with $u_0(x_0)>0$, we have $u(x_0,t)>0$ for all $t>0$. Let $\varepsilon>0$ be fixed, such that $u_0(x)>0$ holds in $[-1,1]\cap[x_0-\varepsilon,x_0+\varepsilon]$. Choose a nonnegative smooth function $\xi(x)$, such that

$$\xi'(\pm1)=0,\quad \xi(x)=1 \quad \text{in } (x_0-\delta,x_0+\delta), \tag{6.42}$$

$$\int_{-1}^1 \xi(x)u_0^{2-n}(x)dx \le C<\infty. \tag{6.43}$$

Multiplying both sides of the equation (6.1) by $\xi G_0'(u_\delta)$, and then integrating over Q_t, we obtain

$$\int_{-1}^1 \xi(x)G_0(u_\delta(x,t))dx-\int_{-1}^1 \xi(x)G_0(u_{\delta0}(x))dx$$
$$-\iint_{Q_t}[D^3u_\delta Du_\delta\xi+D^3u_\delta h(u_\delta)\xi']dxds=0, \tag{6.44}$$

where

$$h(u_\delta) = -u_\delta^n \int_{u_\delta}^A \frac{ds}{s^n} = \frac{1}{n-1} A^{1-n} u_\delta^n - \frac{1}{n-1} u_\delta.$$

Since $Du_\delta(\pm 1, t) = 0$, $\xi'(\pm 1) = 0$, we have

$$-\iint_{Q_t} D^3 u_\delta D u_\delta \xi dxds = \iint_{Q_t} [D^2 u_\delta \xi + D u_\delta D^2 u_\delta \xi'] dxds,$$

and

$$\iint_{Q_t} D^3 u_\delta h(u_\delta) \xi' dxds = -\iint_{Q_t} [h'(u_\delta) D u_\delta D^2 u_\delta \xi' + h(u_\delta) D^2 u_\delta \xi''] dxds.$$

Let $\xi = \zeta^s$, $s \geq 4$. Then

$$\begin{aligned}
&\iint_{Q_t} |D u_\delta D^2 u_\delta \xi'| dxds + \iint_{Q_t} |h'(u_\delta) D u_\delta D^2 u_\delta \xi'| dxds \\
\leq\ & C \left\{ \iint_{Q_t} \zeta^s D^2 u_\delta dxds \iint_{Q_t} \zeta^{s-2} (D u_\varepsilon)^2 dxds \right\}^{1/2} \\
\leq\ & \frac{1}{4} \iint_{Q_t} \xi D^2 u_\delta dxds + C,
\end{aligned}$$

and

$$\begin{aligned}
&\iint_{Q_t} |h(u_\delta) D^2 u_\delta \xi''| dxds \leq C \iint_{Q_t} \zeta^{s-2} |D^2 u_\delta| dxds \\
\leq & C \left\{ \iint_{Q_t} \zeta^s D^2 u_\delta dxds \iint_{Q_t} \zeta^{s-4} dxds \right\}^{1/2} \\
\leq & \frac{1}{4} \iint_{Q_t} \xi D^2 u_\delta dxds + C.
\end{aligned}$$

It follows from (6.44) that

$$\int_{-1}^1 \xi(x) u_\delta(x,t)^{2-n} dx \leq \int_{-1}^1 \xi(x) u_0(x)^{2-n} dx + C \leq C. \tag{6.45}$$

Let $E_\varepsilon = [-1, 1] \cap [x_0 - \varepsilon, x_0 + \varepsilon]$. Then

$$\int_{E_\varepsilon} u_\delta(x,t)^{2-\delta} dx \leq C.$$

Letting $\delta \to 0$, we obtain

$$\int_{E_\varepsilon} u(x,t)^{2-\delta} dx \le C.$$

Since $u(\cdot,t) \in C^{1/2}$, $n \ge 4$, the similar argument as in Theorem 4.6.3 shows that $u(x,t) > 0$ for any $x \in E_\varepsilon$, $t > 0$. Therefore, $\text{supp}u(\cdot,t) \supset \text{supp}u_0$. The proof is complete. □

Theorem 4.6.8 *Let $0 \le u_0 \in H_0^1(-1,1)$ with $\text{supp}u_0 \subset (-1,1)$, and let $0 < n < 2$. Then the solution obtained above has the property of finite speed of disturbances, namely, there exists an increasing interval $[x_1(t), x_2(t)]$ with $x_1(t)$ and $x_2(t)$ being continuous and $\text{supp}u_0 \subset [x_1(0), x_2(0)]$, such that*

$$\text{supp}u(\cdot,t) \subset [x_1(t), x_2(t)].$$

Proof. The proof can be found in [YG] for $0 < n < 1$, and [BE4] for $0 < n < 2$. □

Remark 4.6.1 The result stated in the above theorem can be extended to two dimensional case. The readers may refer to [GY] for a result about such property of radial solutions. We may also discuss other properties of solutions. For example, we refer to [BS] for the discussion of travelling wave solutions. Moreover, there are several recent papers devoted to other properties of solutions, see for example, [BCH], [BF], [BHK], [BKO], [BPU], [FB], [GRU], [HP1], [HS], [HUL], [LAU2], [MW].

4.7 Cahn-Hilliard Equation with Degenerate Mobility

4.7.1 *Models with degenerate mobility*

In §4.4.4 and §4.4.5, we have discussed the Cahn-Hilliard equation

$$\frac{\partial u}{\partial t} + D\left[m(u)(kD^3u - DA(u))\right] = 0 \tag{7.1}$$

with mobility $m(u)$ being a constant or a positive function. In some cases in physics, the mobility may have zero points, in other words, the equation may have some points of degeneracy. A typical case is $m(u) = \dfrac{1}{3}u^3$, see Chapter 4 in [TAY]. In this case, the equation can be used to describe the diffusive process of an oil film spreading over a solid surface. Such equation also appears in the inter diffusion processes for the mixture of two phases, see for example [EG]. In this case, the mobility

$$m(u) = (1-u^2)^p \overline{m}(u), \quad p \geq 1$$

has two zero points $u = \pm 1$.

In more general case, the mobility $m(u)$ can be expressed by

$$m(u) = (1-u)(m_{BB}(u) - m_{AB}(u)) + u(m_{AA}(u) - m_{BA}(u)),$$

where the matrix

$$\begin{pmatrix} m_{AA}(u) & m_{AB}(u) \\ m_{BA}(u) & m_{BB}(u) \end{pmatrix}$$

characterizes physical parameters related to the two phases A and B, which is usually nonnegative definite, see [NS]. An extreme case is that the matrix is zero, namely, $m(u) \equiv 0$. If the matrix is diagonal, then $m(u)$ turns out to be

$$m(u) = 2um_{AA}(u) = 2(1-u)m_{BB}(u),$$

which has at least two zero points $u = 0$ and $u = 1$.

4.7.2 *Definition of physical solutions*

Consider the boundary value problem for the equation (7.1) with the following conditions

$$J\Big|_{x=0,1} \equiv m(u)(kD^3u - DA(u))\Big|_{x=0,1} = 0, \tag{7.2}$$

$$Du\Big|_{x=0,1} = 0, \tag{7.3}$$

$$u(x,0) = u_0(x). \tag{7.4}$$

Since the unknown function u in the equation (7.1) represents a physical quantity such as concentration, it is natural to call the solution u satisfying $0 \leq u(x,t) \leq 1$ a physical solution. To ensure the existence of physical solutions, we assume that

$$m(0) = m(1) = 0, \quad m(s) > 0 \quad \text{for } s \in (0,1). \tag{7.5}$$

We do not require other structure conditions for $m(s)$ and $A(s)$. However, for simplicity, we will assume that

$$m(s) = 0, \quad \text{for } s < 0 \quad \text{or } s > 1,$$

$$|A'(s)| \le \max_{0 \le s \le 1} |A'(s)| \equiv A_0, \quad \forall s \in \mathbb{R}.$$

Notice that these conditions only play an auxiliary role in the proof of the existence of solutions, and are really unnecessary, because the solutions obtained by means of parabolic regularization satisfy $0 \le u(x,t) \le 1$.

Definition 4.7.1 A function $u \in C^\alpha(\overline{Q}_T), \alpha \in (0,1)$ is said to be a generalized solution of the problem (7.1)–(7.4), if the following conditions are fulfilled:

(1) $u \in L^\infty(0,T; H^1(0,1))$, $D^3u \in L^2_{loc}(P)$,

$$\iint_P m(u)(D^3u)^2 dxdt < +\infty,$$

where

$$P = \{(x,t) \in \overline{Q}_T; 0 < u(x,t) < 1\};$$

(2) Du, D^3u are locally Hölder continuous in P, and

$$Du\Big|_{\Gamma \cap P} = D^3u\Big|_{\Gamma \cap P} = 0$$

holds in the usual sense, where

$$\Gamma = \{(0,t); t \in [0,T]\} \cup \{(1,t); t \in [0,T]\};$$

(3) For any $\varphi \in C^1(\overline{Q}_T)$,

$$-\int_0^1 u(x,T)\varphi(x,T)dx + \int_0^1 u_0(x)\varphi(x,0)dx$$
$$+ \iint_{Q_T} u\frac{\partial \varphi}{\partial t} dxdt + + \iint_P m(u)(kD^3u - DA(u))D\varphi dxdt = 0.$$

If u satisfies $0 \le u(x,t) \le 1$, then u is said to be a physical solution.

4.7.3 *Existence of solutions*

Theorem 4.7.1 *Let $m(s) \in C^{1+\alpha}(\mathbb{R})(\alpha \in (0,1))$, $A(s) \in C^1(\mathbb{R})$ and $u_0 \in C^{4+\alpha}(\bar{I})(I = (0,1))$, $D^i u_0(0) = D^i u_0(1) = 0(i = 1,3)$. Then the problem (7.1)–(7.4) admits at least one generalized solution.*

Proof. Let $m_\varepsilon(s)$, $a_\varepsilon(s)$ and u_0^ε be the smooth approximate functions of $m(s)$, $a(s) = A'(s)$ and u_0 respectively, $A_\varepsilon(s) = \int_0^s a_\varepsilon(s)ds$. Consider the regularized equation

$$\frac{\partial u_\varepsilon}{\partial t} + D[(m_\varepsilon(u_\varepsilon) + \varepsilon)(kD^3 u_\varepsilon - DA_\varepsilon(u_\varepsilon))] = 0, \tag{7.6}$$

with initial and boundary value conditions

$$Du_\varepsilon\Big|_{x=0,1} = D^3 u_\varepsilon\Big|_{x=0,1} = 0, \tag{7.7}$$

$$u_\varepsilon(x,0) = u_0^\varepsilon(x). \tag{7.8}$$

From the auxiliary assumption on $A(s)$, we see that there exists some constant $\mu \geq 0$, such that $a(s) \geq -\mu$. So, we may further assume

$$a_\varepsilon(s) \geq -\mu.$$

Thus

$$\begin{aligned} H_\varepsilon(s) \equiv \int_0^s A_\varepsilon(\tau)d\tau = & \int_0^s d\tau \int_0^\tau a_\varepsilon(\sigma)d\sigma \\ = & \int_0^s (s-\sigma)a_\varepsilon(\sigma)d\sigma \geq -\frac{1}{2}s^2\mu. \end{aligned}$$

Although $H_\varepsilon(s)$ does not satisfy (5.32), checking the proof of Theorem 4.5.3 we may find that, for

$$F_\varepsilon(t) = \int_0^1 \left(\frac{k}{2}(Du_\varepsilon)^2 + H_\varepsilon(u_\varepsilon) + \frac{1}{2}\mu u_\varepsilon^2\right) dx,$$

the following estimate holds

$$F_\varepsilon(t) \leq F_\varepsilon(0).$$

So, the conclusion of Theorem 4.5.3 is still valid, i.e., the regularized problem (7.6)–(7.8) admits a smooth solution u_ε. Moreover, an argument similar to that in Section 5 shows that the following estimates hold

$$\sup_{0<t<T}\int_0^1 (Du_\varepsilon(x,t))^2dx \le C, \tag{7.9}$$

$$\iint_{Q_T}(m_\varepsilon(u_\varepsilon)+\varepsilon)(D^3u_\varepsilon)^2dxdt \le C, \tag{7.10}$$

$$\left|\int_0^1 u_\varepsilon(x,t)dx\right| \le C. \tag{7.11}$$

The estimates (7.9) and (7.11) imply

$$\sup_{Q_T}|u_\varepsilon(x,t)| \le C, \tag{7.12}$$

$$|u_\varepsilon(x_1,t)-u_\varepsilon(x_2,t)| \le C|x_1-x_2|^{1/2}. \tag{7.13}$$

In addition, using the method in Section 5, we may also obtain

$$|u_\varepsilon(x,t_1)-u_\varepsilon(x,t_2)| \le C|t_1-t_2|^{1/8}. \tag{7.14}$$

Based on the above estimates and using the approaches in Section 6, we may conclude that the limit function u of a subsequence of $\{u_\varepsilon\}$ is a generalized solution of the problem (7.1)–(7.4). Here, we omit the details. The proof is complete. □

Remark 4.7.1 If we drop out the requirement (2) in the definition of generalized solutions, then the smoothness conditions in Theorem 4.7.1 can be weakened. In fact, the smoothness assumptions on $m(s)$ and u_0 can be replaced by $m(s)\in C^\alpha(\mathbb{R})$, $u_0\in H^1(I)$.

4.7.4 *Physical solutions*

Theorem 4.7.2 *In addition to the assumptions of Theorem 4.7.1, we assume that* $0\le u_0(x)\le 1$. *Then the generalized solution obtained in Theorem 4.7.1 by means of parabolic regularization is a physical solution, namely,* u *satisfies* $0\le u(x,t)\le 1$.

Proof. Let u be the generalized solution obtained in Theorem 4.7.1 by means of parabolic regularization. In what follows, we show that $0\le u(x,t)\le 1$. As an example, we show that $u(x,t)\ge 0$. We argue by contradiction. Assume that the set

$$E=\{(x,t)\in\overline{Q}_T;u(x,t)<0\}$$

is non empty. For convenience, suppose that $u(x,t)$ has been extended continuously to the whole plane, such that $u(x,t) \geq 0$ for $t < 0$, $u(x,t) = 0$ for $t \geq T+1$.

For any fixed $\delta > 0$, choose a C^∞ function $H_\delta(s)$, such that $H_\delta(s) = -\delta$ for $s \geq -\delta$, $H_\delta(s) = -1$ for $s \leq -2\delta$, and $H_\delta(s)$ is nondecreasing. Then we choose $\alpha(s)$ to be the kernel of a one-dimensional mollifier, namely, $\alpha(s) \in C^\infty(\mathbb{R})$, $\operatorname{supp}\alpha = [-1,1]$, $\alpha(s) > 0$ in $(-1,1)$, and

$$\int_{-1}^{1} \alpha(s)ds = 1.$$

For fixed $h > 0, \delta > 0$, set

$$u^h(x,t) = \int_{-\infty}^{\infty} u(x,s)\alpha_h(t-s)ds,$$
$$\beta_\delta(t) = \int_t^\infty \alpha\left(\frac{s-T/2}{T/2-\delta}\right)\frac{1}{T/2-\delta}ds,$$

where $\alpha_h(s) = \dfrac{1}{h}\alpha(\dfrac{s}{h})$.

Substituting

$$\varphi_\delta^h(x,t) \equiv [\beta_\delta(t)H_\delta(u^h)]^h,$$

as a test function, into the integral equality in the definition of generalized solutions, we obtain

$$\begin{aligned} &-\int_0^1 u(x,T)\varphi_\delta^h(x,T)dx + \int_0^1 u_0(x)\varphi_\delta^h(x,0)dx \\ &+\iint_{Q_T} u\frac{\partial}{\partial t}\varphi_\delta^h dxdt + \iint_P km(u)D^3uD\varphi_\delta^h dxdt \\ &-\iint_P m(u)DA(u)D\varphi_\delta^h dxdt = 0. \end{aligned} \tag{7.15}$$

To analyze the above integral equality, we first discuss the properties of the test function $\varphi_\delta^h(x,t)$. From the definition of $\beta_\delta(t)$, we see that

$$\varphi_\delta^h(x,t) = 0, \qquad t \geq T-\delta/2, h < \delta/2. \tag{7.16}$$

By the continuity of $u(x,t)$, there exists $\eta_1(\delta) > 0$, such that

$$u^h(x,t) \geq -\delta/2, \qquad t \leq \eta_1(\delta), 0 \leq x \leq 1, h < \eta_1(\delta). \tag{7.17}$$

Using the definitions of $\beta_\delta(t)$ and $H_\delta(s)$, we also have

$$H_\delta(u^h(x,t)) \ge -\delta, \qquad t \le \eta_1(\delta), 0 \le x \le 1, h < \eta_1(\delta) \tag{7.18}$$

and

$$\varphi_\delta^h(x,t) \ge -\delta, \qquad t \le \frac{1}{2}\eta_1(\delta), 0 \le x \le 1, h < \frac{1}{2}\eta_1(\delta). \tag{7.19}$$

Obviously, for any $f(t), g(t) \in L^2(\mathbb{R})$,

$$\begin{aligned} &\int_{-\infty}^{\infty} f(t)g^h(t)dt = \int_{-\infty}^{\infty} f(t)dt \int_{-\infty}^{\infty} g(s)\alpha_h(t-s)ds \\ = &\int_{-\infty}^{\infty} f(t)dt \int_{-\infty}^{\infty} g(s)\alpha_h(s-t)ds = \int_{-\infty}^{\infty} g(s)ds \int_{-\infty}^{\infty} f(t)\alpha_h(s-t)dt \\ = &\int_{-\infty}^{\infty} f^h(t)g(t)dt. \end{aligned}$$

Now, let $0 < h < \frac{1}{2}\eta_1(\delta)$. From (7.16), (7.18) and (7.19), we have

$$\begin{aligned} &\iint_{Q_T} u\frac{\partial}{\partial t}\varphi_\delta^h dxdt = \int_{-\infty}^{\infty} dt \int_0^1 u\left(\frac{\partial}{\partial t}(\beta_\delta(t)H_\delta(u^h))\right)^h dx \\ = &\iint_{Q_T} u^h \frac{\partial}{\partial t}(\beta_\delta(t)H_\delta(u^h))dxdt. \end{aligned}$$

Integrating by parts, we obtain

$$\begin{aligned} &\iint_{Q_T} u^h \frac{\partial}{\partial t}(\beta_\delta(t)H_\delta(u^h))dxdt \\ = &\int_0^1 u^h(x,T)\beta_\delta(T)H_\delta(u^h(x,T))dx- \\ &-\int_0^1 u^h(x,0)\beta_\delta(0)H_\delta(u^h(x,0))dx \\ &-\iint_{Q_T} \beta_\delta(t)H_\delta(u^h)\frac{\partial u^h}{\partial t}dxdt \\ = &\ \delta\int_0^1 u^h(x,0)dx - \iint_{Q_T} \beta_\delta(t)\frac{\partial}{\partial t}F_\delta(u^h)dxdt, \end{aligned}$$

where $F_\delta(s) = \int_0^s H_\delta(\sigma)d\sigma$. From (7.18), it follows that

$$\begin{aligned}
F_\delta(u^h(x,0)) &= \int_0^{u^h(x,0)} H_\delta(\sigma)d\sigma \\
&= \int_0^1 H_\delta(\lambda u^h(x,0))d\lambda \cdot u^h(x,0) = -\delta u^h(x,0).
\end{aligned}$$

Therefore

$$\begin{aligned}
&\iint_{Q_T} u^h \frac{\partial}{\partial t}(\beta_\delta(t)H_\delta(u^h))dxdt \\
=\ &\delta \int_0^1 u^h(x,0)dx + \int_0^1 \beta_\delta(0)F_\delta(u^h(x,0))dx + \\
&+ \iint_{Q_T} F_\delta(u^h)\beta_\delta'(t)dxdt \\
=\ &-\frac{1}{T/2-\delta}\iint_{Q_T} F_\delta(u^h)\alpha\left(\frac{t-T/2}{T/2-\delta}\right)dxdt.
\end{aligned} \tag{7.20}$$

On the other hand, from (7.16) and (7.19), it is easy to see that

$$-\int_0^1 u(x,T)\varphi_\delta^h(x,T)dx = 0, \tag{7.21}$$

$$\int_0^1 u_0(x)\varphi_\delta^h(x,0)dx = -\delta\int_0^1 u_0(x)dx. \tag{7.22}$$

Substituting (7.20), (7.21) and (7.22) into (7.15), we have

$$\begin{aligned}
&-\frac{2}{T-2\delta}\iint_{Q_T} F_\delta(u^h)\alpha\left(\frac{t-T/2}{T/2-\delta}\right)dxdt + \\
&-\delta\int_0^1 u_0(x)dx + \iint_P km(u)D^3uD\varphi_\delta^h dxdt - \\
&-\iint_P m(u)DA(u)D\varphi_\delta^h dxdt = 0.
\end{aligned} \tag{7.23}$$

On the other hand, from the continuity of $u(x,t)$, there exists $\eta_2(\delta) > 0$, such that

$$u(x,t) \geq -\delta/2, \qquad (x,t) \in P_\delta, \tag{7.24}$$

where

$$P_\delta = \{(x,t); \mathrm{dist}((x,t),P) < \eta_2(\delta)\}.$$

Here, we have used the fact that $u(x,t) > 0$ in P. Therefore

$$H_\delta(u^h(x,t)) = -\delta, \quad (x,t) \in P_{\delta/2}, 0 < h < \frac{1}{2}\eta_2(\delta),$$

where

$$P_{\delta/2} = \{(x,t); \mathrm{dist}((x,t),P) < \frac{1}{2}\eta_2(\delta)\},$$

which shows that, as long as $h < \frac{1}{2}\eta_2(\delta)$, $\varphi_\delta^h(x,t)$ is only a function of t in P. Therefore

$$D\varphi_\delta^h(x,t) = 0, \qquad (x,t) \in P, 0 < h < \frac{1}{2}\eta_2(\delta) \tag{7.25}$$

and (7.23) becomes

$$-\frac{2}{T-2\delta}\iint_{Q_T} F_\delta(u^h)\alpha\left(\frac{t-T/2}{T/2-\delta}\right) dxdt - \delta\int_0^1 u_0(x)dx = 0,\ 0 < h < \eta(\delta),$$

where $\eta(\delta) = \min(\eta_1(\delta), \eta_2(\delta))$. Letting $h \to 0$ leads to

$$-\frac{2}{T-2\delta}\iint_{Q_T} F_\delta(u)\alpha\left(\frac{t-T/2}{T/2-\delta}\right) dxdt - \delta\int_0^1 u_0(x)dx = 0. \tag{7.26}$$

From the definitions of $F_\delta(s)$ and $H_\delta(s)$, it is easy to see that

$$F_\delta(u(x,t)) \to -\chi_E(x,t)u(x,t), \qquad (\delta \to 0).$$

Thus letting $\delta \to 0$ in (7.26), we have

$$\iint_E |u(x,t)|\alpha\left(\frac{2t-T}{T}\right) dxdt = 0,$$

which contradicts the fact that $\alpha\left(\frac{2t-T}{T}\right) > 0$ for $0 < t < T$. Therefore E must be an empty set, and hence $u(x,t) \geq 0$. The proof is complete. □

Remark 4.7.2 The Cahn-Hilliard equation with degenerate mobility in multi-dimensional case has also been discussed by some authors. For example, we refer to [EG] for a study of the existence of weak solutions, [YL1], [LY] for the nonnegativity of solutions and the property of finite speed of disturbances in two dimensional case. For numerical approach, we refer to [BBG].

Bibliography

[AA1] Angenent, S. B. and Aronson, D. G., The focusing problem for the radially symmetric porous medium equation, Comm. Partial Differential Equations, 20(7-8)(1995), 1217–1240.

[AA2] Angenent, S. B. and Aronson, D. G., Self-similarity in the post-focussing regime in porous medium flows, European J. Appl. Math., 7(3)(1996), 277–285.

[AB] Abdulla, U. G., Local structure of solutions of the Dirichlet problem for N-dimensional reaction-diffusion equations in bounded domains, Adv. Differential Equations, 4(2)(1999), 197–224.

[AB1] Abdullaev, U. G., Local structure of solutions of the reaction-diffusion equations, Proceedings of the Second World Congress of Nonlinear Analysts, Part 5 (Athens, 1996), Nonlinear Anal., 30(5)(1997), 3153–3163.

[AB2] Abdullaev, U. G., Instantaneous shrinking of the support of a solution of a nonlinear degenerate parabolic equation, Mat. Zametki, 63(3)(1998), 323–331.

[AB3] Abdullaev, U. G., Instantaneous shrinking and exact local estimations of solutions in nonlinear diffusion absorption, Adv. Math. Sci. Appl., 8(1)(1998), 483–503.

[ABF] Alikakos N., Bates P. W. and Fusco G., Slow motion for the Cahn-Hilliard equation in one space dimension, Journal of Differential Equations, 90(1991), 81–135.

[AC] Aronson D. G. and Caffarelli L. A., The initial trace of a solution of the porous medium equation, Trans. Amer. Math. Soc., 280(1983), 351–366.

[ACK] Aronson D. G., Caffarelli L. A. and Kamin S., How an initially stationary interface begins to move in porous medium flow, SIAM J. Math. Anal., 14(1983), 639–658.

[ACP] Aronson D. G., Caffarelli L. A. and Peletier L. A., Stabilization of solutions of a degenerate nonlinear diffusion problem, Nonlinear Analysis, 6(10)(1982), 1001–1022.

[ACV] Aronson D. G., Caffarelli L. A. and Vázquez J. L., Interfaces with a corner point in one-dimensional porous medium flow, Comm. Pure Appl. Math. 38(1981), 375–404.

[AD] Andreucci D. and DiBenedetto E., On the Cauchy problem and initial traces for a class of evolution equation with strongly nonlinear sources, Estratto dagli Annali della Scuola Normale Superiore di Pisa, Sscienze Fisiche Mathematiche-Serie 6, 18(3)(1991), 363–441.

[ADA] Adams R. A., Sobolev Spaces, Academic Press, New York, 1975.

[AE] Alikakos N. D. and Evans L. C., Continuity of the gradient of the solutions of certain degenerate parabolic equation, J. Math. Pures et Appl., 62(1983), 253–268.

[AM] Alikakos N. D. and McKinney W. R., Remarks on the equilibrium theory for the Cahn-Hilliard equation in one space dimension, in "Reaction Diffusion Equations" (Brown K. J. and Lacey A. A., Eds), pp. 75–93, Oxford Science Publications, 1990.

[ANG] Angenent A., Analyticity of the interface of the porous media equation after waiting time, Research Report of Mathematical Institute of University of Leiden, No. 30, 1985.

[AR1] Aronson D. G., Regularity properties of flows through porous media, SIAM J. Appl. Math., 17(2)(1969), 461–467.

[AR2] Aronson D. G., Regularity properties of flows through porous media: a counterexample, SIAM J. Appl. Math., 19(2)(1970), 299–307.

[AR3] Aronson D. G., Regularity properties of flows through porous media: the interface, Arch. Rat. Mech. Anal., 37(1)(1970), 1–10.

[AV] Aronson, D. G., Gil, O. and Vázquez, J. L., Limit behaviour of focusing solutions to nonlinear diffusions, Comm. Partial Differential Equations, 23(1-2)(1998), 307–332.

[AVA] Aronson D. G. and Vázquez J. L, Eventual C^∞-regularity and concavity for flows in one-dimendsional porous media, Arch. Rat. Mech. Anal., 99(4)(1987), 329–348.

[BA1] Barenblatt G. I., On some unsteady motions of a liquid and a gas in a porous medium, Prikl. Mat. Meh., 16(1952), 67–78.

[BA2] Barenblatt G. I., On a class of exact solutions of the plane one-dimensional problem of unsteady filtration into a porous medium, Prik. Mat. Meh., 17(1953), 739–742.

[BA3] Barenblatt G. I., On limiting self-similar motions in the theory of unsteady filtration of a gas in a porous medium and the theory of the boundary layer, Prikl. Mat. Meh., 18(1954), 409–414.

[BAA] Betelú S. I., Aronson, D. G. and Angenent, S. B., Renormalization study of two-dimensional convergent solutions of the porous medium equation, Phys. D, 138(3-4)(2000), 344–359.

[BAD] Badii, M., Quasilinear parabolic equations with nonlinear flux boundary conditions and semigroups, NoDEA Nonlinear Differential Equations Appl., 5(3)(1998), 265–282.

[BBC] Bénilan Ph., Brézis H. and Crandall M. G., The continuous dependence on of solutions of $u_t - \Delta\varphi(u) = 0$, Indiana Univ. Math. J., 30(1981), 161–177.

[BBD] Boretta E., Bertsch M. and Dal Passo R., Nonnegative solutions of a fourth-order nonlinear degenerate parabolic equation, Arch. Rat. Mech. Anal., 129(1995), 175–200.

[BBG] Barrett, J. W., Blowey, J. F. and Garcke, H., Finite element approximation of the Cahn-Hilliard equation with degenerate mobility, SIAM J. Numer. Anal., 37(1)(1999), 286–318.

[BC] Brézis H. and Crandall M. G., Uniqueness of solutions of the initial value problem for $u_t - \Delta\varphi(u) = 0$, J. Math. Pures et Appl., 58(1979), 153–163.

[BCH] Bayada, G. and Chambat, M., On the various aspects of the thin film equation in hydrodynamic lubrication when the roughness occurs, Applications of multiple scaling in mechanics (Paris, 1986), 14–30, Rech. Math. Appl., 4, Masson, Paris, 1987.

[BCT] Bonafede, S., Cirmi, G. R. and Tedeev, A. F., Finite speed of propagation for the porous media equation with lower order terms, Discrete Contin. Dynam. Systems, 6(2)(2000), 305–314.

[BCP] Bénilan Ph., Crandall M. G. and Pierre M., Solutions of the porous medium equation under optimal conditions on initial values, Indiana Univ. Math. J., 33(1984), 51–87.

[BDG] Budd, C. J., Dold, J. W. and Galaktionov, V. A., Self-similar blow-up for a quasilinear parabolic equation with gradient diffusion and exponential source, Adv. Differential Equations, 2(1)(1997), 85–124.

[BEA] Bear J., Dynamics of fluids in porous media, New York, American Elsevier Publishing Company Inc., 1972.

[BE1] Bernis F., Compactness of the support for some nonlinear elliptic problems of arbitrary order in dimension n, Comm. Partial Differential Equations, 9(1984), 271–312.

[BE2] Bernis F., Qualitative properties for some nonlinear higher order degenerate parabolic equations, Houston J. Math., 14(3)(1988), 319–352.

[BE3] Bernis F., Source-type solutions of fourth order degenerate parabolic equations, Proc. Microprogram Nonlinear Diffusion Equations Equilibrium States, MSRI, Berkley, California (Edited by W. -M. Ni, L. A. Peletier and J. Serrin), Vol. I, pp. 123–146. Springer, New York(1988).

[BE4] Bernis F., Finite speed of propagations and continuity of the inteface for thin viscous flows, Advances in Diff. Eqs., 1(3)(1996), 337–368.

[BE5] Bernis F., The free boundary of thin viscous flows, Proc. ICIAM, 95, Z. Angew. Math. Mech., 1996.

[BEN] Bénilan Ph., Equations d'evolution dans un espaces de Banach quelconque et applications, Thesis, Ossay, 1972.

[BF] Bernis, F. and Ferreira, R., Source-type solutions to thin-film equations: the critical case, Appl. Math. Lett., 12(7)(1999), 45–50.

[BF1] Brézis H. and Friedman A., Nonlinear parabolic equations involving measure as initial conditions, J. Math. Pures et Appl., 62(1983), 73–97.

[BF2] Bernis, F. and Friedman A., Higher order nonlinear degenerate parabolic equations, Journal of Differential Equations, 83(1990), 179–206.

[BFU] Bénilan Ph. and Fujita H., Ed., Japan Society for the Promotion of Science, Tokyo, (1978), 15–53.

[BG] Bénilan, Ph. and Gariepy, R., Strong solutions in L^1 of degenerate parabolic equations, J. Differential Equations, 119(2)(1995), 473–502.

[BG1] Budd, C. and Galaktionov, V., Stability and spectra of blow-up in problems with quasi-linear gradient diffusivity, R. Soc. Lond. Proc. Ser. A, Math. Phys. Eng. Sci., 454(1977)(1998), 2371–2407.

[BHK] Bernis, F., Hulshof, J. and King, J. R., Dipoles and similarity solutions of the thin film equation in the half-line, Nonlinearity, 13(2)(2000), 413–439.

[BK1] Bürger, R. and Karlsen, K. H., Analysis and numerics of strongly degenerate convection-diffusion problems modelling sedimentation-consolidation processes, The mathematics of finite elements and applications, X, MAFELAP 1999 (Uxbridge), 215–224, Elsevier, Oxford, 2000.

[BK2] Bürger, R., Evje, S. and Karlsen, K. H., On strongly degenerate convection-diffusion problems modeling sedimentation-consolidation processes, J. Math. Anal. Appl. 247(2)(2000), 517–556.

[BKO] Boatto, S., Kadanoff, L. P. and Olla, P., Traveling-wave solutions to thin-film equations, Phys. Rev. E(3), 48(6)(1993), 4423–4431.

[BKP] Bertsch M., Kersner R. and Peletier L. A., Positivity versus localization in degenerate diffusion equations, Nonlinear Analysis, TMA, 9(9)(1985), 987–1008.

[BM] Bernis F. and McLeod J. B., Similarity solutions of a higher order nonlinear diffusion equation, Nonlinear Analysis, TMA, 17(1991), 1039–1068.

[BNS] Bandle, C., Nanbu, T. and Stakgold, I., Porous medium equation with absorption, SIAM J. Math. Anal., 29(5)(1998), 1268–1278.

[BP] Bertsch, M., Dal Passo, R., Hyperbolic phenomena in a strongly degenerate parabolic equation, Arch. Rat. Mech. Anal., 117(1992), 349–387.

[BPU] Bertozzi, A. L. and Pugh, M. C., Long-wave instabilities and saturation in thin film equations, Comm. Pure Appl. Math., 51(6)(1998), 625–661.

[BRC] Brooks R. H. and Covey A. T., Properties of porous media affecting fluid flow, Journal of the irrigation and drainage division, ASCE, 92(1966), 61–88.

[BS] Bertozzi, A. L. and Shearer, M., Existence of undercompressive traveling waves in Thin Film equations, SIAM J. Math. Anal., 32(1)(2000), 194–213.

[BU] Bürger, R., Ein Anfangs-Randwertproblem einer quasilinearen entarteten parabolischen Gleichung in der Theorie der Sedimentation mit Kompression, Dissertation, Universität Stuttgart, Stuttgart, 1996, Universität Stuttgart, Stuttgart, 1996, 118 pp.

[BV] Barenblatt, G. I. and Vazquez, J. L., A new free boundary problem for unsteady flows in porous media, European J. Appl. Math., 9(1)(1998), 37–54.

[BW1] Bürger, R. and Wendland, W. L., Entropy boundary and jump conditions in the theory of sedimentation with compression, Math. Methods Appl. Sci., 21(9)(1998), 865–882.

[BW2] Bürger, R. and Wendland, W. L., Existence, uniqueness, and stability of generalized solutions of an initial-boundary value problem for a degenerating quasilinear parabolic equation, J. Math. Anal. Appl., 218(1)(1998), 207–239.

[BWC] Bürger, R., Wendland, W. L. and Concha, F., Model equations for gravitational sedimentation-consolidation processes, ZAMM Z. Angew. Math. Mech., 80(2)(2000), 79–92.

[C1] Cheng, F., Regularity results for a strongly degenerate parabolic equation, J. Partial Differential Equations, 10(3)(1997), 275–283.

[C2] Cheng, F., The hyperbolic properties of a class of strongly degenerate parabolic equations, Acta Math. Appl. Sinica, 21(1)(1998), 22–32.

[CA] Campanato S., Propriety di Holderianita di alcuna classi di funzioni, Ann. Sc. Norm. Sup. Pisa, 17(1963), 175–188.

[CAR] Carrillo, Jos Entropy solutions for nonlinear degenerate problems, Arch. Ration. Mech. Anal., 147(4)(1999), 269–361.

[CB] Concha, F. and Bürger, R., Wave propagation phenomena in the theory of sedimentation: mathematical theory of gravitational solid-liquid separation processes, Numerical methods for wave propagation (Manchester, 1995), 173–196, Fluid Mech. Appl., 47, Kluwer Acad. Publ., Dordrecht, 1998.

[CC] Cho, C., Nonnegative weak solutions of a porous medium equation with strong absorption, Differential Integral Equations, 11(6)(1998), 847–874.

[CD1] Chen Yazhe and DiBenedetto E., Boundary estimates for solutions of nonlinear degenerate parabolic systems, J. Reine Angew Math, 395(1989), 102–131.

[CD2] Chen Yazhe and DiBenedetto E., On the local behavior of solutions of singular parabolic equations, Arch. Rat. Mech. Anal., 103(1988), 319–345.

[CD3] Chen Yazhe and DiBenedetto E., Hölder estimates of solutions of singular parabolic equations with measurable coefficients, Arch. Rat. Mech. Anal., 118(1992), 251–271.

[CD4] Chen Yazhe and DiBenedetto E., On the Harnack inequality for nonnegative solutions of singular parabolic equations, Proc. of Conf. Nonlinear Diffusion in Honour of J. Serrin, Minneapolis, May, 1990.

[CDE] Cortázar, C., del Pino, M. and Elgueta, M., On the blow-up set for $u_t = \Delta u^m + u^m$, $m > 1$, Indiana Univ. Math. J., 47(2)(1998), 541–561.

[CE] Caffarelli L. A. and Evans L. C., Continuity of the temperature for the two phase Stefan problem, Arch. Rat. Mech. Anal., 81(1983), 199–220.

[CF1] Caffarelli L. A. and Friedman A., Continuity of the density of a gas flow in a porous medium, Tran. Amer. Math. Soc., 252(1979), 99–113.

[CF2] Caffarelli L. A. and Friedman A., Regularity of the free boundary of a gas flow in n−dimension porous medium, Indiana Univ. Math. J. 29(1980), 361–391.

[CGS] Carr J., Gurtin M. and Slemrod M., Structured phase transitions on a finite interval, Arch. Rat. Mech. Anal., 86(1984), 317–351.

[CH] Cahn J. W. and Hilliard J. E., Free energy of a nonuniform system, I. Interfacial free energy, J. Chem. Phys., 28(1958), 258–267.

[CH1] Chen Yazhe, Uniqueness of weak solutions of quasilinear degenerate parabolic equations, The Proc. of the 1982 Changchun Sym. on Diff. Geom. and Diff. Eqs., 317–332.

[CH2] Chen Yazhe, Hölder estimates for solutions of uniformly degenerate quasilinear parabolic equations, Chin. Ann. of Math., 5B(4)(1984), 661–678.

[CH3] Chen Yazhe, Hölder continuity of the gradient of the solutions of certain degenerate parabolic equations, Chin. Ann. Math., 8B(3)(1987), 343–356.

[CH4] Chen Yazhe, Hölder continuity of the gradients of solutions of nonlinear degenerate parabolic systems, Acta Math. Sinica, New series, 2(4)(1986), 309–331.

[CH5] Chen Yazhe, On finite diffusing speed for uniformly degenerate quasilinear parabolic equation, Chin. Ann. Math., 7B(3)(1990), 318–329.

[CH6] Chen Yazhe, Global Lipschitz continuity of free boundaries in the one-phase Stefan problem, Ann. Mat. Pura Appl., 142(4)(1985), 197–213.

[CHO] Cho, C. and Choe, H. J., The asymptotic behaviour of solutions of a porous medium equation with bounded measurable coefficients, J. Math. Anal. Appl., 210(1)(1997), 241–256.

[CM] Cohen D. S. and Murray J. D., A generalized diffusion model for growth and dispersal in a population, J. Math. Biology, 12(1981), 237–249.

[CV] Caffarelli, L. and Vazquez, J. Viscosity solutions for the porous medium equation, Differential equations: La Pietra, 1996, (Florence), 13–26, Proc. Sympos. Pure Math., 65, Amer. Math. Soc., Providence, RI, 1999.

[CVW] Caffarelli L. A., Vázquez J. L. and Wolanski N. I., Lipschitz continuity of solutions and interfaces of the N-dimensional porous medium equation, Indiana Univ. Math. J., 36(2)(1987), 373–401.

[CVW1] Chaves, M., Vazquez, J. L. and Walias, M., Optimal existence and uniqueness in a nonlinear diffusion-absorption equation with critical exponents, Proc. Roy. Soc. Edinburgh Sect. A, 127(2)(1997), 217–242.

[CW] Caffarelli L. A. and Wonlanski N. I., $C^{1,\alpha}$ regularity of the free boundary for the N-dimensional porous media equation, Comm. Pure Appl. Math., 43(1990), 885–902.

[CX] Chen, Xinfu, Global asymptotic limit of solutions of the Cahn-Hilliard equation, J. Differential Geom, 44(2)(1996), 262–311.

[CZ] Chen Yazhe and Zhou Shulin, A note on three classes of nonlinear parabolic equations with measurable coefficients, J. Partial Diff. Eqs., 8(4)(1995), 310–320.

[D1] DiBenedetto E., On the local behaviour of solutions of degenerate parabolic equations with measurable coefficients, Ann. Supola Norm. Sup. Pisa Ci Sci. Serie IV, XIII, 3(1986), 487–535.

[D2] DiBenedetto E., Degenerate parabolic equations, Springer-Verlag 1993, New York, Inc.

[D3] DiBenedetto, E., Continuity of weak solutions to a general porous medium equation, Indiana Univ. Math. J., 32(1)(1983), 81–118.

[DA] Daskalopoulos, P., The Cauchy problem for variable coefficient porous medium equations, Potential Anal., 7(1)(1997), 485–516.

[DAD] Da Prato, G. and Debussche, A., Stochastic Cahn-Hilliard equation, Nonlinear Anal., 26(2)(1996), 241–263.

[DD] Diáz G. D. and Diáz I. D., Finite extinction time for a class of nonlinear parabolic equations, Comm. PDE, 8(1982), 124–135.

[DD1] Daskalopoulos, P. and del Pino, M., On the Cauchy problem for $u_t = \Delta \log u$ in higher dimensions, Math. Ann., 313(2)(1999), 189–206.

[DD2] Daskalopoulos, P. and del Pino, M., Recent progress on super-diffusive nonlinear parabolic problems, Reaction diffusion systems (Trieste, 1995), 107–119, Lecture Notes in Pure and Appl. Math., 194, Dekker, New York, 1998.

[DD3] Daskalopoulos, P. and del Pino, M. A., On a singular diffusion equation, Comm. Anal. Geom., 3(3-4)(1995), 523–542.

[DD4] Daskalopoulos, P. and del Pino, M. A., On fast diffusion nonlinear heat equations and a related singular elliptic problem, Indiana Univ. Math. J., 43(2)(1994), 703–728.

[DD5] Daskalopoulos, P. and Del Pino, M., On nonlinear parabolic equations of very fast diffusion, Arch. Rational Mech. Anal., 137(4)(1997), 363–380.

[DF] DiBenedetto E. and Friedman A., Hölder estimates for nonlinear degenerate parabolic systems, J. Reine Angew. Math., 357(1985), 1–22.

[DGLV] Dold, J. W., Galaktionov, V. A., Lacey, A. A. and Vázquez, J. L., Rate of approach to a singular steady state in quasilinear reaction-diffusion equations, Ann. Scuola Norm. Sup. Pisa Cl. Sci.,

[DH1] DiBenedetto E. and Herrero M. A., On the Cauchy problem and initial traces for a degenerate parabolic equation, Trans. Amer. Soc., 314(1)(1989), 187–224.

[DH2] DiBenedetto E. and Herrero M. A., Nonnegative solutions of the evolution p-Laplacian equation, Initial traces and Cauchy problem when

$1 < p < 2$, Arch. Rat. Mech. Anal., 111 (2)(1990), 225–290.

[DHA1] Daskalopoulos, P. and Hamilton, R., Regularity of the free boundary for the porous medium equation, J. Amer. Math. Soc., 11(4)(1998), 899–965.

[DHA2] Daskalopoulos, P. and Hamilton, R., The free boundary for the n-dimensional porous medium equation, Internat. Math. Res. Notices, (17)1997, 817–831.

[DHA3] Daskalopoulos, P. and Hamilton, R., C^∞-regularity of the interface of the evolution p-Laplacian equation, Math. Res. Lett., 5(5)(1998), 685–701.

[DI] Diller, D. J., A remark on the continuous dependence on ϕ of solutions to $U_T - \Delta\phi(U) = 0$, Differential Integral Equations, 11(3)(1998), 425–438.

[DK] Dahlberg B. E. J. and Kenig C. E., Nonnegative solutions of the porous medium equation, Comm. P. D. E., 9(1984), 409–437.

[DO] Dong Guangchang, Nonlinear Partial Differential Equations of Second Order, Tsinghua Univ. Press, Beijing, 1988.

[DP] Díaz, J. I. and Padial, J. F., Uniqueness and existence of solutions in the $\mathrm{BV}_t(Q)$ space to a doubly nonlinear parabolic problem, Publ. Mat., 40(2)(1996), 527–560.

[DS] de Pablo, A. and Sánchez, A., Travelling wave behaviour for a porous-Fisher equation, European J. Appl. Math., 9(3)(1998), 285–304.

[DT] Diaz, J. I. and de ThÉLin, F., On a nonlinear parabolic problem arising in some models related to turbulent flows, SIAM J. Math. Anal., 25(4)(1994), 1085–1111.

[DU1] Dubinskii, Yu. A., Weak convergence of nonlinear ellptic and parabolic equations, Mat. Sb., 67(107)(1965), 609–642.

[DU2] Dubinskii, Yu. A., Quasilinear ellptic-parabolic equations, Mat. Sb., 77(3)(1968), 354–389.

[DUY] Van Duyn C. J. and Ye Qixiao, The flow of two immiscible fluids through a porous medium, J. Nonlinear Analysis, TMA, 8(4)(1984), 353–367.

[DY] Dong Guangchang and Ye Qixiao, On the uniqueness of solutions of nonlinear degenerate parabolic equations, Chin. Ann. Math., 3(1982), 279–284.

[EB] Ebmeyer, C. K., Finiter Elemente für die Poröse-Medien-Gleichung im R^n, Dissertation, Rheinische Friedrich-Wilhelms-Universität zu Bonn, Bonn, 1995. Bonner Mathematische Schriften [Bonn Mathematical Publications], 287. Universität Bonn, Mathematisches Institut, Bonn, 1996. 69 pp.

[EF] Elliott C. M. and French D. A., Numerical studied of the Cahn-Hilliard equation for phase separation, IMA J. Appl. Math., 38(1987), 97–128.

[EFG] Evans, L. C., Feldman, M. and Gariepy, R. F., Fast/slow diffusion and collapsing sandpiles, J. Differential Equations, 137(1)(1997), 166–209.

[EG] Elliott C. M. and Garcke H., On the Cahn-Hilliard equation with degenerate mobility, SIAM J. Math. Anal., 27(2)(1996), 404–423.

[EK1] Evje, S. and Karlsen, K. H., Monotone difference approximations of BV solutions to degenerate convection-diffusion equations, SIAM J. Numer. Anal., 37(6)(2000), 1838–1860.

[EK2] Evje, S. and Karlsen, K. H., Discrete approximations of BV solutions to doubly nonlinear degenerate parabolic equations, Numer. Math., 86(3)(2000), 377–417.

[EK3] Evje, S. and Karlsen, K. H., Degenerate convection-diffusion equations and implicit monotone difference schemes, Hyperbolic problems: theory, numerics, applications, Vol. I (Zürich, 1998), 285–294, Internat. Ser. Numer. Math., 129, Birkhäuser, Basel, 1999.

[EK4] Evje, S. and Karlsen, K. H., Viscous splitting approximation of mixed hyperbolic-parabolic convection-diffusion equations, Numer. Math., 83(1)(1999), 107–137.

[EK5] Evje, S. and Karlsen, K. H., Monotone difference approximation of BV solutions to degenerate convection-diffusion equations, Report, 115. University of Bergen, Department of Applied Mathematics, Bergen, 1998, 21 pp.

[EM] Elliott C. M. and Mikelic A., Existence for the Cahn-Hilliard phase separation model with a nondifferentiable energy, Annali di Matematica pura ed applicata, (IV)(CLVIII)(1991), 181–203.

[ERV] Esteban J. R., Rodriguez A. and Vazquez J. L., A nonlinear heat equation with singular diffusivity, Comm. PDE, 13(8)(1988), 985–1039.

[EV] Evans L. C., Application of nonlinear semigroups theory to certain partial differential equations, in Nonlinear Evolution Equations (ed. M. C. Crandall), Acad. Press, 1978.

[EY] Eyre D. J., Systems of Cahn-Hilliard equations, SIAM J. Appl. Math., 53(6) (1993), 1686–1712.

[EZ] Elliott C. M. and Zheng S. M., On the Cahn-Hilliard equation, Arch. Rat. Mech. Anal., 96(4)(1986), 339–357.

[FB] Ferreira, R. and Bernis, F., Source-type solutions to thin-film equations in higher dimensions, European J. Appl. Math., 8(5)(1997), 507–524.

[FIL] Filo J., Uniform bounds for solutions of a degenerate diffusion equation with nonlinear boundary conditions, Comment. Math. Univ. Carolinae, 30(3)(1989), 485–495.

[FK] Friedman A. and Kamin S., The asymptotic behaviour of gas in an n-dimensional porous medium, Trans. Amer. Math. Soc., 262(1980), 551–563.

[FR1] Friedman A., Partial Differential Equations of Parabolic Type, Prentice-Hall, Inc., 1964.

[FR2] Friedman A., Variational Principles and Free-boundary Problems, A Wiley-Interscience Publication, New York, 1982.

[FU] Fujii, A. and Ohta, M., Asymptotic behavior of blowup solutions of a parabolic equation with the p-Laplacian, Nonlinear evolution equations and their applications, 966(1996), 136–150.

[G] Giaquinta M., Multiple integrals in the calculus of variations and nonlinear elliptic systems, Princeton Univ. Press. Princeton NJ., 1983.

[GG] Gilding, B. H. and Goncerzewicz, J., Localization of solutions of exterior domain problems for the porous media equation with radial symmetry, SIAM J. Math. Anal., 31(4)(2000), 862–893.

[GHY] Gao Hang and Yin Jingxue, On a class of antistrophic diffusion equations, Sys. Sci. Math. Sci., 8(4)(1995), 311–318.

[GI] Gilding, B. H., A nonlinear degenerate parabolic equation, Annali della Scuola Norm. Sup. di Pisa, 4(3)(1977), 393–432.

[GIK] Gilding, B. H. and Kersner, R., A necessary and sufficient condition for finite speed of propagation in the theory of doubly nonlinear degenerate parabolic equations, Proc. Roy. Soc. Edinburgh Sect. A, 126(4)(1996), 739–767.

[GK] Galaktionov, V. A. and King, J. R., On the behaviour of blow-up interfaces for an inhomogeneous filtration equation, IMA J. Appl. Math., 57(1)(1996), 53–77.

[GM] Gmira A., On quasilinear parabolic equations involving measure data, Asymptotic Analysis, 3(1990), 43–56, North-Holland.

[GMC] Greenspan H. P. and McCay B. M., On the wetting of a surface by a very viscous fluid, Stud. Appl. Math., 64(1981), 95–112.

[GNI] Grinfeld, M. and Novick-Cohen, A. The viscous Cahn-Hilliard equation: Morse decomposition and structure of the global attractor, Trans. Amer. Math. Soc., 351(6)(1999), 2375–2406.

[GP] Galaktionov, V. A. and Peletier, L. A., Asymptotic behaviour near finite-time extinction for the fast diffusion equation, Arch. Rational Mech. Anal., 139(1)(1997), 83–98.

[GP1] Gilding B. H. and Peletier L. A., On a class of similarity solutions of the porous media equation, J. Math. Anal. Appl., 55(1976), 351–364.

[GP2] Gilding B. H. and Peletier L. A., The Cauchy problem for an equation in the theory of infiltration, Arch. Rat. Mech. Anal., 61(1976), 127–140.

[GPV] Galaktionov, V. A., Peletier, L. A. and Vazquez, J. L., Asymptotics of the fast-diffusion equation with critical exponent, SIAM J. Math. Anal., 31(5)(2000), 1157–1174.

[GR] Greenspan H. P., On the motion of a small viscous droplet that wets a surface, J. Fluid Mech., 84(1978), 125–143.

[GRA] Gránge O., Sur la Résolution d'une équation et d'une lnéquation paraboliques non linéaires, J. Fun. Anal., 11(1972), 77–92.

[GRU] Grün, G. and Rumpf, M., Nonnegativity preserving convergent schemes for the thin film equation, Numer. Math., 87(1)(2000), 113–152.

[GS] Glushko, V. P. and Savchenko, Yu. B., Higher-order degenerate elliptic equations: Spaces, Operators, Boundary value problems, J. Soviet Math. 39(6)(1987), 3088–3148.

[GSV1] Galaktionov, V. A., Shmarev, S. I. and Vazquez, J. L., Second-order interface equations for nonlinear diffusion with very strong absorption, Commun. Contemp. Math., 1(1)(1999), 51–64.

[GSV2] Galaktionov, V. A., Shmarev, S. I. and Vazquez, J. L., Regularity of solutions and interfaces to degenerate parabolic equations, The intersection comparison method. Free boundary problems: theory and applications (Crete, 1997), 115–130, Chapman & Hall/CRC Res. Notes Math., 409, Chapman & Hall/CRC, Boca Raton, FL, 1999. 35K65

[GSV3] Galaktionov, V. A., Shmarev, S. I. and Vazquez, J. L., Regularity of interfaces in diffusion processes under the influence of strong absorption, Arch. Ration. Mech. Anal., 149(3)(1999), 183–212.

[GU] Guo, J. S. and Sun, I. M., Remarks on a singular diffusion equation, Nonlinear Anal., 27(10)(1996), 1109–1115.

[GV] Gratton, J. and Vigo, C., Evolution of self-similarity, and other properties of waiting-time solutions of the porous medium equation: the case of viscous gravity currents, European J. Appl. Math., 9(3)(1998), 327–350.

[GVA] Gil, O. and Väzquez, J. L., Focusing solutions for the p-Laplacian evolution equation, Adv. Differential Equations, 2(2)(1997), 183–202.

[GW] Guo, Boling and Wang, Youde, Global attractors for the initial value problems of Cahn-Hilliard equations on compact manifolds, Sinica (N.S.), 12(3)(1996), 314–325.

[GY] Gao Wenjie and Yin Jingxue, Existence of physical solutions to the boundary value problem of an equation in modeling oil film spreading over a solid surface, J. Partial Differential Equations, 11(4)(1998), 335–346.

[GYY] Gao Yuhuan and Yin Jingxue, Classical solutions for a nonlinear degenerate parabolic equations, Northeastern Math. J., 9(2)(1993), 267–274.

[HA] Hardy G. H., Littlewood J. E. and Pólya G., Inequalities, Cambridge University Press, Cambridge, 1952.

[HE] Herrero M. A., On the growth of the interfaces of a nonlinear degenerate parabolic equation, in: Research Notes in Math. 89, Pitman, Boston, MA-London 1983, 218–224.

[HK] Ham, Y. and Ko, Y., C-infinity interfaces of solutions for one-dimensional parabolic p-Laplacian equations, Electron. J. Differential Equations, (1)1999, 12 pp.

[HKL] Hazewinkel M., Kaashoek J. F. and Leynse B., Pattern formation for a one dimensional evolution equation based on Thom's river basin model, Report #8519/B, Econometric Institute, Erasmus University (1985).

[HKR] Höllig K. and Kreiss H. O., C^∞-Regularity for the porous medium equation, Math. Z., 192(1984), 217–224.

[HO] Hocking L. M., Sliding and spreading of thin two-dimensional drops, Quart. J. Mech. Appl. Math., 34(1981), 37–55.

[HP] Herrero M. A. and Pierre M., The Cauchy Problem for $u_t = \Delta u^m$

($0 < m < 1$), Trans. Amer. Math. Soc., 291(1985), 145–158.

[HP1] Hastings, S. P. and Peletier, L. A., On self-similar solutions of the thin film equation when $n = 3$, Differential Integral Equations, 11(1)(1998), 1–22.

[HPI] Höllig K. and Pilant M., Regularity of the free boundary for the porous medium equation, Indianan Univ. Math. J., 34(1985), 723–732.

[HS] Hulshof, J. and Shishkov, A. E., The thin film equation with $2 \leq n < 3$: finite speed of propagation in terms of the L^1-norm, Adv. Differential Equations, 3(5)(1998), 625–642.

[HU] Huang Qingbo, Boundary value problem for a class of degenerate and singular nonlinear diffusion equations, Northeastern Math. J., 5(2)(1989), 155–167.

[HU1] Hui, K. M., Existence of solutions of the equation $u_t = \Delta \log u$, Nonlinear Anal., 37(7)(1999), Ser. A: Theory Methods, 875–914.

[HU2] Hui, K. M., Nonexistence of fundamental solutions of the equation $u_t = \Delta \ln u$, J. Math. Anal. Appl., 182(3)(1994), 800–809.

[HU3] Hui, K. M., Nonnegative solutions of the fast diffusion equation with strong reaction term, Nonlinear Anal., 19(12)(1992), 1155–1178.

[HU4] Hui, K. M., Singular limit of solutions of the equation $u_t = \Delta(u^m/m)$ as $m \to 0$, Pacific J. Math., 187(2)(1999), 297–316.

[HU5] Hui, K. M., Singular limit of solutions of the porous medium equation with absorption, Trans. Amer. Math. Soc., 350(11)(1998), 4651–4667.

[HU6] Hui, K. M., Singular limit of solutions of the generalized p-Laplacian equation, Nonlinear Anal., 24(9)(1995), 1327–1345.

[HU7] Hui, K. M., Singular limit of solutions of $u_t = \Delta u^m - A \cdot \nabla(u^q/q)$ as $q \to \infty$, Trans. Amer. Math. Soc., 347(5)(1995), 1687–1712.

[HU8] Hui, K. M., Asymptotic behaviour of solutions of $u_t = \Delta u^m - u^p$ as $p \to \infty$, Nonlinear Anal., 21(3)(1993), 191–195.

[HU9] Hui, K. M., Asymptotic behaviour of solutions of $u_t = \Delta \log u$ in a bounded domain, Differential Integral Equations, 14(2)(2001), 175–188.

[HUL] Hulshof, J., A local analysis of similarity solutions of the thin film equation, Nonlinear analysis and applications, 179–192, GAKUTO Internat. Ser. Math. Sci. Appl., 7, Gakkōtosho, Tokyo, 1996.

[HV] Herrero M. A. and Vázquez J. L., On the propagation properties of a nonlinear degenerate parabolic equation, Comm. PDE, 7(1982), 1381–1402.

[HY] Hayasida, K. and Yamashiro, T., An ill-posed estimate of positive solutions of a degenerate nonlinear parabolic equation, Tokyo J. Math., 19(2)(1996), 331–352.

[I] Ivanov A. V., Hölder estimates for quasilinear doubly degenerate parabolic equations, J. Soviet Math., 56(2)(1991), 2320–2347.

[IR] Irmay S., On the hydraulic conductivity of unsaturated soils, Trans. Amer. Geophys Union, 35(1972), 463–467.

[IV1] Ivanov, A. V., Quasilinear degenerate and nonuniformly elliptic and parabolic equations of second order, American Mathematical Society, Providence, Rhode Island, 1982.

[IV2] Ivanov, A. V., The regularity theory for (M, L)-Laplacian parabolic equation, Zap. Nauchn. Sem. S.-Peterburg. Otdel. Mat. Inst. Steklov. (POMI), 243 (1997), Kraev. Zadachi Mat. Fiz. i Smezh. Vopr. Teor. Funktsii., 28,87–110, 339; translation in J. Math. Sci., 99(1)(2000), 854–869.

[JA] Jäger, W. and Lu, Yunguang, Global regularity of solution for general degenerate parabolic equations in 1-D, J. Differential Equations, 140(2)(1997), 365–377.

[JI] Jiang Lishang, Regularity and Uniqueness of generalized solutions of quasilinear degenerate parabolic equations, Acta Mathematica Sinica, 29(1)(1986), 1–9.

[KA1] Kalashnikov A. S., The nature of the propagation of perturbations in problems of nonlinear heat conduction with absorption, (Russian), Ž. Vyčisl. Mat. i Mat. Fiz., 14(1974), 891–905,

[KA2] Kalashnikov A. S., The differential properties of the generalized solutions of equations of the type of nonstationary filtration (Russian), Collection of articles dedicated to the memory of Ivan Georgievič Petrovskii (Russian) Vestnik Moskov. Univ. Ser. I Mat. Meh., 29(1)(1974), 62–68.

[KA3] Kalashikov A. S., The Cauchy problem in the class of increasing functions for equations of non-stationary filtration type (Russian) Vestnik Moskov. Univ. Ser. I Mat. Meh., 47(6)(1963), 17–27.

[KA4] Kalashnikov, A. S., Cauchy problem for second-order degenerate parabolic equations with nonpower nonlinearities, J. Soviet Math., 33(3)(1986), 1014–1025.

[KAM] Kamin S.,Source-type solutions for equations of nonstationary filtration, J. Math. Anal. Appl., 64(1978), 263–276.

[KAM2] Kamin, S., Heat propagation in an inhomogeneous medium, Progress in partial differential equations: the Metz surveys, 4, 229–237, Pitman Res. Notes Math. Ser., 345, Longman, Harlow, 1996.

[KAM3] Bertsch, M. and Kamin, S., The porous media equation with nonconstant coefficients, Adv. Differential Equations, 5(1-3)(2000), 269–292.

[KAT] Kath R., Waiting and propagating fronts in nonlinear diffusion, Physica, 12D(1984), 375–381.

[KBDE] Karlsen, K. H., Brusdal, K., Dahle, H. K., Evje, S. and Lie, K.-A., The corrected operator splitting approach applied to a nonlinear advection-diffusion problem, Comput. Methods Appl. Mech. Engrg., 167(3-4)(1998), 239–260.

[KE1] Kersner R., Nonlinear heat conduction with absorption: space localization and extinction in finite time, SIAM J. Appl. Math., 43(1983), 1274–1285.

[KE2] Kersner R., Conditions for the localization of heat perturbations in a semibounded moving medium in the presence of absorption (Russian), Vestnik Moskov. Univ. Ser. I Mat. Meh., 31(4)(1976), 52–58.

[KI1] King J. R., Mathematical aspects of semiconductor process modelling, Thesis, University of Oxford, 1986.

[KI2] King J. R., The isolation oxidation of silicon: the reaction-controlled case, SIAM J. Appl. Math., 49(1989), 1064–1080.

[KJ] Kjellberg B., On some inequalities, C. R. Dixième Congrès des Mathématiciens Scandinaves 1946, Jul. Gjellerups Forlag, Copenhagen, 1947, 333–340.

[KM1] Kwak, M., A porous media equation with absorption. II. Uniqueness of the very singular solution, J. Math. Anal. Appl., 223(1)(1998), 111–125.

[KM2] Kwak, M., A porous media equation with absorption. I. Long time behaviour, J. Math. Anal. Appl., 223(1)(1998), 96–110.

[KN] Knerr B. F., The porous medium equation in one dimension, Trans. AMS, 234(2)(1977), 381–414.

[KO] Kosygina, E. R., On the propagation of perturbations in two-dimensional porous media, Vestnik Moskov. Univ. Ser. I Mat. Mekh., (3)1995, 36–40, 102 translation in Moscow Univ. Math. Bull., 50(3)(1995), 32–35.

[KP1] Kamin S. and Peletier L. A., Source type solutions of degenerate diffusion equations with absorption, Israel J. Math., 50(1985), 219–230.

[KP2] Kamin S. and Peletier L. A., Large time behaviour of solutions of the porous media equation with absorption, Israel J. Math, 55(1986), 129–146.

[KRI] Krickeberg K., Distributionen, Funktionen beschränkter variation und Lebesguesher Inhalt nichtparametrischer Flächen., Ann. Mat. pura ed Appl., 44(1957).

[KRU] Kruzhkov, S. N., First order quasilinear equations with several independent variables (Russian) Mat. Sb. (N. S.), 81(2)(1970), 228–255.

[KY1] Ko, Y., $C^{1,\alpha}$ regularity of interfaces for solutions of the parabolic p-Laplacian equation, Comm. Partial Differential Equations, 24(5-6)(1999), 915–950.

[KY2] Ko, Y., $C^{1,\alpha}$ regularity of interface of some nonlinear degenerate parabolic equations, Nonlinear Anal., 42(7)(2000), Ser. A: Theory Methods, 1131–1160.

[L] Ladyzhenskaja O. A., New equation for the description of incompressible fluids and solvability in the large boundary value for them, Proc. Steklov Inst. Math., 102(1967), 95–118. (Russian)

[LA] Laptev, G. I., Solvability of second-order quasilinear parabolic equations with double degeneration, Sibirsk. Mat. Zh., 38(6)(1997), 1335–1355, iii translation in Siberian Math. J., 38(6)(1997), 1160–1177.

[LA1] Lacey A. A., Initial motion of the free boundary for a nonlinear diffusion equation, IMA J. Appl. Math., 42(1982), 113–119.

[LAC] Lacey A. A., The motion with slip of a thin viscous droplet over a solid surface, Stud. Appl. Math., 67 (1982), 217–230.

[LAU1] Laurençot, Ph., A remark on the uniqueness of fundamental solutions to the p-Laplacian equation, $p > 2$, Portugal. Math., 55(4)(1998), 385–389.

[LAU2] Laurençot, Ph., Degenerate Cahn-Hilliard equation as limit of the phase-field equation with non-constant thermal conductivity, Free Boundary Problems, Theory and Applications (Zakopane, 1995), 135–144, Pitman Res. Notes Math. Ser.363, Longman, Harlow, 1996.

[LE] Leoni, G., On very singular self-similar solutions for the porous media equation with absorption, Differential Integral Equations, 10(6)(1997), 1123–1140.

[LH1] Li Huilai, Free boundary problem for degenerate parabolic equations, Lecture Notes in Math., 1306, 102–130, Springer-Verlag, 1988.

[LH2] Li Huilai, Radial solutions of free boundary problems for degenerate parabolic equations, J. PDE, 3(1)(1990), 47–68.

[LH3] Li Huilai, A degenerate Stefan problem with two free boundary, North-eastern Math. J., 11(3)(1995), 263–274.

[LHF1] Li, H. F., The Cauchy problem for a class of nonlinear filtration equations, J. Math. Study, 29(3)(1996), 44–54.

[LHF2] Li, H. F. and Jiang, C.Sh., The Cauchy problem, Harnack inequality, and initial trace for a class of porous medium equations in N dimensions, Nanjing Daxue Xuebao Shuxue Bannian Kan, 13(2)(1996), 231–241.

[LI1] Lieberman, G. M., Strong convergence of bounded sequences of solutions of porous medium equations, Differential Integral Equations, 11(3)(1998), 395–407.

[LI2] Lieberman, G. M., Study of global solutions of parabolic equations via a priori estimates. III, Equations of p-Laplacian type, Singularities and differential equations, (Warsaw, 1993), 199–221, Banach Center Publ., 33, Polish Acad. Sci., Warsaw, 1996.

[LI] Lions J. L., Quelques méthodes de résolution des problèmes aux limites non linéaires, Dunod, Paris, 1969.

[LOT] Lacey A. A., Ockendon J. R. and Taylor A. B., "Waiting-time" solutions of a nonlinear diffusion equation, SIAM J. Appl. Math., 42(1982), 1252–1264.

[LS1] Laurençot, Ph. and Simondon, F., Source-type solutions to porous medium equations with convection, Commun. Appl. Anal., 1(4)(1997), 489–502.

[LS2] Laurençot, Ph. and Simondon, F., Comportement asymptotique pour des équations des milieux poreux avec convection, C. R. Acad. Sci. Paris Sér. I Math., 323(7)(1996), 775–780.

[LS3] Laurençot, Ph. and Simondon, F., Source-type solutions to porous medium equations with convection.

[LS4] Laurençot, Ph. and Simondon, F., Long-time behaviour for porous medium equations with convection, Proc. Roy. Soc. Edinburgh Sect. A, 128(2)(1998), 315–336.

[LSU] Ladyzhenskaja O. A., Solonnikov V. A. and Ural'tzeva N. N., Linear and quasilinear equations of parabolic type, Transl. Math. Mono., 23, AMS. Providence RI, 1968.

[LU1] Ladyzhenskaya, O. A. and Ural'ceva, N. N., Boundary problems for linear and quasilinear parabolic equations I, Izb. A. N. SSSR, Ser. Mat., 26(1)(1962).

[LU2] Ladyzhenskaya, O. A. and Ural'ceva, N. N., Linear and Quasilinear Elliptic Equations, Academic Press, New York(1968).

[LY] Liu Changchun and Yin Jingxue, Finite Speed of Propagation of Perturbations for the Cahn-Hilliard Equation with Degenerate Mobility, Journal of Partial Differential Equations, 14(3)(2001).

[LZ] Li, Desheng and Zhong, Chenkui, Global attractor for the Cahn-Hilliard system with fast growing nonlinearity, J. Differential Equations, 149(2)(1998), 191–210.

[MA] Moulay, M. S. and Ali Ziane, T. Régularité des solutions d'un problème d'évolution dégénéré en dimension N, Maghreb Math. Rev., 3(1)(1994), 89–96.

[ME] Mikelić A. and Elezović, On the stochastic Cahn-Hilliard equation, Nonlinear Analysis, TMA, 16(12)(1991), 1169-1200.

[MP1] Martinson L. K. and Paplov K. B., Unsteady shear flows of a conducting fluid with a theological power law, Marnit. Gidrodenamika, 2(1970), 50–58.

[MP2] Martinson L. K. and Paplov K. B., The effect of magnetic plasticity in non-Newtonian fluids, Magnit. Gidrodinamika, 3(1969), 69–75.

[MP3] Martinson L. K. and Paplov K. B., On the spatial localization of thermal perturbations in the theory of nonlinear heat conductions, Zh. Vychisl. Matem. i Matem. Fiz., 12(4)(1972), 1048–1053.

[MSW] Maier-Paape, S., Stoth, B. and Wanner, T., Spinodal decomposition for multicomponent Cahn-Hilliard systems, J. Statist. Phys., 98(3-4)(2000), 871–896.

[MU] Mossino, J. and Ughi, M., On the regularity at shrinking points for the porous media equation, NoDEA Nonlinear Differential Equations Appl., 7(1)(2000), 1–21.

[MU1] Muskat M., The flow of homogeneous fluids through porous media, McGraw-Hill, New York, 1937.

[MW] Melkonian, S. and Winternitz, P., Symmetry properties and solutions of nonlinear dispersive thin-film equations in three dimensions, J. Math. Phys., 32(11)(1991), 3213–3222.

[NC] Novick-Cohen A., The nonlinear Cahn-Hilliard equation: Transition from spinodal decomposition to nucleation behavior, J. Statist. Phys., 38(1985), 707–723.

[NI] Nirenberg L., On elliptic partial differential equations, Ann. Scuola

Norm. Sup. Pisa, 13(1959), 115–162.

[NS] Novick-Cohen A. and Segel L. A., Nonlinear aspects of the Cahn-Hilliard equation, Physica 10(D)(1984), 277–298.

[OK] Okamoto, K., A kinetic approach to nonlinear degenerate parabolic equations, Hiroshima Math. J., 23(1993), 577–606.

[OKZ] Oleinik O. A., Kalashnikov A. S., Chou Yulin, Cauchy problem and boundary value problem for the equation of nonstationary filtration, Uzv. AN. SSSR. Ser. Mat., 22(5)(1958), 667–704.

[OR] Oleinik, O. A., Radkevič, E. V., Second Order Differential Equations with Nonnegative Characteristic Form, American Mathematical Society, Rhode Island and Plenum Press, New York, 1973.

[OR1] Osher M. and Ralston J., L^1-stability of travelling waves with applications to convective porous media flow, Comm. Pure Appl. Math., 35(1982), 737–749.

[PA] Passo, R. D., Uniqueness of the entropy solution of a strongly degenerate parabolic equation, Comm. PDE, 18(1&2)(1993), 265–279.

[PE] Pego R. L., Front migration in the nonlinear Cahn-Hilliard equation, Proc. R. Soc. Lond. A422(1989), 261–278.

[PE1] Peletier L. A., On the existence of an interface in nonlinear diffusion processes, Lecture Notes in Math., 415, Springer-Verlag, Berlin, 1974, 412–416.

[PE2] Peletier L. A., A necessary and sufficient condition for the existence of an interface in flows through porous media, Arch. Rat. Mech. Anal., (56)(1974), 183–190.

[PE3] Peletier L. A., The porous media equation, in: Applications of nonlinear analysis in the physical sciences, Pitman, London 1981, 229–241.

[PIE] Pierre M., Uniqueness of the solutions of $u_t - \Delta\varphi(u) = 0$ with initial data on a measure, Nonlinear Analysis, TMA, 6(1982), 175–187.

[PIM] Pimbley G. H., Wave solutions travelling along quadratic paths for the equation $\dfrac{\partial u}{\partial t} - (k(u)u_x)_x = 0$, Quart. Appl. Math., 35(1977), 129–138.

[PT] Peletier L. A. and Terman D., A very singular solution of the porous media equation with absorption, J. Diff. Eqs., 65(1986), 396–410.

[PW] Peletier L. A. and Wang Junyu, A very singular solution of a quasilinear degenerate diffusion equation with absorption, Trans. Amer. Math. Soc., 307(1988), 813–826.

[PZ1] Peletier L. A. and Zhao Junning, Source-type solutions of the porous media equation with absorption: The fast diffusion case, Nonlinear Analysis, TMA, 14(2)(1990), 107–121.

[PZ2] Peletier L. A. and Zhao Junning, Large time behaviour of solution of the porous media equation with absorption: the fast diffusion case, Nonlinear Analysis, TMA, 17(10)(1991), 991–1009.

[QF] Qian, L. and Fan, W., A remark of the Hölder estimate of solutions of some degenerate parabolic equations, Acta Math. Sci. (English Ed.), 19(4)(1999), 463–468.

[QI] Qi, Y. W. and Wang, M., The global existence and finite time extinction of a quasilinear parabolic equation, Adv. Differential Equations, 4(5)(1999), 731–753.

[QV] Quirós, F. and Vázquez, J. L., Asymptotic behaviour of the porous media equation in an exterior domain, Ann. Scuola Norm. Sup. Pisa Cl. Sci.,

[QW] Qi, Y. W. and Wang, M. X., Singular solutions of doubly singular parabolic equations with absorption, Electron. J. Differential Equations, (67)2000, 22 pp.

[RHM] Renson I., Hornberger G. M. and Molz F. J., Numerical Methods in Subsurface Hydrology with an Introduction to the Finite Element Method, New York, Wiley-Interscience, 1971.

[RW] Redheffer R. and Walter W., Inequalities involving derivatives, Pacific J. Math., 85(1979), 165–178.

[SA1] Sacks P. E., The initial and boundary value problem for a class of degenerate parabolic equations, Comm. PDE, 8(1983), 693–733.

[SA2] Sacks P. E., Continuity of solutions of a singular parabolic equation, Nonlinear Anal., TMA, 7(1983), 387–409.

[SAK1] Sakaguchi, S., The number of peaks of nonnegative solutions to some nonlinear degenerate parabolic equations, J. Math. Anal. Appl., 203(1)(1996), 78–103.

[SAK2] Sakaguchi, S., Spatial critical points of nonnegative solutions of the evolution p-Laplacian equation: the fast diffusion case, Differential Integral Equations, 10(6)(1997), 1049–1063.

[SBC] Sander, G. C., Braddock, R. D., Cunning, I. F., Norbury, J. and Weeks, S. W., Source solutions for the nonlinear diffusion-convection equation, J. Austral. Math. Soc. Ser. B, 39(1)(1997), 28–45.

[SC] Schatzman M., Stationary solutions and asymptotic behaviour of an equation like the porous media equation, Research Notes in Math. 89., Pitman, Boston, London 1983, 243–263.

[SI] Simondon F., Effect régularisant local pour $u_t = (\varphi(u))_{xx}$, C. R. Acad. Sci. Paris Sér. 299(1)(1984), 969–972.

[SIV] Sivashinsky G. I., On cellular instability in the solidification of a dilute binary alloy, Physica 8D, (1983), 243–248.

[SO1] Soltanov, K. N., On the solvability of some nonlinear parabolic problem (Russian), Diff. Eqs., 16(1980), 882–887.

[SO2] Soltanov, K. N., Some embedding theorem and its application to nonlinear equations (Russian), Diff. Eqs., 20(12)(1982), 2181–2184.

[ST] Segura de León, S. and Toledo, J., Regularity for entropy solutions of parabolic p-Laplacian type equations, Publ. Mat., 43(2)(1999), 665–683.

[SU1] Su, N., Multidimensional degenerate diffusion problem with evolutionary boundary condition: existence, uniqueness, and approximation, Flow in porous media (Oberwolfach, 1992), 165–178, Internat. Ser. Numer. Math., 114, Birkhäuser, Basel, 1993.

[SU2] Su, N., Mathematical problems on the fluid-solute-heat flow through porous media.Partially saturated case

[SU3] Su, N., Extinction in finite time of solutions to degenerate parabolic equations with nonlinear boundary conditions, J. Math. Anal. Appl., 246(2)(2000), 503–519.

[SUR] Suzuki, R., Existence and nonexistence of global solutions to quasilinear parabolic equations with convection, Hokkaido Math. J., 27(1)(1998), 147–196.

[SW] Stein E. and Weiss G., Introduction to Fourier Analysis on Euclidean Spaces, Princeton University Press, Princeton, N. J., 1971.

[SWA] Swartzendruber D., The flow of water in unsaturated soils, Flow Through Porous Media, Edited by R. J. M. Deviet, 215–292, New York, Academic Press, 1969.

[TAR] Tartar, L., Compensated compactness and applications to partial differential equations, in Nonlinear Analysis and Mechanics, Heriot-Watt Symposium, IV, 136–192, Research Notes in Mathematics, 1979.

[TAY] Tayler A. B., Mathematical Models in Applied Mechanics, Clarendon, Oxford, 1986.

[TK] Tayler A. B. and King J. R., Free boundaries in semi-conductor fabrication, in Free Boundary Problem: Theory and Applications, Irsee, Bavaria, Germany, 1987, Ed. by Hoffmann K. H. and Sprekels J., Vol. I, Pitman Research Notes in Mathematics 185, Longman, Harlow, 1990, 243–259.

[TY] Takeuchi, S. and Yamada, Y., Asymptotic properties of a reaction-diffusion equation with degenerate p-Laplacian, Nonlinear Anal., 42(1)(2000), Ser. A: Theory Methods, 41–61.

[UG] Ughi J. L., Initial values of nonnegative solutions of filtration equation, J. Diff. Equat. 47(1983), 107 132.

[UR] Urbano, J. M., A free boundary problem with convection for the p-Laplacian, Rend. Mat. Appl., (7), 17(1)(1997), 1–19.

[VA1] Vázquez J. L., The interfaces of one-dimensional flows in porous media, Trans. Amer. Math. Soc. 285(1984), 717–737.

[VA2] Vázquez J. L., Behaviour of the velocity of ono-dimensional flows in porous media, Trans. Amer. Math. Soc. 286(1984), 787–802.

[VE] Veron L., Effect regularisants de semigroups nonlineaires dans espaces de Banach, Ann. Fac. Sc. Toulouse.

[VH1] Vol'pert, A. I. and Hudjaev, S. I., On the problem for quasilinear degenerate parabolic equations of second order (Russian), Mat. Sb., 3(1969), 374–396.

[VH2] Vol'pert, A. I., and Hudjaev, S. I., Analysis of a class of discontinuous functions and the equations of mathematical physics (Russian), Izda. Nauka Moskwa, 1975.

[VI] Vishik, K. N., On the solvability of the boundary value problem for quasilinear equations of higher order (Russian), Mat. Sb., 59(1962), 289–325.

[VIT] Vitillaro, E., Blow-up for the porous media equation with source term and positive initial energy, J. Math. Anal. Appl., 247(1)(2000), 183–197.

[VO] Vol'pert, A. I., *BV* spaces and quasilinear equations, Mat. Sb., 73(2)(1967), 255–302.

[WA] Wang Rouhuai, On the Schauder type theory about the general parabolic boundary value problem, Acta Scientiarum Naturlium Universitatis Jilinensis, (2)(1964), 35–64.

[WAW] Wang, Shoutian and Wang, Bixiang, Approximate inertial manifolds to the Cahn-Hilliard equation, Pure Appl. Math., 12(2)(1996), 37–42.

[WC] Wang, M. X. and Chen, Sh. X., Convective porous medium equations with nonlinear forcing at the boundary, J. Math. Anal. Appl., 211(1)(1997), 49–70.

[WD] Wu Dequan, Uniqueness of the weak solution of quasilinear degenerate parabolic equations, Proc. of the Beijing Sym. on Diff. Geom. and Diff. Eqs., 1980(Abstract), Acta Math. Sinica, 25(1982).

[WEW1] Wei, Juncheng and Winter, M., Stationary solutions for the Cahn-Hilliard equation, Ann. Inst. H. Poincare Anal. Non Lineaire, 15(4)(1998), 459–492.

[WEW2] Wei, Juncheng and Winter, M., On the stationary Cahn-Hilliard equation: bubble solutions, SIAM J. Math. Anal., 29(6)(1998), 1492–1518.

[WEW3] Wei, Juncheng and Winter, M., On the stationary Cahn-Hilliard equation: interior spike solutions, J. Differential Equations, 148(2)(1998), 231–267.

[WJ1] Wu Dequan, Jiang Lishang, Generalized solutions of degenerate quasilinear parabolic equations in multi-dimension, Chin. Ann. Math., 6A(6)(1985), 737–750.

[WJ2] Wu Dequan, Jiang Lishang, First boundary value problem of degenerate quasilinear parabolic equations in multi-dimension, Northeastern Math. J., 1(1)(1985), 54–67.

[WM1] Wang Mingxin, Nonlinear Parabolic Equations, Science Press, Beijing, 1993.

[WM2] Wang Mingxin, Travelling wave solutions of egenerate parabolic equations I, Chin. Ann. Math., 12A(5)(1991), 627–635.

[WM3] Wang Mingxin, Travelling wave solutions of a class of degenerate and singular parabolic equations I, System Science and Mathematical Sciences, 12(2)(1992), 127–135.

[WM4] Wang Mingxin, Travelling wave solutions of a class of degenerate and singular parabolic equations II, System Science and Mathematical Sciences, 12(4)(1992), 341–349.

[WN] Weiss, G. S. and Nishiura, Y., The singular limit of the Cahn-Hilliard equation with a nonlocal term, Variational problems and related topics Kyoto, 1995, 62–68.

[WW] Wu Zhuoqun and Wang Junyu, The first boundary value problem for quasilinear degenerate parabolic equations, Proceedings of International Beijing Symposium on Differential Geometry and Partial Differential Equations, 1981.

[WWX1] Wang, S., Wang, M. X. and Xie, C. H., A nonlinear degenerate diffusion equation not in divergence form, Z. Angew. Math. Phys., 51(1)(2000), 149–159.

[WWX2] Wang, S., Wang, M. X. and Xie, C. H., Parabolic equation of the m-Laplacian with nonlinear boundary condition, Chinese Sci. Bull., 43(11)(1998), 905–908.

[WWY] Wang, Yifu, Wu, Zhuoqun and Yin, Jingxue, Periodic solutions of evolution p-Laplacian equations with weakly nonlinear sources, Int. J. Math. Game Theory Algebra, 10(1)(2000), 67–77.

[WY1] Wu Zhuoqun and Yin Jingxue, Some properties of functions in BV_x and their applications to the uniqueness of solutions for degenerate quasilinear parabolic equations, Northeastern Math. J., 5(4)(1989), 395–422.

[WY2] Wu Zhuoqun and Yin Jingxue, Uniqueness of solutions of the Cauchy problem for the system of dynamics of biological groups, Northeastern Math. J., 9(1)(1993), 134–142.

[WYW] Wang, Yifu, Yin, Jingxue and Wu, Zhuoqun, Periodic solutions of evolution p-Laplacian equations with nonlinear sources, J. Math. Anal. Appl., 219(1)(1998), 76–96.

[WZ1] Wu Zhuoqun and Zhao Junning, The first boundary value problem for quasilinear degenerate parabolic equations of second order in several space variables, Chin. Ann. of Math., 4B(1)(1983), 57–76.

[WZ2] Wu Zhuoqun and Zhao Junning, Some general results on the first boundary value problem for quasilinear degenerate parabolic equations, Chin. Ann. of Math., 4B(3)(1983), 319–328.

[WZQ1] Wu Zhuoqun, A free boundary problem for degenerate quasilinear parabolic equations, Nonlinear Analysis, TMA, 9(9)(1985), 937–951.

[WZQ2] Wu Zhuoqun, A note on the first boundary value problem for quasilinear degenerate parabolic equations, Acta Mathematica Scientia.

[XU1] Xu, Xiangsheng, On the Cauchy problem for a singular parabolic equation, Pacific J. Math., 174(1)(1996), 277–294.

[XU2] Xu, Xiangsheng, Some results on functional capacity and their applications to p-Laplacian problems involving measure data, Nonlinear Anal., 27(1)(1996), 17–36.

[YA] Yang, S. X., Extinction of solutions for evolution p-Laplacian equations with a gradient term, Xiamen Daxue Xuebao Ziran Kexue Ban, 38(6)(1999), 811–815.

[YG] Yin Jingxue and Gao Wenjie, Solutions with compact support to the cauchy problem of an equation modeling the motion of viscous droplets, ZAMP, 47(5)(1996), 659–671.

[YI1] Yin Jingxue, On a class of quasilinear parabolic equations of second order with double-degeneracy, J. PDE, 3(4)(1990), 49–64.

[YI2] Yin Jingxue, On the uniqueness and stability of BV solutions for nonlinear diffusion equations, Comm. PDE, 15(12)(1990), 1671–1683.

[YI3] Yin Jingxue, A uniqueness theorem for solutions of nonlinear diffusion equations, Northeastern Math. J., 7(3)(1991), 366–370.

[YI4] Yin Jingxue, The compensated compactness method applied to quasilinear parabolic equations of higher order with doubly strong degeneration, Chin. Ann. of Math., 12B(4)(1991), 500–519.

[YI5] Yin Jingxue, Properties of finite speed of propagation of perturbations of solutions of nonlinear diffusion equations, Acta Mathematica Sinica, 34(3)(1991), 360–364.

[YI6] Yin Jingxue, Solutions with compact support for nonlinear diffusion equations, Nonlinear Analysis, TMA, 19(4)(1992), 309–321.

[YI7] Yin Jingxue, Continuous solutions of the first boundary value problem for nonlinear diffusion equations, Acta Mathematica Sinica, 10(1)(1994), 64–91.

[YI8] Yin Jingxue, On the Cahn-Hilliard equation with nonlinear principal part, Journal of Partial Differential Equations, 7(1)(1994), 77–96.

[YI9] Yin Jingxue, On the existence of nonnegative continuous solutions of the Cahn-Hilliard equation, Journal of Differential Equations, 97(2)(1992), 310–327.

[YI10] Yin Jingxue, A uniqueness theorem for a class of degenerate quasilinear parabolic equations of fourth order, J. Partial Differential Equations, 2(4)(1989), 89–95.

[YI11] Yin Jingxue, On a class of degenerate quasilinear parabolic equations of fourth order, Northeastern Math. J., 4(3)(1988), 253–270.

[YI12] Yin Jingxue, On the classical solutions of degenerate quasilinear parabolic equations of fourth order, J. Partial Differential Equations, 2(2)(1989), 39–52.

[YI13] Yin Jingxue, Finite speed of propagation of solutions for some quasilinear higher order parabolic equations with doubly strong degeneration, Acta Mathematica Scientia, 11(2)(1991), 164–174.

[YI14] Yin Jingxue, A note on the uniqueness of solutions for double degenerate nonlinear parabolic equations, J. Partial Differential Equations, 5(2)(1992), 94–96.

[YI15] Yin Jingxue, Parabolic equations with both nonconvexity nonlinearity and strong degeneracy, Acta Mathematica Scientia, 14(4)(1994), 451–460.

[YL] Yin Jingxue and Li Yong, Classical radial solutions for singular and degenerate elliptic equations, Nonlinear Analysis, TMA, 24(11)(1995), 1619–1626.

[YL1] Yin Jingxue and Liu Changchun, Radial symmetric solutions of the Cahn-Hilliard equation with degenerate mobility, Electronic Journal of Qualitative Theory of Differential Equations, (2)(2001), 1–14.

[YU1] Yuan Hongjun, Hölder continuity of weak solutions for quasilinear doubly degenerate parabolic equations of second order, Acta Scientiarum Naturalium Universitatis Jilinensis, 2(1991), 36–52.

[YU2] Yuan Hongjun, Finite velocity of the propagation of perturbations for general porous medium equations with strong degeneracy, Nonlinear Analysis, TMA, 23(6)(1994), 721–729.

[YU3] Yuan Hongjun, Höder continuity of interfaces for the porous medium equation with absorption, Comm. PDE, 18(5&6)(1993), 965–976.

[YU4] Yuan Hongjun, The Cauchy problem for a quasilinear degenerate parabolic system, Nonlinear Analysis, TMA, 23(2)(1994), 155–164.

[YU5] Yuan Hongjun, Regularity of the free boundary for certain degenerate parabolic equations, Chin. of Math., 15A(1)(1994), 89–97.

[YU6] Yuan Hongjun, Localization Condition for a Class of Nonlinear Diffusion Equations, Chin. Ann. Math., 17(A)(1)(1996), 1–12.

[YU7] Yuan Hongjun, Regularity of the free boundary for general filtration equations, Northeastern Math. J., 9(2)(1993), 197–210.

[YU8] Yuan Hongjun, Solutions to the equation of non-Newtonian polytropic filtration under optimal condition on initial values, Northeastern Math. J., 10(4)(1994), 1–8.

[YU9] Yuan Hongjun, Localization problem for general filtration equations, Northeastern Math. J., 11(4)(1995), 387–400.

[YU10] Yuan, Hongjun, Extinction and positivity for the evolution p-Laplacian equation, J. Math. Anal. Appl., 196(2)(1995), 754–763.

[YZ] Yang Jinshun and Zhao Junning, The asymptotic behavior of solutions of some doubly degenerate nonlinear parabolic equations, Northeastern Math. J., 11(2)(1995), 241–252.

[ZE] Zheng Songmu, Asymptotic behavior of solution to the Cahn-Hilliard equation, Appl. Anal., 23(1986), 165–184.

[ZH] Zhou Shulin, Hölder continuity of solutions for quasilinear singular parabolic equations of second order, Doctoral Thesis of Beijing University.

[ZH1] Zhao Junning, Uniqueness of solutions of the first boundary value problem for quasilinear degenerate parabolic equation, Northeastern Math. J., 1(1)(1985), 153–165.

[ZH2] Zhao Junning, The Cauchy problem of p-Laplacian equation with absorption, Nonlinear Analysis and Microlocal Analysis, Proceedings of the International Conference at Nankai Inst. of Math. 1991, 226–235.

[ZH3] Zhao Junning, Source-type solutions of a quasilinear degenerate parabolic equation with absorption, Chin. Ann. of Math., 15B(1)(1994), 89–104.

[ZH4] Zhao Junning, The asymptotic behaviour of solutions of a quasilinear degenerate parabolic equation, J. Diff. Eqs., 102(1)(1993), 33–52.

[ZH5] Zhao Junning, Existence and nonexistence of solutions for $u_t = \operatorname{div}(|\nabla u|^{p-2}\nabla u) + f(\nabla u, u, x, t)$, J. Math. Anal. Appl., 172(1)(1993),

130–146.

[ZH6] Zhao Junning, On the Cauchy problem and initial traces for evolution p-Laplacian equations with strongly nonlinear sources, J. Diff. Eqs., 121(1995), 329–383.

[ZH7] Zhao Junning, The Cauchy problem for $u_t = \text{div}(|\nabla u|^{p-2}\ \nabla u)$ when $\frac{2N}{N+1} < p < 2$, Nonlinear Analysis, TMA, 24(5)(1995), 615–630.

[ZH8] Zhao Junning, Some properties of solutions of quasilinear degenerate parabolic equations and quasilinear degenerate elliptic equations, Northeastern Math. J., 2(3)(1986), 281–302.

[ZH9] Zhao Junning, Continuity of solutions for a class of quasilinear degenerate parabolic equations, Northeastern Math. J., 7(3)(1991), 356–365.

[ZH10] Zhao Junning, Uniqueness of solutions for higher dimensional quasilinear degenerate parabolic equation, Chin. Ann. of Math., 13B(2)(1992), 129–136.

[ZH11] Zhao Junning, Source-type solution of degenerate quasilinear parabolic equations, J. Diff. Eqs., 92(2)(1991), 179–198.

[ZH12] Zhao, Junning, Singular solutions for a convection diffusion equation with absorption, Acta Math. Sci., 15(4)(1995), 431–441.

[ZHO] Zhou Yulin, Initial value problems for nonlinear degenerate systems of filtration type, Chin. Ann. Math., 5B(4)(1984), 633–652.

[ZHU] Zhu, Ning, On the Cauchy problem and initial trace for nonlinear filtration type with singularity, J. Partial Differential Equations, 9(2)(1996), 129–138.

[ZL] Zhao Junning and Liu Huizhao, The Cauchy problem of the porous medium equation with absorption, J. Partial Diff. Eqs., 7(3)(1994), 231–247.

[ZR] Zeldovich Ya. B. and Raizer Yu. R., Physics of Schock Waves and High-Temperature Hydrodynamic Phenomena, V. II. Academic Press, New York, London, 1969.

[ZX] Zhao Junning and Xu Zhonghai, Cauchy problem and initial traces for a doubly nonlinear degenerate parabolic equation, Science in China, 39A(7)(1996), 673–684.

[ZY1] Zhao Junning and Yuan Hongjun, The Cauchy problem of some doubly nonlinear degenerate parabolic equations, Chin. of Math., 16(A)(2)(1995), 179–194.

[ZY2] Zhao Junning and Yuan Hongjun, Uniqueness of the solutions of $u_t = \Delta u^m$ and $u_t = \Delta u^m - u^p$ with initial datum a measures: the fast diffusion case, Journal of PDE, 7(2)(1994), 143–159.

[ZY3] Zhao Junning and Yuan Hongjun, Lipschitz continuity of solutions and interfaces of the evolution p-Laplacian equation, Northeastern Math. J., 8(1)(1992), 21–37.